Regenerative Medicine: Principles and Clinical Applications

Regenerative Medicine: Principles and Clinical Applications

Edited by Mark Walters

hayle medical

New York

Hayle Medical,
750 Third Avenue, 9th Floor,
New York, NY 10017, USA

Visit us on the World Wide Web at:
www.haylemedical.com

ISBN: 978-1-63241-515-8

The publisher's policy is to use permanent paper from mills that operate a sustainable forestry policy. Furthermore, the publisher ensures that the text paper and cover boards used have met acceptable environmental accreditation standards.

Printed in China.

Cataloging-in-Publication Data

Regenerative medicine : principles and clinical applications / edited by Mark Walters.
p. cm.
Includes bibliographical references and index.
ISBN 978-1-63241-515-8
1. Regenerative medicine. 2. Medicine. 3. Regeneration (Biology). I. Walters, Mark.
R856 .R44 2018
610.28--dc23

Table of Contents

Preface

The world is advancing at a fast pace like never before. Therefore, the need is to keep up with the latest developments. This book was an idea that came to fruition when the specialists in the area realized the need to coordinate together and document essential themes in the subject. That's when I was requested to be the editor. Editing this book has been an honour as it brings together diverse authors researching on different streams of the field. The book collates essential materials contributed by veterans in the area which can be utilized by students and researchers alike.

Regenerative medicine studies the processes and methods by which damaged cells, tissues and organs can be medically restored. This field seeks to treat diseases and disorders through cell therapy as well as through transplantation of tissues and organs. Regenerative medicine is used in reconstructive surgery, organ transplant, endodontics, biomedicine, etc. This book traces the progress of this field and highlights some of its key concepts and applications. It will be of great help to students in the fields of tissue engineering, developmental biology and biomedicine.

Each chapter is a sole-standing publication that reflects each author´s interpretation. Thus, the book displays a multi-facetted picture of our current understanding of application, resources and aspects of the field. I would like to thank the contributors of this book and my family for their endless support.

Editor

Nicotine Promotes Acquisition of Stem Cell and Epithelial-to-Mesenchymal Properties in Head and Neck Squamous Cell Carcinoma

Michael Andrew Yu[1,¶], Alan Kiang[1,¶], Jessica Wang-Rodriguez[2,¶], Elham Rahimy[1], Martin Haas[3], Vicky Yu[1], Lesley G. Ellies[2], Jing Chen[4], Jian-Bing Fan[4], Kevin T. Brumund[1], Robert A. Weisman[1], Weg M. Ongkeko[1]*

1 Division of Otolaryngology-Head and Neck Surgery, Department of Surgery, University of California San Diego, La Jolla, California, United States of America, **2** Veterans Administration Medical Center and Department of Pathology, University of California San Diego, La Jolla, California, United States of America, **3** Moores Cancer Center, University of California San Diego, La Jolla, California, United States of America, **4** Illumina Inc., San Diego, California, United States of America

Abstract

The ability of nicotine to enhance the malignancy of cancer cells is known; however, the possibility that nicotine could regulate a cancer stem cell phenotype remains to be well-established. In this study we sought to determine whether long-term exposure to nicotine could promote cancer stem cell-like properties in two head and neck squamous cell carcinoma cell lines, UMSCC-10B and HN-1. Nicotine treatment induced epithelial-to-mesenchymal transition (EMT) in both cell lines by repressing E-cadherin expression, and led to the induction of stem cell markers Oct-4, Nanog, CD44 and BMI-1, which was reversed upon ectopic re-expression of E-cadherin. Nicotine-treated cells formed spheres at a higher efficiency than non-treated cells, formed larger tumors when injected into mice, and formed tumors with 4-fold greater efficiency compared to control cells when injected at limiting doses. Consistent with previous literature, nicotine-treated cells demonstrated a greater capacity for survival and also a higher tendency to invade. Comparison of microRNA profiles between nicotine and control cells revealed the upregulation of miR-9, a repressor of E-cadherin, and the downregulation of miR-101, a repressor of EZH2. Taken together, these results suggest that nicotine may play a critical role in the development of tobacco-induced cancers by regulating cancer stem cell characteristics, and that these effects are likely mediated through EMT-promoting, microRNA-mediated pathways. Further characterization of such pathways remains a promising avenue for the understanding and treatment of tobacco-related cancers.

Editor: Zhongjun Zhou, The University of Hong Kong, Hong Kong

Funding: This work was supported by the American Academy of Otolaryngology-Head and Neck Surgery Young Investigator Award (to WMO); ARMY/41482, Department of Defense/CDMRP (to WMO). The corresponding author had full access to all the data in the study and takes responsibility for the integrity of the data and the accuracy of the data analysis. The funders had no role in study design, data collection and analysis, decision to publish, or preparation of the manuscript.

Competing Interests: The authors have the following interests: JC and JBF are employed by Illumina, Inc. Technology by Illumina was used in this study to characterize cell lines. There are no further patents, products in development or marketed products to declare.

* E-mail: wongkeko@ucsd.edu

¶ These authors contributed equally to this work.

Introduction

Current theory suggests that cancers are maintained by a subpopulation of tumor initiating cells, or cancer stem cells (CSC). Such a cell has the capacity to give rise to whole tumors due to two fundamental properties: the ability to self-renew, and the ability to differentiate into multiple cell types [1,2]. Critical genes for self-renewal and pluripotency in human embryonic stem cells, such as Oct-4 and Nanog, are likewise elevated in cancer stem cells [3,4]. In addition, the surface glycoprotein CD44 and oncogene BMI1 have been identified as key markers of the CSC subpopulation in breast cancer and head and neck squamous cell carcinoma (HNSCC) [5]. When compared to CD44- cells, CD44+ cells form tumors in mice much more efficiently, show increased invasion, and exhibit markedly higher resistance to chemotherapy drugs, all hallmarks of cancer stem cells [5–7]. In addition, cancer stem cells form spheres in non-adherent conditions in greater size and number than differentiated cancer cells [7,8].

It is generally proposed that cancer stem cells originate either from adult stem cells that have lost control of proliferation, or progenitor cells that have acquired the ability to self-renew. However, several studies seem to support the theory that CSCs arise from differentiated tumor cells that have undergone a process of dedifferentiation to become more stem-like. Most notably, differentiated breast cancer cells that were forced to undergo epithelial-to-mesenchymal transition (EMT) were found to exhibit a more stem cell-like phenotype [9,10]. In addition, retinal pigment epithelial cells were found to undergo a program of transdifferentiation accompanied by EMT following retinal detachment [11].

EMT is a developmental program characterized by loss of cell adhesion, increase in cell mobility, repression of E-cadherin, and upregulation of mesenchymal markers such as N-cadherin and vimentin. EMT can be induced either by the activation of the transcription factors Snail and Twist, or by repression of E-cadherin [12]. Due to its effects on cell adhesion and mobility, EMT is often activated during cancer invasion and metastasis [13,14].

Approximately 47,000 new cases of head and neck squamous cell carcinoma (HNSCC) arise in the US each year [15]. Statistics show that tobacco use and alcohol abuse are the primary risk factors for HNSCC, associated with over 80% of documented cases [16]. Despite the well-established link between tobacco use and cancer, there is lack of understanding of the molecular mechanisms that govern tobacco-induced carcinogenesis. The majority of studies to date have focused on DNA repair or tumor suppressor pathways in response to tobacco use, while little is known about the effect of tobacco on pathways that regulate stem cell self-renewal and pluripotency.

Studies have revealed that nicotine, the major addictive component of tobacco, promotes proliferation, invasion, and angiogenesis in tumor cells while inhibiting apoptosis [17,18]. More intriguingly, nicotine has also been shown to induce EMT in breast and lung cancer [19]. It was recently shown that tobacco smoke can promote lung tumorigenesis by triggering inflammation, which was long believed to play a critical role in cancer development [20]. Correspondingly, nicotine exposure has been shown in several cases to lead to the activation of NF-kappaB, a key regulator of inflammatory responses [21–23]. From these lines of evidence it is clear that nicotine can help promote cancer development; however, its role along with that of tobacco use in general remains largely uninvestigated in context of the cancer stem cell model. Previous studies have shown that nicotine upregulates FOXM1, an oncogene and stem cell gene [24] and that it is capable of increasing the proportion of ALDH+ breast cancer cells [25]. However, neither of these studies investigated the ability of nicotine to increase sphere formation in vitro or tumor-forming efficiency at limiting dilutions in vivo, arguably some of the most important metrics in determining cancer stem cell phenotype. The relationship between nicotine and classical stem cell genes such as CD44, BMI-1, Oct-4 or Nanog, is also unknown. In this study, we sought to determine whether nicotine, through activation of EMT, could alter the phenotype of cancer cells into a more stem cell-like state. We first showed that nicotine indeed promotes EMT in HNSCC cell lines. We then used a combination of gene expression and functional assays to show whether the cells developed properties of cancer stem cells after exposure to nicotine, and whether these properties were mediated via EMT. Most importantly, we investigated whether nicotine could increase the tumor-initiating capability of HNSCC cells in vitro, solidifying its role in promoting the cancer stem cell phenotype. The results of these experiments are essential to a better understanding of the development and progression of tobacco-induced cancer from a stem cell model perspective.

Materials and Methods

Cell Culture and Nicotine Treatment

The HNSCC cell lines used were UMSCC10B, a gift from Dr. Tom Carey, University of Michigan and HN-1, a gift from Dr. J.S. Gutkind, National Institute for Dental and Craniofacial Research. HN-1 was derived from a tongue squamous carcinoma [26]. UMSCC10B was derived from a metastatic lymph node from a different patient [27]. Cells were cultured in low glucose DMEM supplemented with 10% fetal bovine serum, 2% L-glutamine and 2% pen-strep. For nicotine treatment, an appropriate amount of nicotine hemisulfate salt solution (Sigma-Aldrich, St Louis, MO) was directly added to the culture dish. The nicotine along with the medium was replaced every three to four days, and cells were passaged at 90% confluence. Cells were not serum-starved for any experiments in this paper.

E-cadherin Transfection

In order to test whether the expected increase in stem cells gene expression is mediated via induction of EMT, E-cadherin was transiently transfected into these cells using Lipofectamine 2000 (Invitrogen, Carlsbad, CA), following the manufacturer's instructions. The vector used in this study was based on the pWZL-blast backbone. Cells were harvested 72 hours later and qRT-PCR was performed to verify the induction of e-cadherin as well as determine the expression of genes of interest.

Quantitative RT-PCR

Cells were harvested two days after being passaged at about 70–80% confluence. Total cell lysate was collected and RNA was extracted using an RNeasy kit (Qiagen). cDNA was synthesized using Superscript III Reverse Transcriptase (Invitrogen, Carlsbad, CA) as per the manufacturer's instructions. Real-time PCR reaction mixes were prepared using Power SYBR Green (Applied Biosystems, Foster City, CA), and run on the 7300 Real-time PCR System (Applied Biosystems) using the following program: 95°C for 10 min, 95°C for 30 s, and 60°C for 1 min, for 40 cycles. Results were analyzed using the ddCt method. Experiments were done in technical triplicates and were repeated at least twice independently. GAPDH gene expression was measured as endogenous control. Primers were custom ordered (Eurofins MWG Operon, Huntsville, AL) using the following sequences: Snail forward 5′- TCTGAGTGGGTCTGGAGGTG-3′, reverse 5′- CTCTAGGCCCTGGCTGCTAC-3′; Twist forward 5′-GGGCCGGAGACCTAGATGTCATTG-3′, reverse 5′-GAATGCAGAGGTGTGAGGATGGTG-3′; GAPDH forward 5′-CTTCGCTCTCTGCTCCTCC-3′, reverse 5′-CAATACGACCAAATCCGTTG -3′. N-cadherin forward 5′-AGCTTCTCACGGCATACACC-3′, reverse 5′- GTGCATGAAGGACAGCCTCT-3′. Vimentin forward 5′-GGAAATGGCTCGTCACCTTCGT-3′, reverse 5′-AGAAATCCTGCTCTCCTCGCCT-3′. Oct-3/4 forward 5′-GCAAAGCAGAAACCCTCGTGC-3′ reverse 5′- ACCACACTCGGACCACATCCT-3′. Nanog forward 5′-GATTTGTGGGCCTGAAGAAA-3′ reverse 5′-TTGGGACTGGTGGAAGAATC-3′. CD44 forward 5′-ACACCACGGGCTTTTGACCAC-3′ reverse 5′- AG-GAGTTGCCTGGATTGTGCTTG-3′. BMI-1 forward 5′-TCCACAAAGCACACACATCA-3′ reverse 5′-CTTTCATTGTCTTTTCCGCC-3′.

Matrigel Invasion Assay

Invasion of 10B and HN-1 cells was measured using a Matrigel invasion assay (Becton Dickinson, Bedford, MA). Transwell inserts of 8 μm pore size were coated with a final concentration of 1 mg/mL of Matrigel in cold serum-free DMEM. Cells were trypsinized, and 500 uL of cell suspension (1×10^5 cells/mL) were added in triplicate wells. The lower chamber of the transwell was filled with 750 μl of culture media containing 0.5% serum as a chemoattractant, along with the treatment of nicotine and allowed to incubate at 37°C for 48 hours. Invading cells on the lower surface that passed through the filter were fixed and stained using crystal violet in gluteraldehyde and photographed. The number of the stained nuclei was counted in a predetermined and consistent section of each well.

Cell Survival (Clonogenic) Assay

In clonal growth assays all populations were plated in triplicate at 500 cells per 60×15 mm culture plate and cultured in a Mitomycin-C treated J2-3T3 feeder layer with media supplemented with 0.5% fetal bovine serum. Colonies were fixed and stained with a crystal violet/formalin solution and counted after 7–10 days using AlphaImager 2200 and 1220 software (Alpha Innotech Co., San Leandro, CA) such that colonies containing at least 50 cells will be counted as positive.

Western Blotting

Nicotine-treated 10B and HN-1 were plated in complete medium and grown as normal until they reached 80% confluence. Cells were lysed following 2 days after the last nicotine treatment. Cell lysates were separated on a 4–12% SDS-polyacrylamide protein gel (Invitrogen, Carlsbad, CA, USA) and transferred onto a nitrocellulose membrane. The membrane was blocked overnight at 4°C with 5% BSA in TBS-T (TBS with 0.1% Tween-20) and incubated with primary antibodies at a 1:1,000 dilution. After a 3×10 minute wash with TBS-T, the membrane was incubated with the appropriate secondary antibody at 1:10,000 dilution at room temperature. After three additional washes, the protein-antibody complexes were detected using enhanced chemiluminescence (Pierce, Rockford, IL).

Immunofluorescence

10B and HN1 cells were cultured on cover slips under the nicotine and non-nicotine conditions previously stated. The cells were fixed with 4% paraformaldehyde and blocked in goat serum in Dulbecco's phosphate buffered saline at room temperature prior to incubation with mouse monoclonal to anti-human vimentin (Sigma Aldrich, St. Louis, MO). Cells were then incubated with a goat anti-mouse FITC conjugated secondary antibody (Chemicon, Temecula, CA) and counterstained with DAPI. Finally, SlowFade Gold antifade reagent (Invitrogen, Carlsbad, CA) was used to mount the cover slips onto slides. Fluorescent images were obtained at 40X using Leica DMIRE2 inverted fluorescence microscope and computer program Simple PCI was used for image capture.

Sphere Formation Assay

Cells were seeded in a 24-well low adhesion plate (Corning Inc, Lowell, MA) at a density of 1500 cells/well and an initial volume of 500 ul. Spheres were grown in DMEM F12 medium (Gibco) supplemented with Vitamin B27, 20 ng/ml EGF, 20 ng/ml FGF, and 4 ug/ml of heparin. EGF and FGF were replenished every three days and 500 ul of fresh media with or without nicotine was added every four days. Aside from the addition of heparin, cells were agitated daily to minimize clumping. After two weeks, cells were counted and photographed.

Tumorigenicity Assay in SCID Mice

Control and nicotine-treated HN-1 cells were trypsinized, pelleted and resuspended at the appropriate concentration in a 1:10 solution of matrigel (Sigma-Aldrich, St. Louis, MO). 200 ul of cells (5000 or 200,000 cells) were then injected subcutaneously into SCID mice. After 12 weeks, the mice were sacrificed and the tumors were compared between control and nicotine-treated samples. Mice were not administered nicotine throughout the experiment. All animal work was approved and conducted in accordance with UCSD IRB policy and procedures (protocol #S07410).

MicroRNA profiling

MicroRNA was isolated from treated and non-treated cells using the mirVana miRNA isolation kit (Ambion, Austin, TX), following the manufacturer's instructions. Samples were run on the Illumina MicroRNA Array Profiling platform [28]. Analyses were performed using BRB-ArrayTools developed by Dr. Richard Simon and BRB-ArrayTools development team. Clustering algorithms were performed by Cluster 3.0 and visualized with TreeView (Eisen Lab, Stanford University).

Results

Nicotine Treatment Induces Epithelial-to-mesenchymal Transition in HNSCC Cell Lines

In the process of determining whether nicotine can promote a more stem cell-like phenotype, we first assessed its ability to induce epithelial-to-mesenchymal transition in the HNSCC cell lines UMSCC10B and HN-1. 10B and HN-1 cells were cultured in the presence of 3 mM and 1 mM nicotine, respectively, for 6 weeks. Such dosages are within physiological limits of nicotine concentrations found in the saliva of tobacco users [18]. In addition, long-term treatment was intended to mimic the conditions of carcinogenesis in habitual tobacco users. At 6 weeks, expression of the EMT markers Snail, Twist, and Vimentin were assayed by qRT-PCR in both cell lines (Fig. 1A, 1B, 1C). At a sufficient dose of nicotine (which is cell-line dependent), nicotine-treated cells expressed significantly higher levels of mesenchymal markers and lower levels of epithelial markers, suggesting that long-term nicotine treatment is capable of inducing EMT in these two HNSCC cell lines. To further confirm the acquisition of an EMT phenotype, western blots were performed to determine the protein levels of Snail and E-cadherin (Fig. 1D). Nicotine-treated HN-1 cells displayed a fibroblast-like morphology, indicating mesenchymal phenotype (Fig. 1E) which was further confirmed by an immunofluorescence experiment showing increased expression of vimentin (Fig. 1G). Together, these results show that nicotine induces EMT under our experimental conditions, and are consistent with previous studies performed on breast and lung cancer. As a control, we also performed an experiment to test the effect of nicotine on EMT markers in normal oral keratinocytes (Fig. 1F).

Nicotine-treated Cells Exhibit Higher Levels of Stem Cell Markers

CD44 and BMI-1 are critical markers of self-renewal in HNSCC, and their expression has been implicated in tumorigenic potential as well as metastatic ability. Oct-4 and Nanog play essential roles in pluripotency and self-renewal in human embryonic stem cells. We therefore reasoned that the increased expression of these markers is indicative of a more stem cell-like state. The expression of these markers was compared between control and nicotine-treated cells using qRT-PCR (Fig. 2A–D). The expression of stem cell markers was upregulated in nicotine-treated cells, indicating possible regulation of CSCs by nicotine. As a control, we also performed an experiment to see whether nicotine could induce similar changes in normal human keratinocytes (Fig. 2E).

Nicotine Exposure Increases Survival, Sphere Formation, Tumor Size, and Multipotency

Nicotine treated cells exhibited a greater capacity to survive, as determined by a clonal growth assay (Fig. 3A). Both CSCs and normal stem cells have been demonstrated previously to form

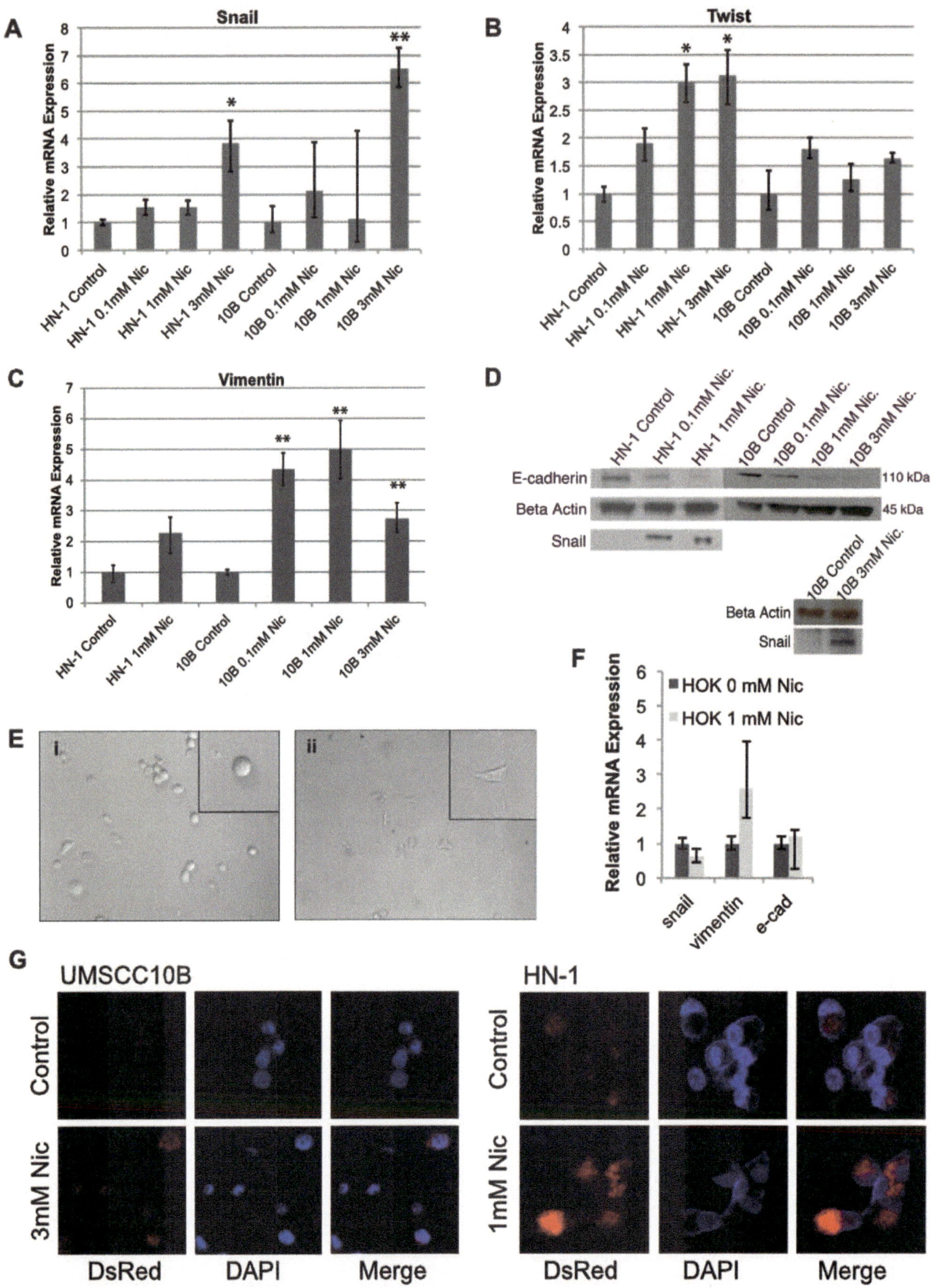
A
Snail
Relative mRNA Expression
HN-1 Control
HN-1 0.1mM Nic
HN-1 1mM Nic
HN-1 3mM Nic
10B Control
10B 0.1mM Nic
10B 1mM Nic
10B 3mM Nic
B
Twist
C
Vimentin
D
E-cadherin
Beta Actin
Snail
110 kDa
45 kDa
HN-1 Control
HN-1 0.1mM Nic.
HN-1 1mM Nic.
10B Control
10B 0.1mM Nic.
10B 1mM Nic.
10B 3mM Nic.
E
i
ii
F
HOK 0 mM Nic
HOK 1 mM Nic
snail
vimentin
e-cad
G
UMSCC10B
HN-1
Control
3mM Nic
1mM Nic
DsRed
DAPI
Merge

Figure 1. Long term treatment of nicotine alters mRNA levels of epithelial to mesenchymal transition expression in head and neck squamous cell carcinoma cell lines UMSCC10B and HN-1. A) Snail, B) Twist, C) Vimentin were all induced corresponding to an increased EMT phenotype. D) Western blot showing increased EMT expression in Snail and decreased E-cadherin in UMSCC10B and HN1 following long-term exposure to nicotine. E) Nicotine-treated HN-1 cells (ii) had a more mesenchymal morphology compared to non-treated cells (i). F) Effect of nicotine on expression of EMT markers in normal oral keratinocytes. G) EMT is confirmed by immunofluorescence, with nicotine treated cells displaying higher expression of the mesenchymal marker vimentin. All images were photographed under 40X. *p<0.05, **p<0.01.

spheres at much higher efficiency compared to differentiated cells [29]. We sought to determine the effect of long term nicotine exposure on sphere formation, as it could be an indicator of a CSC-like phenotype. Nicotine-treated HN-1 cells were seeded in a 24-well low adhesion plate format (Corning Inc, Lowell, MA) at a density of 1500 cells per well. After two weeks, spheres were counted. We found that long term nicotine treated cells formed a greater number of spheres compared to control (Figure 3B right). In addition, nicotine exposed cells formed spheres larger in diameter compared to control cells, which seemed to form aggregates rather than true spheres (Fig. 3B left). This shows that nicotine-exposed cells have a higher capacity to self-renew and proliferate in suspension, which is a critical property of both normal and cancer stem cells [29]. To perform a preliminary survey on the effect of nicotine on tumor formation, we injected both control and 1 mM nicotine-treated HN-1 cells (500,000 cells) into SCID mice and observed that nicotine-treated cells formed significantly larger tumors compared to control cells (Figure 3C). To evaluate multipotency, we looked at the effect of nicotine treatment on the expression of INHBA, a regulator of multipotency in mesenchymal stem cells (Figure 3). This experiment revealed that nicotine upregulates INHBA, indicating increased multipotency. Together these results suggest that long-term exposure to nicotine could promote the acquisition of some functional properties of cancer stem cells.

Nicotine-treated Cells are More Invasive Compared to Control Cells

CSCs are characteristically more invasive compared to normal tumor cells. To determine whether long term nicotine treatment results in a higher level of cellular invasion, we performed a matrigel invasion assay comparing treated 10B and HN-1 to control 10B and HN-1 cells. Our results showed 2–3 fold higher invasion in nicotine treated cells, which is possibly a consequence of the EMT induced by nicotine (Figure 4).

Nicotine Enhances In Vivo Tumor Growth and Tumorigenicity

To confirm and strengthen our preliminary findings in mice, we performed two sets of mouse experiments, one demonstrating that nicotine-treated cells form bigger tumors in nude mice compared to control cells, and that nicotine-treated cells also form tumors more efficiently at limiting dilutions. HN-1 control and HN-1 1 mM long term nicotine cells were first made to overexpress red fluorescent protein (RFP) using a viral-based vector. After an RFP-expressing population was confirmed by fluorescent microscopy,

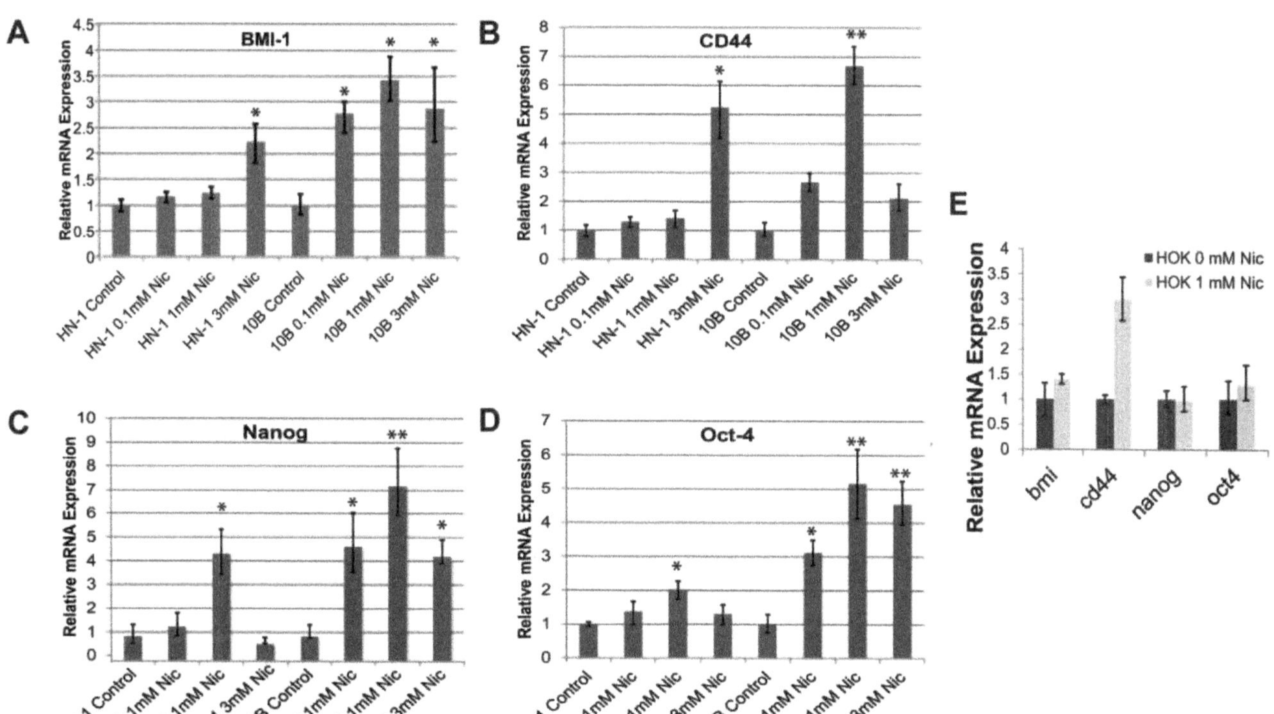

Figure 2. Long term treatment of nicotine increases mRNA expression of stem cell genes in head and neck squamous cell carcinoma cell lines UMSCC10B and HN-1. A) CD44, B) Bmi1, C) Oct4, D) Nanog. E) Effect of nicotine on expression of stem cell markers in normal oral keratinocytes. *p<0.05, **p<0.01.

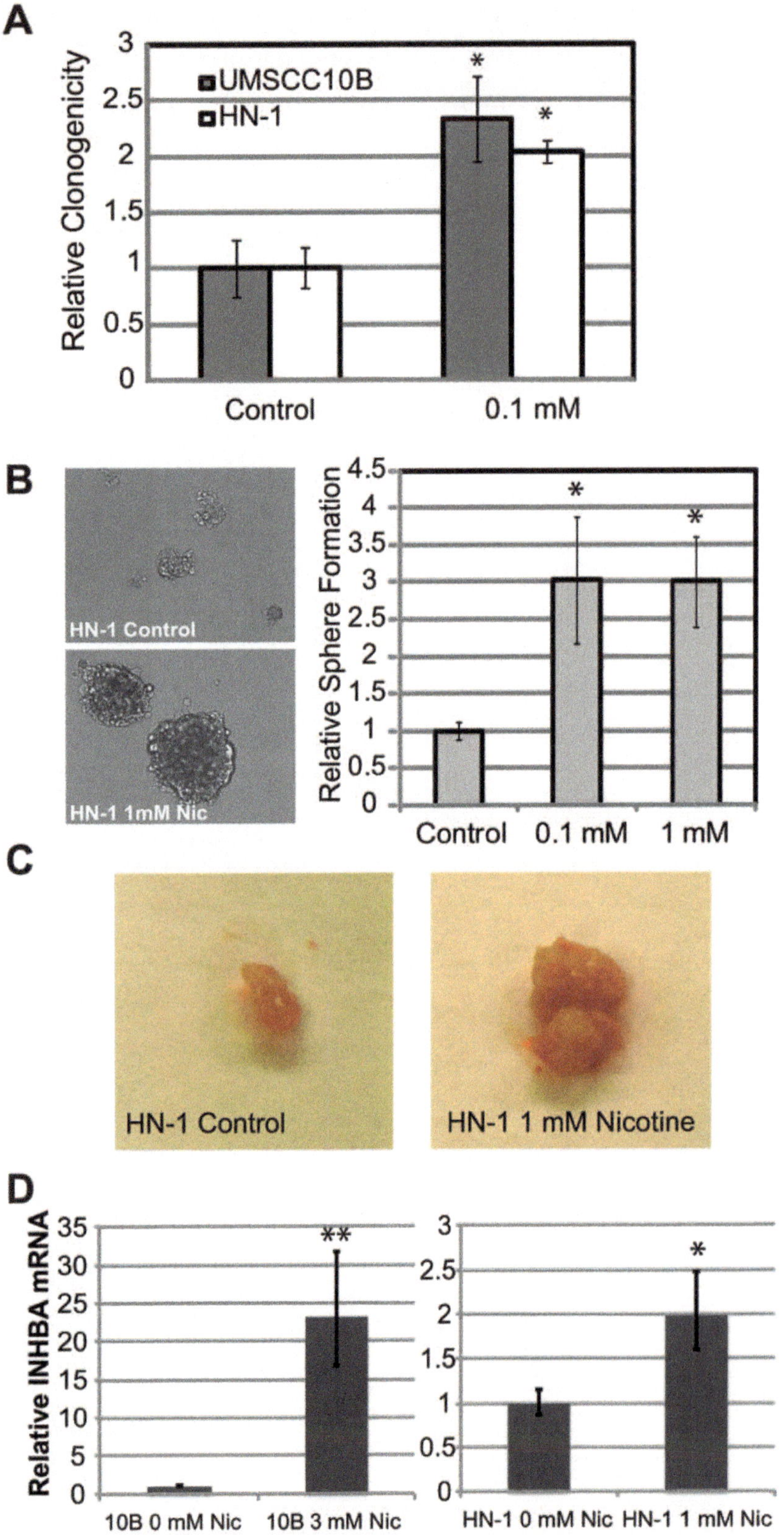

Figure 3. Increased cell survival, sphere formation and tumor size after exposure to nicotine. A) Nicotine enhanced the ability of 10B and HN-1 cells to form colonies under low serum conditoins, suggesting increased survival capacity. B) Photos on left are representative images of spheres formed by nicotine-treated or control HN-1 cells. Graph on right indicates the significant increase in number of spheres formed by nicotine-treated HN-1 cells compared to untreated cells. C) Representative photos comparing tumor sizes between 1 mM-nicotine-treated HN-1 cells and untreated HN-1 cells. D) Relative expression of the multipotency regulator INHBA between nicotine-treated cells and control cells. *$p<0.05$, **$p<0.01$.

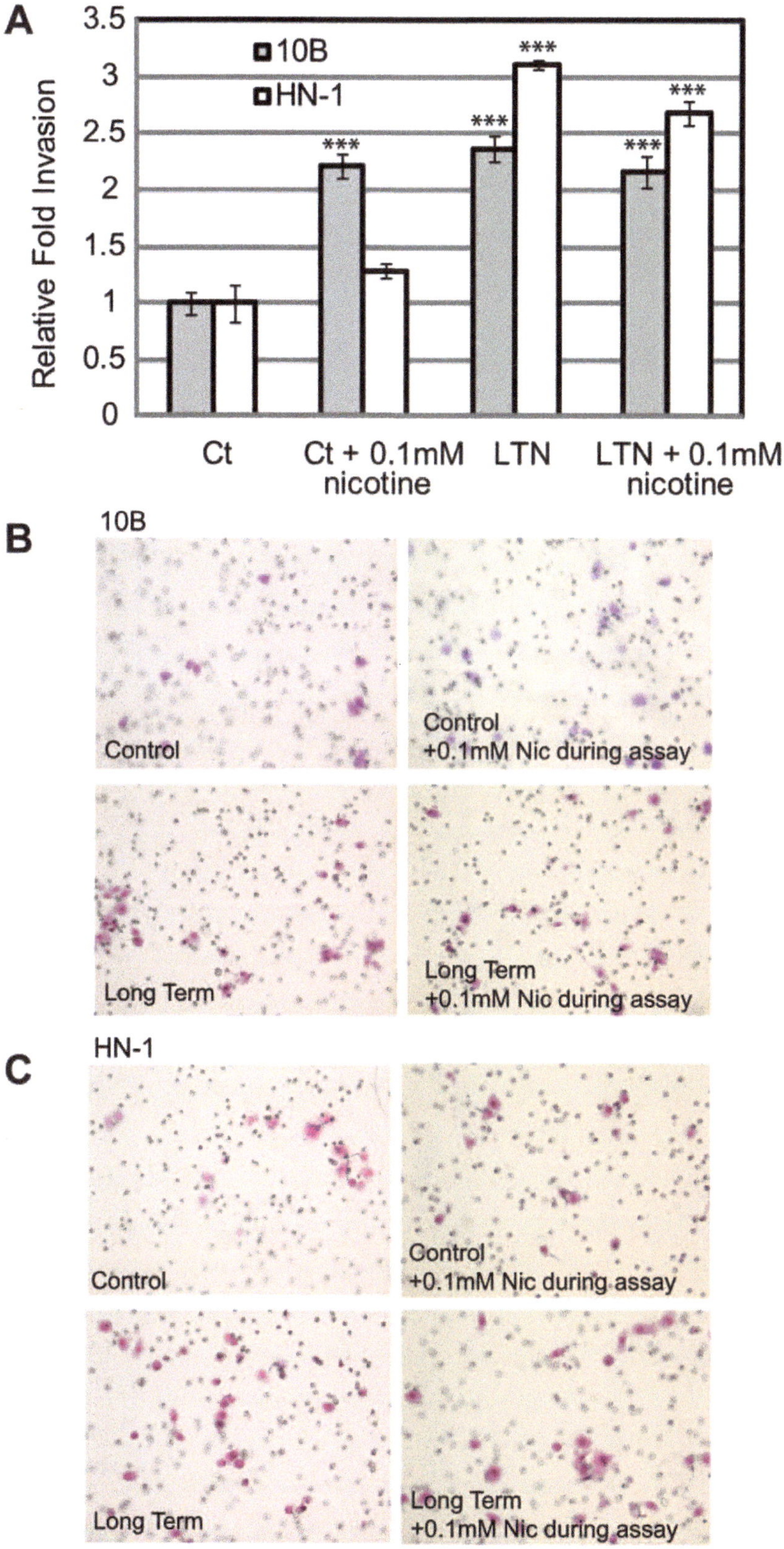

Figure 4. Long-term nicotine exposure increases invasiveness of 10B and HN-1. A) Bar graph indicating the results of a matrigel invasion assay. B) Representative wells of invaded 10B cells. Permutations include control, control with addition of 0.1 mM nicotine during the invasion assay, cells treated long term in 0.1 mM nicotine without addition of nicotine during assay, cells treated long term in 0.1 mM nicotine with addition of 0.1 mM nicotine during assay. C) Representative wells of invaded HN-1 cells, with same permutations as above. $^{*}p<0.05$, $^{**}p<0.01$, $^{***}p<0.001$.

these cells were injected subcutanouesly at two different doses (5 k and 500 k) into the flanks of nude mice. Mice were not administered nicotine throughout this experiment. Mice that were injected with 500 k cells were sacrificed on the 47th day, while those injected with 5 k cells were sacrificed a week later. Results indicated that at a dose of 500 k cells, nicotine-treated HN-1 cells formed significantly larger tumors, with an average weight of 0.49 g compared to 0.185 g for control cells (Figure 5). At a dose of 5 k cells, nicotine-treated HN-1 cells were 4 times more tumorigenic compared to control cells, forming tumors at 4 out of 6 of the injection sites versus 1 out of 6 for control (Figure 6).

Long-term Nicotine Exposure Results in Differential Expression of Critical miRNAs

To address the possibility that the effects of nicotine treatment are miRNA-mediated, we performed an array analysis of global miRNA expression using Illumina's MicroRNA Profiling assay. A handful of miRNAs were shown to be significantly up- or down-regulated in both cell lines after long-term exposure to nicotine (Figure 7A). These included several genes which have been shown previously to play roles in EMT and other oncogenic properties. Upregulated genes included miR-9, which was previously shown to inhibit cancer metastasis by suppressing e-cadherin [30], and

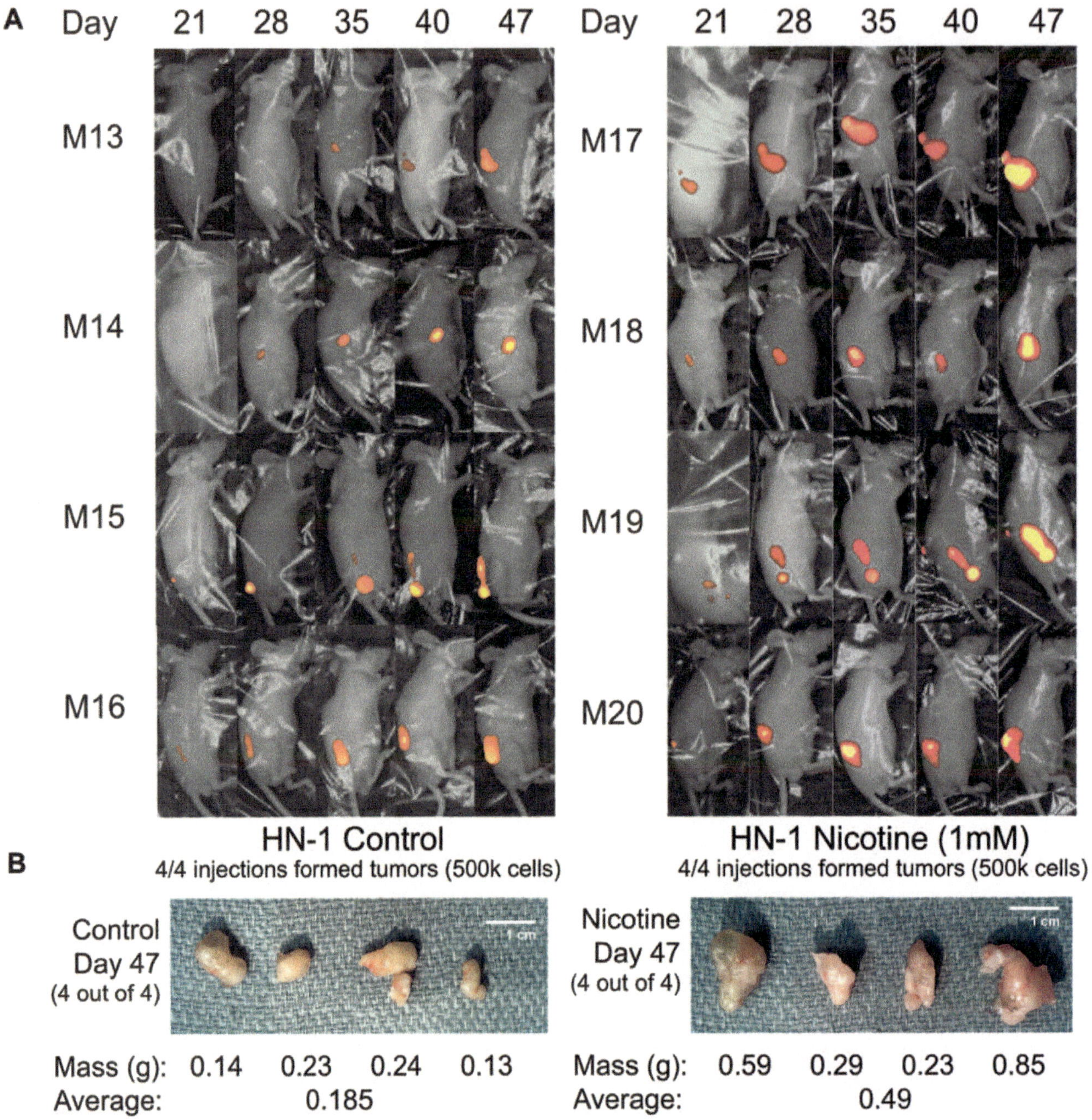

Figure 5. Long-term nicotine exposure increases size of tumors formed in nude mice. A) Fluorescent whole-body imaging of nude mice each injected with 500 k of either control or 1 mM-nicotine-treated HN-1 cells overexpressing RFP. B) Photographs and weights of tumors dissected from mice 47 days post-injection.

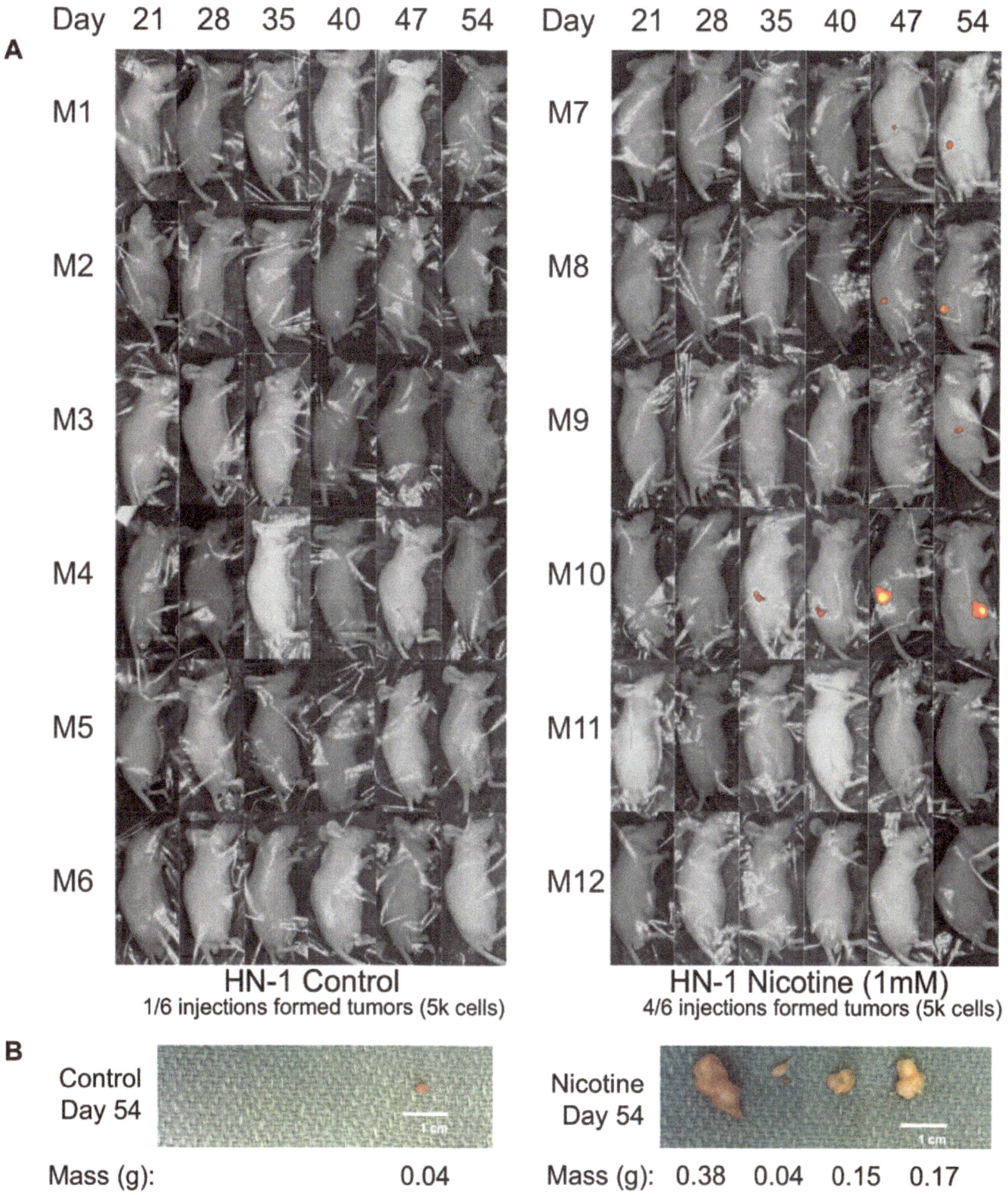

Figure 6. Long-term nicotine exposure increases tumorigenicity. A) Fluorescent whole-body imaging of nude mice each injected with 5 k of either control or 1 mM-nicotine-treated HN-1 cells overexpressing RFP. B) Photographs and weights of tumors dissected from mice 54 days post-injection.

miR-221 [31], which contributes to liver tumorigenesis. miR-135a/b was also highly upregulated and is known to be a potent inducer of Wnt pathway activity [32]. Downregulated miRNAs included miR-101, which is a tumor suppressor and inhibitor of EZH2 [33]. The expression of these two microRNAs were confirmed and validated by qPCR (Figure 7B).

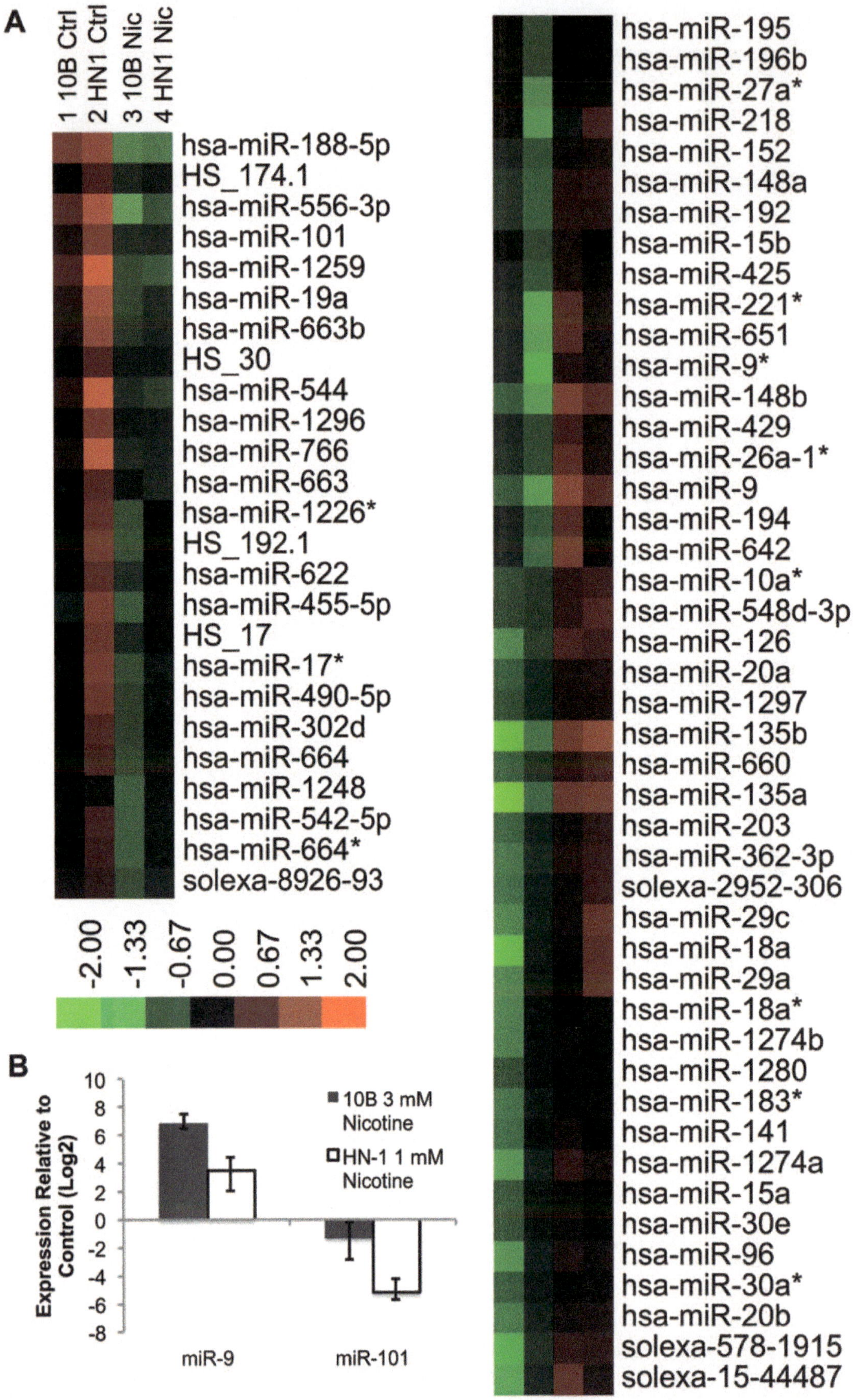
A
1 10B Ctrl
2 HN1 Ctrl
3 10B Nic
4 HN1 Nic
hsa-miR-188-5p
HS_174.1
hsa-miR-556-3p
hsa-miR-101
hsa-miR-1259
hsa-miR-19a
hsa-miR-663b
HS_30
hsa-miR-544
hsa-miR-1296
hsa-miR-766
hsa-miR-663
hsa-miR-1226*
HS_192.1
hsa-miR-622
hsa-miR-455-5p
HS_17
hsa-miR-17*
hsa-miR-490-5p
hsa-miR-302d
hsa-miR-664
hsa-miR-1248
hsa-miR-542-5p
hsa-miR-664*
solexa-8926-93
-2.00
-1.33
-0.67
0.00
0.67
1.33
2.00
hsa-miR-195
hsa-miR-196b
hsa-miR-27a*
hsa-miR-218
hsa-miR-152
hsa-miR-148a
hsa-miR-192
hsa-miR-15b
hsa-miR-425
hsa-miR-221*
hsa-miR-651
hsa-miR-9*
hsa-miR-148b
hsa-miR-429
hsa-miR-26a-1*
hsa-miR-9
hsa-miR-194
hsa-miR-642
hsa-miR-10a*
hsa-miR-548d-3p
hsa-miR-126
hsa-miR-20a
hsa-miR-1297
hsa-miR-135b
hsa-miR-660
hsa-miR-135a
hsa-miR-203
hsa-miR-362-3p
solexa-2952-306
hsa-miR-29c
hsa-miR-18a
hsa-miR-29a
hsa-miR-18a*
hsa-miR-1274b
hsa-miR-1280
hsa-miR-183*
hsa-miR-141
hsa-miR-1274a
hsa-miR-15a
hsa-miR-30e
hsa-miR-96
hsa-miR-30a*
hsa-miR-20b
solexa-578-1915
solexa-15-44487
B
Expression Relative to Control (Log2)
10
8
6
4
2
0
-2
-4
-6
-8
10B 3 mM Nicotine
HN-1 1 mM Nicotine
miR-9
miR-101

Figure 7. MicroRNA profile of nicotine-treated vs non-treated cells. A) Log expression values are computed based on fold expression from the median value within each row. Genes were excluded based on the criteria of having at least 20% of expression values showing a 1.5 fold increase or decrease relative to the median. Downregulated genes that clustered with miR-101 and upregulated genes that clustered with miR-9 were magnified within the figure. B) qPCR validation of miR-9 and miR-101 expression.

Transfecting e-cadherin in Nicotine-treated Cells Partially Reverts the Previously Observed Induction of Stem Cell and EMT Markers

To determine whether the induction of stem cell markers were conferred via EMT signaling, we transfected mouse e-cadherin into long term nicotine treated 10B and HN-1. Cells transfected with e-cadherin showed reduction of these stem cell markers (Figure 8). These results suggest that nicotine-induced expression of pluripotency markers Oct-4 and Nanog are at least partially mediated through EMT. It remains to be clarified whether there exists a pathway aside from EMT or a more direct manner in which nicotine may modulate the expression of Oct-4 and Nanog. The overexpression of e-cadherin was greater than ten-fold in all of these experiments.

Reversal of Nicotine-induced Gene Expression Occurs Following Withdrawal of Nicotine

To determine whether the effects of nicotine could be reversed by withdrawal of nicotine, we looked at expression of stem cell and EMT markers in nicotine-treated cells after a week-long withdrawal from nicotine. Our results revealed that the expression levels of these markers reverted to basal levels within one week of nicotine withdrawal (Figure 9). Re-addition of nicotine for 24 hours slightly induced these markers above basal levels in some cases, but for the most part did not elicit a significant response,

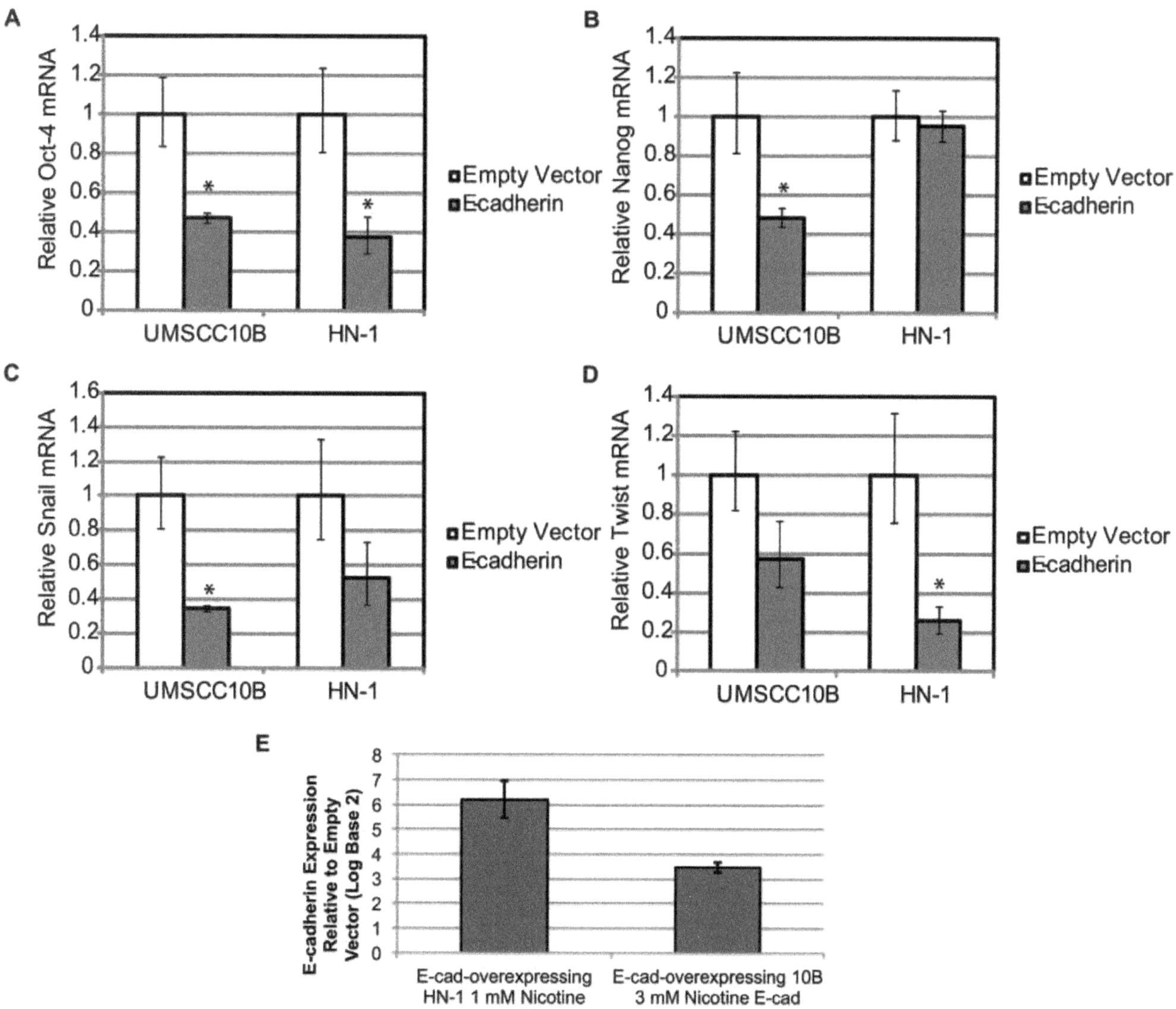

Figure 8. Overexpression of E-cadherin decreases expression of (A) Oct-4, (B) Nanog, (C) Snail, and (D) Twist, indicating a reversal in phenotype induced by long term nicotine treatment. Expression values are compared against an empty vector control. E) indicates the overexpression level of e-cadherin.

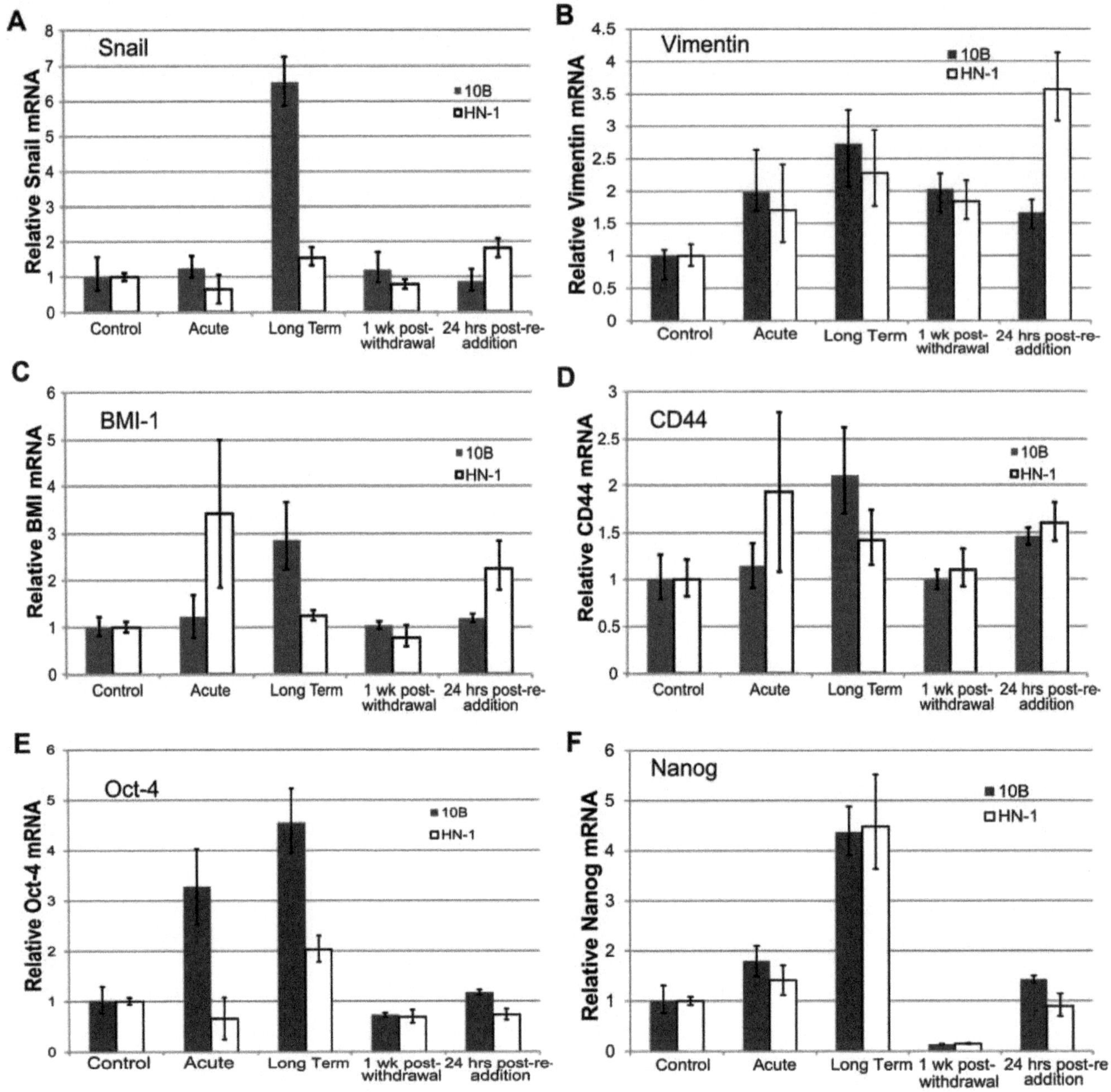

Figure 9. Effect of nicotine withdrawal and subsequent reintroduction on long term nicotine-treated cells. Bar graphs showing expression of (A) Snail, (B) Vimentin, (C) BMI1, (D) CD44, (E) Oct-4, (F), Nanog under various conditions.

suggesting that long-term nicotine treatment does not confer a "memory" effect whereby cells become more sensitized to future nicotine exposure.

Discussion

Here we have shown that long-term nicotine exposure generates cells with properties that are reminiscent of cancer stem cells. Our findings may be important for understanding tobacco-induced carcinogenesis in the context of the cancer stem cell paradigm. Knowing that nicotine or tobacco enables the acquisition of CSC-like properties, one can speculate on the implications of this finding in regard to tobacco-induced carcinogenesis and appropriate treatment modalities. First, most smoking-cessation programs currently rely on the chronic administration of nicotine. Our findings suggest that such treatments can potentially increase the risk for cancer by promoting self-renewal and EMT in premalignant cells even after tobacco use has stopped. Second, the potential ability of nicotine to bring about a malignant stem cell state implies that the CSC model is relevant in tobacco-induced carcinomas. Because nicotine induces a subpopulation of CSCs, treatments specifically aimed at targeting CSCs could be both justified and necessary in the treatment of tobacco-induced cancers. While nicotine typically inhibits apoptosis, it would be interesting to determine whether long-term nicotine exposure can increase cell sensitivity towards CSC-specific inhibitors, such as salinomycin or metformin [34,35]. Third, the way in which nicotine contributes to the development of cancer can potentially be described by a model in which vital tumor suppressors are first rendered inoperative by various carcinogens found in tobacco. In some fashion, this event would greatly potentiate the ability of nicotine to activate EMT or other dedifferentiation pathways and drive the development of the cancer into an aggressive and metastatic phenotype. Another possibility is that if given enough time, nicotine can act upon somatic stem cells or progenitor cells to induce a cancer stem cell phenotype perhaps through the persistent activation of inflammatory cascades and through sustained oxidative stress.

We have shown that overexpressing E-cadherin in nicotine-treated cells partially reverses the induction of the stem cell genes Oct-4 and Nanog, suggesting that the stem cell properties conferred by nicotine are mediated at least in part by EMT. Our findings seem to be consistent with the observations of Dasgupta et al, who suggested that nicotine-induced EMT is primarily mediated through the repression of ECM proteins such as E-cadherin. Future experiments must characterize the mechanisms by which nicotine represses E-cadherin and induces EMT, as well as other targets of nicotine which may contribute to the acquisition of a CSC-like state. Our profiling experiments revealed a set of differentially expressed microRNAs between treated and non-treated cells, suggesting a role of miRNAs in mediating the effects of nicotine. Most notably, miR-9, which was upregulated in nicotine-treated cells, has been shown to promote metastasis in breast cancer by repressing E-cadherin [30]. miR-9 has also been shown to control the migration and proliferation of progenitor cells derived from hESCs [36]. The roles of miR-9 and other differentially expressed miRNAs remain to be fully established in the context of nicotine-induced EMT, and should be addressed by future experiments.

Studies indicate that epigenetic mechanisms may play an essential role in regulating EMT [37–39]. However, little is currently known about the effect of nicotine on expression and activity of DNA methyltransferases and HDACs, although this information may be critical in describing the mechanisms of our findings. Possible targets include DNMT1, whose expression has been linked to the repression of E-cadherin [40,41]. In addition, E-cadherin has been shown to be downregulated by a transcriptional repressor complex consisting of Snail, HDAC1 and HDAC2 [42].

Aside from EMT, pathways that are induced by nicotine and may also contribute to the acquisition of stem cell characteristics include the well-studied PI3K/Akt and MAPK signaling pathways. Nicotine was found to reduce the senescence of endothelial progenitor cells by increasing telomerase activity via Akt signaling [43]. In addition, the anti-apoptotic effects of nicotine have been shown to be partly mediated by PI3K/Akt and MAPK in nasal epithelial cancer [44]. It is possible that nicotine, which has been shown to target NF-kappaB [21], may also work through pro-inflammatory pathways to promote stem cell properties. Takahashi et al recently described a model of lung tumorigenesis in which tobacco smoke acts as a tumor promoter, causing increased proliferation of chemically (NNK) and genetically (K-Ras activation) induced lung cancer cells *in vivo* through IKKβ- and JNK1-mediated inflammatory signaling. As tobacco smoke comprises a myriad of agents, it would be useful from a molecular perspective to examine the isolated role of nicotine in that context.

Removal of nicotine seems to allow the expression of stem cell and EMT markers to revert to their original levels within a week, suggesting that maintenance of EMT and stem cell characteristics may require sustained exposure to nicotine. There is also little evidence of sensitization towards nicotine that is brought on by long-term nicotine treatment, as acute reintroduction of nicotine to cells that have undergone withdrawal did not seem to elicit a response that was significantly different from control cells. Even though the effect of nicotine on stem cell and EMT gene expression seems to wear off easily, nicotine-treated cells were still more tumorigenic in mice that did not receive nicotine throughout the course of the experiment. Thus, even the brief timeframe in which nicotine enhances the stem cell phenotype of the cancer cell may make the difference in terms of whether it is able to form a tumor.

The exact conditions required for nicotine-induced stemness have yet to be identified and future studies must determine which genetic mutations, if any, are prerequisites for nicotine to induce EMT and dedifferentiation in tumor cells. It must also be examined whether nicotine or tobacco can activate similar pathways of EMT and dedifferentiation in normal adult progenitor cells, as this would have enormous implications in cancer stem cell biology, specifically in helping to understand the origin of cancer stem cells. As of now, our findings imply that nicotine, due to its ability to regulate EMT and stem cell properties, could hold a much more essential role in cancer development than previously thought. Preliminary studies in the cancer stem cell field have shown that selective targeting of CSCs can significantly reduce the rate of tumor formation as well as metastasis in mice [35]. Further studies to characterize the pathways through which nicotine acts to promote stem cell properties could therefore greatly contribute to our understanding of the initiation and progression of tobacco-induced cancers, and may lead the way into more novel and effective treatment modalities for these diseases.

Author Contributions

Conceived and designed the experiments: WMO MAY AK. Performed the experiments: MAY AK JC JBF ER VY. Analyzed the data: MAY AK JWR MH LGE KTB RAW WMO. Contributed reagents/materials/analysis tools: MH JC JBF. Wrote the paper: AK WMO.

References

1. Reya T, Morrison SJ, Clarke MF, Weissman IL (2001) Stem cells, cancer, and cancer stem cells. Nature 414: 105–111.
2. Lobo NA, Shimono Y, Qian D, Clarke MF (2007) The biology of cancer stem cells. Annu Rev Cell Dev Biol 23: 675–699.
3. Trosko JE (2006) From adult stem cells to cancer stem cells: Oct-4 Gene, cell-cell communication, and hormones during tumor promotion. Ann N Y Acad Sci 1089: 36–58.
4. Hombach-Klonisch S, Paranjothy T, Wiechec E, Pocar P, Mustafa T, et al. (2008) Cancer stem cells as targets for cancer therapy: selected cancers as examples. Arch Immunol Ther Exp (Warsz) 56: 165–180.
5. Prince ME, Sivanandan R, Kaczorowski A, Wolf GT, Kaplan MJ, et al. (2007) Identification of a subpopulation of cells with cancer stem cell properties in head and neck squamous cell carcinoma. Proc Natl Acad Sci U S A 104: 973–978.
6. Besancon R, Valsesia-Wittmann S, Puisieux A, de Fromentel CC, Maguer-Satta V (2009) Cancer stem cells: the emerging challenge of drug targeting. Curr Med Chem 16: 394–416.
7. Okamoto A, Chikamatsu K, Sakakura K, Hatsushika K, Takahashi G, et al. (2008) Expansion and characterization of cancer stem-like cells in squamous cell carcinoma of the head and neck. Oral Oncol.
8. Zhou XD, Wang XY, Qu FJ, Zhong YH, Lu XD, et al. (2009) Detection of cancer stem cells from the C6 glioma cell line. J Int Med Res 37: 503–510.
9. Mani SA, Guo W, Liao MJ, Eaton EN, Ayyanan A, et al. (2008) The epithelial-mesenchymal transition generates cells with properties of stem cells. Cell 133: 704–715.
10. Santisteban M, Reiman JM, Asiedu MK, Behrens MD, Nassar A, et al. (2009) Immune-Induced Epithelial to Mesenchymal Transition In vivo Generates Breast Cancer Stem Cells. Cancer Res 69: 2887–2895.
11. Saika S, Kono-Saika S, Tanaka T, Yamanaka O, Ohnishi Y, et al. (2004) Smad3 is required for dedifferentiation of retinal pigment epithelium following retinal detachment in mice. Lab Invest 84: 1245–1258.

12. Lee JM, Dedhar S, Kalluri R, Thompson EW (2006) The epithelial-mesenchymal transition: new insights in signaling, development, and disease. J Cell Biol 172: 973–981.
13. Gupta GP, Massague J (2006) Cancer metastasis: building a framework. Cell 127: 679–695.
14. Thiery JP (2002) Epithelial-mesenchymal transitions in tumour progression. Nat Rev Cancer 2: 442–454.
15. Jemal A, Siegel R, Ward E, Hao Y, Xu J, et al. (2008) Cancer statistics, 2008. CA Cancer J Clin 58: 71–96.
16. Sturgis EM, Cinciripini PM (2007) Trends in head and neck cancer incidence in relation to smoking prevalence: an emerging epidemic of human papillomavirus-associated cancers? Cancer 110: 1429–1435.
17. Martin JW, Mousa SS, Shaker O, Mousa SA (2009) The multiple faces of nicotine and its implications in tissue and wound repair. Exp Dermatol 18: 497–505.
18. Wright SC, Zhong J, Zheng H, Larrick JW (1993) Nicotine inhibition of apoptosis suggests a role in tumor promotion. FASEB J 7: 1045–1051.
19. Dasgupta P, Rizwani W, Pillai S, Kinkade R, Kovacs M, et al. (2009) Nicotine induces cell proliferation, invasion and epithelial-mesenchymal transition in a variety of human cancer cell lines. Int J Cancer 124: 36–45.
20. Takahashi H, Ogata H, Nishigaki R, Broide DH, Karin M (2010) Tobacco smoke promotes lung tumorigenesis by triggering IKKbeta- and JNK1-dependent inflammation. Cancer Cell 17: 89–97.
21. Crowley-Weber CL, Dvorakova K, Crowley C, Bernstein H, Bernstein C, et al. (2003) Nicotine increases oxidative stress, activates NF-kappaB and GRP78, induces apoptosis and sensitizes cells to genotoxic/xenobiotic stresses by a multiple stress inducer, deoxycholate: relevance to colon carcinogenesis. Chem Biol Interact 145: 53–66.
22. Nakao S, Ogata Y, Sugiya H (2009) Nicotine stimulates the expression of cyclooxygenase-2 mRNA via NFkappaB activation in human gingival fibroblasts. Arch Oral Biol 54: 251–257.
23. Lau PP, Li L, Merched AJ, Zhang AL, Ko KW, et al. (2006) Nicotine induces proinflammatory responses in macrophages and the aorta leading to acceleration of atherosclerosis in low-density lipoprotein receptor(−/−) mice. Arterioscler Thromb Vasc Biol 26: 143–149.
24. Gemenetzidis E, Bose A, Riaz AM, Chaplin T, Young BD, et al. (2009) FOXM1 upregulation is an early event in human squamous cell carcinoma and it is enhanced by nicotine during malignant transformation. PLoS One 4: e4849.
25. Hirata N, Sekino Y, Kanda Y (2010) Nicotine increases cancer stem cell population in MCF-7 cells. Biochem Biophys Res Commun 403: 138–143.
26. Schwachofer JH, Crooijmans RP, Hoogenhout J, Kal HB, Theeuwes AG (1990) Effectiveness of cis-platin and carboplatin in the chemotherapy of squamous cell carcinoma grown as multicellular spheroids. Anticancer Res 10: 805–811.
27. Wu W, Tang X, Hu W, Lotan R, Hong WK, et al. (2002) Identification and validation of metastasis-associated proteins in head and neck cancer cell lines by two-dimensional electrophoresis and mass spectrometry. Clin Exp Metastasis 19: 319–326.
28. Chen J, Lozach J, Garcia EW, Barnes B, Luo S, et al. (2008) Highly sensitive and specific microRNA expression profiling using BeadArray technology. Nucleic Acids Res 36: e87.
29. Dontu G, Wicha MS (2005) Survival of mammary stem cells in suspension culture: implications for stem cell biology and neoplasia. J Mammary Gland Biol Neoplasia 10: 75–86.
30. Ma L, Young J, Prabhala H, Pan E, Mestdagh P, et al. (2010) miR-9, a MYC/MYCN-activated microRNA, regulates E-cadherin and cancer metastasis. Nat Cell Biol 12: 247–256.
31. Pineau P, Volinia S, McJunkin K, Marchio A, Battiston C, et al. (2010) miR-221 overexpression contributes to liver tumorigenesis. Proc Natl Acad Sci U S A 107: 264–269.
32. Nagel R, le Sage C, Diosdado B, van der Waal M, Oude Vrielink JA, et al. (2008) Regulation of the adenomatous polyposis coli gene by the miR-135 family in colorectal cancer. Cancer Res 68: 5795–5802.
33. Friedman JM, Liang G, Liu CC, Wolff EM, Tsai YC, et al. (2009) The putative tumor suppressor microRNA-101 modulates the cancer epigenome by repressing the polycomb group protein EZH2. Cancer Res 69: 2623–2629.
34. Hirsch HA, Iliopoulos D, Tsichlis PN, Struhl K (2009) Metformin selectively targets cancer stem cells, and acts together with chemotherapy to block tumor growth and prolong remission. Cancer Res 69: 7507–7511.
35. Gupta PB, Onder TT, Jiang G, Tao K, Kuperwasser C, et al. (2009) Identification of selective inhibitors of cancer stem cells by high-throughput screening. Cell 138: 645–659.
36. Delaloy C, Liu L, Lee JA, Su H, Shen F, et al. (2010) MicroRNA-9 coordinates proliferation and migration of human embryonic stem cell-derived neural progenitors. Cell Stem Cell 6: 323–335.
37. Vrba L, Jensen TJ, Garbe JC, Heimark RL, Cress AE, et al. (2010) Role for DNA methylation in the regulation of miR-200c and miR-141 expression in normal and cancer cells. PLoS One 5: e8697.
38. Tryndyak VP, Beland FA, Pogribny IP (2010) E-cadherin transcriptional down-regulation by epigenetic and microRNA-200 family alterations is related to mesenchymal and drug-resistant phenotypes in human breast cancer cells. Int J Cancer 126: 2575–2583.
39. Dumont N, Wilson MB, Crawford YG, Reynolds PA, Sigaroudinia M, et al. (2008) Sustained induction of epithelial to mesenchymal transition activates DNA methylation of genes silenced in basal-like breast cancers. Proc Natl Acad Sci U S A 105: 14867–14872.
40. Benton G, Crooke E, George J (2009) Laminin-1 induces E-cadherin expression in 3-dimensional cultured breast cancer cells by inhibiting DNA methyltransferase 1 and reversing promoter methylation status. FASEB J 23: 3884–3895.
41. Shieh YS, Shiah SG, Jeng HH, Lee HS, Wu CW, et al. (2005) DNA methyltransferase 1 expression and promoter methylation of E-cadherin in mucoepidermoid carcinoma. Cancer 104: 1013–1021.
42. von Burstin J, Eser S, Paul MC, Seidler B, Brandl M, et al. (2009) E-cadherin regulates metastasis of pancreatic cancer in vivo and is suppressed by a SNAIL/HDAC1/HDAC2 repressor complex. Gastroenterology 137: 361–371, 371 e361–365.
43. Junhui Z, Xiaojing H, Binquan Z, Xudong X, Junzhu C, et al. (2009) Nicotine-reduced endothelial progenitor cell senescence through augmentation of telomerase activity via the PI3K/Akt pathway. Cytotherapy 11: 485–491.
44. Shen T, Le W, Yee A, Kamdar O, Hwang PH, et al. (2010) Nicotine induces resistance to chemotherapy in nasal epithelial cancer. Am J Rhinol Allergy 24: e73–77.

Multi-Kinase Inhibitor C1 Triggers Mitotic Catastrophe of Glioma Stem Cells Mainly through MELK Kinase Inhibition

Mutsuko Minata[1]¶, Chunyu Gu[1,7]¶, Kaushal Joshi[1], Mariko Nakano-Okuno[2], Christopher Hong[1], Chi-Hung Nguyen[4], Harley I. Kornblum[5], Annie Molla[6], Ichiro Nakano[1,3]*

1 Department of Neurological Surgery, The Ohio State University Medical Center, Columbus, Ohio, United States of America, **2** Department of Internal Medicine, The Ohio State University Medical Center, Columbus, Ohio, United States of America, **3** James Comprehensive Cancer Center, The Ohio State University Medical Center, Columbus, Ohio, United States of America, **4** Pharmaco-chemistry, UMR 176 CNRS-Institut Curie, Orsay, France, **5** Departments of Psychiatry and Molecular and Medical Pharmacology, University of California Los Angeles, Los Angeles, California, United States of America, **6** Institut Albert Bonniot, Université Joseph Fourier, Grenoble Cedex 9, France, **7** Departments of Neurosurgery, Beijing Sanbo Brain Hospital, Capital Medical University, Beijing, China

Abstract

Glioblastoma multiforme (GBM) is a highly lethal brain tumor. Due to resistance to current therapies, patient prognosis remains poor and development of novel and effective GBM therapy is crucial. Glioma stem cells (GSCs) have gained attention as a therapeutic target in GBM due to their relative resistance to current therapies and potent tumor-initiating ability. Previously, we identified that the mitotic kinase maternal embryonic leucine-zipper kinase (MELK) is highly expressed in GBM tissues, specifically in GSCs, and its expression is inversely correlated with the post-surgical survival period of GBM patients. In addition, patient-derived GSCs depend on MELK for their survival and growth both *in vitro* and *in vivo*. Here, we demonstrate evidence that the role of MELK in the GSC survival is specifically dependent on its kinase activity. With *in silico* structure-based analysis for protein-compound interaction, we identified the small molecule Compound 1 (C1) is predicted to bind to the kinase-active site of MELK protein. Elimination of MELK kinase activity was confirmed by *in vitro* kinase assay in nano-molar concentrations. When patient-derived GSCs were treated with C1, they underwent mitotic arrest and subsequent cellular apoptosis *in vitro*, a phenotype identical to that observed with shRNA-mediated MELK knockdown. In addition, C1 treatment strongly induced tumor cell apoptosis in slice cultures of GBM surgical specimens and attenuated growth of mouse intracranial tumors derived from GSCs in a dose-dependent manner. Lastly, C1 treatment sensitizes GSCs to radiation treatment. Collectively, these data indicate that targeting MELK kinase activity is a promising approach to attenuate GBM growth by eliminating GSCs in tumors.

Editor: Anita B. Hjelmeland, Cleveland Clinic, United States of America

Funding: This work was supported by the Start-up budget for I. Nakano at the Ohio State University, Department of Neurological Surgery; and the American Cancer Society Grant for I. Nakano (MRSG-08-108-01). HIK is supported by the NIH, NS0525630. IN is also supported by the NIH, R01 NS083767, 1R21CA175875, 1R01NS087913; and Vincent J. Sgro, The American Brain Tumor Association, National Brain Tumor Foundation. The funders had no role in study design, data collection and analysis, decision to publish, or preparation of the manuscript.

Competing Interests: The authors have declared that no competing interests exist.

* E-mail: Ichiro.Nakano@osumc.edu

¶ These authors contributed equally to this work.

Introduction

Glioblastoma multiforme (GBM) is the most common and lethal primary brain tumor in adults, and therefore, there is an urgent need to develop novel therapeutic strategies that effectively target therapy-resistant GBM cells. Among heterogeneous GBM cells glioma stem cells (GSCs) represent a subpopulation of highly tumorigenic cells that possess stem cell characteristics. While our understanding of GSCs is evolving, there is a great deal of evidence supporting the hypothesis that GSCs drive GBM propagation and promote resistance to conventional therapies such as radiation and chemotherapy [1–9].

Maternal embryonic leucine zipper kinase (MELK) is a serine/threonine kinase that is highly expressed in various organ-specific stem cells and cancers [10,11]. Furthermore, MELK expression is correlated with a poor prognosis of a variety of cancers, including GBM [10–13]. We previously demonstrated that MELK is abundantly expressed in GBM with preferential expression in GSCs and that targeting MELK-mediated pathways disrupt cell cycle progression of GSCs *in vitro* and tumor growth *in vivo*, suggesting that MELK is a clinically relevant molecular target for GBM therapy [10,14–17]. To gain insights in the mechanisms of action, we recently identified that MELK forms a protein complex with the oncogenic transcription factors c-JUN and FOXM1 in GSCs but not in non-GSCs or normal stem/progenitor cells [18,19]. Further, both of these protein interactions are specifically dependent on the MELK kinase domain [18]. These results suggest that inhibition of the kinase activity of MELK could disrupt key interactions with pivotal oncogenes in cancer cells, while relatively sparing normal cells. In this study, we sought to identify a novel small molecule that potently inhibits MELK kinase activity.

Materials and Methods

Ethics

Experiments using de-identified human tissue-derived materials were carried out under the approved Institutional Review Board at University of California, Los Angeles (UCLA) or Ohio State University (OSU). Microarray studies were carried out at UCLA. Primary samples collected at UCLA were de-identified and sent to OSU for further studies. The OSU Institutional Review Board approved this research study and waived the need for further written informed consent from the participants. The name of this protocol is Investigating Novel Therapeutic Strategies for Brain Tumor Treatment and the approval number at OSU is 2005C0075. I Nakano serves as the Principal Investigator for this approved protocol. All animal experimentation was performed at OSU with the approval of the OSU Animal Research Committee, following NIH guidelines.

Tissue Culture

Cells derived from 3 samples of GBM surgical tissues were established in Dr. Harley Kornblum's laboratory at UCLA and were cultured as previously described [20]. Neurosphere cultures derived from these 3 samples were designated as GBM146, GBM157 and GBM206. GBM1600 cells were kindly provided by Dr. Paul Mischel at UCLA and cultured in DMEM/F12 with 10% fetal bovine serum (FBS) (Sigma-Aldrich, MO)[16]. U87 and U251 were obtained from ATCC (VA) and maintained in DMEM (Life technologies, NY) with 10% FBS (Life technologies, NY).

cDNA Microarray

RNA was extracted from GBM sphere samples (GBM146, GBM157, and GBM206) treated with 1 μM Siomycin A or control (DMSO) for 24 hours with RNeasy Mini Kit according to the manufactur's protocol (Qiagen). RNA samples were subjected to cluster (A) and canonical pathway analyses (D) by Ingenuity software (Ingenuity Systems, www.ingenuity.com). The GEO submission number for this microarray is GSE50227.

Flow Cytometry and Cell Sorting

Flow cytometry and cell sorting of CD133(+) and CD133(−) cells from GBM spheres were performed using CD133 antibody (clone: AC133) according to manufacturer's protocol (Miltenyi Biotec, CA) and as described previously [16,17].

In silico Docking Model and In vitro Kinase Assay

Using the structure-based virtual screening method, C1 was identified as a potential MELK inhibitor from readily available half a million commercial compounds. C1 compound was subsequently validated via experimental enzyme assay as previously described [21,22]. Briefly, in order to account for the correct binding of the ligands, crystal structural complexes of kinases for each MELK inhibitor in the Protein Data Bank were sought and similarity analysis based on SMILES was carried through the National Center for Biotechnology Information. In total, 16 groups of templates, 4 structurally distinct MELK, were selected for induced fit MELK conformational modeling on the basis of backbone root-mean-square deviation of the binding site residues, visual inspection of the p-loop, and individual docking performance. Furthermore, the templates were selected using a MELK domain sequence with the Basic Local Alignment Search Tool. Using these conformers and known inhibitors, all the docking and virtual screening calculations were performed with the Virtual Screening Workflow script and selected the final 3 protein models. A total of 30 compounds of varying potency with at least 25% inhibition of MELK activity at 1 μM were collected and downloaded the structures from readily available half a million commercial compounds. With the screening to the ATP binding pocket of all 3 selected conformers using Glide HTVS docking, the top 10% of the compounds were carried forward by the more exhaustive Glide SP docking algorithm. The most highly scored 10% of the SP docked compounds were narrowed down and finally the 3 compounds, showed a pair of hydrogen bonds with the hinge residues, were selected. Subsequently, the 3 compounds were validated via experimental enzyme assays, C1 was the most selective (K_d = 18 μM), which showed no or minimal activity to the other kinases. Similarity search to the Chemical Abstracts Service database was performed in order to check the novelty of this computationally discovered MELK inhibitor candidate.

GBM Slice Culture

GBM surgical tissues of 2 patients were received immediately after surgery from the Department of Pathology at OSU and they were histopathologically diagnosed as GBM by the assigned neuro-pathologists. Serial sections of the surgical specimens were cut to create tumor blocks (10 mm in diameter) and these blocks were transferred into 6 well plates as described previously [23]. Tumor blocks were then injected with either DMSO (5%) or C1 (2.5 nM) and incubated for 16 hours at 37°C in humidified air containing 5% CO_2. After incubation we confirmed that the tumor slice cultures retain the histopathological characteristics of GBM. These treated tissues were fixed with 10 mL of 10% v/v formalin for 24 hours and processed for paraffin-embedded sections (4 μm thickness) for immunohistochemistry.

Xenograft

Ten thousand GBM157 sphere cells in 5 μl of phosphate buffered saline (PBS) were injected intracranially into immuno-compromised mice (n = 16) (Athymic NCr-nu/nu; National Cancer Institute, Strain Code 01B74) according to the methods described previously [19,23]. At day 7 after transplantation, varying doses of C1 (2.5 pmol: n = 3 25 pmol: n = 4, 250 pmol: n = 5) or DMSO (n = 4) were injected into tumor cavities. Three days following C1 or DMSO injection, we sacrificed 3 treated mice (DMSO: n = 1, 25 pmol: n = 1, 250 pmol: n = 1) and stained the brains with the proliferation marker Ki-67. For analysis of tumor growth, 13 mice were sacrificed at 8 weeks after transplantation. The tissues were fixed with ice cold 4% paraformaldehyde in PBS overnight, sunk in 20% sucrose in PBS, and stored at −80°C until use. Sections were subsequently cryoprotected, sectioned at 20 μm and stained with the human-specific Nestin for measuring the tumor size according with the same protocol that we used in our previous study [16,25].

Immunocytochemistry and Immunohistochemistory

Studies were performed as described previously [16,17,19]. The primary antibody for MELK (1:200, Sigma-Aldrich, Missouri) was used to visualize the fluorescent signals using the following secondary antibodies: Alexa 488 or Alexa 555 (1:1000, Cell Signaling Technology, MA). Specificity was determined using no-primary control slides. For immunohistochemistry, the following primary antibodies were used: Nestin (anti-Nestin, clone 10C2, 1:200, mouse monoclonal antibody, MAB5326, MA) and Ki67 (anti-Human Ki-67, clone MIB-1, 1:1, mouse monoclonal antibody, Dako, Denmark). The Envision system (Dako) followed by Diaminobenzidine (DAB) method was used for detection of primary antibody according the manufacturer's protocol. For paraffin-embedded slides, hematoxylin was used as a nuclear counterstain. IHC scoring was performed using automated digital image analysis (ImageJ).

Time-lapsed Microscopy

U251 cells were transfected with the vector E-GFP-N1 using lipofectamine (Invitrogen) according to the manufacturer's protocol. Cells were selected by gentamicin (100 µg/ml), seeded on a 2-wells Lab-Tek chambered coverglass (Nalge Nunc International), and maintained under standard culture conditions (37°C, 5% CO_2) for 22 to 24 hours. One μM of C1 or DMSO was added to the cell culture just before imaging. To avoid drug combinations, cells were only synchronized by trypsination around 24 hours before imaging. Images were acquired on a Zeiss dynascope confocal microscope (LSM 710) equipped for alive cells (37°C, 5% CO_2) by using a Plan-Apochromat 40 X-water immersion objective. The positions of the mitotic cells on the stage were registered and therefore, these cells were imaged continuously. None of the followed mitotic cells divided in two daughter cells. Control and treated cells were imaged simultaneously. Then, the Lab-Teck was kept under normal culture conditions and cells were fixed upon 72 hours of treatment by 4% paraformaldehyde and nuclei stained by Hoechst 3342. Images were recorded with the same equipment as above but at room temperature and analyzed with the Zen software provided by Zeiss. Only one cell line was imaged since the behavior of the cells upon C1 treatment is similar to what was previously observed in HeLa, Hek or H358 cells [26,27]. Three independent experiments were conducted and 10 to 15 fields were followed in each.

Apoptosis Assay

U251 cells treated with either C1 or DMSO for 48 hours was analyzed by flow cytometry with Annexin V antibody and Propidium Iodide (Life technologies, NY) using the Apoptosis Detection Kit (R&D Systems, MN) according to the manufacturer's instructions. Data was confirmed by 3 independent experiments.

Neurosphere Formation Assay and Radiosensitivity Assay

Neurosphere formation assay and radiosensitivity assay were performed as described previously [16,17,19].

Statistical Analysis

Statistical analysis was performed using the SPSS17 Statistics software (IBM Corporation, NY) using one-way ANOVA and student's T test. A probability of $p<0.05$ was considered to be significant. All the data are shown in mean ± standard error of the mean (SEM).

Results

Siomycin a Treatment of GSCs Results in Downregulation of Genes in the DNA Damage-induced Repair Pathway

Previously we demonstrated that the thiazole antibiotic Siomycin A attenuates a MELK-mediated signaling, thereby diminishing GSC growth *in vitro* and *in vivo* [16]. Here we first sought to determine the downstream pathways in GSCs that are suppressed by Siomycin A treatment. We performed cDNA microarray with 3 well-characterized GBM neurosphere samples (GBM146, GBM157, and GBM206) [10] treated with either 1 µM of Siomycin A or vehicle (DMSO) for 48 hours. Unbiased cluster analysis separated these 3 samples into 2 groups; either DMSO-treated or Siomycin A-treated GBM neurospheres (Fig. 1A). Consistent with our previously published quantitative reverse transcriptase (qRT)-PCR [19], Siomycin A significantly downregulated MELK as well as its binding partner FOXM1 (Fig. 1B). The other downregulated genes included mitotic genes such as Aurora A/B and Survivin (Fig. 1B and C). Pathway analysis using Ingenuity indicated that the most downregulated pathway is the DNA damage-induced ATM/ATR pathway; a pathway that regulates the G2/M checkpoint (Fig. 1D). In addition, the transcriptional p53 signaling pathway was also significantly downregulated by Siomycin A treatment (Fig. 1D), supporting our previous study that demonstrated MELK action is mediated, at least in part, through inhibition of p53 pathway [18].

Kinase Activity of MELK is Essential for GSC Survival

Next we investigated whether the function of MELK in GSCs specifically depends on its kinase activity. To address this question, we combined lentiviral infection of MELK shRNA vector, to downregulate endogenous MELK protein, with overexpression of MELK wild type or the kinase dead version of MELK (MELK D150A) [28]. Flow cytometry with Annexin V (AV) and Propidium Iodide (PI) demonstrated that GBM1600 spheres infected with MELK shRNA have more cells in early (AV(+), PI(−)) and late (AV(+), PI(+)) stages of apoptosis, compared to the control shRNA infected cells (Fig. 2A). We then introduced MELK wild type or D150A cDNA into these infected cells. As a control, we used GFP-overexpression vector. When wildtype MELK was restored in MELK shRNA infected GBM1600 spheres, we observed a partial reversal of the effects on apoptosis (AV(+), PI(+) cells; 30.7% in the control samples vs. 15.4% in MELK wild type samples) [18]. In contrast, overexpression of MELK D150A failed to rescue the effects of MELK knockdown (AV(+); PI(+) cells; 37.0% in MELK D150A sample), indicating that this mutant MELK lacks the ability to recover MELK elimination-induced cell death. Given that our previous observation that the D150 residue of MELK is required for the interaction of MELK protein with the oncogenic transcriptional factors c-JUN and FOXM1 in a cancer-specific manner [18,19], it is likely that the kinase domain is essential for MELK-driven GSC survival.

Identification of C1 as an Inhibitor for Mitotic Kinases Including MELK

The above data raised a possibility that the kinase domain of MELK is a potential therapeutic target for GBM. We therefore sought to discover small molecules that specifically inhibit its kinase activity. To this end, we performed an *in silico* screening of small molecules and identified a benzo[e]pyridoindole, C1 (Fig. 2B), as a multi-kinase inhibitor with significant activity against the mitotic kinases, MELK and Aurora B. Effects of C1 on other kinases exhibited substantially lower potency [21]. Computer-based molecular structure analysis supported the predicted docking of C1 to the ATP-binding site of MELK protein (Fig. 2C). The inhibition of MELK kinase activity by C1 was further validated, as we found that compound C1 inhibited the kinase activity of recombinant MELK protein with an IC50 of 42 nM *in vitro* (Fig. 2D).

C1 Treatment Inhibits GSCs to a Greater Extent than Non-GSCs *In vitro*

Next, we sought to assess the sensitivity of GSCs to C1 *in vitro*. First, we compared the effects of C1 treatment on neurosphere formation from patient-derived GBM cells and normal neural progenitors [17]. We incubated the 3 GSC samples (GBM146, GBM157, and GBM206) and normal neural progenitors (16wf) with varying concentrations of C1 to measure the impact on

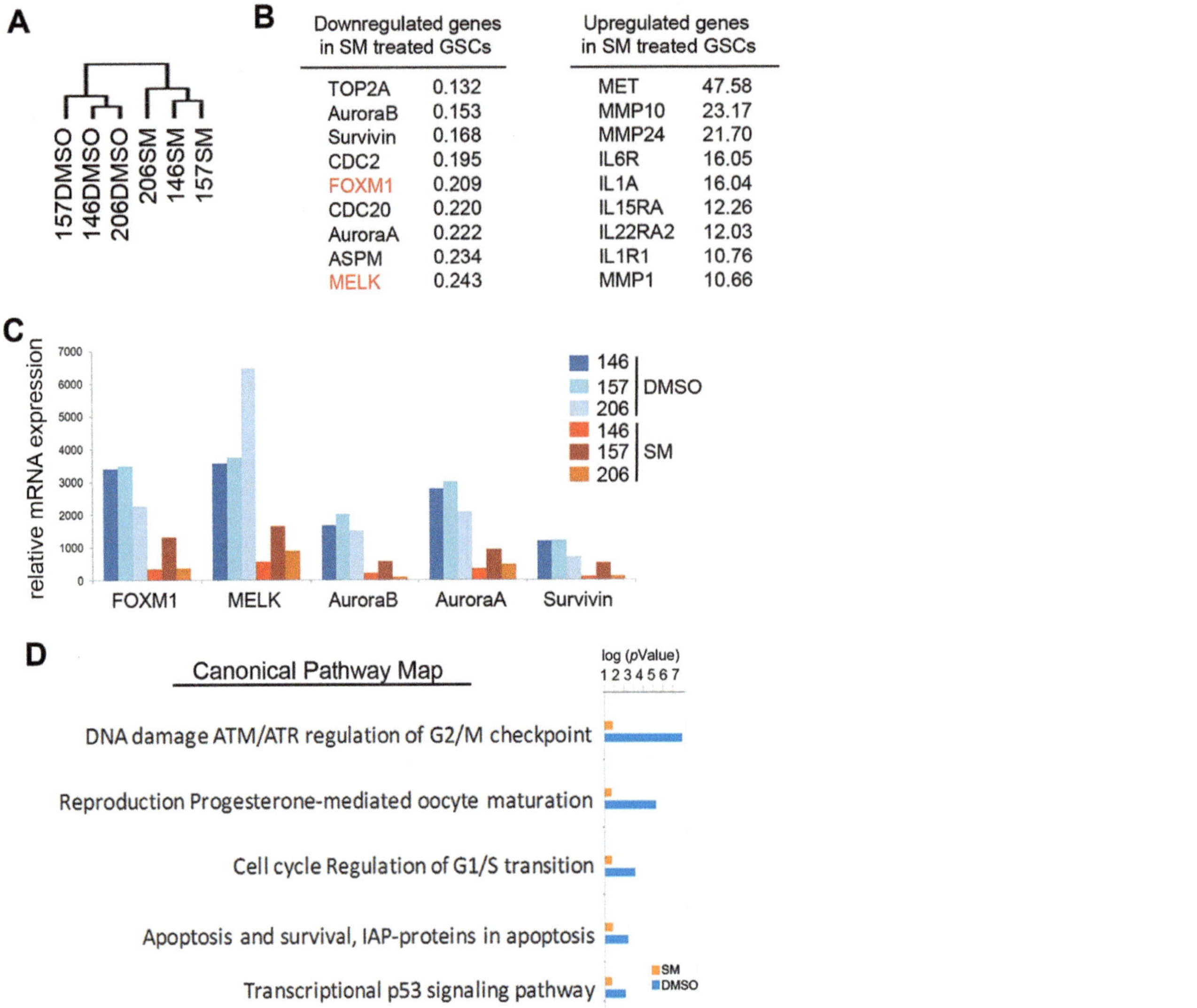

Figure 1. Genes in the DNA damage-induced response pathway are downregulated in Siomycin A-treated GSCs. cDNA microarray of GBM146, GBM157, and GBM206 samples treated with 1 μM Siomycin A or control (DMSO) were subjected to cluster (A) and canonical pathway analyses (D) using Ingenuity software. Log (*p*Value) of most significantly downregulated pathways are shown ($p<0.05$). The most downregulated and upregulated genes in Siomycin A-treated GSCs are shown in (B) and (C), respectively. Expression of FOXM1, MELK, Aurora A/B, and Survivin were significantly decreased by Siomycin A treatment compared with DMSO treatment.

neurosphere formation. C1 treatment attenuated neurosphere formation of all 3 GBM samples at substantially lower doses (GBM146: 440 nM; GBM157: 370 nM; GBM206: 370 nM) than normal progenitors (16wf: 790 nM)(Fig. 3A).

We then performed FACS analysis with GSCs treated with either C1 or DMSO, as the expression of the cell surface CD133 is well-recognized as a surrogate, but not definitive, marker for GSCs [24,29,30]. Following separation of GBM157 cells into CD133(+) and CD133(−) populations by cell sorter, individual cell populations were separately treated with 1μM of C1 or DMSO under the identical serum-free conditions for 48 hours. CD133(+) tumor cells with C1 treatment showed significant reduction of the sphere number compared with the DMSO-treated cells. On the contrary this C1-induced cell number reduction was not apparently observed in CD133(−) tumor cells ($p<0.01$; Fig. 3B). Collectively, C1 has more potent inhibitory effect on GSC growth compared to the growth of non-GSCs or normal progenitors *in vitro*.

C1 Induces Apoptotic Cell Death in Organotypic Slice Cultures of GBM Surgical Specimens

Organotypic slice culture of tumor tissues is one experimental model readily available to assess the influence of drug treatment on tumor cells *in situ* without significantly destroying the tumor microenvironment [31]. We performed intratumoral injection of C1 or DMSO using 2 patient-derived slice cultures of GBM tumors (Fig. 3C). Immunohistochemistry for Ki-67 exhibited the presence of abundant proliferating tumor cells in the DMSO treated samples but not in the C1 treated ones ($p<0.0026$; Fig. 3D), suggesting that C1 treatment suppressed GBM cell proliferation *in situ*. We then assessed the effect of C1 injection on sphere-forming GSCs within these slice cultures. After the C1-treated or DMSO-treated slice cultures were dissociated into single cells, we performed neurosphere formation assay. As shown in Fig. 3E, we found the significant reduction both of the number of neurospheres and the total cell numbers by C1 treatment in both cases

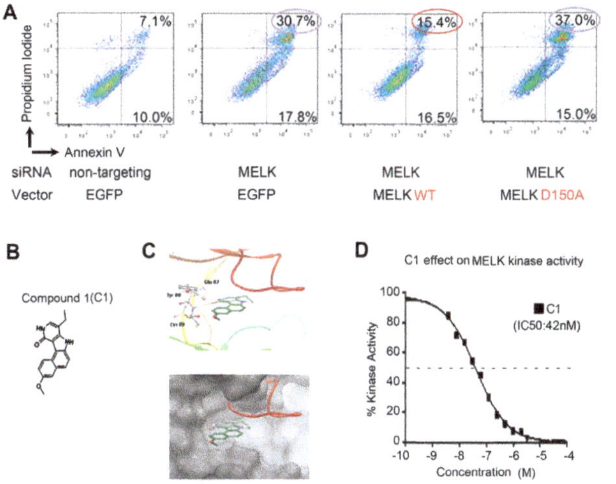

Figure 2. The kinase activity plays an essential role in the action of MELK on GSCs and identification of C1 as a MELK kinase inhibitor. A, GBM1600 spheres transfected with MELK wild type vector or MELK kinase dead (MELK D150A) vector were subjected to transfection of shRNA for MELK to eliminate endogenous MELK. Non-targeting shRNA and GFP vector were used as controls. The proportion of apoptosis at 48 hours post-transfection was analyzed by flow cytometry using Annexin V antibody and Propidium Iodide. B, Schema showing the chemical structure of the benzo[e]pyridoindole, Compound 1 (C1). C, *In silico* modeling of C1 predicted binding to MELK through hydrogen bonding with the backbone residues of the hinge region of the ATP binding pocket (Glu87, Tyr88, Cys89) by computational analysis. A pair of hinge residue hydrogen bonds makes favorable hydrophobic interactions. D, A ten-dose curve allows the determination of the *in vitro* efficiency of C1 towards the recombinant kinase (IC50: concentration of C1 that inhibits kinase activity by 50%). Solid black squares indicate individual data points.

($p<0.01$). These data suggest that C1 has a potent inhibitory effect on the GBM cell proliferation and GSC growth under the condition of a relatively preserved microenvironment for GBM tumors.

C1 Inhibits GBM Cell Growth *In vivo*

Given the potent inhibitory effects of C1 on GSCs *in vitro* and *ex vivo*, we next investigated the effect of C1 treatment on mouse models of GBM *in vivo*. Following transplantation of GBM157 cells into immunocompromised mouse brains, we injected C1 at day 7 with quantities of 2.5 pmol, 25 pmol, and 250 pmol into the tumor cavities (Fig. 3F). At day 3 post-C1 injection, we observed a substantial decrease in proliferating GBM157 cells, as evidenced by far fewer cells labeled with Ki-67(+) in C1-treated (25 pmol and 250 pmol) tumors than DMSO-treated (40% in DMSO-treated tumors vs. 32% in C1-treated tumors (25 pmol); Fig. 3G).

With the other set of mice harboring GBM157-derived intracranial tumors, we measured tumor sizes at 8 weeks post-transplantation. While there was only a marginal difference in tumor sizes between the control and C1 treatment at 2.5 pmol, tumors treated with C1 at both 25 pmol and 250 pmol exhibited a 3-fold decrease in size compared to the control ($n = 7$, $p<0.0001$; Fig. 3H and I). These results suggest that intra-tumoral treatment with C1 diminishes the *in vivo* growth of GSC-derived tumors in mouse brains.

C1 Treatment Induces Early Mitotic Exit and Subsequent Mitotic Catastrophe and Cell Death of GSCs

Next, we investigated the mechanism of cell cycle arrest induced by MELK kinase inhibition in 2 GBM-derived cell lines, U87 and U251. We first confirmed that C1 decreases their growth *in vitro*. With both cell lines, the total viable cells were significantly decreased by C1 treatment compared to DMSO-treated cells (U87: $p<0.001$, U251: $p<0.0001$; Fig. 4A). We then performed time-lapsed immunocytochemistry with C1-treated U251 cells expressing GFP. The top panel in Fig. 4B (left side is treated with DMSO, right side with C1) displayed mitotic cells readily dividing and giving rise to two daughter cells (arrows and arrowheads). C1-treated mitotic U251 cells could not undergo anaphase and prematurely escaped from mitosis, as evidenced by the round cell spread on the substrate at 1 hour 40 min (cell focused from former field) post-treatment. As showed in the bottom panel of Fig. 4B, C1-treated cells at 72 hours formed polyploid cells with multiple abnormal micronuclei within single cells, unlike the control cells exhibiting normal, euploid nuclei. This phenomenon is known as "*mitotic catastrophe*," which irreversibly drives cells into apoptosis, necrosis, or senescence [32]. In contrast, when U251 cells were treated with temozolomide (TMZ), the current first-line chemotherapy for GBM, cell growth was attenuated without affecting mitotic division (data not shown). Collectively these data suggest that the mechanism of cell death by C1 is distinct from that of TMZ.

Since C1 is a multi-kinase inhibitor, we then sought to determine the major molecular target of C1 in GSCs. We postulated that if C1 attenuates GSC survival mainly through MELK kinase inhibition, the phenotype of C1-treated GSCs should be identical to that of shRNA-mediated MELK elimination. Previously, we found that MELK knockdown prevents GSCs from proper mitotic progression, resulting in accumulation of tumor cells in the G2/M phase, generation of polyploid cells with multiple micronuclei, and subsequent apoptotic cell death [17,33]. We monitored C1-treated GBM1600 cells from pre-mitosis phase until mitotic completion. As expected, GBM1600 cells at pre-mitosis phase express high levels of MELK protein (Fig. 4C, left panel). We then followed the effects of C1 through karyokinesis and cytokinesis of GBM1600 cells. These C1 treated GBM1600 cells demonstrated micronucleated, chromatin condensation (signs of apoptosis) and asymmetric division (Fig. 4C, middle and right panels). DMSO-treated cells, on the other hand, showed round chromatin and homogenous cell size and figures. To support these data, cell cycle analysis with PI showed the accumulation of C1-treated GBM1600 GSCs at the G2/M phase (62.7% with C1 treatment vs. 19.3% with DMSO treatment; Fig. 4D). To determine the subsequent cell fate of C1 treated cells, we performed FACS analysis with U251 cells using Annexin V and PI (Fig. 4E). Both early and late apoptotic fractions were significantly increased in C1-treated cells in a dose-dependent manner ($p<0.001$; Fig. 4E). Taken together, these data indicate that C1 treatment induces failure of mitotic progression to undergo anaphase and G2/M arrest of GSCs. These effects subsequently promote early mitotic exit, metaphase-associated mitotic catastrophe, and subsequent apoptotic cell death.

C1 Sensitizes GSCs to Radiation Treatment

Irradiation is the current first line post-surgical therapy for GBM patients. Nonetheless, the survival benefit for GBM patients of radiation treatment is no greater than 3 months [34,35]. One potential reason for the limited efficacy of this treatment is the rapid induction of the DNA damage repair genes and proteins in tumor cells [4]. In particular, GSCs are known to upregulate these

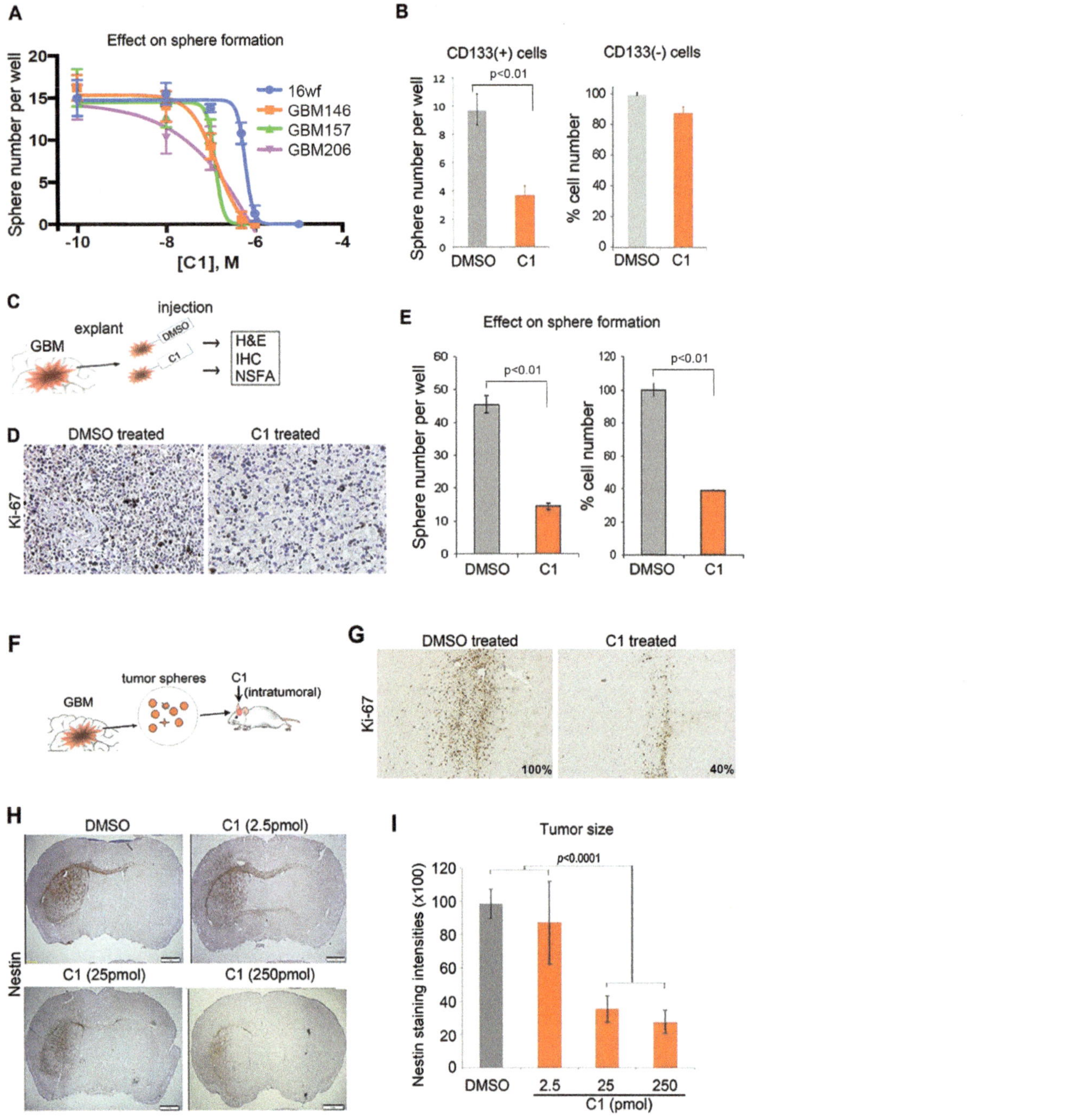

Figure 3. C1 treatment inhibits GSC proliferation *in vitro* and *in vivo*. A, Graph of neurosphere forming assay indicating the relative neurosphere numbers of C1-treated patient-derived GBM samples (GBM146, GBM157, and GBM206) and normal neural progenitors (16wf). B, CD133(+) and (−) cells, separated from GBM157-derived sphere cultures, were treated with 1 μM C1 or DMSO (control) under the identical serum-free conditions for 48 hours. The effect on CD133(+) cells was assessed by the neurosphere number per well, and the effect on CD133(−) cells was assessed by the % change of the total cell number in comparison to the control sample. C, Schematic showing organotypic slice cultures explanted GBM tissues and treated with C1 or DMSO (control) for 16 hours and evaluated with H&E, Ki67, and Nestin staining. D, Immunohistochemistry of C1- or DMSO-treated GBM slice cultures with anti-Ki-67 monoclonal antibody (Original magnification, ×200). E, Graph indicating the numbers of neurospheres (left) or total cells (right) in serum-free medium derived from C1- or DMSO-treated slice cultures for 16 hours. F, Schematic drawing of the effect of C1 treatment for the mouse intracranial GBM models derived from GSCs. Cells from GBM157 spheres were injected intracranially into immunocompromised mice (C1 mice: n=4, control mice: n=12). At day 7 post transplantation, C1 was injected intratumorally at quantities of 2.5 pmol (n=3), 25 pmol (n=4), or 250 pmol (n=5). G, Representative images for immunohistochemistry with Ki-67 staining of GBM slice cultures treated with 25 pmol C1 or DMSO at day 10 of treatment. Ki-67 positive cells in each group were analyzed automated digital image analysis (Original magnification, ×200). H, Representative images for immunohistochemistry with human-specific Nestin antibody using GBM157-derived mouse intracranial tumors treated with varying doses of C1 or DMSO intratumoral injection (bar: 1 mm). I, Graph indicates tumor sizes in each group as determined by Nestin staining intensities analyzed using automated digital image analysis.

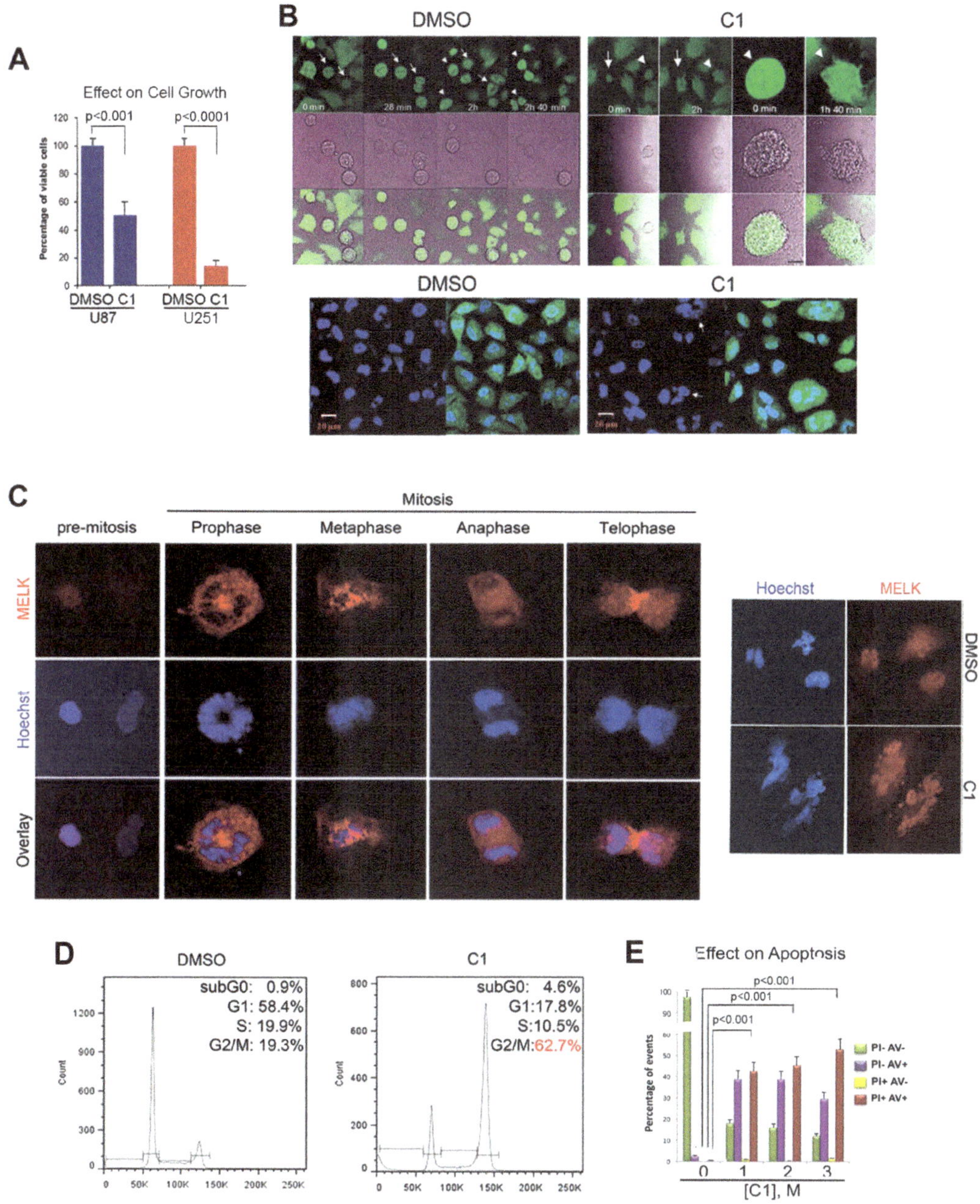

Figure 4. C1 treatment accumulates GSCs in G2/M and triggers subsequent mitotic catastrophe. A, Proliferation assays on two glioblastoma cell lines (U87 and U251). U87 and U251 cells were treated with 5.7 μM C1 or DMSO. Cells were trypsinized and estimated by counting, in duplicate, after 72 h of treatment. Two different experiments were conducted with similar results. B, Time-Lapse on mitotic U251 cells stably expressing GFP was performed in the absence (DMSO) or in the presence of C1 (at 1 μM). The compound was added to the cell culture just before imaging and then cells were continuously imaged. Three independent experiments were conducted and 10 to 15 fields were followed in each. None of the followed mitotic cells divided in two daughter cells. Representative field is imaged, DNA is in blue and the merge shows GFP and DNA. Several polyploid cells were present in the image. Arrows and arrowheads in upper panels of DMSO and C1- treated cell indicate the same cells through time-lapse. The elapse times are indicated on each photo, in some assays, a zoom of one cell is shown (the red bar represents 5 μm) this cell is present on the former field and labelled with an arrow). Images in bottom panels show DNA only (left) and DNA overlap with GFP (right) after 72-hour treatment with DMSO or C1 (the red bars on each panel represent 20 μm). Arrows in the bottom panel of C1-treated cells indicate mitotic catastrophe by C1 treatment. C, Pictures demonstrate pre-mitotic phase (left panel), mitotic (mid panel) process by full karyokinesis and cytokinesis and after cell division (right panel). MELK expression was determined with immunocytochemistry of GBM1600 cells with anti-MELK antibody (red), chromatin staining with Hoechst stain (blue). Picture of pre-mitosis shows GBM1600 cells highly expressed MELK at pre-mitosis phase (400× magnification). Then GBM1600 cells were treated with 5 μM C1 or control and were subjected to immunocytochemistry 3 days later with anti-MELK and chromatin staining (640× magnification). Data were confirmed by three independent experiments. C1 treated cells are micronucleated at metaphase and followed multinuclear chromatin condensation (mid panel). Right panel show multinuclear asymmetric divided chromatin of C1 treated cell compared with DMSO treated cell. D, Flow cytometric analysis of C1- and DMSO-treated GBM1600 cells with Propidium Iodide at 3 days after treatment shows 62.7% of C1-treated cells resulted in the G2/M arrest, whereas the control cells have 19.3% of the G2/M arrested cells.E, Graph indicating the proportions of live, early apoptotic, and late apoptotic U251 cells with varying doses of C1 or DMSO.

DNA repair genes more efficiently than non-GSCs, which may partially explain their pronounced radioresistance [4,36]. Since we found that inhibition of MELK-mediated pathway potently suppressed the DNA damage repair pathway in GSCs (Fig. 1D), we hypothesized that C1 treatment combined with radiation would have a greater efficacy over radiation alone. To address this question, we used radiation treatment at sub-lethal doses (2Gy and 4Gy) for GSCs with or without C1 treatment (Fig. 5). While radiation alone did not noticeably affect GSC survival at the indicated doses, the combination with 1μM of C1 treatment resulted in a significant reduction in the GSC growth ($p<0.0001$), indicating that C1 treatment sensitizes GSCs to radiation-induced cell death *in vitro*.

Discussion

In this study, we demonstrated that MELK acts on GSC survival through its kinase activity. We performed the computational structure analysis of MELK protein to determine the ATP binding region of this kinase. Using this information, we identified C1 as a kinase inhibitor with substantial potency against MELK. C1 treatment not only induced G2/M arrest and subsequent mitotic catastrophe of GSCs *in vitro*, but also inhibited the growth of GSC-derived tumors *in vivo*. Finally, we found that C1 treatment sensitizes GSCs to radiation-induced cell death, supporting C1 as an attractive molecule-targeting therapy to combine with current standard protocols of GBM treatment.

Our study contributes several novel findings to the current understanding of the GBM pathophysiology. To our knowledge, we present the first pre-clinical data for MELK kinase inhibition-mediated suppression of GBM cell growth *in vivo* using a small molecule kinase inhibitor. Previously, we found that the oxo-group of C1 is necessary for kinase inhibitory activity [27]. Our *in silico* model predicted that the oxo-group of C1 forms a strong hydrogen bond with the Tyr88 residue of the ATP-binding triplet domain, confirming our earlier report [21]. Our study therefore illustrates that pharmacological inhibition of kinase activity at the ATP binding site has very attractive therapeutic potential and strongly warrants further study and drug development. However, similarity of the structure of the ATP binding site has been noticed with various protein kinases. In fact, our prior study exhibited that C1 inhibits multiple protein kinases with potent effects on not only MELK but also Aurora kinases and Chk2 with the highest affinities towards Aurora B and MELK [21]. With the current dataset, we cannot draw a definitive conclusion that the potent effect on GSCs by C1 treatment is solely due to MELK inhibition. Both Aurora B and MELK are mitotic kinases, and the relative inhibition of each by C1 is indistinguishable. However, the selectivity towards CD133(+) cells in this study may suggest that MELK inhibition predominantly contributes to C1 efficacy in GSCs [10,16]. The individual effects of C1 on these kinases aside, we found that inhibition of the MELK-mediated pathway also suppresses expression of Aurora kinases (Fig. 1B and C). In addition, we recently revealed that the oncogenic transcription factor FOXM1 is a substrate for MELK specifically in GSCs. Given that Aurora kinases are known downstream targets of FOXM1, multi-kinase inhibition would provide better efficacy with lower doses to avoid unwanted toxic effects on normal orgains [19]. Further work is definitely needed to determine the contributions of Aurora kinases and other protein kinases to GBM pathogenesis and to clarify the full scope of C1 target molecules.

Given that mutations in the kinase domain or in other molecular elements of MELK do not appear to be a frequent event in cancers, the distinct function of MELK in normal and oncogenic stem cells is likely to be epigenetic. Several recent studies, including our own, have implicated MELK in cell cycle regulation [37,38], successful cell division [39], and suppression of apoptosis [16,40,41], making it entirely possible that MELK contributes to tumor initiation and propagation through molecular interactions with deregulated oncogenes and/or tumor suppressor genes. In support of this notion, we recently demonstrated that a JNK pathway-driven interaction of MELK with another transcription factor/oncoprotein c-JUN is essential for GSC survival, proliferation, and radioresistance in a p53 dependent manner [18]. Introducing a point mutation in MELK protein at the D150 residue, which is required for proper kinase activity [28], attenuated the protein complex formation with c-JUN. Furthermore, this interaction with c-JUN was unique to GSCs and was not found in normal neural progenitors. Collectively, it is possible that C1 interrupts the oncogenic JNK signaling cascade through inhibition of MELK kinase activity and the resulting interaction with c-JUN. Given that JNK signaling orchestrates a variety of cellular processes, pharmacological inhibition of MELK, a more downstream and possibly cancer-specific protein, may lead to fewer off-target effects and greater specificity in targeting cancer cells. Further studies are required to elucidate this possibility.

The potent radioresistance of GSCs has been partly attributed to upregulation of the ATM/ATR DNA damage response pathway [42,43]. In this study, we found that the greatest effect of MELK signaling inhibition was on the ATM/ATR DNA damage response pathway and C1 treatment radiosensitizes GBM cells at least *in vitro*. Recently, Golding *et al.* reported that ATM inhibition effectively radiosensitizes GBM cells without harming normal neural progenitor cells [44]. Further, Raso *et al.* demonstrated that radiosensization through ATM inhibition occurs preferentially in GSCs but not in non-GSCs [45]. We previously demonstrated that treatment of GSCs with Siomycin A reduces GSC-derived tumor growth *in vivo* without causing a noticeable harmful effect on normal brain cells [16]. Taken together, MELK inhibition may attenuate radiation-induced ATM/ATR activation in GSCs that are essential for their role in the DNA damage repair and survival.

Regarding the clinical application of C1 for GBM therapeutics, some open questions remain. In fact, the efficacy of chemotherapy of brain malignancies is often hampered by the presence of the blood-brain barrier (BBB). From the point of molecular weight, the size calculated from the structure of C1 is 293 Da, which is

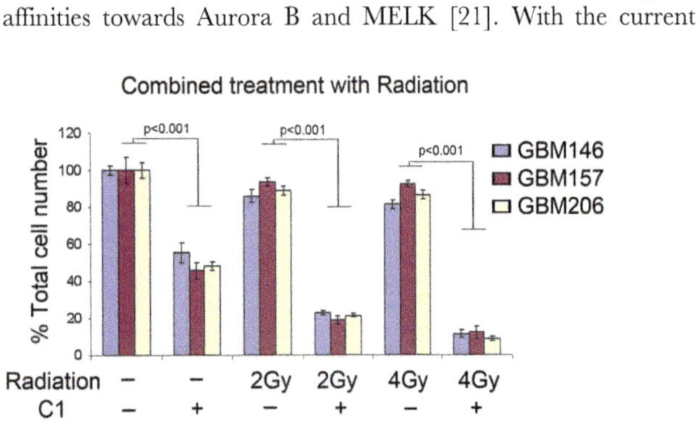

Figure 5. C1 sensitizes GBM cells to radiation-induced cell death, *in vitro*. Graph indicating the relative total cell numbers after monotherapy (1 μM C1 or radiation) or after combination therapy. Cells treated with combination therapy were radiated (2Gy or 4Gy) at 24 hours post-C1 treatment and the subsequent total cell numbers were counted at 3 days after irradiation. Data are shown from 3 separate experiments.

presumably small enough to penetrate the BBB. However, the permeability of the BBB is not solely dependent on the molecular size but also affected by several kinds of drug property and circumstances. Given the potent effect of C1 treatment on mouse GBM-like tumor models *in vivo*, it is attempted to evaluate the permeability of the BBB and bioavailablity/stability of C1 *in vivo*.

In conclusion, our data indicate that C1 is a novel inhibitor for protein kinases with substantial inhibitory effect on MELK. This study suggests that pharmacological inhibition of MELK kinase activity represents an attractive therapeutic approach for GBM that may overcome the resistance seen after current, standard treatment protocols. We postulate that C1 may also effectively treat a variety of cancers with elevated activation of MELK.

Acknowledgments

We thank Dr. Jeremy Rich for constructive criticism for this study. We also thank Dr. Chenglong Li for help on protein structure analysis in this study.

Author Contributions

Conceived and designed the experiments: IN. Performed the experiments: CG CH KJ CHN AM. Analyzed the data: HIK AM IN. Contributed reagents/materials/analysis tools: CHN. Wrote the paper: CH HIK AM IN.

References

1. Filiano AN, Fathallah-Shaykh HM, Fiveash J, Gage J, Cantor A, et al. (2011) Gene expression analysis in radiotherapy patients and C57BL/6 mice as a measure of exposure to ionizing radiation. Radiat Res 176: 49–61.
2. Grunda JM, Fiveash J, Palmer CA, Cantor A, Fathallah-Shaykh HM, et al. (2010) Rationally designed pharmacogenomic treatment using concurrent capecitabine and radiotherapy for glioblastoma; gene expression profiles associated with outcome. Clin Cancer Res 16: 2890–2898.
3. Sulman E, Aldape K, Colman H (2008) Brain tumor stem cells. Curr Probl Cancer 32: 124–142.
4. Bao S, Wu Q, McLendon RE, Hao Y, Shi Q, et al. (2006) Glioma stem cells promote radioresistance by preferential activation of the DNA damage response. Nature 444: 756–760.
5. Hambardzumyan D, Squatrito M, Carbajal E, Holland EC (2008) Glioma formation, cancer stem cells, and akt signaling. Stem Cell Rev 4: 203–210.
6. Furnari FB, Fenton T, Bachoo RM, Mukasa A, Stommel JM, et al. (2007) Malignant astrocytic glioma: genetics, biology, and paths to treatment. Genes Dev 21: 2683–2710.
7. Fan QW, Cheng C, Hackett C, Feldman M, Houseman BT, et al. (2010) Akt and autophagy cooperate to promote survival of drug-resistant glioma. Sci Signal 3: ra81.
8. Ziegler DS, Wright RD, Kesari S, Lemieux ME, Tran MA, et al. (2008) Resistance of human glioblastoma multiforme cells to growth factor inhibitors is overcome by blockade of inhibitor of apoptosis proteins. J Clin Invest 118: 3109–3122.
9. Zhong D, Liu X, Khuri FR, Sun SY, Vertino PM, et al. (2008) LKB1 is necessary for Akt-mediated phosphorylation of proapoptotic proteins. Cancer Res 68: 7270–7277.
10. Nakano I, Masterman-Smith M, Saigusa K, Paucar AA, Horvath S, et al. (2008) Maternal embryonic leucine zipper kinase is a key regulator of the proliferation of malignant brain tumors, including brain tumor stem cells. J Neurosci Res 86: 48–60.
11. Hebbard LW, Maurer J, Miller A, Lesperance J, Hassell J, et al. (2010) Maternal embryonic leucine zipper kinase is upregulated and required in mammary tumor- initiating cells in vivo. Cancer Res 70: 8863–8873.
12. Gray D, Jubb AM, Hogue D, Dowd P, Kljavin N, et al. (2005) Maternal embryonic leucine zipper kinase/murine protein serine-threonine kinase 38 is a promising therapeutic target for multiple cancers. Cancer Res 65: 9751–9761.
13. Kuner R, Falth M, Pressinotti NC, Brase JC, Puig SB, et al. (2013) The maternal embryonic leucine zipper kinase (MELK) is upregulated in high-grade prostate cancer. J Mol Med (Berl) 91: 237–248.
14. Lee J, Son MJ, Woolard K, Donin NM, Li A, et al. (2008) Epigenetic-mediated dysfunction of the bone morphogenetic protein pathway inhibits differentiation of glioblastoma-initiating cells. Cancer Cell 13: 69–80.
15. Visnyei K, Onodera H, Damoiseaux R, Saigusa K, Petrosyan S, et al. (2011) A molecular screening approach to identify and characterize inhibitors of glioblastoma stem cells. Mol Cancer Ther 10: 1818–1828.
16. Nakano I, Joshi K, Visnyei K, Hu B, Watanabe M, et al. (2011) Siomycin A targets brain tumor stem cells partially through a MELK-mediated pathway. Neuro Oncol 13: 622–634.
17. Nakano I, Paucar AA, Bajpai R, Dougherty JD, Zewail A, et al. (2005) Maternal embryonic leucine zipper kinase (MELK) regulates multipotent neural progenitor proliferation. J Cell Biol 170: 413–427.
18. Gu C, Banasavadi-Siddegowda YK, Joshi K, Nakamura Y, Kurt H, et al. (2013) Tumor-specific activation of the C-JUN/MELK pathway regulates glioma stem cell growth in a p53-dependent manner. Stem Cells 31: 870–881.
19. Joshi K, Banasavadi-Siddegowda Y, Mo X, Kim SH, Mao P, et al. (2013) MELK- dependent FOXM1 phosphorylation is essential for proliferation of glioma stem cells. Stem Cells 31: 1051–1063.
20. Hemmati HD, Nakano I, Lazareff JA, Masterman-Smith M, Geschwind DH, et al. (2003) Cancerous stem cells can arise from pediatric brain tumors. PNAS 100: 15178–83.
21. Hoang TM, Favier B, Valette A, Barette C, Nguyen CH, et al. (2009) Benzo[e]pyridoindoles, novel inhibitors of the aurora kinases. Cell Cycle 8: 765–772.
22. Mahasenan KV, Li C (2012) Novel inhibitor discovery through virtual screening against multiple protein conformations generated via ligand-directed modeling: a maternal embryonic leucine zipper kinase example. J Chem Inf Model 52: 1345–1355.
23. Joshi K, Demir H, Yamada R, Miyazaki T, Ray-Chaudhury A, et al. (2011) Method for novel anti-cancer drug development using tumor explants of surgical specimens. J Vis Exp.
24. Mao P, Joshi K, Li J, Kim SH, Li P, et al. (2013) Mesenchymal glioma stem cells are maintained by activated glycolytic metabolism involving aldehyde dehydrogenase 1A3. Proc Natl Acad Sci U S A 110: 8644–8649.
25. Miyazaki T, Pan Y, Joshi K, Purohit D, Hu B, et al. (2012) Telomestatin impairs glioma stem cell survival and growth through the disruption of telomeric G-quadruplex and inhibition of the proto-oncogene, c-Myb. Clin Cancer Res 18: 1268–1280.
26. Delacour-Larose M, Vu HL, Molla A (2009) Aurora B kinase, an immobile passenger! Cell Cycle 8: 3600–3601.
27. Le LT, Vu HL, Naud-Martin D, Bombled M, Nguyen CH, et al. (2013) Hydrosoluble benzo[e]pyridoindolones as potent inhibitors of aurora kinases. ChemMedChem 8: 289–296.
28. Hanks SK, Hunter T (1995) Protein kinases 6. The eukaryotic protein kinase superfamily: kinase (catalytic) domain structure and classification. FASEB J 9: 576–596.
29. Lottaz C, Beier D, Meyer K, Kumar P, Hermann A, et al. (2010) Transcriptional profiles of CD133+ and CD133- glioblastoma-derived cancer stem cell lines suggest different cells of origin. Cancer Res 70: 2030–2040.
30. Beier D, Rohrl S, Pillai DR, Schwarz S, Kunz-Schughart LA, et al. (2008) Temozolomide preferentially depletes cancer stem cells in glioblastoma. Cancer Res 68: 5706–5715.
31. Hu B, Kong LL, Matthews RT, Viapiano MS (2008) The proteoglycan brevican binds to fibronectin after proteolytic cleavage and promotes glioma cell motility. J Biol Chem 283: 24848–24859.
32. Vitale I, Galluzzi L, Castedo M, Kroemer G (2011) Mitotic catastrophe: a mechanism for avoiding genomic instability. Nat Rev Mol Cell Biol 12: 385–392.
33. Ianzini F, Mackey MA (1997) Spontaneous premature chromosome condensation and mitotic catastrophe following irradiation of HeLa S3 cells. Int J Radiat Biol 72: 409–421.
34. Louis DN, Ohgaki H, Wiestler OD, Cavenee WK, Burger PC, et al. (2007) The 2007 WHO classification of tumours of the central nervous system. Acta Neuropathol 114: 97–109.
35. Stupp R, Hegi ME, Mason WP, van den Bent MJ, Taphoorn MJ, et al. (2009) Effects of radiotherapy with concomitant and adjuvant temozolomide versus radiotherapy alone on survival in glioblastoma in a randomised phase III study: 5-year analysis of the EORTC-NCIC trial. Lancet Oncol 10: 459–466.
36. Tamura K, Aoyagi M, Wakimoto H, Ando N, Nariai T, et al. (2010) Accumulation of CD133-positive glioma cells after high-dose irradiation by Gamma Knife surgery plus external beam radiation. J Neurosurg 113: 310–318.
37. Mirey G, Chartrain I, Froment C, Quaranta M, Bouche JP, et al. (2005) CDC25B phosphorylated by pEg3 localizes to the centrosome and the spindle poles at mitosis. Cell Cycle 4: 806–811.
38. Davezac N, Baldin V, Blot J, Ducommun B, Tassan JP (2002) Human pEg3 kinase associates with and phosphorylates CDC25B phosphatase: a potential role for pEg3 in cell cycle regulation. Oncogene 21: 7630–7641.
39. Chartrain I, Couturier A, Tassan JP (2006) Cell-cycle-dependent cortical localization of pEg3 protein kinase in Xenopus and human cells. Biol Cell 98: 253–263.
40. Jung H, Seong HA, Ha H (2008) Murine protein serine/threonine kinase 38 activates apoptosis signal-regulating kinase 1 via Thr 838 phosphorylation. J Biol Chem 283: 34541–34553.

41. Lin ML, Park JH, Nishidate T, Nakamura Y, Katagiri T (2007) Involvement of maternal embryonic leucine zipper kinase (MELK) in mammary carcinogenesis through interaction with Bcl-G, a pro-apoptotic member of the Bcl-2 family. Breast Cancer Res 9: R17.
42. Frosina G (2009) DNA repair and resistance of gliomas to chemotherapy and radiotherapy. Mol Cancer Res 7: 989–999.
43. Ropolo M, Daga A, Griffero F, Foresta M, Casartelli G, et al. (2009) Comparative analysis of DNA repair in stem and nonstem glioma cell cultures. Mol Cancer Res 7: 383–392.
44. Golding SE, Rosenberg E, Adams BR, Wignarajah S, Beckta JM, et al. (2012) Dynamic inhibition of ATM kinase provides a strategy for glioblastoma multiforme radiosensitization and growth control. Cell Cycle 11: 1167–1173.
45. Raso A, Vecchio D, Cappelli E, Ropolo M, Poggi A, et al. (2012) Characterization of glioma stem cells through multiple stem cell markers and their specific sensitization to double-strand break-inducing agents by pharmacological inhibition of ataxia telangiectasia mutated protein. Brain Pathol 22: 677–688.

3

Mesenchymal Stem Cell-Cardiomyocyte Interactions under Defined Contact Modes on Laser-Patterned Biochips

Zhen Ma[1¤], Huaxiao Yang[1], Honghai Liu[1], Meifeng Xu[2], Raymond B. Runyan[3], Carol A. Eisenberg[4], Roger R. Markwald[5], Thomas K. Borg[5], Bruce Z. Gao[1*]

1 Department of Bioengineering, Clemson University, Clemson, South Carolina, United States of America, **2** Department of Pathology and Laboratory Medicine, University of Cincinnati Medical Center, Cincinnati, Ohio, United States of America, **3** Department of Cellular and Molecular Medicine, University of Arizona, Tucson, Arizona, United States of America, **4** New York Medical College/Westchester Medical Center Stem Cell Laboratory, New York Medical College, Valhalla, New York, United States of America, **5** Department of Regenerative Medicine and Cell Biology, Medical University of South Carolina, Charleston, South Carolina, United States of America

Abstract

Understanding how stem cells interact with cardiomyocytes is crucial for cell-based therapies to restore the cardiomyocyte loss that occurs during myocardial infarction and other cardiac diseases. It has been thought that functional myocardial repair and regeneration could be regulated by stem cell-cardiomyocyte contact. However, because various contact modes (junction formation, cell fusion, partial cell fusion, and tunneling nanotube formation) occur randomly in a conventional coculture system, the particular regulation corresponding to a specific contact mode could not be analyzed. In this study, we used laser-patterned biochips to define cell-cell contact modes for systematic study of contact-mediated cellular interactions at the single-cell level. The results showed that the biochip design allows defined stem cell-cardiomyocyte contact-mode formation, which can be used to determine specific cellular interactions, including electrical coupling, mechanical coupling, and mitochondria transfer. The biochips will help us gain knowledge of contact-mediated interactions between stem cells and cardiomyocytes, which are fundamental for formulating a strategy to achieve stem cell-based cardiac tissue regeneration.

Editor: Maurizio Pesce, Centro Cardiologico Monzino, Italy

Funding: This work was partially supported by National Institutes of Health (NIH) (SC COBRE P20RR021949, Career Award 5k25hl088262-04, and 5R01 HL085847); National Science Foundation (NSF) (MRI, CBET-0923311and SC EPSCoR RII EPS-0903795 through SC GEAR program); and Guangdong Provincial Department of Science and Technology, China (2011B050400011). BZG would also like to acknowledge the support from the grant established by the State Key Laboratory of Precision Measuring Technology and Instruments (Tianjin University). ZM would like to acknowledge his Siebel Institute Postdoctoral Fellowship (41523-31595-44-OYZHMA-IQKEH). The funders had no role in study design, data collection and analysis, decision to publish, or preparation of the manuscript.

Competing Interests: The authors have declared that no competing interests exist.

* E-mail: zgao@clemson.edu

¤ Current address: Department of Bioengineering, University of California, Berkeley, California, United States of America

Introduction

Cardiovascular disease, a pervasive clinical problem afflicting more than five million Americans, has an annual mortality rate of more than 20% [1]. The pathology of diseases, such as myocardial infarction, involves death of cardiomyocytes and leads to dysfunctional tissue. Transplantation of exogenous stem cells to the heart has been proposed to prevent or reverse heart failure [2–4]. It has been shown that not only can the transplanted stem cells transdifferentiate into cardiac phenotypes, they can also protect native cardiomyocytes. The protective effect from stem cell benefits the infarcted myocardium in a paracrine manner by secreting multiple soluble factors, which may act through reduction in infiltration of inflammatory neutrophils, inactivation of fibrogenic cells and scarring, stimulation of angiogenesis and vascularization, or recruitment and activation of resident cardiac stem cells [5,6]. Besides the protective paracrine effect, contact-mediated intercellular interactions have been demonstrated to benefit myocardial repair and regeneration through three mechanisms that induce (1) cardiomyogenic differentiation of stem cells [7,8]; (2) functional integration of the stem cells with host cardiomyocytes [9,10]; and (3) delivery of molecules or even subcellular organelles from stem cells to enhance cardiomyocyte vitality and function [11]. However, a therapeutic procedure, which is heavily dependent on understanding signaling pathways and the structural and functional interactions between the transplanted stem cells and the host cardiomyocytes, has yet to be established.

The structural and functional integrations between cells are closely related to their particular contact-modes, including junction formation, tunneling nanotube connection, and cell fusion. In junction-formation mode, junctional proteins (e.g., connexins and cadherins) are distributed at the contact area between stem cells and cardiomyocytes: Connexins play an important role in electrical coupling, and cadherins do so for mechanical coupling [12]. Stem cells can also interact with cardiomyocytes by partial or full cell fusion process [13,14]: Fused cells exhibit both stem cell and cardiomyocyte characteristics. A newly discovered mode of intercellular interaction between stem cells and cardiomyocytes is formation of thin-membrane channels (tunneling nanotubes). These nanotubular structures contain actin and microtubules to establish cytosolic connectivity and facilitate intercellular transmission of various cellular components [15]. Our knowledge of contact-mediated, in vitro stem cell-cardiomyocyte interactions is based mainly on conventional cell-culture models, which contain (1) an undefined number of contacting cells (one cell contacts multiple cells simultaneously); (2) an undefined population of contacting cells (homotypic and heterotypic cell contacts); and (3) undefined contact modes. These undefined cellular contacts make it difficult in conventional cell-culture models to interpret and recognize the functional interactions associated with one specific contact mode formed between stem cells and cardiomyocytes.

To address this issue, we designed laser-patterned biochips to allow only one contact mode to form between stem cells and cardiomyocytes to systematically study their intercellular interactions. In our biochips, two types of microenvironment (contact-promotive and contact-preventive microwells) were created by lithographic microfabrication methods. Individual stem cells and cardiomyocytes were laser-patterned into the microwells using the laser-guided cell micropatterning technique (LGCM) [16], which provides high spatiotemporal resolution for single-cell studies. Biochips such as these, with well-defined cell arrangements, can be used to form specific contact modes so that contact-mediated stem cell-cardiomyocyte interaction corresponding to each specific contact mode can be quantified systematically.

Materials and Methods

Ethics Statement

All experiments that involved animal use were performed in compliance with relevant laws and institutional guidelines. These experiments were approved by Clemson University's Institutional Animal Care and Use Committee through protocol AUP2010-032.

Cell culture

Neonatal cardiomyocytes were isolated from three-day neonatal rats using a two-day protocol and cultured using high-glucose Dulbecco's Modified Eagle's Medium (DMEM) supplemented with 20% fetal bovine serum (FBS) and 1% penicillin streptomycin [17]. In previous characterizations, our isolation protocol yielded approximately 95% cardiomyocytes. Because of the size difference between cardiomyocytes and nonmyocyte cells, we were able to select and deposit only cardiomyocytes into the microwells during the laser-patterning procedure. Commercial rat mesenchymal stem cells (rMSCs) from bone marrow were purchased from ScienCell^{TM} research laboratories and characterized by the company with antibodies to CD73, CD90, CD105, and oil red staining after adipocyte differentiation. The rMSCs were cultured using mesenchymal stem-cell medium (MSCM) from the same company and used for cell patterning before the fifth passage. The cardiomyocyte-culture medium was used as the coculture medium for the biochips during and after the laser-patterning procedure. Characterization of the rMSCs and cardiomyocytes used in our culturing system has been recently published [18]. To further characterize the rMSCs and cardiomyocytes in the microwells after laser-patterning procedure, MSCs were stained for CD90 and CD105 in contact-promotive microwells, whereas cardiomyocytes were stained for cardiac troponin T and myosin heavy chain in contact-preventive microwells (Figure S1). The geometric restriction within the microwells did not affect the expression of these cell-type specific markers.

Biochip design and construction

The basic design concept was to create a microwell with defined geometrical restrictions that contained one rMSC and one cardiomyocyte. We produced two types of microwell: contact-promotive and contact-preventive. For each biochip, identical 9×9 microwells were created by attaching an elastomeric membrane with through holes onto a glass coverslip. The distance between two neighboring microwells was 200 μm. The microwell that promoted cellular contact was designed in a rectangular shape (50 μm long and 25 μm wide) that allowed the confined cell bodies (a single rMSC and a single cardiomyocyte) to form physical contact. The microwell that prevented cellular contact was designed in a dumbbell shape, with two circles (30 μm in diameter) connected by one channel (30 μm long and 15 μm wide). The curvature between the circular shape and the relatively narrow channel deterred formation of broad contact between a cell on one side of the channel and a cell on the other side. The microwell depth was 40 μm to restrict the cell bodies inside the microwells. Elastomeric membranes were microfabricated with polydimethylsiloxane (PDMS) using the standard procedure of photolithography and soft-lithography. The biochip assembly, treatment, and reuse are described in our previous publication [19]. Briefly, the biochips were cleaned, activated, sterilized by ethanol wash, oxygen plasma, and UV exposure, and coated with fibronectin before use. The biochips were placed inside the cell-deposition chambers, one biochip for each chamber, where individual rMSCs and cardiomyocytes were laser-patterned into the microwells.

Laser-guided cell micropatterning (LGCM)

Consecutive placement of single rMSCs and cardiomyocytes into the microwells on the biochip is critical to this research. The high spatiotemporal resolution required in multiple-cell-type positioning was achieved using LGCM, which is thoroughly described in our previous publication [20]. After placing one biochip into the cell-deposition chamber, we used a nanopump system to inject nanoliters of cell suspension into the chamber above the biochip. An individual cell was selected and guided to the microwell at a speed of 150 μm/s horizontally and 50 μm/s vertically by the optical force generated from a weakly focused Gaussian laser beam (Movie S1).

To create heterotypic cell pairs, cardiomyocytes were first laser-patterned into the left side of the microwells on the biochip, one in each microwell, and then rMSCs were patterned into the right side of the microwells, each of which already contained a cardiomyocyte, shown in Figure 1. The orientation of the microwell was determined by the triangular features established outside the central region, and thus cardiomyocytes were distinguished from rMSCs in the experiments that followed. Immediately after the laser-patterning procedure, the biochip was transferred from the cell-deposition chamber to a commercial 35 mm Petri dish containing cardiomyocyte culture medium. All biochips were incubated at 37°C in a humidified atmosphere of 95% air and 5% CO_2, and the culture medium was changed every two days.

Immunocytochemistry

On Day 5 (four days after laser-patterning), immunostaining was performed on the biochips. The cells were fixed, permeabilized, and blocked in 4% paraformaldehyde (10 min), 0.1% Triton X-100 (15 min), and 2% bovine serum albumin (BSA) with 4% donkey or goat serum. Next, the cells were labeled with the primary antibodies (mouse anti-sarcomericα-actinin (1:400, Sigma-Aldrich), rabbit anti-connexin 43 (1:200, Sigma-Aldrich), mouse anti-N-cadherin (1:200, Sigma-Aldrich), and mouse anti-

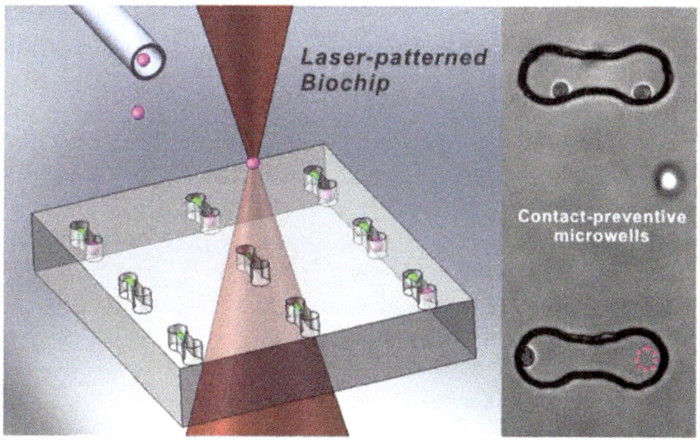

Figure 1. Schematic of the laser-patterning procedure. An rMSC is pumped out of the hollow fiber, trapped by the focused laser beam, and guided into the right side of a contact-preventive microwell (dotted pink circle) containing a cardiomyocyte (left side of the microwell). (pink: rMSCs; green: cardiomyocytes).

CD105 (1:200, Santa Cruz Biotechnology Inc.)) at 4°C overnight. Excess primary antibodies were removed by a triple wash in PBS, and cells were stained with secondary antibodies (Alexa Fluor 488 donkey-anti-mouse IgG (1:200, Invitrogen) and Oregon Green 488 goat-anti-rabbit (1:200, Invitrogen)) at room temperature for 2 hrs. After three washes with PBS, the slides were prepared with ProLong antifade kit mounting medium (Invitrogen Inc.) including cell-nuclei-staining DAPI. The DAPI-labeled nuclei can be used to identify the cell position in the microwells. The nonspecific control groups (one group without primary antibodies and one group without secondary antibodies) were conducted simultaneously with the experimental groups. Immunostained cells were imaged using a confocal fluorescent microscope (Eclipse Ti, Nikon) equipped with a high-sensitivity quantitative monochrome camera (CoolSnap HQ2, Photometrics) and NIS-Elements imaging software (Nikon).

Live cell-membrane and mitochondria labeling

In cell fusion studies, rMSCs and cardiomyocytes were labeled by DiI and DiO (Vybrant Multicolor Cell-Labeling Kit, Invitrogen), respectively, before the laser-patterning procedure. The serum culture medium was removed from the rMSC culture flask. The rMSCs were washed with warm PBS solution and then dissociated with 0.25% Trypsin-EDTA. The cells were collected in a 15 mL tube with serum culture medium and centrifuged into a pellet. The cells were resuspended by 2 mL serum-free medium (DMEM, 1% penicillin/streptomycin). Then, 10 μL DiI solution was added to the tube, and the cells were incubated in the dark at 37°C for 30 min. The cells were centrifuged into a pellet to remove the DiI solution and washed three times with serum-free medium. Then fresh serum culture medium was added to the tube for laser-patterning. Primary isolated cardiomyocytes were labeled with DiO using, except for the trypsin treatment, the protocol for rMSCs DiI-labeling.

To visualize mitochondria transfer from rMSCs to cardiomyocytes, the mitochondria of rMSCs were live-labeled by MitoTracker (MitoTracker Red CMXRos, Invitrogen). The lyophilized MitoTracker was dissolved in high-quality anhydrous dimethylsulfoxide (DMSO) to a final concentration of 1 mM as the stock solution. The labeling protocol was the same as that for cell-membrane labeling except for the use of a final MitoTracker concentration of 400 nM in serum-free medium.

Statistical analysis

The occurrence of each contact mode was statistically analyzed according to our immunostaining results. First, each biochip was observed under bright field, and microwells containing round-shape cells, which were considered to be dead cells, were not used in the confocal examination. The percentage of occurrence of each contact mode was calculated by equation below (Eq. 1) for one biochip. Totally, 10 identical biochips were analyzed to collect the statistical results. Data were expressed as mean ± SD.

$$\text{Percentage of contact mode} = \frac{\text{Number of cell pairs in the contact mode}}{\text{Total number of viable cell pairs}} \quad (1)$$

Results

Laser-patterned biochips

The biochips with heterotypic cells were created by laser-patterning rMSCs and cardiomyocytes into the contact-promotive and -preventive microwells, shown in Figure 2 (A, B); the images were taken immediately after the laser-patterning procedure using a Zeiss phase-contrast microscope. The cardiomyocytes were on the left side of each microwell, and the rMSCs were on the right side. Typically, after being cultured 48 hours, the surviving cell pairs spread inside the fibronectin-coated microwells. Surviving heterotypic cell pairs were 58±3.1% of all laser-patterned cell pairs, slightly lower than that of conventional culture, which is consistent with the reports emphasizing the decreased viability on microwell-based single-cell analysis [21,22]. In the following experiments, only the surviving cell pairs were studied. In Figure 2 (C, E), the DAPI-stained (blue) nuclei indicate the locations of the two cells inside one microwell, and the α-actinin staining (green) indicates cardiomyocyte structure. In contact-promotive biochips, 96±3.8% of the viable cell pairs formed broad contact between rMSCs and cardiomyocytes, as shown in Figure 2 (C, D); the remaining pairs, which did not contact each other (Figure S2 (A)), were not used in the statistical analysis. In contact-preventive biochips, cellular contact was impeded. Therefore, only 15±6.1% viable cell pairs were able to form broad cell contact by cell migration across the narrow channel. Further, 55±8.7% viable cell pairs did not form any type of cell contact as shown in Figure 2 (E, F), and the other cell pairs formed long-distance connections.

Junction formation

Junction formation between contacted cells was determined through imaging data on junctional protein expression: When the fluorescence image of junctional protein expression featured a thin line along the border of two cells in contact rather than a diffusive distribution, junction formation was counted. Two types of junctional proteins, connexin 43 and N-cadherin, were examined because of their importance to electrical and mechanical coupling between cardiac cells. For the heterotypic cell pairs, 84±3.9% were observed with connexin 43 expression, among which 28±6.5% exhibited junction formation, as shown in Figure 3 (A, B). The others exhibited only diffusive distribution throughout the cell bodies. In addition, 67±4.4% of cell pairs were observed with N-cadherin expression, among which 33±4.2% had junction formation, as shown in Figure 3 (C, D). The junctional-protein expression did not always cover the entire broad-contact area. In more than 90% of junction-formation mode, only a portion of the

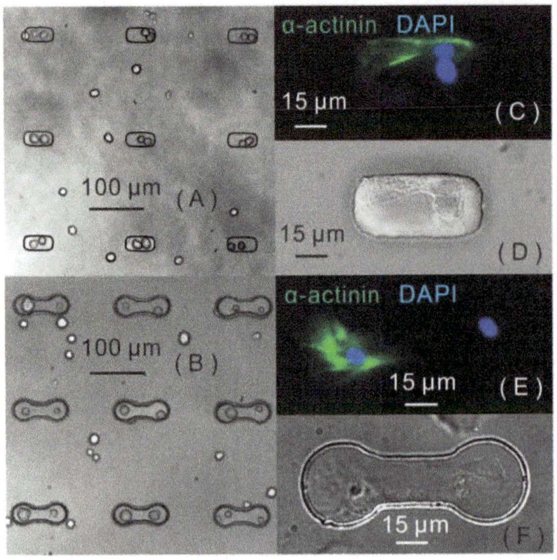

Figure 2. Laser-patterned biochips. Portions of laser-patterned biochips shown with (A) contact-promotive microwells and (B) contact-preventive microwells. After 48 hours culturing, the heterotypic cells (C, D) formed broad cellular contact in contact-promotive microwells; the heterotypic cells (E, F) did not form broad cellular contact in contact-preventive microwells. (α-actinin positive: cardiomyocytes, α-actinin negative: rMSCs, DAPI: nuclei).

cellular contact area was found with junctional distribution of connexin 43 and N-cadherin. Figure 3E demonstrates the formation of connexin 43 between PKH26-labeled cardiomyocytes and a GFP-positive rMSC in our coculture dish. It is also important to note that the diffusely expressed junctional proteins inside the cytoplasm are not likely to support intercellular coupling; rather they represent newly produced, trafficking, or internalized connexin and cadherin molecules [12].

Cell fusion

In general, cell fusion is a natural process present from the time that a spermatozoid fuses with an ovule, and it is closely related to tissue formation, viral infection, and immune response [23]. The importance of cell fusion to regenerative medicine is increasingly recognized. For example, bone marrow-derived stem cells are able to fuse with mature somatic cells, such as Purkinje neurons, cardiomyocytes, and hepatocytes [24]. Between rMSCs and cardiomyocytes in contact, this phenomenon is an infrequently occurring form of intercellular interaction: Less than 1% occurrence was reported in previous coculture studies [25,26]. To investigate the cell-fusion phenomenon at the single-cell level, we respectively labeled rMSCs and cardiomyocytes with DiI and DiO and laser-patterned them into the contact-promotive biochips. We found several double-nuclei cells with mixed labels (DiI and DiO), which indicated fused hybrid cells [11], as shown in Figure 4 (A–D). No clear contact area on the fused cell was observed in the bright-field image, which indicated full cell fusion accompanied by membrane reorganization. The ratio of cell fusion in our biochip-based coculture model varied from 0–6.3% among 15 contact-promotive biochips. Controlling cell contact in our biochips, we were able to evaluate the efficiency of cell fusion between rMSCs and cardiomyocytes at the single-cell level based on independent measurements of fluorescence mixing, nuclei centralization, and membrane reorganization.

In addition to full cell fusion, we observed partial cell fusion, in which two connected cells exchanged membrane and organelle components at their contact area [27]. In our research, mitochondrial transfer was used to identify the partial cell fusion within the contact-promotive biochips. The percentage of partial cell fusion was as high as 28±5.8% among contacted cell pairs in our biochips. We found that dye-tracked mitochondria from rMSCs migrated and accumulated in a contacted cardiomyocyte around its nucleus, as shown in Figure 4 (E). Unlike the membrane reorganization observed in the full cell fusion, the contact area between rMSCs and cardiomyocytes at their cell membranes was still identified, as shown in Figure 4 (F). The interacting cells maintained their stable cell-membrane structure, but fused in small areas, which resulted in direct cytoplasmic contact between the cells.

Tunneling nanotube

Long, thin connections have been characterized in cultured PC12 cells [28]. These tunneling, or membrane, nanotubes have been found to provide novel cell-to-cell communication, including organelle transfer [29] and electrical coupling [30,31], over long distances. We used contact-preventive biochips to mimic the long-distance-communication condition; thus nanotube formation

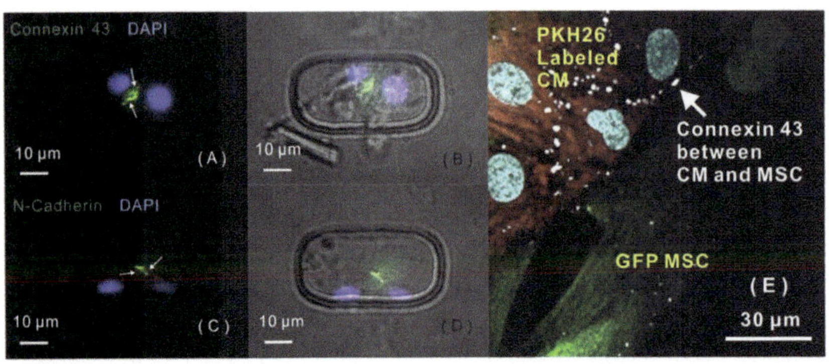

Figure 3. Junction formation between rMSCs and cardiomyocytes. Junctional distribution of (A, B) connexin 43 and (C, D) N-cadherin expressed at the contact area between rMSCs and cardiomyocytes in the contact-promotive microwells. (White arrows point to the junctional distribution of protein expression), (E) tetra-stained confocal image of cardiomyocyte (CM) and mesenchymal stem cell (MSC) coculture. The cell membranes of CM (red) and MSC (green) were labeled by different markers. Connexin43 was expressed at the junctional area between the CM and MSC membranes.

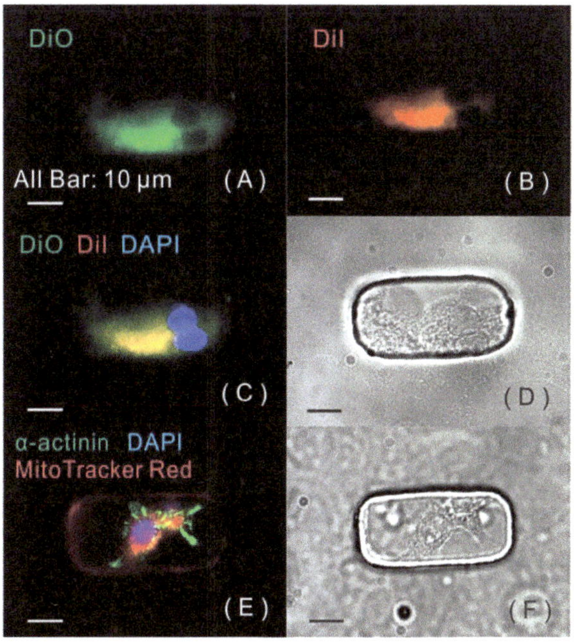

Figure 4. Cell fusion between rMSCs and cardiomyocytes. Full cell fusion (A–D) was observed by a DiO-labeled cardiomyocyte (A, green) fused with a DiI-labeled MSC (B, red) into double-nuclei, mixed labels (observed in fluorescent image, C), and membrane reorganization (observed in bright-field image, D). Partial cell fusion (E, F) was observed by rMSC mitochondrial transfer; labeled mitochondria (red) migrate across the membrane at contact area and accumulate near the cardiomyocyte nucleus.

could be studied in an isolated model at the single-cell level. We observed that a beating cardiomyocyte in one contact-preventive microwell was connected by the nanotube extended from an rMSC (Movie S2). Time-lapse microscopy images, as shown in Figure 5 (A–D), demonstrated the process of nanotube formation in three hours: A nanotube extended from an rMSC to contact the cardiomyocyte in the microwell; it extended 20 μm in three hours.

The membrane protrusion extending from rMSCs was highlighted by staining rMSCs with specific membrane marker CD105, as shown in Figure 6 (A, B). In addition, the contacted cardiomyocyte was stained by α-actinin, as shown in Figure 6 (C,

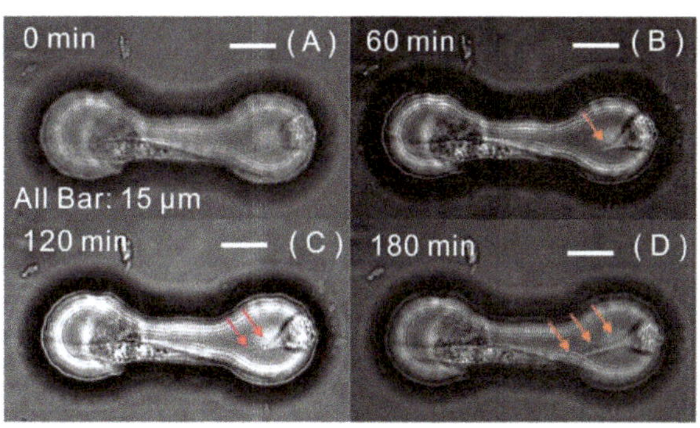

Figure 5. Tunneling nanotube extension from one rMSC. Time-lapse images (A–D) show the nanotube extension from an rMSC making contact with the cardiomyocyte in 180 minutes within the same microwell. (Red arrows point to the nanotube).

D). Our results showed that in a microwell, a tunneling nanotube developed from the membrane protrusion initiated from only rMSCs and extended towards the cardiomyocytes. Although some cardiomyocytes developed a filopodium-like structure that contacted distant rMSCs, its size was too large to be a nanotube. In the contact-preventive microwells, 18±4.1% of surviving, non-contacting heterotypic cell pairs were formed tunneling nanotubes from rMSCs, and 12±6.7% of cardiomyocytes exhibited a filopodia-like structure.

By tracking labeled mitochondria from rMSCs, we observed that mitochondria (red) were transferred from rMSCs to distant cardiomyocytes through tunneling nanotubes, as shown in Figure 6 (E). We also found that labeled rMSC mitochondria accumulated in the nucleus area of paired cardiomyocytes through filopodia-like extensions from the cardiomyocyte, as shown in Figure 6 (F). Although the total percentage of long-distance communication (including tunneling nanotubes and filopodia-like structures) was as high as 30% in contact-preventive biochips, the percentage of mitochondrial transfer was only 11±2.3% in contact-preventive biochips. Time-lapse microscopic recording was conducted to visualize the process of mitochondrial transfer, as shown in Figure 7 (A–D). The velocities of mitochondrial propagation along the tunneling nanotube were calculated by the digital image-correlation method. The transfer velocities were highly dependent on the size and shape of mitochondria (0.3–1 μm thick and up to 50 μm long [32]). The dot-shape mitochondria had higher transfer velocities than the rod-shape mitochondria, as shown in Figure 7(E).

Discussion

Using our biochips, we isolated different contact modes between MSCs and cardiomyocytes at the single-cell level and investigated MSC-cardiomyocyte interactions through junctional coupling, cell fusion, partial cell fusion, and nanotubular connections. Although directly extrapolating the results to actual tissues is unfeasible, these data demonstrate that the single-cell models, each with a defined contact mode, are significant in providing basic interactional data at the molecular level, such as the direction of mitochondria transfer and the relationship between the molecular transfer rate and the extent of cell fusion. These data cannot be obtained systematically under a complicated cell-cell interactional environment. Our results suggest that 1) cell fusion, observed in microwell studies between stem cells and cardiomyocytes, may be an additional mechanism by which grafted cells can improve the infarcted myocardium; and 2) partial cell fusion and tunneling nanotubes can facilitate transfer of rMSC mitochondria to cardiomyocytes. Mitochondrial transfer from stem cells has been reported to rescue aerobic respiration in mammalian cells and may also promote cardiomyocyte reprogramming back to a progenitor-like state.

Among different culture models, means of cell fusion have varied widely, and therefore extensive cell fusion as an *in-vitro* artifact cannot be ruled out [33,34]. Our concept of a defined microenvironment with single-cell resolution provides a tool to identify the cell-fusion phenomenon and quantitatively study its occurrence rate. Using a confocal fluorescence microscope, we found double-nuclei cells with mixed labels in contact-promotive biochips with a maximal percentage of 6.3%, which is higher than the average (0.7%) previously reported for cell fusion between stem cells and cardiomyocytes [26,35]. Since the cell-fusion process requires that (at least) two interacting cells have a large contact area, there are two possible reasons for the high occurrence of cell fusion in our biochips: (1) Our 1:1 rMSC-to-

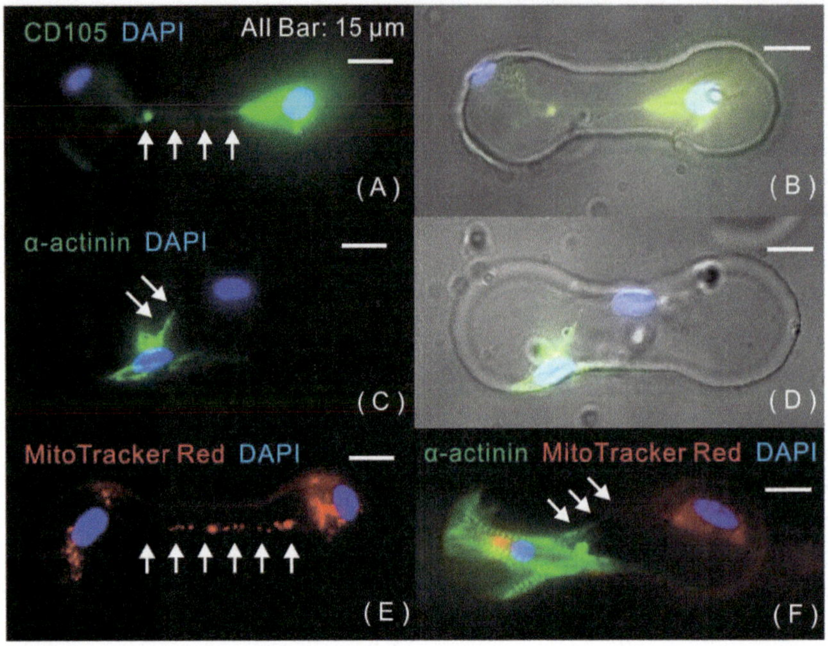

Figure 6. Long-distance communication between rMSCs and cardiomyocytes. Long distance connections between rMSCs and cardiomyocytes in contact-preventive microwells through either (A, B) rMSC-origin tunneling nanotube (determined by MSC surface marker, CD105) or (C, D) cardiomyocyte-origin filopodium-like structure (determined by cardiac marker, α-actinin). (E): Mitochondria transfer from one rMSC to its contacting cardiomyocyte. (F): Mitochondria transfer through the cardiomyocyte-origin filopodia and accumulate around the cardiomyocyte's nucleus. (White arrows point to the nanotubes).

cardiomyocyte ratio was much higher than the seeding ratio of 1:40 typical of conventional coculture [36]; (2) Our contact-promotive microenvironment increased the probability of broad cellular contact formation up to 95% in the limited space of a microwell. The cell-fusion phenomenon was also observed by patterning one rMSC and one cardiomyocyte in a larger circular microwell (Figure S2 (C, D)), but the occurrence decreased to 4.2%.

Partial cell fusion is characterized as fusion of the heterocellular plasma membranes and resulted in direct cytoplasmic contact and organelle transfer between the cells. In vivo partial cell fusion has been confirmed in the border zone of a rabbit myocardial infarction, where close contact between cardiac fibroblasts and dedifferentiated cardiomyocytes is accompanied by disruption of the basal lamina [27]. In our study, the partial cell fusion phenomenon was identified by mitochondrial transfer from rMSCs to contacted cardiomyocytes. Surprisingly, about 30% of heterotypic cell pairs with a broad contact area were found to have mitochondrial transfer; thus the occurrence of partial cell fusion was much higher than full cell fusion and approximately

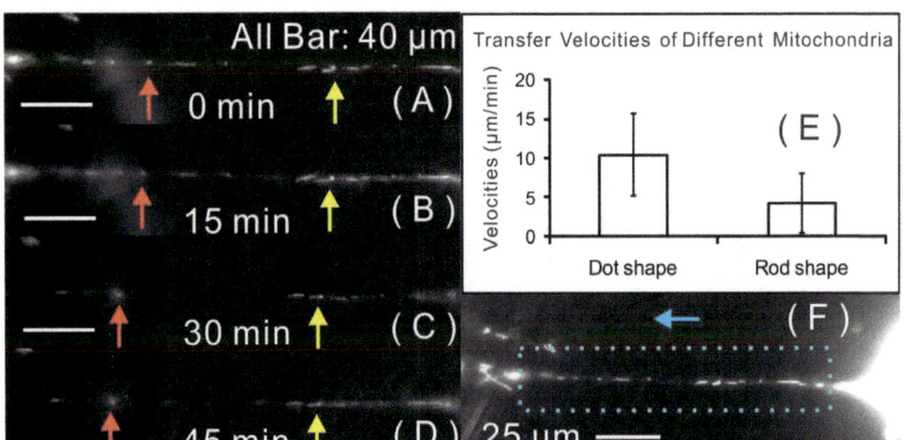

Figure 7. Mitochondria propagation through a nanotube. Time-lapse images (Zeiss Fluorescent Microscope, 63×, NA = 1.4) of mitochondrial propagation with a time interval of 15 minutes (A–D); red arrows point to a dot-shape (small) mitochondria, and yellow arrows point to a rod-shape (large) mitochondria. (E): The transfer velocities are highly dependent on mitochondrial size and shape; larger mitochondria have lower velocities. (F): rMSC and cardiomyocyte cell bodies connected by a nanotube, which is enlarged in the left column (A–D).

comparable to the rate of junction formation. This observation of partial cell fusion suggests that engrafted stem cells can contribute to the infarcted myocardium through partial cell fusion along with electrical and mechanical coupling through connexin 43 and N-cadherin. Partial cell fusion provides a unique opportunity for transfer of only cytoplasmic components without incorporation of the donor nucleus into the acceptor cells [37]. Although the benefit of partial cell fusion to neonatal cardiomyocytes has not been fully elucidated, it has been observed that rMSCs and adult-cardiomyocyte fused cells look like a type of cardiac progenitor: They demonstrate proliferative potential and express early cardiac transcriptional factors, but they lack contractile proteins [38]. Partial cell fusion opens a path for intercellular interactions to transport the cargo vesicle, mitochondria, growth factors and cytokines. It has been reported that MSCs synthesize and secrete a broad spectrum of growth factors, and cytokines such as VEGF, FGF, MCP-1, HGF, IGF-I, SDF-1, and thrombopoietin. Many of these factors have been demonstrated to produce beneficial effects on the heart, including neovascularization, attenuation of ventricular wall thinning and increased angiogenesis. Although partial cell fusion associated with mitochondrial transfer has been reported to facilitate cardiac reprogramming to progenitor-like cells, there might be other trafficking components that cross the fusion area that play key roles in the reprogramming procedure. It has been shown that a combination of three developmental transcription factors (i.e., Gata4, Mef2c, and Tbx5) rapidly and efficiently reprogram postnatal cardiac or dermal fibroblasts directly into differentiated cardiomyocyte-like cells [39,40].

Tunneling nanotubes, a novel biological phenomenon in cell-to-cell communication over long distance, allow for transfer of cargo vesicles, mitochondria, and small molecules such as calcium [41,42]. Actin filaments and microtubules are known to support the nanotubes between rMSCs and cardiomyocytes. In vivo nanotubular connection has been confirmed in mouse-heart tissue, in which fibroblasts extended membrane nanotubes between adjacent muscle bundles [43]. To study this long-distance communication phenomenon, we designed contact-preventive biochips to separate one rMSC and one cardiomyocyte inside two compartments connected by one narrow channel. In our study, the nanotubes were extended only from rMSCs. The other type of long-distance connection from cardiomyocytes resembled as filopodium-like structures protruding from the membrane. The overall amount of long-distance connections, including nanotubes and filopodia, was about 30% in our contact-preventive biochips, lower than that typically observed in conventional culture. A similar conclusion was made by Dr. Rustom's group, who found that nanotube formation among neuronal cells on a microstructured surface was 6% lower than on a conventional glass substrate [44].

We found that mitochondria from rMSCs can be transferred into cardiomyocyte cytoplasm either through rMSC-origin nanotubes or cardiomyocyte-origin filopodia. The mitochondria from rMSCs accumulated around the cardiomyocyte nucleus after transfer, which may modulate transcription-factor activities in cardiomyocytes through mitochondria-to-nucleus retrograde signaling. Recently, it has been reported that, after being transferred into cardiomyocytes, stem-cell mitochondria may cause somatic reprogramming [38]. In this report by Acquistapace and coworkers, although correlation between reprogramming and mitochondria transfer was confirmed using mitochondrial-depleted human MSCs, a genetic mechanism of reprogramming was not investigated. However, mitochondrial-retrograde signaling has been hypothesized to participate in the physiology and pathology of multiple cell types. For example, in skeletal myoblasts, mitochondrial stress caused upregulation of a number of genes involved in Ca^{2+} transport and storage, including Ryanodine Receptor I or II (RyR1 or RyR2), calreticulin, and calsequestrin [45]. Furthermore, mitochondrial-retrograde signaling has been hypothesized to induce phenotypic changes and progression in tumors. Increased expression of cathepsin L, a target gene involved in retrograde signaling, is an important factor in the invasive behavior of tumor cells [46]. Based on these facts, it could be hypothesized that, after being transferred into the cardiomyocytes, the MSC mitochondria would activate a mitochondrial-retrograde signaling pathway to induce dedifferentiation of the cardiomyocytes because of higher cytoplasmic Ca^{2+} concentration. According to the study on Zebrafish heart regeneration, it has been proposed that heart regeneration can be mediated by cardiomyocyte dedifferentiation and proliferation [47]. These findings suggest a regenerative pathway that is an alternative to well-accepted candidates, including differentiation, cell fusion, and paracrine effects. In particular, transfer of stem-cell mitochondria may explain why functional cardiac improvements are observed despite the fact that few of the donor cells are engrafted long-term.

In our contact-promotive biochips, connexin 43 and N-cadherin expression between two interacting cells was analyzed. The diffusive distribution of these two proteins was frequently observed in our biochips, but junctional distribution remained at one-third of the entire sample. Similar research has been conducted by Dr. Bursac's group [12], who found upregulation of expression of connexin 43 and N-cadherin in skeletal muscle cells and mesenchymal stem cells that were cocultured with cardiomyocytes. However, their results were obtained on only homotypic/heterotypic cell pairs with broad cellular contact. In the study we reported here, connexin 43 and N-cadherin were also expressed on noncontact rMSC-cardiomyocyte cell pairs in the contact-preventive biochips at a rate of approximately 50% but only in a diffusive manner (Figure S2 (B)). These results indicate that cellular contact not only upregulated protein expression, but also relocated expressed proteins to the contact area to facilitate cellular interaction through the junction mode. Although a diffusive distribution of connexin 43 and N-cadherin has been reported in the infarct zone [48] and in regions with implanted cells, further studies are necessary to investigate the precise distribution of junctional proteins at heterocellular interfaces within the heart.

Conclusions

In our study, two types of biochips (contact-promotive and -preventive), which were created using laser-guided cell-micropatterning and microfabrication techniques, were used to study contact-mediated cellular interactions between rMSCs and cardiomyocytes. Four in vivo-relevant intercellular interaction modes, including junction formation, cell fusion, partial cell fusion, and tunneling nanotube, were modeled by our biochip systems. Investigation of these contact modes and functional integration of stem cells inside the heart may enhance our understanding of the mechanisms for cardiac-cell therapies.

Supporting Information

Movie S1 One cell was being trapped and guided during laser-patterning procedure.

Movie S2 One cardiomyocyte connected with one rMSC through tunneling nanotube was beating within a contact-preventive microwell.

Figure S1 The cells in the microwells were characterized to demonstrate that the geometric restriction within the microwells did not affect the expression of these cell-type specific markers. The cardiomyocytes were characterized by (A, B) myosin heave chain and (C, D) cardiac troponin T, and rMSCs were characterized by (E, F) CD105 and (G, H) CD90.

Figure S2 (A) noncontact cell pair in a contact-promotive microwell (α-actinin positive: cardiomyocytes, α-actinin negative: rMSCs, DAPI: nuclei), (B) diffusive connexin 43 distribution on noncontact cell pair in a contact-preventive microwell, (C, D) cell fusion in a circle-shape microwell with double-nuclei, mixed labels, and membrane reorganization.

Author Contributions

Conceived and designed the experiments: ZM TKB BZG. Performed the experiments: ZM HY. Analyzed the data: ZM HY HL MX. Contributed reagents/materials/analysis tools: MX RBR CAE RRM BZG. Wrote the paper: ZM MX BZG.

References

1. Oettgen P, Boyle AJ, Schulman SP, Hare JM (2006) Cardiac stem cell therapy. Need for optimization of efficiency and safety monitoring. Circulation 114: 353–358.
2. De Angelis A, Piegari E, Cappetta D, Marino L, Filippelli A, et al. (2010) Anthracycline cardiomyopathy is mediated by depletion of the cardiac stem cell pool and is rescued by restoration of progenitor cell function. Circulation 121: 276–292.
3. Hosoda T, D'Amario D, Cabral-Da-Silva MC, Zheng H, Padin-Iruegas ME, et al. (2009) Clonality of mouse and human cardiomyogenesis in vivo. Proc Natl Acad Sci U S A 106: 17169–17174.
4. Orlic D, Kajstura J, Chimenti S, Bodine DM, Leri A, et al. (2003) Bone marrow stem cells regenerate infarcted myocardium. Pediatr Transplant 7: 86–88.
5. Abarbanell AM, Coffey AC, Fehrenbacher JW, Beckman DJ, Herrmann LJ, et al. (2009) Proinflamatory cytokine effects on mesenchymal stem cell therapy for the ischemic heart. Ann Thorac Surg 88: 1036–1043.
6. Rogers TB, Pati S, Gaa S, Riley D, Khakoo AY, et al. (2011) Mesenchymal stem cells stimulate protective genetic reprogramming of injured cardiac ventricular myocytes. J Mol Cell Cardiol 50: 346–356.
7. Wang T, Xu Z, Jiang W, Ma A (2006) Cell-to-cell contact induced mesenchymal stem cell to differentiate into cardiomyocyte and smooth muscle cell. Int J Cardiol 109: 74–81.
8. Xu M, Wani M, Dai YS, Wang J, Yan M, et al. (2004) Differentiation of bone marrow stromal cells into the cardiac phenotype requires intercellular communication with myocytes. Circulation 110: 2658–2665.
9. Pijnappels DA, Schalij MJ, Ramkisoensing AA, van Tuyn J, de Vries AA, et al. (2008) Forced alignment of mesenchymal stem cells undergoing cardiomyogenic differentiation affects functional integration with cardiomyocyte cultures. Circ Res 103: 167–176.
10. Valarmathi MT, Fuseler JW, Goodwin RL, Davis JM, Potts JD (2011) The mechanical coupling of adult marrow stromal stem cells during cardiac regeneration assessed in a 2D co-culture model. Biomaterials 32: 2834–2850.
11. Cselenyak A, Pankotai E, Horvath EM, Kiss L, Lacza Z (2010) Mesenchymal stem cells rescue cardiomyoblasts from cell death in an in vitro ischemic model via direct cell-to-cell connections. BMC Cell Biol 11: 29.
12. Pedrotty DM, Klinger RY, Badie N, Hinds S, Kardashian A, et al. (2008) Structural coupling of cardiomyocytes and noncardiomyocytes: quantitative comparisons using a novel micropatterned cell pair assay. Am J Physiol Heart Circ Physiol 295: H390–400.
13. Avitabile D, Crespi A, Brioschi C, Parente V, Toietta G (2011) Human cord blood CD34+ progenitor cells acquire functional cardiac properties through a cell fusion process. Am J Physiol Heart Circ Physiol 300: H1875–1884.
14. Song YH, Pinkernell K, Alt E (2011) Stem cell induced cardiac regeneration: fusion/mitochondria exchange and/or transdifferetiation? Cell Cycle 10: 2281–2286.
15. Plotnikov EY, Khryapenkova TG, Vasileva AK, Marey MV, Galkina SI, et al. (2008) Cell-to-cell cross-talk between mesenchymal stem cells and cardiomyocytes in co-culture. J Cell Mol Med 12: 1622–1631.
16. Ma Z, Pirlo RK, Wan Q, Yun JX, Yuan X, et al. (2011) Laser-guidance-based cell deposition microscope for heterotypic single-cell micropatterning. Biofabrication 3: 034107.
17. Liu H, Qin W, Shao Y, Ma Z, Ye T, et al. (2011) Myofibrillogenesis in live neonatal cardiomyocytes observed with hybrid two-photon excitation fluorescence-second harmonic generation microscopy. J Biomed Opt 16: 126012.
18. Ma Z, Liu Q, Liu H, Yang H, Yun JX, et al (2012) Cardiogenic regulation of stem-cell electrical property in a laser-patterned biochip. Cell Mol Bioeng 5: 327–336
19. Ma Z, Liu Q, Liu H, Yang H, Yun JX, et al. (2012) Laser-patterned stem-cell bridges in a cardiac muscle model for on-chip electrical conductivity analyses. Lab Chip 12: 566–573.
20. Pirlo RK, Ma Z, Sweeney A, Liu H, Yun JX, et al. (2011) Laser-guided cell micropatterning system. Rev Sci Instrum 82: 013708.
21. Ma Z, Gao BZ (2012) Quantitatively analyzing the protective effect of mesenchymal stem cells on cardiomyocytes in single-cell biochips. Biotechnol Lett 34:1385–1391.
22. Lindstrom S, Andersson-Svahn H (2011) Miniaturization of biological assays – overview on microwell devices for single-cell analyses. Biochim Biophys Acta 1810: 308–316.
23. Alvarez-Dolado M, Martinez-Losa M (2011) Cell fusion and tissue regeneration. Adv Exp Med Biol 713: 161–175.
24. Alvarez-Dolado M, Pardal R, Garcia-Verdugo JM, Fike JR, Lee HO, et al. (2003) Fusion of bone-marrow-derived cells with Purkinje neurons, cardiomyocytes and hepatocytes. Nature 425: 968–973.
25. Alvarez-Dolado M (2007) Cell fusion: biological perspectives and potential for regenerative medicine. Front Biosci 12: 1–12.
26. Nygren JM, Jovinge S, Breitbach M, Sawen P, Roll W, et al. (2004) Bone marrow-derived hematopoietic cells generate cardiomyocytes at a low frequency through cell fusion, but not transdifferentiation. Nat Med 10: 494–501.
27. Driesen RB, Dispersyn GD, Verheyen FK, van den Eijnde SM, Hofstra L, et al. (2005) Partial cell fusion: a newly recognized type of communication between dedifferentiation cardiomyocytes and fibroblasts. Cardiovasc Res 68: 37–46.
28. Rustom A, Saffrich R, Markovic I, Walther P, Gerdes HH (2004) Nanotubular highways for intercellular organelle transport. Science 303: 1007–1010.
29. Gerdes HH, Bukoreshtliev NV, Barroso JF (2007) Tunneling nanotubes: a new route for exchange of components between animal cells. FEBS Lett 581: 2194–2201.
30. Wang X, Gerdes HH (2012) Long-distance electrical coupling via tunneling nanotubes. Biochim Biophys Acta 8: 2082–2086.
31. Wang X, Veruki ML, Bukoreshtliev NV, Hartveit E, Gerdes HH (2010) Animal cells connected by nanotubes can be electrically coupled through interposed gap-junction channels. Proc Natl Acad Sci U S A 107: 17194–17199.
32. Bereiter-Hahn J (1990) Behavior of mitochondria in the living cell. Int Rev Cytol 122: 1–63.
33. Garbade J, Schubert A, Rastan AJ, Lenz D, Walther T, et al. (2005) Fusion of bone marrow-derived stem cells with cardiomyocytes in a heterologous in vitro model. Eur J Cardiothorac Surg 28: 685–691.
34. Kajstura J, Rota M, Whang B, Cascapera S, Hosoda T, et al. (2005) Bone marrow cells differentiate in cardiac cell lineages after infarction independently of cell fusion. Circ Res 96: 127–137.
35. Zhang S, Wang D, Estrov Z, Raj S, Willerson JT, et al. (2004) Both cell fusion and transdifferentiation account for the transformation of human peripheral blood CD34-positive cells into cardiomyocytes in vivo. Circulation 110: 3803–3807
36. He XQ, Chen MS, Li SH, Liu SM, Zhong Y, et al. (2010) Co-culture with cardiomyocytes enhanced the myogenic conversion of mesenchymal stromal cells in a dose-dependent manner. Mol Cell Biochem 339: 89–98
37. Prochiantz A, Joliot A (2003) Can transcription factors function as cell-cell signaling molecules? Nat Rev Mol Cell Biol 4: 814–819.
38. Acquistapace A, Bru T, Lesault PF, Figeac F, Coudert AE, et al. (2011) Human mesenchymal stem cells reprogram adult cardiomyocytes toward a progenitor-like state through partial cell fusion and mitochondria transfer. Stem Cells 29: 812–824.
39. Ieda M, Fu JD, Delgado-Olguin P, Vedantham V, Hayashi Y, et al. (2010) Direct reprogramming of fibroblasts into functional cardiomyocytes by defined factors. Cell 142: 375–86
40. Qian L, Huang Y, Spencer IC, Foley A, Vedantham V, et al. (2012) In vivo reprogramming of murine cardiac fibroblasts into induced cardiomyocytes. Nature 485: 593–8
41. Onfelt B, Nedvetzki S, Benninger RK, Purbhoo MA, Sowinski S, et al. (2006) Structurally distinct membrane nanotube between human macrophages support long-distance vesicular traffic or surfing of bacteria. J Immunol 177: 8476–8483.
42. He K, Luo W, Zhang Y, Liu F, Liu D, et al. (2010) Intercellular transportation of quantum dots mediated by membrane nanotubes. ACS Nano 4: 3015–3022.
43. He K, Shi X, Zhang X, Dang S, Ma X, et al. (2011) Long-distance intercellular connectivity between cardiomyocytes and cardiofibroblasts mediated by membrane nanotubes. Cardiovasc Res 92: 39–47.
44. Abel MP, Riese SR, Schlicker O, Bukoreshtliev NV, Gerdes HH, et al. (2011) Microstructured platforms to study nanotube-mediated long-distance cell-to-cell connections. Biointerphases 6: 22–31.
45. Biswas G, Adebanjo OA, Freedman BD, Anandatheerthavarada HK, Vijayasarathy C, et al. (1999) Retrograde Ca2+ signaling in C2C12 skeletal

myocytes in response to mitochondrial genetic and metabolic stress: a novel mode of inter-organelle crosstalk. EMBO J 18: 522–533.
46. Amuthan G, Biswas G, Ananadatheerthavarada HK, Vijayasarathy C, Shephard HM, et al. (2002) Mitochondrial stress-induced calcium signaling, phenotypic changes and invasive behavior in human lung carcinoma A549 cells. Oncogene 21: 7839–49.
47. Jopling C, Sleep E, Raya M, Marti M, Raya A, et al. (2010) Zebrafish heart regeneration occurs by cardiomyocyte dedifferentiation and proliferation. Nature 464: 606–609.
48. Kieken F, Mutsaers N, Dolmatova E, Virgil K, Wit AL, et al. (2009) Structural and molecular mechanisms of gap junction remodeling in epicardial border zone myocytes following myocardial infarction. Circ Res 104: 1103–1112.

4

Metformin: A Potential Therapeutic Agent for Recurrent Colon Cancer

Pratima Nangia-Makker[1,2,3], Yingjie Yu[1,3], Anita Vasudevan[1,3], Lulu Farhana[1,3], Sindhu G. Rajendra[1], Edi Levi[1], Adhip P. N. Majumdar[1,2,3]*

1 Veterans Affairs Medical Center, Wayne State University, Detroit, Michigan, United States of America, **2** Karmanos Cancer Institute, Wayne State University, Detroit, Michigan, United States of America, **3** Department of Internal Medicine, Wayne State University, Detroit, Michigan, United States of America

Abstract

Accumulating evidence suggests that metformin, a biguanide class of anti-diabetic drugs, possesses anti-cancer properties. However, most of the studies to evaluate therapeutic efficacy of metformin have been on primary cancer. No information is available whether metformin could be effectively used for recurrent cancer, specifically colorectal cancer (CRC) that affects up to 50% of patients treated by conventional chemotherapies. Although the reasons for recurrence are not fully understood, it is thought to be due to re-emergence of chemotherapy-resistant cancer stem/stem-like cells (CSCs/CSLCs). Therefore, development of non-toxic treatment strategies targeting CSCs would be of significant therapeutic benefit. In the current investigation, we have examined the effectiveness of metformin, in combination with 5-fluorouracil and oxaliplatin (FuOx), the mainstay of colon cancer therapeutics, on survival of chemo-resistant colon cancer cells that are highly enriched in CSCs/CSLCs. Our data show that metformin acts synergistically with FuOx to (a) induce cell death in chemo resistant (CR) HT-29 and HCT-116 colon cancer cells, (b) inhibit colonospheres formation and (c) enhance colonospheres disintegration. *In vitro* cell culture studies have further demonstrated that the combinatorial treatment inhibits migration of CR colon cancer cells. These changes were associated with increased miRNA 145 and reduction in miRNA 21. Wnt/β-catenin signaling pathway was also down-regulated indicating its pivotal role in regulating the growth of CR colon cancer cells. Data from SCID mice xenograft model of CR HCT-116 and CR HT-29 cells show that the combination of metformin and FuOX is highly effective in inhibiting the growth of colon tumors as evidenced by ~50% inhibition in growth following 5 weeks of combination treatment, when compared with the vehicle treated controls. Our current data suggest that metformin together with conventional chemotherapy could be an effective treatment regimen for recurring colorectal cancer (CRC).

Editor: Shrikant Anant, University of Kansas School of Medicine, United States of America

Funding: This study was supported by grants from the NIH (AG014343) and the Department of Veteran Affairs (I101BX001927). The funders had no role in study design, data collection and analysis, decision to publish, or preparation of the manuscript.

Competing Interests: The authors have declared that no competing interests exist.

* E-mail: a.majumdar@wayne.edu

Introduction

Recent understanding of the heterogeneous makeup of the cancer cells in a tumor has revealed the presence of CSCs/CSLCs [1,2], which exhibit self-renewing characteristics, ability to initiate tumor from a small number of cells that are highly chemo-resistant [3–7]. Carcinoma recurrence is in part due to fact that conventional chemotherapy only targets the rapidly dividing cells that form bulk of the tumor, but spares the CSCs/CSLCs [8]. The proportion of CSCs is reported to increase after conventional chemotherapy [9]. Thus, presence of chemotherapy resistant CSCs/CSLCs in the primary tumor may in part be responsible for a failure of complete eradication of tumor resulting in its recurrence at the primary and secondary sites. Development of novel therapeutic strategies, which specifically target CSCs/CSLCs is, therefore, warranted.

Metformin, (1,1-dimethylbiguanide hydrochloride) a FDA approved biguanide anti-diabetic drug, is derived from French lilac (*Galega officinalis*), a plant used in folk medicine for several centuries. In addition to its function as a gluconeogenesis suppressor, metformin has recently been shown to possess strong anti-cancer properties. In type 2 diabetic patients metformin has been shown to suppress carcinogenesis in breast, pancreas and lung, and to diminish cancer related mortality (reviewed in [10,11]. In a number of preclinical studies, metformin reduced proliferation, induced apoptosis, caused cell cycle arrest, and reduced incidence and growth of experimental tumors *in vivo* [12–18]. Some reports also indicate that metformin improved the response of human breast tumor xenografts to conventional chemotherapy by eradicating CSCs in the tumor [19,20].

Many investigators have reported the pivotal role of Wnt/β-catenin signaling pathway in the regulation of epithelial stem cell self-renewal [21,22], and its dysregulation has been implicated in many malignancies including colorectal cancer (CRC) [23,24]. In CRC, as a result of inactivation of APC gene, the coordinated phosphorylation and destruction of β-catenin is disrupted. As a result, β-catenin accumulates in the cytoplasm, complexes with DNA-binding proteins of TCF/LEF family and translocates to the nucleus [25,26]. Once there, it activates the transcription of its target genes like cyclin D1, c-myc, MMP-7, MT1-MMP, axin1 etc. [27]. Many of these target genes are implicated in intestinal stem cell proliferation and carcinogenesis [28]. Recently we showed that Wnt/β-catenin pathway plays a critical role in growth and maintenance of colonospheres, which are highly enriched in CSC/CSL cells [29]. Increased levels of β-catenin in colonospheres were associated with induction of transcriptional activation of TCF/LEF, which was decreased when β-catenin was silenced using siRNA [29].

A novel class of short non-coding RNAs, so called miRNAs has emerged as a regulator of mRNA translation. A miRNA can act as a tumor suppressor or an oncogene depending on its targets in different tissues and cell types [30]. Comprehensive analyses of miRNA-expression patterns in human cancers have revealed that different cancer types have distinct miRNA expression patterns [31].We have reported that miR21, which has been shown to be up-regulated in CRC [32], induces stemness in chemo-resistant (CR) colon cancer cells [33]. miR21 has been shown to regulate the growth, migration, invasion and apoptosis related properties of cancer cells [34]. Down-regulation of miR21 by CDF, an analog of curcumin [35], resulted in normalization of PTEN/Akt axis in CR HCT-116 and CR HT-29 cells and reduced growth [36]. Over-expression of miR21in HCT-116 cells resulted in an increased β-catenin activity [33].

The present investigation was undertaken to examine whether metformin could be used in conjunction with conventional chemotherapy to inhibit the growth of chemo-resistant colon cancer cells and the regulation of this process. Herein, we demonstrate that metformin acts synergistically with FuOx, a combination of 5-FU and oxaliplatin, the mainstay in colon cancer chemotherapy to inhibit the growth of colon CR cells *in vitro* and *in vivo* as well as their migration via down-regulating miR21and inhibiting the Wnt/β-catenin signaling pathway.

Materials and Methods

Cell lines and Reagents

Human colon cancer cells HT-29 and HCT-116 were obtained from the American Type Culture Collection (ATCC, Rockville, MD). The cells were maintained in Dulbecco's modified Eagle's medium (4.5 g/I D-glucose) supplemented with 10% fetal bovine serum (Invitrogen, Grand Island, NY) and 1% antibiotic/antimycotic in humidified incubator at 37°C in an atmosphere of 95% air and 5% carbon diaoxide. 5-Fluorouracil + Oxaliplatin (FuOx) resistant cells (CR cells) were generated as described earlier [37–39] in our laboratory. The CR cells were maintained in normal culture medium containing 2× FuOx (50 μM 5 FU + 1.25 μM Ox). Endothelial cells were a gift from Dr Dipak Banerjee, University of Puerto Rico and maintained in Earle's Minimum Essential medium supplemented with 10% fetal bovine serum (Hyclone) as previously described [40]. The medium was changed two times a week, and cells were passaged using 0.05% trypsin/EDTA (Invitrogen). The use of cell lines was approved by the Human Investigation Committee, Wayne State University, Detroit, MI. Metformin hydrochloride was purchased from Sigma Chemical Co. A 2 M solution was prepared in sterile distilled water and stored at −20°C. FU and Ox were obtained from Sigma Chemical Co.

Determination of cellular growth

The growth of colon cancer cells was assessed by mitochondrial-dependent reduction of 3-(4,5-dimethylthiazol-2yl)-2, 5-diphenyltetrazolium bromide (MTT) (Sigma) to formazan as described previously [41]. Briefly, the cells (5×10^3) were seeded in quadruplicates onto 24 well culture dishes. After 24 hr, fresh medium containing various concentrations of metformin and/or FuOx was added. After 72 hr, cell proliferation was determined by MTT assay. Briefly, the medium was removed and the cells were incubated at 37°C with MTT (0.5 mg/mL) for 4 hr. The medium was aspirated and the cells were solubilized in 0.04 N HCl in isopropanol. The optical density (OD) was measured at 570 and 630 nm.

Analysis of interaction between metformin and FuOx

Combination indices (CI) method adapted for *in vitro* drug testing was employed to determine the nature of interaction between metformin and FuOx. This method utilizes multiple drug effect equation originally derived from enzyme kinetics method, where the output is represented as CI and/or isobologram analysis.CI analysis was performed by utilizing Calcusyn software (Biosoft, Ferguson, MO). Based on CI values, extent of synergism/anatogonism is determined. In general, CI values below 1 suggest synergy, whereas CI values above 1 indicate antagonism between the drugs. CI values in the range of 0.9–1.10 would mainly indicate additive effects: those between 0.9–0.85 suggest slight synergy, and values in the range of 0.7–0.3 are indicative of moderate synergy. Any values less than 0.3 would suggest strong synergistic interactions between the drugs.

Formation and differentiation of colonospheres

The ability of cells to form spheres in suspension was evaluated as described by Liu et al [42] with slight modifications [29,43]. Briefly, colonospheres were generated by incubating a limited number of parental and CR HCT-116 and HT-29 cells at a concentration of 100 cells per 200 μL in serum -free stem cell medium (SCM) containing DMEM/F12 (1:1) supplemented with B27 (Life technologies, Gaithersberg, MD) 20 ng/mL epidermal growth factor (Sigma, St. Louis, MO), 10 ng/mL fibroblast growth factor (Sigma) and antibiotic/antimycotic in 24 well plates in the presence of metformin alone or in combination with FuOx. The colonospheres formed after 8 days were evaluated for their size and number by light microscopy. In one set of experiments, the colonospheres formed in normal medium were treated with metformin and/or FuOx to analyze their effect on sphere differentiation and attachment.

Extreme limiting dilution analysis

Extreme limiting dilution analysis (ELDA) was performed as described by Hu and Smyth [44]. Briefly, single cell suspension obtained from adherent cells pretreated with metformin and/or FuOx for 72 hr were plated at a concentration of 100, 10 and 1 cell per 100 μLSCM (24 well for each dilution) in 96 well plates and incubated for 8 days. At the end of 8 days, the number of wells showing formation of colonospheres was counted. The frequency of sphere forming cells was determined using ELDA webtool at http://bioinf.wehi.edu.au/software/elda.

Cell migration assay

This assay was performed using a Boyden chamber (Neuroprobe Inc., Cabin John, MD) as described earlier [41,45]. In the lower chamber 100 μg/mL Matrigel (BD Biosciences, Bedford, MA) alone or mixed with metformin and/or FUOX was added. HT-29 or CR HT-29 cells (5×10^4) cells were loaded in the upper chamber. The two chambers were separated by a polycarbonate filter of 8 μ pore size and incubated in a 37°C tissue culture incubator for 16 hr, after which the filter was removed, the cells on top of the filter were wiped off and the migrated cells were fixed, stained using Protocol Hema 3 stain set (Fisher Scientific Company, Pittsburgh, PA). The migrated cells were photographed using Olympus 45 microscope supporting a Spot Idea camera and the migrated cells per field were counted. Each assay was carried out thrice and 6 wells were used for each treatment.

In the next set of experiments, migration of colon cancer cells towards endothelial cells was measured. In this investigation, enodothelial cells or HT-29/CR HT-29 cells (2.4×10^4) were prelabeled with viable stain DiO or DiI respectively (Invitrogen), washed twice with complete medium and seeded in each chamber of the cell culture insert (ibidi GmbH). After 24 hr, the cell culture insert was removed and the migration of the co-cultures towards each other for wound healing was observed. In some cultures, metformin and/or FuOx was added at the time of removal of the insert. The cells were photographed at 0, 24 and 48 hr after removal of the insert.

Western blot analysis

Western blot analysis was performed according to the standard protocol [46]. Briefly, the cells were solubilized in lysis buffer (50 mM Tris HCl, pH 7.4, 150 mM NaCl, 5 mM EDTA, 1% Triton X100, 0.5% NP40, 25 µg/mL each of aprotinin, leupeptin and pepstatin, 1× phosphatase inhibitor cocktail (Sigma). After clarification at 10,000 g for 30 min, the supernatant was used for protein analysis. Protein concentration determined by the Bio-Rad protein assay kit ((Bio-Rad, Hercules, CA) and aliquots containing 25 µg protein were separated by SDS-polyacrylamide gel electrophoresis. After electrophoresis, proteins were transferred to a polyvinylidene difluoride membrane (Millipore) by electroblotting and subjected to western blot analysis with the recommended dilution of primary antibody. Following washes the blots were reacted with secondary antibody mix containing 1:2500 dilutions of the appropriate horseradish peroxidase-conjugated secondary antibody (Amersham Biosciences). Protein bands were visualized using a commercially available enhanced chemiluminiscence kit (Amersham Biosciences). Blots were also immuno-reacted with a 1:5000 dilution of anti-actin mouse monoclonal antibody (Santa Cruz Biotechnology Inc., Santa Cruz, CA) to normalize for variation in protein loading.

Isolation of RNA and Quantitative Polymerase Chain Reaction analysis

Total RNA was extracted from parental and CR cells treated with metformin and/or FuOx for 48 hr using TRIZOL reagent (Invitrogen) according to manufacturer's instructions. Total RNA was treated with DNase1 to remove contaminating genomic DNA, subsequently purified using miRNAeasy Mini Kit (Qiagen, Valencia, CA). RNA concentration was measured at an optical density of 260 nm.

To quantitate miR-21 and miR-145, first the cDNA synthesis was carried out using Taqman MicroRNA Reverse transcription kit (Applied Biosystems, Foster City, CA). The miRNA RT-PCR primers for miR21, miR145 and endogenous control RNU6B were purchased from Applied Biosystems. Real time qRT-PCR analysis was carried out using Applied Biosystems 7500 real time PCR System. The PCR mix containing Taqman 2× Universal PCR Master Mix were processed as follows: 95°C for 10 min followed by 40 cycles of 95°C for 15 sec and 60°C for 60 sec. Signal was collected at the endpoint of each cycle. The gene expression ΔC_T values for each sample were calculated by normalizing with internal control RNU6B, and relative quantitation values were plotted.

Flow cytometric analysis

Single cell suspensions of metformin and/or FuOx treated or untreated CR HT-29 cells were subjected to direct immunofluorescence staining followed by flow cytometric analysis according to standard protocol [38]. Briefly, the cells were harvested and washed with PBS. Half a million cells were suspended in 90 µl PBS containing 0.5% BSA. After 10 min incubation at room temperature, 10 µl fluorophore (PE-Cy7 or PerCP-Cy5) conjugated ant-human CD44 or CD166 antibody was added and incubated for 30 min in dark at room temperature. The samples were then washed and analyzed using a FACS DiVa (BD, San Jose, CA). The cells stained with IgG2b (isotype-negative control) served as gating control. The proportion of $CD44^+/CD166^{+/low}$ cells was determined on the basis of fluorescence intensity-spectra.

Tumor growth in SCID mice

To determine whether the combination of metformin and FuOx would be an effective therapeutic strategy for colon tumors, SCID mice xenograft model of colon tumor was utilized. To generate colon tumors, four weeks old female SCID mice, purchased from Taconic Laboratory, were injected s.c. either with 1×10^6 CR HCT-116 or CR HT-29 cells suspended in 100 µl Matrigel. They were then divided into 2 subgroups of 4 mice and treatment with metformin was started 7 days after inoculation of the cells in one subgroup. Metformin (RIOMET 500 mg/5 ml; Ranbaxy Laboratories, Princeton, NJ) was dissolved in 100 mL drinking water to attain the dosage of 200 mg/kg body weight. The water was changed daily and measured for water intake. Metformin treatment was continued for 5 weeks until the mice were sacrificed. Metformin group was also injected IP with a mixture of 25 mg/kg 5- fluorouracil and 2 mg/kg oxaliplatin once a week for 3 weeks. Tumors were measured once a week and tumor volumes were calculated using the formula: tumor volume = length × width × width/2. Mice were regularly monitored for any signs of discomfort. All animal experiments were performed according to the Wayne State University's Institutional Animal Care and Use Committee (IACUC) approved protocol # A02-02-13. Animal Welfare Assurance # A3310-01.

Single cell isolation from the xenograft

Small portions (5–10 mg) of the tumor, generated in SCID mice by CR HCT-116 and CR HT-29 cells, as described below were washed extensively in 1× PBS containing 10% antibiotic/antimycotic (AB/AM) and subsequently incubated overnight in Dulbecco's Minimum Essential Media (DMEM/F12) containing 5% antibiotic/antimycotic at 4°C. The tissue was cut into fine pieces using sterile scalpel and then digested with 1.5 mg/ml collagenase I (Sigma-Aldrich) and 20 µg/ml hyaluronidase I (Sigma Aldrich) under gentle agitation for 2–3 hr at 37°C. The digested tissue was filtered through 40 µ filter and centrifuged at 1200 rpm for 5 min. The supernatant (containing dead cells as well as the fat cells) was discarded and the cells (pellet) were washed three times with DMEM/F12 media containing 5% AB/AM. The cells were suspended in previously defined stem cell media.

Statistical analysis

All *in vitro* experiments were repeated thrice in triplicates. Statistical analysis was performed using Microsoft Excel. The p-values were calculated using 2 sample t-test, assuming unequal variances. Values <0.05 were considered statistically significant.

Results

Metformin and FuOx combination therapy inhibits growth of chemo-resistant HT-29 and HCT-116 cells synergistically

To examine whether metformin acts synergistically with FuOx to inhibit the growth of chemo-resistant cells, the cells were treated with incremental doses of metformin and FuOx, each alone or in combination. After 72 hr, the cell growth was determined by MTT assay. Dose response curves Fig 1A & B show that while 10 mM metformin or 8× (200 μM FU and 5 μM Oxaliplatin) FuOx alone caused ~40 and 60% inhibition, together they caused an inhibition of ~70% and 80% in CR HT-29 and CR HCT-116 cells respectively. The results show that the combination of metformin and FuOx acts synergistically to inhibit the growth of both CR HCT-116 and CR HT-29 cells. All subsequent experiments were carried out using one of the chemo-resistant colon cancer cell lines.

The fractions of cells affected in response to each treatment, was utilized to perform synergy analysis with Calcusyn software. The CI as formulated by the software, revealed values indicating moderate synergistic interaction between the two agents in CR HCT-116 cells at metformin doses 5 mM and higher and a strong synergy at 10 and 20 mM metformin in the CR HT-29 cells (Table 1). No synergism was seen in the parental cells (data not shown). As the chemo-resistant cells are maintained in 2×FuOx (50 μM FU, 1.25 μM oxaliplatin), next we studied the effect of doses of metformin ranging from 0.6 to 20 mM in combination with 2×FuOx on the growth of CR HT-29 cells. The results show that at all the concentrations used, the combination therapy was more effective in inhibiting cell growth than metformin alone (Fig. 1C). The chemo-resistant cells were more resistant to the FuOx and more sensitive to combination therapy compared to the parental cells (data not shown).

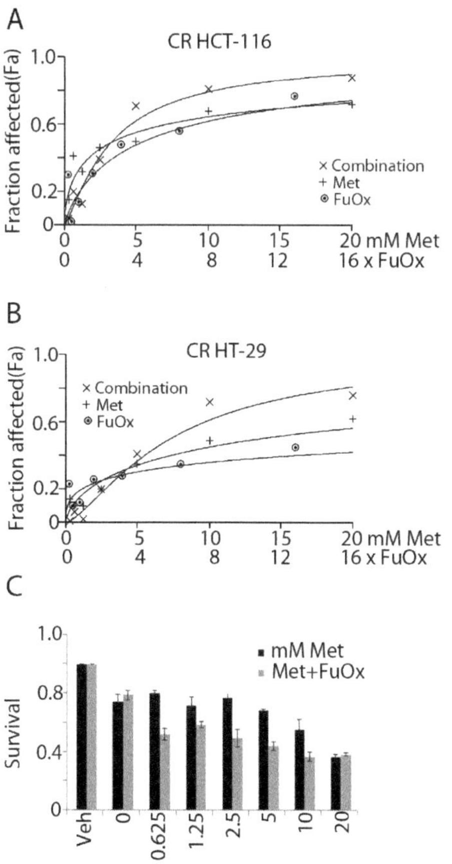

Figure 1. A&B Dose response curves for metformin and/or FuOx in CR HCT-116 (A) and CR HT-29 (B) cells produced by fixed ratio method. Fraction of cells affected by combination of the two drugs (fixed ratio) was higher than either agent alone. Fa represents the fraction of cells affected in response to the treatment. Fa values were used to conduct synergy analysis by CalcuSyn software as described in Materials and Methods. **C.** Relative survival of CR HT-29 cells in the presence of increasing doses of metformin with and without 2×FuOx. Combination treatment is more effective at all the concentrations. Veh: vehicle alone. Each treatment was performed in quadruplicates, bars represent mean ± standard deviation. * p value<0.05.

Metformin and FuOx combination therapy inhibits colonosphere formation (number of stem cell/stem like cells)

It was reported that FuOx surviving cells are enriched in CSCs/CSLCs that grow as large, round unattached floating spheroid colonies (colonospheres) when cultured in serum free stem cell medium at a relatively low density [38] [29]. Figs. 2A and B (left panel) show that in the presence of 2×FuOx, increasing concentrations of metformin decreased the number and size of colonospheres. Metformin also induced attachment and disintegrated the colonospheres (Fig. 2A and B (right panel)) When the colonospheres were trypsinized and seeded in the presence of metformin+FuOx, increasing number of spheres lost their stem-like properties as they are seen to attach to the bottom of the dish.

We also evaluated the cancer stem cell properties of the treated cells by performing an extreme limiting dilution assay (ELDA). CR HT-29 cells were treated with metformin+FuOx for 72 hr and seeded in normal stem cell medium. While metformin alone reduced the frequency to form colonospheres by 1.5–2 folds, the combination treatment reduced it by 7–8 folds that of the untreated control (Table 2).

To determine whether and to what extent the current treatment strategies would affect the proportion of CSCs/CSLCs, flow cytometric analysis was performed following metformin and/or FuOx treatment of CR HT-29 cells. The results revealed a reduction (~65%) in $CD44^{high}/CD166^{low}$ phenotype cell population, when compared with FuOx treated controls. In contrast, the proportion of $CD44^{high}/CD166^{low}$ cells increased from 5.12 to 7.34% in the presence of FuOx, which could, in part, be due to death of FuOx sensitive cells. Combination treatment reduced the proportion of $CD44^{high}/CD166^{low}$ cells to 2.8% (Fig 2C). Taken together, these results suggest that the combination of metformin and FuOx not only decreases CR colon CSC/CSLC population, but also diminishes their stem cell properties.

Metformin and FuOx combination therapy inhibits cell migration

Cell migration was determined by Boyden chamber (Fig 3A) and wound healing (Fig 3B) assays. Boyden chamber assay indicates that while parental HT-29 cells show limited migration, CR HT-29 cells exhibit strong migration towards Matrigel. However, metformin was found to inhibit migration in a dose dependent manner, and the combination therapy caused further reduction: at 10 mM metformin+FuOx, ~6 fold inhibition in migration was observed in CR HT-29 cells (Fig 3A).

Wound healing assay was performed to study the migration of parental and CR HT-29 cells towards endothelial cells. Fig. 3B showed CR HT-29 and endothelial cells migrate towards each

Table 1. Extreme limiting dilution analysis of colonospheres forming frequency of metformin treated CR HT-29 cells.

Number of Cells/well	Number of wells plated	Number of wells showing colonospheres					
		Control	FuOx	5 mM Met	10 mM Met	5 mM Met ±FuOx	10 mM Met ± FuOx
100	24	24	24	24	24	21	21
10	24	17	13	15	14	5	5
1	24	7	2	0	5	2	0
Sphere forming frequency (95%CI)		1/6(1/10–1/4)	1/12(1/21–1/8)	1/11(1/19–1/7)	1/9(1/15–1/6)	1/43(1/67–1/29)	1/48(1/74–1/31)

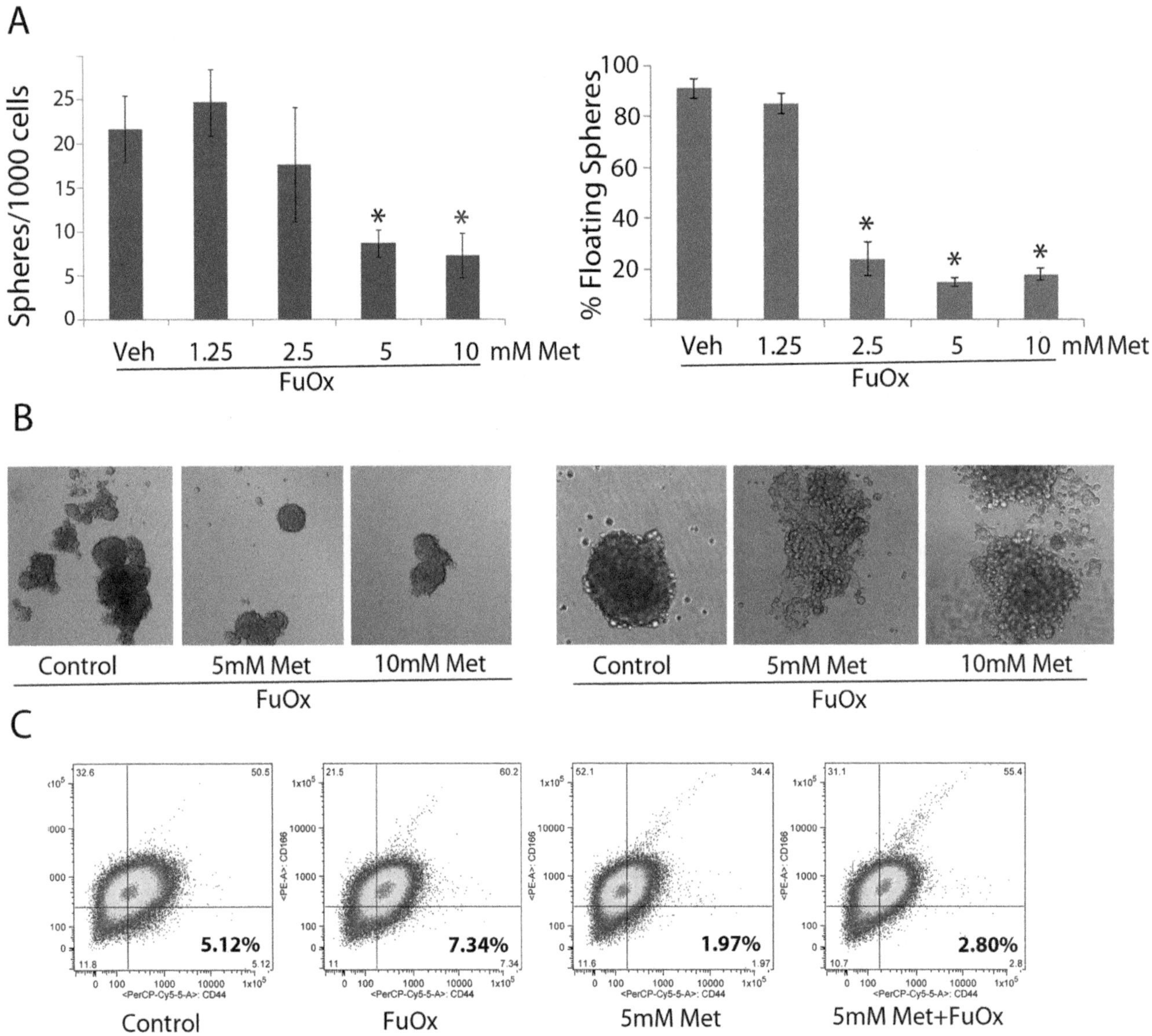

Figure 2. Decreased stem cell characteristics in metformin/FuOx treated CR HT-29 cells. A&B: Formation of primary colonospheres in the presence of metformin and FuOx (left panel). Induction of differentiation and attachment of colonospheres in the presence of metformin/FuOx (right panel). A: graphic representation of the numbers B: photomicrographs of representative spheres. C: Flow cytometric analysis of CD44 and CD166 positive cells after metformin/FuOx treatment.

Table 2. Combination indices for metformin and FuOx combination therapy, as computed by Calcusyn for CR HCT-116 and CR HT-29 colon cancer cells.

Combination Therapy		Combination Index (CI)	
Metformin mM	**FUOX (x)**	**CR HCT-116**	**CR HT-29**
0.31	0.25	26.818	1544.930
0.62	0.5	2.646	19.756
1.25	1	12.488	961.606
2.5	2	2.351	3.746
5	4	0.602	0.931
10	8	0.537	0.194
20	16	0.499	0.277

other. By 48 hr, the wound was completely healed by colon CR cells, while parental cells showed only ~46% wound closure (Fig.3B). Treatment with metformin+FuOx slowed down the wound healing process. While 54% and complete wound closure was observed in untreated CR HT-29 cells at 24 and 48 hr respectively, metformin+FuOx treatment resulted in a 46 and 77% wound closure at 24 and 48 hr respectively. These results further confirm that metformin+FuOx combination is effective in restraining the tumorigenic/migratory/invasive properties of chemo-resistant colon cancer cells.

It has been reported that Akt is an essential down-stream target of PI3K for the remodeling of actin filaments to increase cell migration [47]. This prompted us to analyze Akt/pAkt expression in metformin and FuOx treated cells. Western blot analysis revealed decreased levels of pAkt in treated cells, compared to controls (Fig. 3). We suggest that PI3K/Akt signaling pathway may play a role in regulating the metformin/FuOx-induced inhibition of cell migration.

Metformin and FuOx combination therapy decreases miRNA21 and increases miR145 expression

In view of the notion that miR-21 is oncogenic, while miR-145 is tumor suppressive, we have analyzed their levels in chemo-resistant colon cancer cells. We found the levels of miR21 to be higher in CR colon cancer cells [33] and miR145 to be lower, when compared with the corresponding parental controls (unpublished data). Results of the quantitative real time PCR analysis showed, while metformin either alone or together with FuOx caused a marked reduction of miR-21expression, it induced

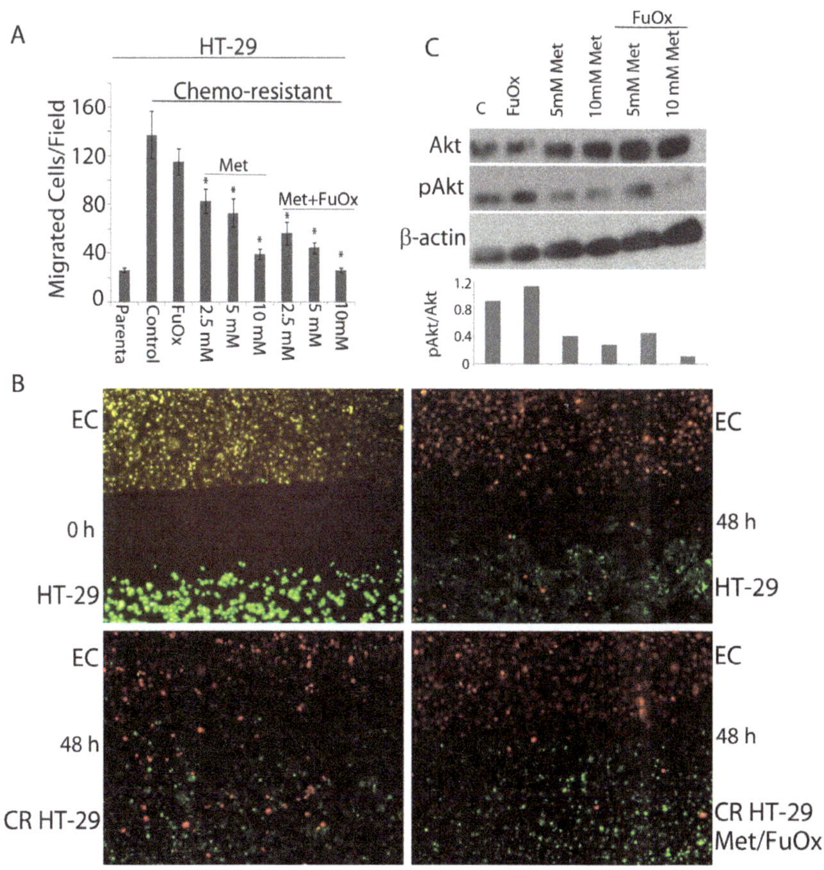

Figure 3. Boyden chamber analysis of migration of parental and CR HT-29 cells. Metformin/FuOx treatment reduced the migration of CR HT-29 cells. Each point represents an average of 6 readings. Bars represent mean ± standard deviation (A). Wound healing assay using endothelial (top/red) and CRC (bottom/green). Top left panel: wound at 0 hr; bottom left panel: complete wound healing at 48 hr in CR HT-29 cells; top right panel: partial wound healing in HT-29 cells after 48 hr; bottom right panel: incomplete wound healing in CR HT-29 cells in the presence of 5 mM metformin/FuOx at 48 hr (B). Western blot analysis of metformin/FuOx treated CR H29 cells indicating decreased pAkt levels (C); relative changes in phospho-Akt (pAkt) were calculated as a ratio of pAkt/Akt after normalization to β-actin, which was used as loading control. *$p<0.05$.

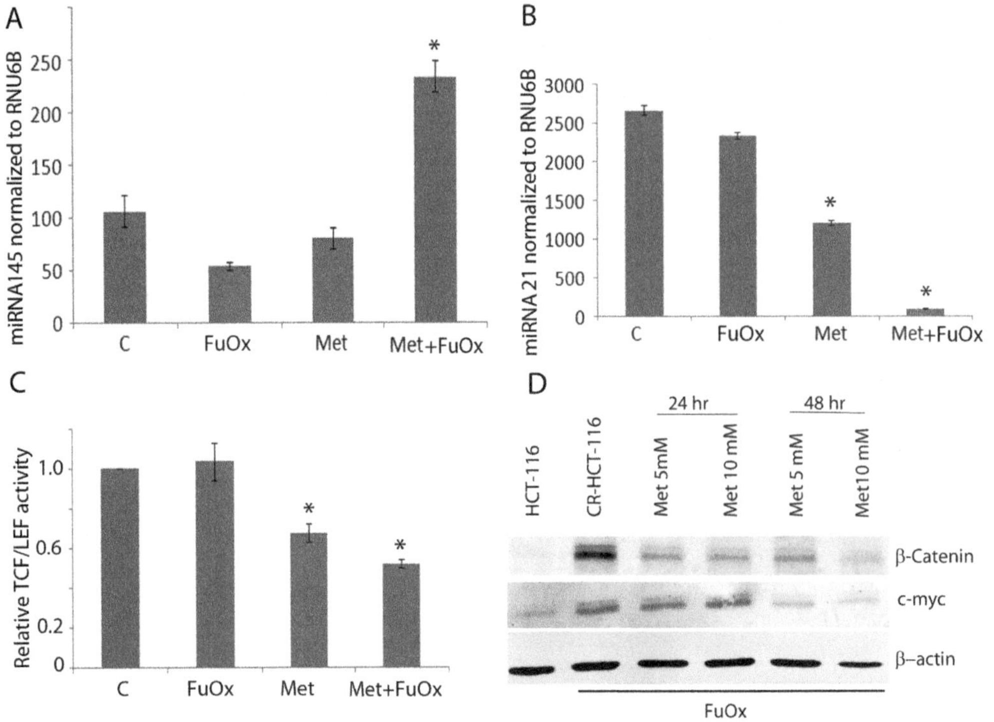

Figure 4. Quantitative real-time PCR showing the expression of miRNA145 (A) and miRNA21(B) normalized to RNU6B in metformin and/or FuOx treated CR HT-29 cells. C: metformin/FuOx treatment of CR HCT-116 cells inhibits transcriptional activity of TCF/LEF. D: Western blot analysis shows a reduced expression of β-catenin and c-myc in metformin/FuOx treated CR HCT-116 cells. β-actin was used as loading control. * $p<0.05$.

miR-145, when compared to their respective controls (Fig. 4). Suffice it to mention that the qRT-PCR values for miR-21 and miR-145 were normalized to the levels of RNU6B.

As the Wnt/β-catenin signaling plays a critical role in the regulation of colon cancer stem cell proliferation, we next studied the effect of metformin treatment on β-catenin activity, its expression and the levels of its target protein c-myc. The combination treatment of metformin and FuOX caused ~50% decrease in the transcriptional activity of TCF/LEF in CR HCT-116 cells, compared to the untreated controls (Fig 4C). Western blot analysis showed a marked decrease in the levels of total β-catenin as well as c-myc expression in CR HCT-116 cells. Taken together, the results suggest an inhibition of Wnt/β-catenin signaling in chemo-resistant colon cancer cells in response to the combination therapy of metformin and FuOx.

Metformin and FuOx combination therapy retards tumor growth in SCID mice

To examine the therapeutic effectiveness of the combination of metform and FuOx, the SCID mice xenograft model of colon tumor was utilized. CR HCT-116 or CR HT-29 cells were injected subcutaneously in the flank region of SCID mice. After palpable tumors were formed, one group was fed metformin in the drinking water and was also injected (i.p) with FuOx, once a week for 3 weeks. Tumor growth was measured by calculating tumor volume for 34 days. Xenografts formed by CR HCT-116 cells showed a linear growth till 27 days, after which the tumor growth slowed down. A similar growth pattern was seen in metformin/FuOx treated mice, albeit the growth was significantly lower compared to controls. By day 34 post-injection, there was a statistically significant (almost 50%) reduction in tumor size in metformin/FuOx treated mice (Fig 5A left panel). CR HT-29 cells injected mice showed a slow tumor growth rate initially. However, a growth spurt was seen in control animals after 27 days. On the other hand, metformin/FuOx treated mice showed a rapid decline in tumor volume after 27 days.

The tumors were harvested, and single cell suspensions were cultured in stem cell medium. Real-time qPCR analysis was performed, which showed a significant decrease in CD44 and increase in CK20 mRNA levels in tumor cells form metformin/FuOx-treated mice (Fig 5B). Metformin/FuOx-treated xenografts also had a reduced number of stem/stem like cells as evidenced by the colonosphere formation (Fig 5C).

Discussion

The current investigation was performed in an attempt to develop new therapeutic strategies for recurrent/metastatic CRC, where the effectiveness of metformin in combination with the conventional chemotherapy was analyzed. We found that metformin acts synergistically with FuOx to inhibit the growth of chemo-resistant HCT-116 and HT-29 cells.

Earlier *in vitro* studies have demonstrated an inhibition of proliferation of colon cancer cells by metformin [48]. *In vivo* studies have demonstrated that metformin delays tumor onset in mouse model for *p53* mutant colon cancer [49] and inhibits colon carcinoma growth stimulated by a high-energy diet [50]. Metformin was also shown to significantly suppress colonic epithelial proliferation by inhibiting the mammalian target of rapamycin (mTOR) pathway [18,51]. In a randomized clinical trial among non-diabetic patients, metformin treatment decreased the number of aberrant crypt foci significantly compared to the

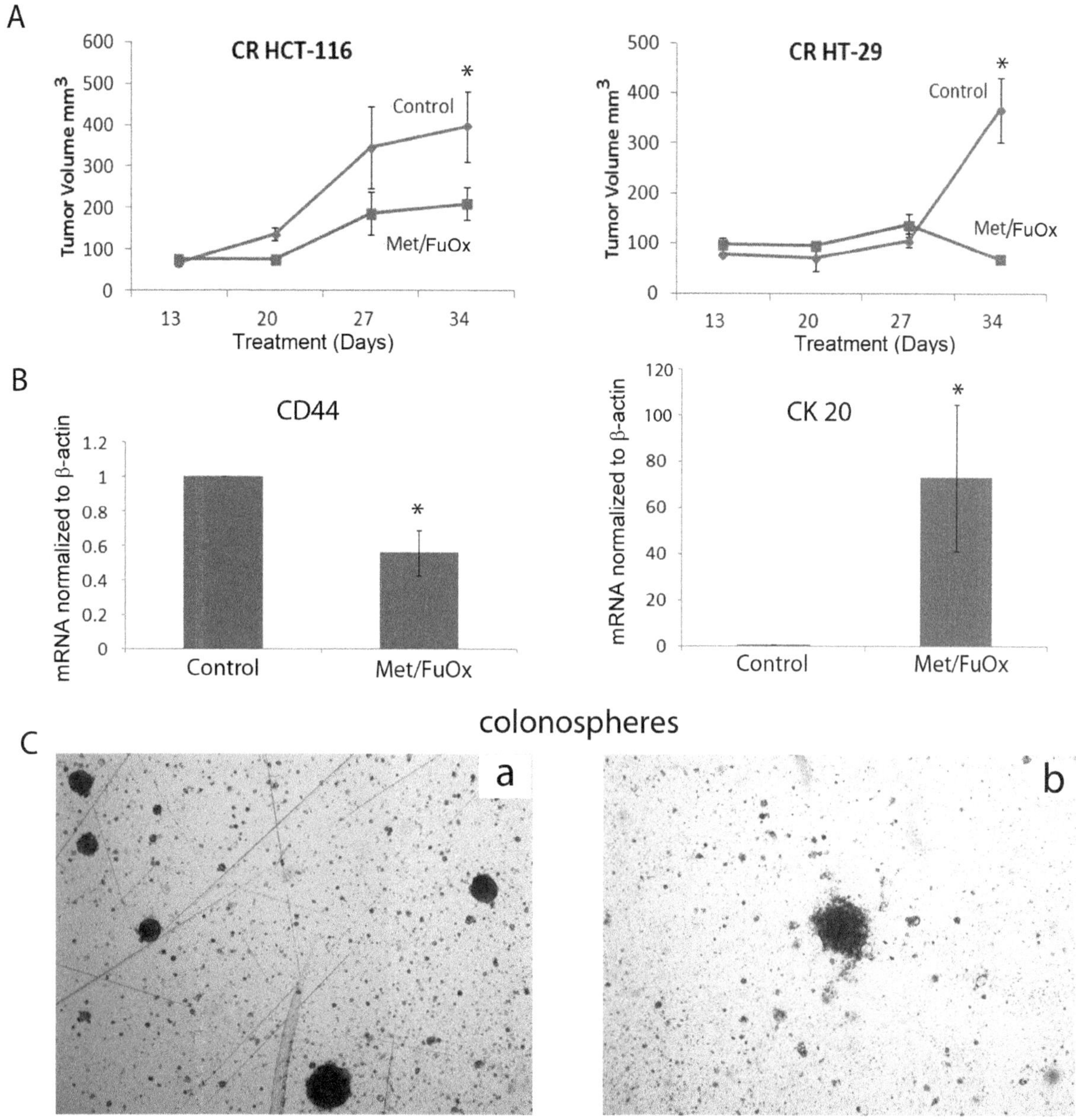

Figure 5. Tumor growth in metformin/FuOx treated SCID mice injected with CR HCT-116 (left pane) or CR HT-29 cells (right panel) (A). Bars represent mean ± standard error. * p >0.05. Real time quantitative PCR on RNA extracted from cells isolated from the CR HT-29 tumor (B). CD44 levels decreased and CK20 levels increased in metformin/FuOx treated xenografts. Colonosphere formation in the cells isolated from CR-HT-29 xenografts (C).

un-treated group with a follow-up of 1 month [52]. Another study performed on diabetic patients with colorectal cancer showed decreased cancer-related mortality in patients using metformin [53]. In breast cancer xenografts Hirsch *et al* [19] showed that metformin treatment specifically eliminated $CD44^{+}/CD24^{-/low}$ stem cells synergistically with doxorubicin. Similarly, in prostate and lung adenocarcinomas, metformin inhibited growth of cancer stem cells [54]. None of these studies however, focused on the recurrent colorectal cancer, which is resistant to the conventional chemotherapy.

Although the reasons for CRC recurrence are not fully understood, it is thought to be related to inability of the conventional chemotherapy to target CSCs/CSLCs [9]. Recent characterization of CSC/CSLCs has led to the identification of key cellular events that among others include epithelial to mesenchymal transition (EMT), drug efflux capability, anti-apoptotic mechanisms and induction of differentiation [55]. Stem cell signaling pathways could, therefore, be targeted to increase the vulnerability of CSCs/CSLCs to therapeutic regimen. Indeed, data from our *in vitro* cell culture studies show an increased cell death in CR colon cancer cells in response to the combination of metformin and FuOx. A similar growth inhibitory effect of metformin/FuOx treatment was also observed *in vivo*, as evidenced by the reduction in tumor volume of xenografts of CR HCT-116 and CR HT-29 cells in SCID mice. Clearly, these results suggest a therapeutic effectiveness of metformin and FuOx combination in tumors caused by chemo-resistant colon cancer cells.

Earlier, we reported a predominance of CSCs/CSLCs in CR colon cancer cells that show increased ability to exclude drugs, and elevated expression of CSC markers such as CD44, CD166, CD133 and ALDH [38,43,56]. Our observation that inhibition of tumor growth in SCID mice by the combination of metformin and FuOx is associated with reduction in CD44 levels and colonosphere forming ability in cells derived from xenograft of CR cells indicates reduction in CSCs/CSLCs. Although the mechanisms for reduction in CSCs/CSLCs are not fully known, our observation of a marked increase in CK-20 expression in colon tumors of metformin/FuOx-treated SCID mice suggests differentiation of CSCs/CSLCs.

Further support for the observation that metformin/FuOx treatment effectively targets colon CSCs/CSLCs and eventually recurrent CRC, comes from the *in vitro* data, which demonstrates that this therapeutic combination inhibits not only the expression of colon CSC markers, but also colonosphere formation and induces their disintegration. Drug resistance and the emergence of CSCs in tumor cells have been attributed in part to the phenomenon of EMT [57–61], that confer on it survival and migratory advantages [57,62] resulting in invasive and metastatic properties (reviewed in [10]). Inhibition of migration in CR cells further indicates that the current combinatorial treatment reduces not only the growth, but also cell migration associated with EMT and invasive properties of CR CSCs/CSLCs. The fact that combination of metformin and FuOx markedly inhibits TCF/LEF activity and decreases β-catenin and c-myc levels suggests that Wnt/β-catenin signaling, which is greatly activated in many malignancies, including CRC [23], plays a pivotal role in regulating the tumor growth.

Recent evidence suggests a role for miRNAs in the regulation of proliferation, apoptosis, differentiation, metabolism and invasion by simultaneously regulating the post-transcriptional expression of numerous genes [63], thereby also playing an important role in CSC function, which involves a complex interplay between multiple pathways. Chemo-resistance in HCT-116 and HT-29 cells was reported to be associated with increased expression of the miR21, an oncomiR [38,43] and reduced expression of the tumor suppressor miR145 (unpublished data). Recently metformin was reported to enhance the cytotoxic effects of 5-FU in HCT-116 cells via regulating miR21 [64]. Our observation that metformin and FuOx treatment reduced the levels of miR21 and increased miR145 in CR HT-29 cells indicates the reversal of chemo-resistant/CSC/CSLC characteristics in colon cancer by modulating miR-21 and miR-145.

Metformin is emerging as a multi-faceted drug. Suppression of mTOR through AMPK activation is believed to constitute the major mechanism underlying its anti-cancer activities [12,17,48,65–68]. The results from the current investigation indicate that metformin acts synergistically with FuOx and inhibits cell proliferation, migration and tumor growth of chemo-resistant colorectal cancer cells partly via affecting the viability of CSCs/CSLCs by modulatig miRNA 21, miRNA 145, and Wnt/β-catenin signaling. In summary, to the best of our knowledge this is the first investigation that shows the therapeutic effectiveness of metformin and FuOx combination therapy on recurring colorectal cancer and demonstrates that the current treatment strategy could be effective against chemo-resistant colon cancer cells that are highly enriched in CSCs/CSLCs.

Acknowledgments

The endothelial cells were a gift from Dr. Dipak Banerjee, University of Puerto Rico, Riomet was a gift from Dr. Ramandeep Rattan, Henry Ford Hospital.

Author Contributions

Conceived and designed the experiments: PNM APNM. Performed the experiments: PNM YY AV SR LF. Analyzed the data: PNM YY. Contributed reagents/materials/analysis tools: EL LF. Wrote the paper: PNM APNM.

References

1. Al-Hajj M, Wicha MS, Benito-Hernandez A, Morrison SJ, Clarke MF (2003) Prospective identification of tumorigenic breast cancer cells. Proc Natl Acad Sci U S A 100: 3983–3988.
2. Bonnet D, Dick JE (1997) Human acute myeloid leukemia is organized as a hierarchy that originates from a primitive hematopoietic cell. Nat Med 3: 730–737.
3. Charafe-Jauffret E, Ginestier C, Iovino F, Wicinski J, Cervera N, et al. (2009) Breast cancer cell lines contain functional cancer stem cells with metastatic capacity and a distinct molecular signature. Cancer Res 69: 1302–1313.
4. Marcato P, Dean CA, Pan D, Araslanova R, Gillis M, et al. (2011) Aldehyde dehydrogenase activity of breast cancer stem cells is primarily due to isoform ALDH1A3 and its expression is predictive of metastasis. Stem Cells 29: 32–45.
5. Pang R, Law WL, Chu AC, Poon JT, Lam CS, et al. (2010) A subpopulation of CD26+ cancer stem cells with metastatic capacity in human colorectal cancer. Cell Stem Cell 6: 603–615.
6. Schulenburg A, Bramswig K, Herrmann H, Karlic H, Mirkina I, et al. (2010) Neoplastic stem cells: current concepts and clinical perspectives. Crit Rev Oncol Hematol 76: 79–98.
7. Tanei T, Morimoto K, Shimazu K, Kim SJ, Tanji Y, et al. (2009) Association of breast cancer stem cells identified by aldehyde dehydrogenase 1 expression with resistance to sequential Paclitaxel and epirubicin-based chemotherapy for breast cancers. Clin Cancer Res 15: 4234–4241.
8. Dean M, Fojo T, Bates S (2005) Tumour stem cells and drug resistance. Nat Rev Cancer 5: 275–284.
9. Jordan CT, Guzman ML, Noble M (2006) Cancer stem cells. N Engl J Med 355: 1253–1261.
10. Del Barco S, Vazquez-Martin A, Cufi S, Oliveras-Ferraros C, Bosch-Barrera J, et al. (2011) Metformin: multi-faceted protection against cancer. Oncotarget 2: 896–917.
11. Dowling RJ, Goodwin PJ, Stambolic V (2011) Understanding the benefit of metformin use in cancer treatment. BMC Med 9: 33.
12. Alimova IN, Liu B, Fan Z, Edgerton SM, Dillon T, et al. (2009) Metformin inhibits breast cancer cell growth, colony formation and induces cell cycle arrest in vitro. Cell Cycle 8: 909–915.
13. Anisimov VN, Berstein LM, Egormin PA, Piskunova TS, Popovich IG, et al. (2005) Effect of metformin on life span and on the development of spontaneous mammary tumors in HER-2/neu transgenic mice. Exp Gerontol 40: 685–693.
14. Ben Sahra I, Laurent K, Loubat A, Giorgetti-Peraldi S, Colosetti P, et al. (2008) The antidiabetic drug metformin exerts an antitumoral effect in vitro and in vivo through a decrease of cyclin D1 level. Oncogene 27: 3576–3586.
15. Cantrell LA, Zhou C, Mendivil A, Malloy KM, Gehrig PA, et al. (2010) Metformin is a potent inhibitor of endometrial cancer cell proliferation–implications for a novel treatment strategy. Gynecol Oncol 116: 92–98.
16. Isakovic A, Harhaji L, Stevanovic D, Markovic Z, Sumarac-Dumanovic M, et al. (2007) Dual antiglioma action of metformin: cell cycle arrest and mitochondria-dependent apoptosis. Cell Mol Life Sci 64: 1290–1302.
17. Liu B, Fan Z, Edgerton SM, Deng XS, Alimova IN, et al. (2009) Metformin induces unique biological and molecular responses in triple negative breast cancer cells. Cell Cycle 8: 2031–2040.
18. Tomimoto A, Endo H, Sugiyama M, Fujisawa T, Hosono K, et al. (2008) Metformin suppresses intestinal polyp growth in ApcMin/+ mice. Cancer Sci 99: 2136–2141.
19. Hirsch HA, Iliopoulos D, Tsichlis PN, Struhl K (2009) Metformin selectively targets cancer stem cells, and acts together with chemotherapy to block tumor growth and prolong remission. Cancer Res 69: 7507–7511.
20. Song CW, Lee H, Dings RP, Williams B, Powers J, et al. (2012) Metformin kills and radiosensitizes cancer cells and preferentially kills cancer stem cells. Sci Rep 2: 362.
21. Brabletz S, Schmalhofer O, Brabletz T (2009) Gastrointestinal stem cells in development and cancer. J Pathol 217: 307–317.
22. Korkaya H, Paulson A, Charafe-Jauffret E, Ginestier C, Brown M, et al. (2009) Regulation of mammary stem/progenitor cells by PTEN/Akt/beta-catenin signaling. PLoS Biol 7: e1000121.

23. Kolligs FT, Bommer G, Goke B (2002) Wnt/beta-catenin/tcf signaling: a critical pathway in gastrointestinal tumorigenesis. Digestion 66: 131–144.
24. Morin PJ, Sparks AB, Korinek V, Barker N, Clevers H, et al. (1997) Activation of beta-catenin-Tcf signaling in colon cancer by mutations in beta-catenin or APC. Science 275: 1787–1790.
25. Aoki K, Taketo MM (2007) Adenomatous polyposis coli (APC): a multi-functional tumor suppressor gene. J Cell Sci 120: 3327–3335.
26. Polakis P (2007) The many ways of Wnt in cancer. Curr Opin Genet Dev 17: 45–51.
27. Fearon ER (2011) Molecular genetics of colorectal cancer. Annu Rev Pathol 6: 479-507.
28. Ishimoto T, Oshima H, Oshima M, Kai K, Torii R, et al. (2010) CD44+ slow-cycling tumor cell expansion is triggered by cooperative actions of Wnt and prostaglandin E2 in gastric tumorigenesis. Cancer Sci 101: 673–678.
29. Kanwar SS, Yu Y, Nautiyal J, Patel BB, Majumdar AP (2010) The Wnt/beta-catenin pathway regulates growth and maintenance of colonospheres. Mol Cancer 9: 212.
30. Calin GA, Croce CM (2006) MicroRNA signatures in human cancers. Nat Rev Cancer 6: 857–866.
31. Lu J, Getz G, Miska EA, Alvarez-Saavedra E, Lamb J, et al. (2005) MicroRNA expression profiles classify human cancers. Nature 435: 834–838.
32. Volinia S, Calin GA, Liu CG, Ambs S, Cimmino A, et al. (2006) A microRNA expression signature of human solid tumors defines cancer gene targets. Proc Natl Acad Sci U S A 103: 2257–2261.
33. Yu Y, Kanwar SS, Patel BB, Oh PS, Nautiyal J, et al. (2012) MicroRNA-21 induces stemness by downregulating transforming growth factor beta receptor 2 (TGFbetaR2) in colon cancer cells. Carcinogenesis 33: 68–76.
34. Lu Z, Liu M, Stribinskis V, Klinge CM, Ramos KS, et al. (2008) MicroRNA-21 promotes cell transformation by targeting the programmed cell death 4 gene. Oncogene 27: 4373–4379.
35. Padhye S, Yang H, Jamadar A, Cui QC, Chavan D, et al. (2009) New difluoro Knoevenagel condensates of curcumin, their Schiff bases and copper complexes as proteasome inhibitors and apoptosis inducers in cancer cells. Pharm Res 26: 1874–1880.
36. Roy S, Yu Y, Padhye SB, Sarkar FH, Majumdar AP (2013) Difluorinated-curcumin (CDF) restores PTEN expression in colon cancer cells by down-regulating miR-21. PLoS One 8: e68543.
37. Nautiyal J, Kanwar SS, Yu Y, Majumdar AP (2011) Combination of dasatinib and curcumin eliminates chemo-resistant colon cancer cells. J Mol Signal 6: 7.
38. Yu Y, Kanwar SS, Patel BB, Nautiyal J, Sarkar FH, et al. (2009) Elimination of Colon Cancer Stem-Like Cells by the Combination of Curcumin and FOLFOX. Transl Oncol 2: 321–328.
39. Yu Y, Sarkar FH, Majumdar AP (2013) Down-regulation of miR-21 Induces Differentiation of Chemoresistant Colon Cancer Cells and Enhances Susceptibility to Therapeutic Regimens. Transl Oncol 6: 180–186.
40. Martinez JA, Torres-Negron I, Amigo LA, Banerjee DK (1999) Expression of Glc3Man9GlcNAc2-PP-Dol is a prerequisite for capillary endothelial cell proliferation. Cell Mol Biol (Noisy-le-grand) 45: 137–152.
41. Nangia-Makker P, Tait L, Shekhar MP, Palomino E, Hogan V, et al. (2007) Inhibition of breast tumor growth and angiogenesis by a medicinal herb: Ocimum gratissimum. Int J Cancer 121: 884–894.
42. Wicha MS, Liu S, Dontu G (2006) Cancer stem cells: an old idea–a paradigm shift. Cancer Res 66: 1883–1890; discussion 1895–1886.
43. Kanwar SS, Yu Y, Nautiyal J, Patel BB, Padhye S, et al. (2011) Difluorinated-curcumin (CDF): a novel curcumin analog is a potent inhibitor of colon cancer stem-like cells. Pharm Res 28: 827–838.
44. Hu Y, Smyth GK (2009) ELDA: extreme limiting dilution analysis for comparing depleted and enriched populations in stem cell and other assays. J Immunol Methods 347: 70–78.
45. Nangia-Makker P, Wang Y, Raz T, Tait L, Balan V, et al. (2010) Cleavage of galectin-3 by matrix metalloproteases induces angiogenesis in breast cancer. Int J Cancer 127: 2530–2541.
46. Nangia-Makker P, Raz T, Tait L, Shekhar MP, Li H, et al. (2013) Ocimum gratissimum retards breast cancer growth and progression and is a natural inhibitor of matrix metalloproteases. Cancer Biol Ther 14: 417–427.
47. Qian Y, Corum L, Meng Q, Blenis J, Zheng JZ, et al. (2004) PI3K induced actin filament remodeling through Akt and p70S6K1: implication of essential role in cell migration. Am J Physiol Cell Physiol 286: C153–163.
48. Zakikhani M, Dowling RJ, Sonenberg N, Pollak MN (2008) The effects of adiponectin and metformin on prostate and colon neoplasia involve activation of AMP-activated protein kinase. Cancer Prev Res (Phila) 1: 369–375.
49. Buzzai M, Jones RG, Amaravadi RK, Lum JJ, DeBerardinis RJ, et al. (2007) Systemic treatment with the antidiabetic drug metformin selectively impairs p53-deficient tumor cell growth. Cancer Res 67: 6745–6752.
50. Algire C, Amrein L, Zakikhani M, Panasci L, Pollak M (2010) Metformin blocks the stimulative effect of a high-energy diet on colon carcinoma growth in vivo and is associated with reduced expression of fatty acid synthase. Endocr Relat Cancer 17: 351–360.
51. Hosono K, Endo H, Takahashi H, Sugiyama M, Uchiyama T, et al. (2010) Metformin suppresses azoxymethane-induced colorectal aberrant crypt foci by activating AMP-activated protein kinase. Mol Carcinog 49: 662–671.
52. Hosono K, Endo H, Takahashi H, Sugiyama M, Sakai E, et al. (2010) Metformin suppresses colorectal aberrant crypt foci in a short-term clinical trial. Cancer Prev Res (Phila) 3: 1077–1083.
53. Lee JH, Kim TI, Jeon SM, Hong SP, Cheon JH, et al. (2012) The effects of metformin on the survival of colorectal cancer patients with diabetes mellitus. Int J Cancer 131: 752–759.
54. Iliopoulos D, Hirsch HA, Struhl K (2011) Metformin decreases the dose of chemotherapy for prolonging tumor remission in mouse xenografts involving multiple cancer cell types. Cancer Res 71: 3196–3201.
55. Roy S, Majumdar AP (2012) Signaling in colon cancer stem cells. J Mol Signal 7: 11.
56. Oh PS, Patel VB, Sanders MA, Kanwar SS, Yu Y, et al. (2011) Schlafen-3 decreases cancer stem cell marker expression and autocrine/juxtacrine signaling in FOLFOX-resistant colon cancer cells. Am J Physiol Gastrointest Liver Physiol 301: G347–355.
57. Hollier BG, Evans K, Mani SA (2009) The epithelial-to-mesenchymal transition and cancer stem cells: a coalition against cancer therapies. J Mammary Gland Biol Neoplasia 14: 29–43.
58. Klarmann GJ, Hurt EM, Mathews LA, Zhang X, Duhagon MA, et al. (2009) Invasive prostate cancer cells are tumor initiating cells that have a stem cell-like genomic signature. Clin Exp Metastasis 26: 433–446.
59. Mani SA, Guo W, Liao MJ, Eaton EN, Ayyanan A, et al. (2008) The epithelial-mesenchymal transition generates cells with properties of stem cells. Cell 133: 704–715.
60. Morel AP, Lievre M, Thomas C, Hinkal G, Ansieau S, et al. (2008) Generation of breast cancer stem cells through epithelial-mesenchymal transition. PLoS One 3: e2888.
61. Santisteban M, Reiman JM, Asiedu MK, Behrens MD, Nassar A, et al. (2009) Immune-induced epithelial to mesenchymal transition in vivo generates breast cancer stem cells. Cancer Res 69: 2887–2895.
62. Thiery JP, Acloque H, Huang RY, Nieto MA (2009) Epithelial-mesenchymal transitions in development and disease. Cell 139: 871–890.
63. Bartel DP (2004) MicroRNAs: genomics, biogenesis, mechanism, and function. Cell 116: 281–297.
64. Feng YH, Wu CL, Shiau AL, Lee JC, Chang JG, et al. (2012) MicroRNA-21-mediated regulation of Sprouty2 protein expression enhances the cytotoxic effect of 5-fluorouracil and metformin in colon cancer cells. Int J Mol Med 29: 920–926.
65. Dowling RJ, Zakikhani M, Fantus IG, Pollak M, Sonenberg N (2007) Metformin inhibits mammalian target of rapamycin-dependent translation initiation in breast cancer cells. Cancer Res 67: 10804–10812.
66. Kalender A, Selvaraj A, Kim SY, Gulati P, Brule S, et al. (2010) Metformin, independent of AMPK, inhibits mTORC1 in a rag GTPase-dependent manner. Cell Metab 11: 390–401.
67. Memmott RM, Mercado JR, Maier CR, Kawabata S, Fox SD, et al. (2010) Metformin prevents tobacco carcinogen–induced lung tumorigenesis. Cancer Prev Res (Phila) 3: 1066–1076.
68. Rocha GZ, Dias MM, Ropelle ER, Osorio-Costa F, Rossato FA, et al. (2011) Metformin amplifies chemotherapy-induced AMPK activation and antitumoral growth. Clin Cancer Res 17: 3993–4005.

5

Knockdown of the HDAC1 Promotes the Directed Differentiation of Bone Mesenchymal Stem Cells into Cardiomyocytes

Dong-feng Lu[1]*, Ying Wang[1], Zi-zhuo Su[1], Zhao-hua Zeng[1], Xiao-wen Xing[1], Zhi-yu He[1], Chunxiang Zhang[2]*

1 Department of Cardiology, the First Affiliated Hospital of Guangzhou Medical University, Guangzhou, Guangdong, China, **2** Department of Pharmacology and Cardiovascular Research Center, Rush Medical college of Rush University, Chicago, Illinois, United States of America

Abstract

Failure of the directed differentiation of the transplanted stem cells into cardiomyocytes is still a major challenge of cardiac regeneration therapy. Our recent study has demonstrated that the expression of histone deacetylase 1 (HDAC1) is decreased in bone mesenchymal stem cells (BMSCs) during their differentiation into cardiomyocytes. However, the potential roles of HDAC1 in cardiac cell differentiation of BMSCs, as well as the mechanisms involved are still unclear. In current study, the expression of HDAC1 in cultured rat BMSCs is knocked down by lentiviral vectors expressing HDAC1-RNAi. The directed differentiation of BMSCs into cardiomyocytes is evaluated by the expression levels of cardiomyocyte-related genes such as GATA-binding protein 4 (GATA-4), Nirenberg, Kim gene 2 homeobox 5 (Nkx2.5), cardiac troponin T (CTnT), myosin heavy chain (MHC), and connexin-43. Compared with that in control BMSCs, the expression of these cardiomyocyte-related genes is significantly increased in these HDAC1 deficient stem cells. The results suggest that HDAC1 is involved in the cardiomyocyte differentiation of BMSCs. Knockdown of the HDAC1 may promote the directed differentiation of BMSCs into cardiomyocytes.

Editor: Gangjian Qin, Northwestern University, United States of America

Funding: This research was supported by the Scientific Foundation of Guangdong, China (No. 2009B030801358; No. 10151009504000008; No. 2012A030400028; No. 7003080). The funders had no role in study design, data collection and analysis, decision to publish, or preparation of the manuscript.

Competing Interests: The authors have declared that no competing interests exist.

* E-mail: chunxiang_zhang@rush.edu (CZ); 497440817@qq.com (DFL)

Introduction

Bone mesenchymal stem cells (BMSCs) are non-hematopoietic stem cells in bone marrow. BMSCs are ideal cell-based regeneration therapy in the treatment of cardiovascular disease because these cells exhibit various advantageous traits, including ready accessibility, ease of amplification, low immunogenicity, and amenability to the introduction of exogenous genes and other genetic modifications [1], [2].

Despite the current optional approaches, failure of the directed differentiation of the transplanted stem cells into cardiomyocytes is still a major challenge of cardiac regeneration therapy [3]. This has highlighted the importance and urgency of studying the novel mechanisms of this critical cellular process and exploring new therapeutic options to improve our cardiac regeneration therapy.

Recently, an increasing number of studies have revealed that the epigenetic modification of histone acetylation may play an important role in the transdifferentiation processes of adult stem cells including their directed cardiac cell differentiation [4], [5], [6], [7]. In a pilot study, we have identified that the expression of histone deacetylase 1 (HDAC1) in BMSCs is significantly decreased during their differentiation into cardiomyocytes. However, until now, the potential role of HDAC1 in the directed differentiation of the transplanted stem cells into cardiomyocytes is still unclear. In current study, the expression of HDAC1 in cultured rat BMSCs is knocked down by lentiviral vectors expressing HDAC1-RNAi and the directed differentiation of BMSCs into cardiomyocytes is evaluated.

Materials and Methods

1.1. Experimental animals

One-month-old male Sprague-Dawley (SD) rats, weighing 80 to 110 g, were purchased from the Guangdong Medical Laboratory Animal Center. All protocols were approved by the Institutional Animal Care and Use Committee at the Guangzhou Medical University and were consistent with the Guide for the Care and Use of Laboratory Animals (NIH publication 85-23, revised 1985).

1.2. The isolation, culture and identification of BMSCs

SD rat BMSCs were isolated and cultured using the whole bone marrow adherence method described previously [8]. Briefly, SD rats were sacrificed, and their femora and tibiae were rapidly stripped. A 5 ml syringe loaded with an appropriate quantity of complete BMSC culture medium (Dulbecco's modified Eagle's medium/Ham's F12 nutrient mixture (DMEM/F12) with 10% FBS, 100 U/ml penicillin, and 100 U/ml streptomycin) was inserted into the metaphysis of each bone, and was utilised to flush bone marrow cells with culture medium. The isolated BMSCs were centrifuged at 1000 rpm for 10 minutes, resuspended in complete medium, and cultured. The culture medium was changed every 2–3 days. After reaching 80% confluence, cultured adherent cells were trypsinised with a 0.25% trypsin solution and passaged. The percentages of cells that were positive for the BMSC surface markers of CD34, CD45, CD29, CD44, and CD90 (Santa Cruz) were analysed by flow cytometry. For all experiments, rat BMSCs from passages 3 to 5 (P3 to P5 cells) were used.

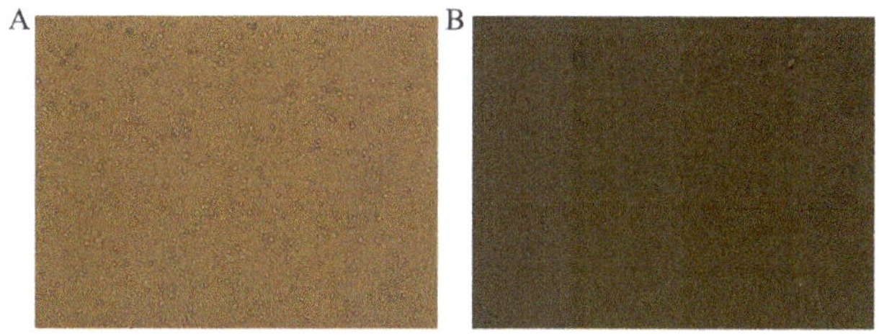

Figure 1. Morphological observations of BMSCs. A. Primary cultured BMSCs (100× magnification). B. BMSCs at passage 3 (P3) (100× magnification).

1.3. The construction and screening of the optimal HDAC1-RNAi lentiviral vector

Based on the sequence information of the HDAC1 gene, four short hairpin RNA (shRNA) sequences and one negative control (NC) sequence were designed. The shRNA expression sequences targeting HDAC1 were ligated into the lentiviral vector pGCSIL-GFP to obtain recombinant pGCSIL-GFP-shHDAC1 plasmids. After the plasmid sequences had been verified, the recombinant plasmids and the lentiviral packaging plasmids pHelper 1.0 and pHelper 2.0 were cotransfected into 293T cells using Lipofectamine 2000. Virus was collected from the culture supernatant, and the viral titre of this solution was determined by serial dilution.

The packaged pGCSIL-GFP-NC lentivirus was used to infect BMSCs at MOIs of 0, 1, 10, and 100. Fluorescence microscopy was utilised to detect the expression of fluorescent proteins after 12 h, 24 h, 48 h, and 72 h of infection. Flow cytometry (at a wavelength of 488 nm) was used to determine infection efficiency and thereby screen for the MOI value and infection time that produced maximal infection efficiency.

Five groups of packaged viruses were used to infect BMSCs, and the BMSCs were collected after 48 hours of infection. Uninfected BMSCs were utilised as a normal control group. To screen for the virus with the best efficiency at silencing HDAC1, real-time quantitative reverse transcription polymerase chain reaction (RT-qPCR) was used to determine the HDAC1 mRNA expression levels for each group, and Western blotting was used to determine the HDAC1 protein expression levels for each group.

1.4. RT-PCR

TRIzol was used to extract total RNA samples from cells. Subsequently, a TaKaRa transcription kit was utilised for the reverse transcription (RT) of these samples; in this reaction, the samples were incubated in a PCR thermocycler at 37°C for 15 minutes and then at 85°C for 5 seconds. The RT-PCR amplifications were performed with the TaKaRa PrimeScript II 1st Strand cDNA Synthesis Kit (D6210A), using glyceraldehyde 3-phosphate dehydrogenase (GAPDH) as an internal reference. The amplification included the following reaction stages: stage I (initial denaturation), which involved an incubation at 95°C for 30 sec; stage II (30 cycles of PCR amplification), which involved 30 cycles of incubation at 95°C for 5 sec, 60°C for 3 sec, and 72°C for 30 s; and stage III (melting curve analysis), which involved an incubation at 72°C for 10 min followed by an incubation at 16°C for 10 min. The primers for various genes were designed using the Primer 5.0 software package and synthesised by TaKaRa Biotechnology. In particular, the following specific primers were synthesised: HDAC1, 149 bp, F: 5′-TCA CCG AAT CCG AAT GAC TCA TAA-3′, R: 5′- CTG GGC GAA TAG AAC GCA AGA-3′; GAPDH, 143 bp, F: 5′- GGC ACA GTC AAG GCT GAG AAT G-3′, R: 5′- ATG GTG GTG AAG ACG CCA GTA-3′;GATA binding protein 4 (GATA-4), 180 bp, F: 5′- AAC GGA AGC CCA AGA ATC TGA-3′, R: 5′- AGC TGC TGT GCC CAT AGT GAG-3′; Nirenberg and Kim gene 2 homeobox 5 (Nkx2.5), 178 bp, F: 5′- GTC TCA ACG CCT ACG GCT AC-3′, R: 5′- CTC TGC ACC GTG TTC AAG TC-3′;cardiac troponin T(CTnT), 162 bp, F: 5′- AGA TTG CGA AGC AGG AGA TG -3′,R: 5′- CAT CCA CTT TGT CCA CAC GA -3′;myosin heavy chain (MHC), 151 bp, F: 5′- GAG CAG GAG CTG ATC GAG AC -3′, R: 5′- CCT CTG CGT TCC TAC ACT CC -3′;and connexin-43, 165 bp, F: 5′- TCG TGT CTG TGC CCA CCC TC -3′, R:5′- TCC CGT ACT TGA ACT TCT TGA TTT -3′. Based on the amplification results, the 2-ΔΔCT method was used to calculate the relative multiple of the starting copy number that existed in the template from each experimental group and thereby indicate the relative differences in mRNA levels among the experimental groups.

1.5. Western blotting

At least 1×10^7 cells were collected for the extraction of nucleoproteins, which was performed as described in the DBI nucleoprotein extraction kit. Protein concentrations were determined with the bicinchoninic acid (BCA) assay. Samples of 16 μg of nucleoproteins were separated on a 10% sodium dodecyl sulphate-polyacrylamide gel electrophoresis (SDS-PAGE) gel and subsequently electrotransferred onto polyvinylidene difluoride (PVDF) membranes for 90 min at 4°C and 290 mA. After these membranes were blocked in 5% non-fat dry milk for 90 min, they were incubated with a diluted anti-HDAC1 primary antibody (1:500 dilution, Abcam) (in a solution that contained 2% non-fat dry milk) and incubated at 4°C overnight. Subsequently, the membranes were washed with phosphate-buffered saline (PBS) or Tris-buffered saline with Tween 20 (TBST), using sufficient saline solution to cover the PVDF membrane for each wash. The first wash step was performed over the course of 15 min, whereas each of the next three washes were 5 min in duration. Following these washes, the membranes were incubated with 5 ml of 1:5000 diluted goat anti-rabbit secondary antibody (in a solution that contained 2% non-fat dry milk) for 1 h at room temperature. An enhanced chemiluminescence (ECL) solution was prepared; the membranes were incubated in this ECL solution in a darkroom for 5 min, and images of the membranes were then digitally developed. The relative expression level of a target protein was quantified by dividing the greyscale value of the target protein band by the greyscale value of the band corresponding to GAPDH, the protein that was utilised as an internal reference.

1.6. Statistical analysis

Statistical analyses were performed using the SPSS 13.0 statistical software package, and the data were expressed as means ± standard deviation. Comparisons between two groups were performed using the independent samples t-test, and comparisons among multiple groups were performed using the one-way analyses of variance; $P<0.05$ was regarded as statistically significant.

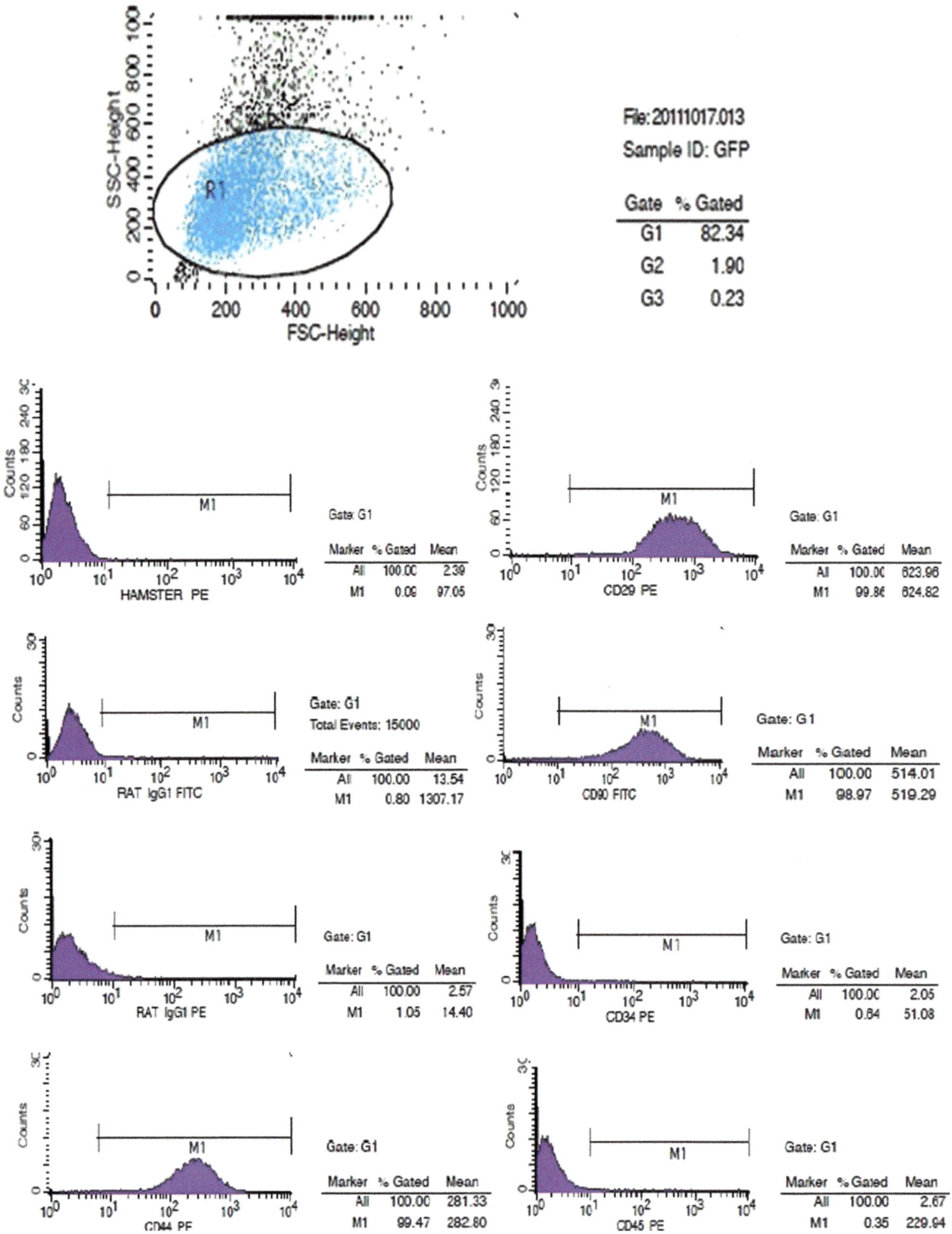

Figure 2. Flow cytometry observations of BMSC surface markers. The expressed levels of the hematopoietic markers CD34 and CD45, and the mesenchymal cell surface markers, CD29, CD44, and CD90 in BMSCs.

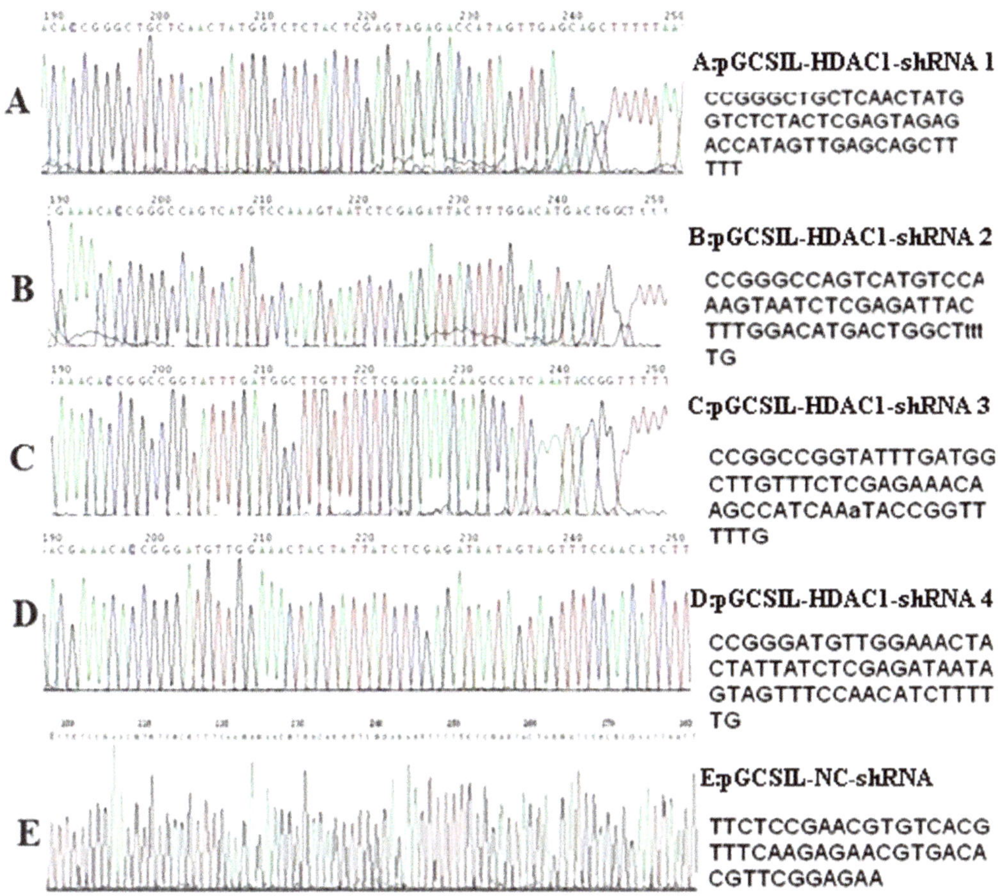

Figure 3. Colour sequencing maps of the recombinant plasmids. The positive clones were designated pGCSIL-HDAC1-shRNA1 through pGCSIL-HDAC1-shRNA4, and the NC vector was known as pGCSIL-NC-shRNA.

Results

1. The morphological characteristics and identification of BMSCs in vitro

At the early stage of primary BMSC cultures, the cells were round. After 24 h of culture, the BMSCs had become adherent and had elongated to adopt fusiform or triangular shapes. After 3–4 d, cells had entered the logarithmic growth phase and exhibited clonal proliferation. These BMSC colonies fused after 9–10 d of culture; at this time, most of the cells exhibited the morphology of flattened long spindles, whereas a small proportion of cells were either triangle- or polygon-shaped. Cells became adherent relatively rapidly after they were passaged. Passaged cells demonstrated a fibroblast-like morphology and an increased volume relative to primary cells. The passaged cells underwent rapid clonal proliferation, requiring only 3–4 d to reach 80% confluence; at this stage, the cells were either re-passaged or cryopreserved. As the number of passages of cells increased, cells became increasingly purified, and the observed cell morphologies became more uniform. Moreover, the BMSCs exhibited a polar arrangement that was swirling in nature or reminiscent of schools of fish (Figure 1). BMSCs at three passages were examined by flow cytometry. The results of these analyses indicated that these BMSCs expressed low levels of the hematopoietic markers CD34 (0.64%) and CD45 (0.35%) but high levels of various stromal and mesenchymal cell surface markers, including CD29 (99.86%), CD44 (99.47%), and CD90 (98.97%) (Figure 2).

2. The construction and screening of the optimal HDAC1-RNAi lentiviral vector

2.1. Sequencing verification of recombinant plasmids. The recombinant vectors of this experiment were sequenced by Shanghai GeneChem. The positive clones were designated pGCSIL-HDAC1-shRNA1 through pGCSIL-HDAC1-shRNA4, and the NC vector was known as pGCSIL-NC-shRNA. The sequencing results indicated that the shRNA expression template was successfully constructed based on the pGCSIL-GFP vector and that the sequences of interest in each vector were completely correct and identical to the designed target sequence (Figure 3).

2.2. The infection of target cells: preliminary experiments and screening for optimal MOI conditions. Expression of enhanced green fluorescent protein (EGFP) could be observed at 12 h after BMSCs had been infected with the pGCSIL-GFP-NC lentivirus; the EGFP expression levels peaked at 48 h after infection, although strong green fluorescence was sustained through 72 h after infection. In particular, compared with the other examined groups, the ENi.S+5 μg/ml polybrene group exhibited a significantly higher number of fluorescent cells. Under optimal infection conditions, at an MOI of between 1 and 10, an infection efficiency of less than 70% was

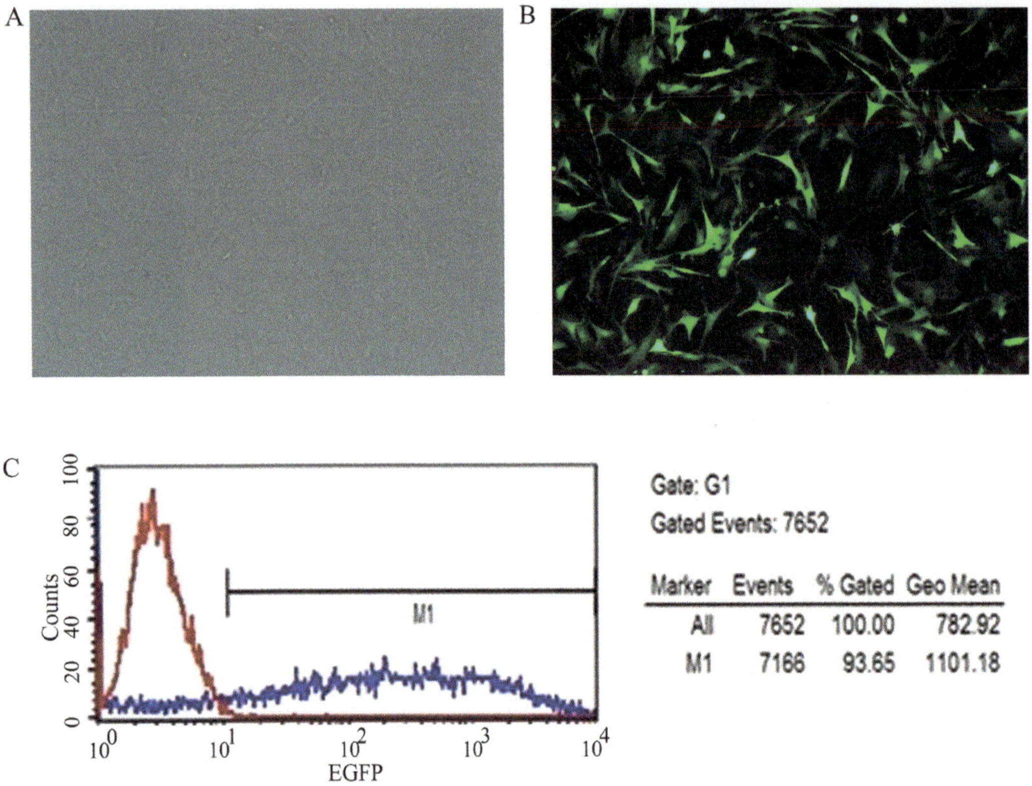

Figure 4. Observations of the efficiency of the lentiviral infection of BMSCs. A, Inverted microscopy of BMSCs infected with lentivirus for 48 h (100× magnification). B, Fluorescence microscopy of BMSCs infected with lentivirus for 48 h (100× magnification). C, Flow cytometric measurements of BMSC infection efficiency (93.65%).

observed. By contrast, at an MOI of 100, an infection efficiency of 90% or more was observed; moreover, cells exhibited strong growth and no significant differences in cell morphology relative to the normal control group. Flow cytometry revealed an EGFP-positive rate of 93.65% among these infected cells (Figure 4).

2.3. HDAC1 mRNA expression determined by RT-PCR. RT-PCR results revealed that although HDAC1 mRNA could be detected in samples from all groups, significantly lower HDAC1 mRNA expression levels were detected in the groups infected with an interference virus compared with those in either the normal control group or the NC group ($P<0.01$, Figure 5, Figure 6). In particular, the normal control group exhibited an HDAC1 mRNA expression level that was 32, 11, 13, and 96 times higher than the HDAC1 mRNA expression levels of the LV-HDAC1-shRNA1, LV-HDAC1-shRNA2, LV-HDAC1-shRNA3, and LV-HDAC1-shRNA4 interference groups, respectively. No significant differences in HDAC1 mRNA expression were found between the normal control group and the NC group. Although all the four vectors that were able to successfully inhibit the HDAC expression, the inhibitory effect of LV-HDAC1-shRNA1 and LV-HDAC1-shRNA4 was more efficient compared with that of LV-HDAC-shRNA2 and LV-HDAC1-shRNA3.

2.4. HDAC1 expression determined by Western blotting. The results of western blotting indicated that although HDAC1 protein was expressed in samples from all groups, significantly lower HDAC1 protein expression levels were detected in all the groups infected with an interference virus compared with those of the normal control group or the NC group ($P<0.01$, Figure 7). In particular, the normal control group exhibited an HDAC1 protein expression level that was 1.66 times, 1.53 times, 1.50 times, and 1.70 times higher than those in LV-HDAC1-shRNA1, LV-HDAC1-shRNA2, LV-HDAC1-shRNA3, and LV-HDAC1-shRNA4 interference groups, respectively. Thus, the inhibitory efficiencies of these interference groups at the protein level were 40%, 35%, 32%, and 41%, respectively. Based on the results, LV-HDAC1-shRNA4 vector was selected to be used in our subsequent experiments.

3. RT-PCR detection of genes related to myocardial development and structure

The expression levels of genes related to myocardial development and structure were detected in BMSCs infected with LV-HDAC1-shRNA4. The genes detected in this study included Nkx2.5, GATA-4, MHC, connexin-43, and CTnT. As shown in Table 1, the expression of these 5 genes was significantly higher in BMSCs infected with the lentivirus compared with that in cells from either the normal control group or the NC group ($P<0.05$). No significant difference was found between the normal control group and the NC group with respect to the expression of the 5 detected genes.

Discussion

In our recent pilot study, we have found that the expression level of HDAC1 is significantly decreased in BMSCs after co-cultured with cardiomyocytes (Data not shown). Moreover, nascent cardiomyocytes exhibite low levels of HDAC1 expression. These initial results suggest that HDAC1 gene expression might be related to the directed differentiation of BMSCs into cardiomyocytes. We thus hypothesize that the HDAC1 might be a

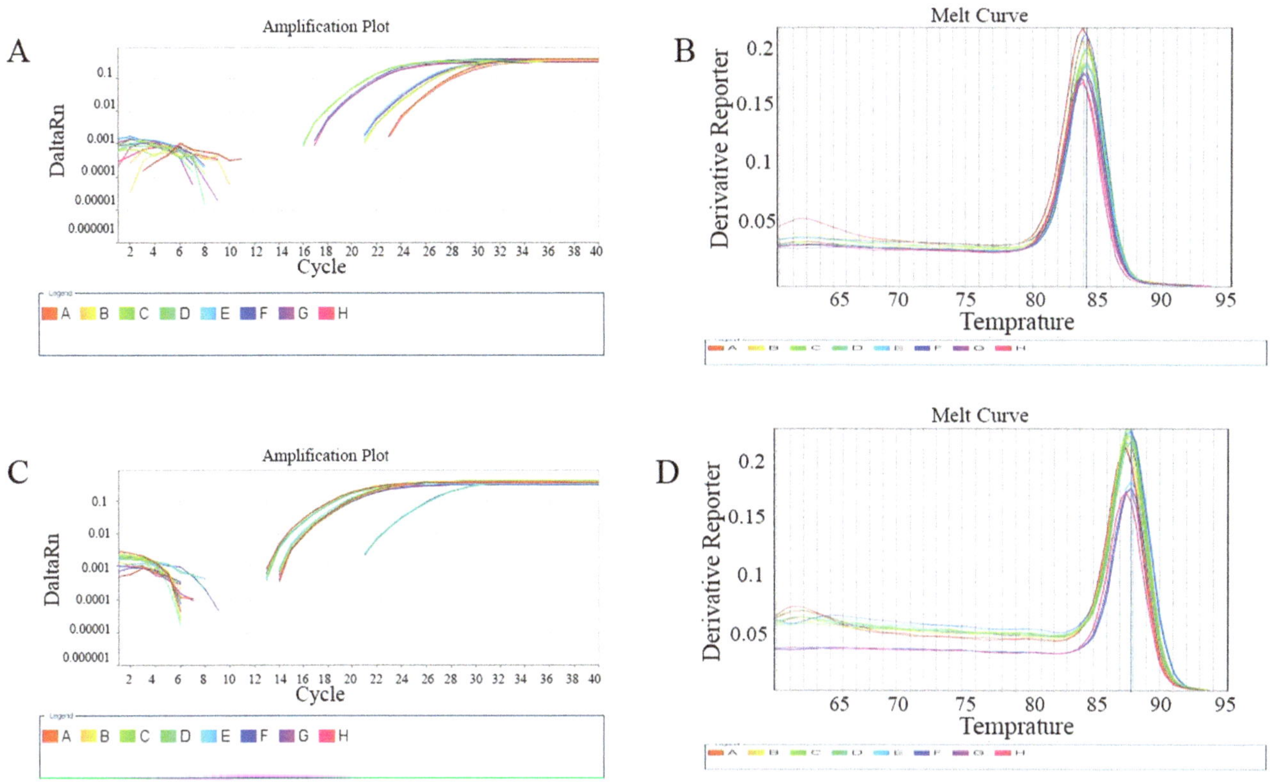

Figure 5. Amplification and melting curves for HDAC1 and GAPDH. A, HDAC1 amplification curve. B, HDAC1 melting curve. C, GAPDH amplification curve. D, GAPDH melting curve.

negatively regulator in cardiac cell differentiation from BMSCs. To test our hypothesis, we constructed a shRNA eukaryotic expression vector that specifically silenced the HDAC1 gene by RNAi. We found that HDAC1-shRNA could significantly reduce HDAC1 expression in BMSCs; this reduction in HDAC1 expression in BMSCs significantly increased the expression of cardiac-specific genes, such as Nkx2.5, GATA-4, MHC, connexin-43, and CTnT. Thus, the specific inhibition of HDAC1 gene in BMSCs is able to increase the differentiation of BMSCs into cardiac cell phenotype.

Directed differentiation of stem cells is a complex process in which multiple genes and signalling pathways are involved [9].

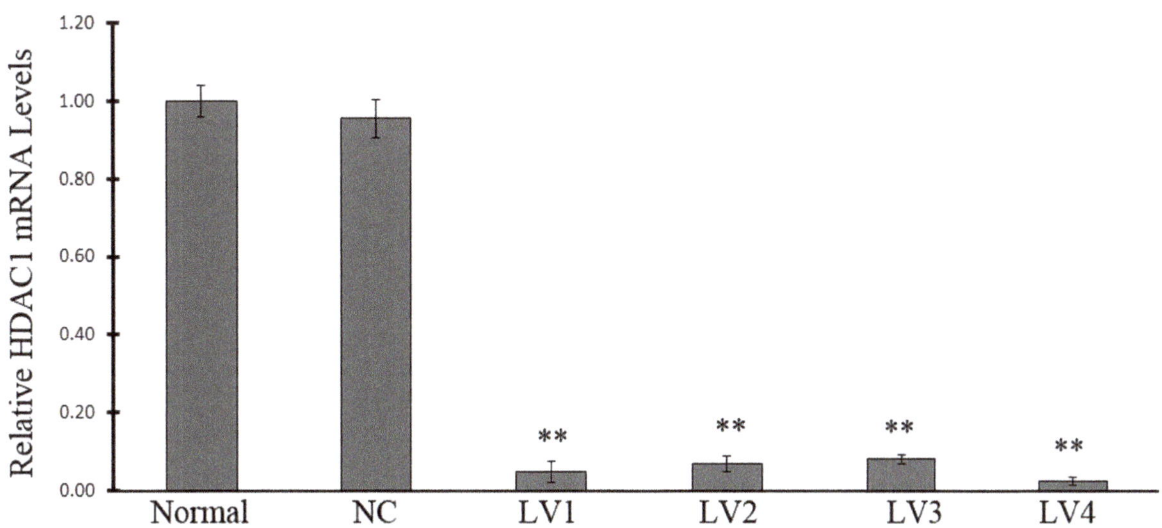

Figure 6. The relative HDAC1 mRNA levels. The cells were infected with vehicle (normal control), NC vector (NC, pGCSIL-NC-shRNA), LV-HDAC1-shRNA1 (LV1), LV-HDAC1-shRNA2 (LV2), LV-HDAC1-shRNA3 (LV3), or LV-HDAC1-shRNA4 (LV4). Note: n = 6; ** $P<0.01$, compared with that in NC group.

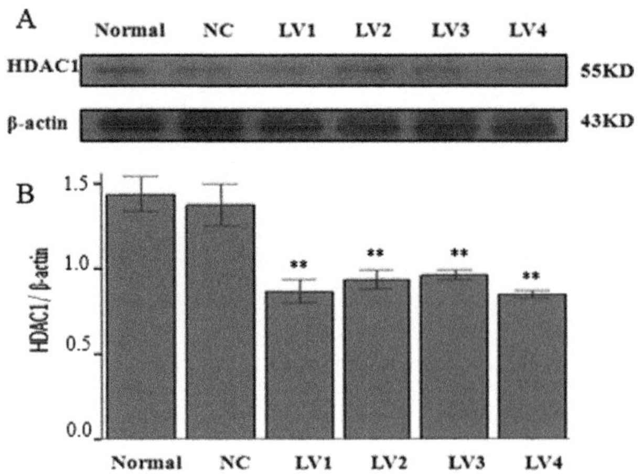

Figure 7. The HDAC1 protein levels. The cells were infected with vehicle (normal control), NC vector (NC, pGCSIL-NC-shRNA), LV-HDAC1-shRNA1 (LV1), LV-HDAC1-shRNA2 (LV2), LV-HDAC1-shRNA3 (LV3), or LV-HDAC1-shRNA4 (LV4). Note: n = 6; ** $P<0.01$, compared with that in NC group.

Recently, epigenetic modification of histone acetylation has been identified to play impartment roles in this stem cell process [4], [5], [6], [7]. Indeed, rat BMSCs treated with the HDAC inhibitors such as suberoylanilide hydroxamic acid (SAHA) and trichostatin A (TSA) exhibit a significant increase in the expression of cardiomyocyte-specific genes, including GATA4, Nkx2.5, and Mef2c [5], [10]. However, both SAHA and TSA are non-specific inhibitors of HDACs that could suppress multiple HDAC subtypes in addition to HDAC1. In addition, SAHA and TSA treatments promote not only the differentiation of BMSCs into cardiomyocyte-like cells but also the differentiation of BMSCs into endothelial cells [11], hepatocytes [12], chondrocytes [13], and adipocytes [14], among other cell types. However, the detailed information regarding the roles of HDAC subtypes in the differentiation of BMSCs into myocardial phenotypes has not yet been reported. In this study, we identified, via the shRNA approach, that HDAC1 is a critical regulator for the directed differentiation of BMSCs into myocardial phenotypes. The results of our study are consistent with the studies in other stem cell lines. For example, Liu et al. [15] and Dovey et al. [16] have reported that the in vitro induction of specific deficiencies in HDAC1

Table 1. Myocardial markers expression levels in experimental groups ($\overline{x}\pm$sd).

Markers	Normal	LV-NC-shRNA	LV-HDAC4-shRNA4
Nkx2.5	1.000±0.023	0.9107±0.094	4.898±1.754*
GATA-4	1.000±0.026	0.9913±0.039	76.455±26.624*
MHC	1.000±0.021	0.900±0.059	52.101±9.366*
Connexin-43	1.000±0.025	1.102±0.032	37.131±3.589*
CTnT	1.000±0.016	1.076±0.076	41.183±13.193*

Note: n = 6; * $P<0.05$, compared with that in LV-NC-shRNA group.

expression in P19CL6 cells and embryonic stem cells could promote the differentiation of stem cells into myocardial phenotypes.

The molecular mechanisms involved in HDAC1-mediated effect on the directed differentiation of BMSCs into myocardial phenotypes are currently unclear. In embryonic stem cells, Hoxha et al. [17] reported that HDAC1 regulates the phenotypic differentiation of embryonic stem cells into cardiomyocytes by controlling the expression of sex-determining region Y-box 17 (SOX17) and bone morphogenetic protein 2 (BMP2). However, because of ethical concerns and other issues, embryonic stem cells and induced pluripotent stem cells might not be used for clinical stem cell transplantation in the near future. The elucidation of the downstream pathways involved in the HDAC-induced the phenotypic differentiation of BMSCs into myocardial phenotypes should be identified in future study. Moreover, the roles of other HDAC subtypes in cardiac cell differentiation from BMSCs merit additional investigation in the future.

In summary, we have identified that the specific inhibition of HDAC1 gene expression in BMSCs by RNAi causes an increase in the mRNA expression levels of genes related to myocardial development and structure. HDAC1 may play an important role in the directed differentiation of BMSCs into cardiomyocytes.

Author Contributions

Conceived and designed the experiments: DFL CZ. Performed the experiments: YW ZZS ZHZ XWX ZYH. Analyzed the data: YW ZZS ZHZ XWX ZYH. Contributed reagents/materials/analysis tools: XWX ZYH. Wrote the paper: DFL CZ.

References

1. Choi SH, Jung SY, Kwon SM, Baek SH (2013) Perspectives on stem cell therapy for cardiac regeneration. Advances and challenges. Circ J 76:1307–1312.
2. Bianco P, Robey PG, Simmons PJ (2008) Mesenchymal stem cells:revisiting history,concepts,and assays. J Cell Stem Cell 2:313–319.
3. Zhang Y, Liang X, Lian Q, Tse HF (2013) Perspective and challenges of mesenchymal stem cells for cardiovascular regeneration. Expert Rev Cardiovasc Ther 11:505–517.
4. Muscari C, Bonafè F, Carboni M, Govoni M, Stanic I, et al. (2008) Difluoromethylornithine Stimulates Early Cardiac Commitment of Mesenchymal Stem Cells in a Modelof Mixed Culture With Cardiomyocytes. J Cell Biochem 103:1046–1052.
5. Feng C, Zhu J, Zhao L, Lu T, Zhang W, et al. (2009) Suberoylanilide hydroxamic acid promotes cardiomyocyte differentiation of rat mesenchymal stem cells. Exp Cell Res 315:3044–3051.
6. Li L, Zhu J, Tian J, Liu X, Feng C (2010) A role for Gcn5 in cardiomyocyte differentiation of rat mesenchymal stem cells. Mol Cell Biochem 345:309–316.
7. Collas P (2009) Epigenetic states in stem cells. Biochimica et Biophysica Acta 1790: 900–905.
8. Huang XP, Sun Z, Miyagi Y, McDonald Kinkaid H, Zhang L, et al. (2010) Differentiation of allogeneic mesenchymal stem cells induces immunogenicity and limits their long-term benefits for myocardial repair. Circulation 122:2419–2429.
9. Burridge PW, Zambidis ET (2013) Highly efficient directed differentiation of human induced pluripotent stem cells into cardiomyocytes. Methods Mol Biol 997:149–161.
10. Yang G, Tian J, Feng C, Zhao LL, Liu Z, et al. (2012) Trichostatin a promotes cardiomyocyte differentiation of rat mesenchymal stem cells after 5-azacytidine induction or during coculture with neonatal cardiomyocytes via a mechanism independent of histone deacetylase inhibition. Cell Transplant 21:985–996
11. Mahpatra S, Firpo MT, Bacanamwo M (2010) Inhibition of DNA methyltransferases and histone deacetylases induces bone marrow-derived multipotent adult progenitor cells to differentiate into endothelial cells. Ethn Dis 20:S1–S60.
12. Snykers S, Vanhaecke T, De Becker A, Papeleu P, Vinken M, et al. (2007) Chromatin remodeling agent trichostatin A: a key-factor in the hepatic differentiation of human mesenchymal stem cells derived of adult bone marrow. BMC Dev Biol 7:24–24.
13. Wang JP, Wen MH, Chen YT, Lee HH, Chiang ER, at al. (2011) Trichostatin A inhibits TGF-β1 induced in vitro chondrogenesis of hMSCs through Sp1 suppression. Differentiation 81:119–126.
14. Haberland M, Carrer M, Mokalled MH, Montgomery RL, Olson EN (2010) Redundant control of adipogenesis by histone deacetylases 1 and 2. J Biol Chem 285:14663–14670.

15. Liu Z, Li T, Liu Y, Jia Z, Li Y, et al. (2009) WNT signaling promotes Nkx2.5 expression and early cardiomyogenesis via downregulation of Hdac1. Biochim Biophys Acta 1793:300–311.
16. Dovey OM, Foster CT, Cowley SM (2010) Histone deacetylase 1 (HDAC1), but not HDAC2, controls embryonic stem cell differentiation. Proc Natl Acad Sci U S A 107:8242–8247.
17. Hoxha E, Lambers E, Wasserstrom JA, Mackie A, Ramirez V, et al. (2012) Elucidation of a novel pathway through which HDAC1 controls cardiomyocyte differentiation through expression of SOX-17 and BMP2. PLoS One 7:e45046.

Ectopic Pregnancy-Derived Human Trophoblastic Stem Cells Regenerate Dopaminergic Nigrostriatal Pathway to Treat Parkinsonian Rats

Tony Tung-Yin Lee[1]⁹, Cheng-Fang Tsai[2]⁹, Tsung-Hsun Hsieh[3], Jia-Jin Jason Chen[3], Yu-Chih Wang[1], Mi-Chun Kao[1], Ruey-Meei Wu[4], Sher Singh[5], Eing-Mei Tsai[1,2]*, Jau-Nan Lee[1]*

1 Department of Obstetrics and Gynecology and Center of Excellence for Environmental Medicine, Kaohsiung Medical University Hospital, Kaohsiung, Taiwan, **2** Graduate Institute of Medicine, Kaohsiung Medical University College of Medicine, Kaohsiung, Taiwan, **3** Institute of Biomedical Engineering, National Cheng Kung University, Tainan, Taiwan, **4** Department of Neurology, National Taiwan University Hospital, Taipei, Taiwan, **5** Department of Life Science, National Taiwan Normal University, Taipei, Taiwan

Abstract

Background: Stem cell therapy is a potential strategy to treat patients with Parkinson's disease (PD); however, several practical limitations remain. As such, finding the appropriate stem cell remains the primary issue in regenerative medicine today. We isolated a pre-placental pluripotent stem cell from the chorionic villi of women with early tubal ectopic pregnancies. Our objectives in this study were (i) to identify the characteristics of hTS cells as a potential cell source for therapy; and (ii) to test if hTS cells can be used as a potential therapeutic strategy for PD.

Methods and Findings: hTS cells expressed gene markers of both the trophectoderm (TE) and the inner cell mass (ICM). hTS cells exhibited genetic and biological characteristics similar to that of hES cells, yet genetically distinct from placenta-derived mesenchymal stem cells. *All-trans* retinoic acid (RA) efficiently induced hTS cells into trophoblast neural stem cells (tNSCs) in 1-day. Overexpression of transcription factor Nanog was possibly achieved through a RA-induced non-genomic c-Src/Stat3/Nanog signaling pathway mediated by the subcellular c-Src mRNA localization for the maintenance of pluripotency in tNSCs. tNSC transplantation into the lesioned striatum of acute and chronic PD rats not only improved behavioral deficits but also regenerated dopaminergic neurons in the nigrostriatal pathway, evidenced by immunofluorescent and immunohistological analyses at 18-weeks. Furthermore, tNSCs showed immunological advantages for the application in regenerative medicine.

Conclusions: We successfully isolated and characterized the unique ectopic pregnancy-derived hTS cells. hTS cells are pluripotent stem cells that can be efficiently induced to tNSCs with positive results in PD rat models. Our data suggest that the hTS cell is a dynamic stem cell platform that is potentially suitable for use in disease models, drug discovery, and cell therapy such as PD.

Editor: Patrick Ching-Ho Hsieh, Institute of Clinical Medicine, National Cheng Kung University, Taiwan

Funding: This work was partly supported by grants of KMUH99-9I04 and NSC 99-2628-B-037-009-MY3. The funders had no role in study design, data collection and analysis, decision to publish, or preparation of the manuscript.

Competing Interests: The authors have declared that no competing interests exist.

* E-mail: jaunanlee@hotmail.com (JNL); tsaieing@yahoo.com (EMT)

⁹ These authors contributed equally to this work.

Introduction

Parkinson's disease (PD) is caused by the dysfunction or the loss of dopaminergic neurons in the nigrostriatal pathway of the midbrain, making it the second most common neurodegenerative disorder in humans. Current pharmacological drugs only provide symptomatic relief but do not retard the disease progression. Consequently, cell replacement by using human fetal mesencephalic tissue [1], [2] or embryonic stem (ES) cell-derived dopaminergic neurons [3], [4], [5] remains an important therapeutic strategy; however, several practical limitations exist, such as shortage of cell sources, variations in outcomes, adverse effects, and socio-ethical issues [6]. Placenta-derived mesenchymal stem (PDMS) cells have also shown promise; but outcomes remain uncertain [7]. In the search for a suitable cell source, we isolated and identified the ectopic pregnancy-derived human trophoblast stem (hTS) cells from the early chorionic villi of a tubal ectopic pregnancy. The characteristics of hTS cells suggest it as a potential alternative source of pluripotent stem cells for the treatment of PD and of other neurodegenerative diseases.

We investigated the microenvironmental factors that affect pluripotency and proliferation of hTS cells. In women, fertilization occurs in the fallopian tubes, where the distinction between the inner cell mass (ICM) and the trophectoderm (TE) [8] and the switch from totipotency to pluripotency takes place during embryogenesis [9]. When an ectopic pregnancy occurs in the fallopian tube, the cellular processes may continue until clinical intervention. Before intervention, however, the microenvironmental factors might affect cell differentiation. For example, the pleiotropic cytokine LIF expresses significantly higher levels in the fallopian tubes than that in the endometrium [10], descending from the ampulla to the isthmus [11]. LIF levels may elevate up to 2- to 4-fold in ectopic pregnancies [11]. Functionally, LIF activates transcription factors, Oct4 and Nanog, for the maintenance of pluripotency in ES cells [12], [13]. On withdrawal of LIF, cell proliferation continues but the caudal-related homeobox transcription factor Cdx2 is activated, driving ES cells differentiation into trophectoderm fate [14]. Nevertheless, the significance of LIF on trophoblastic development is largely unknown. In this study, we clarified the inter-relationships of LIF, Oct4, Nanog and Cdx2 on pluripotency and proliferation of hTS cells.

Next, we investigated if retinoic acid (RA) can effectively induce hTS cell differentiation to a neural cell. RA is a well-recognized signaling molecule, involved in the development, regeneration, and maintenance of the nervous system [15]. RA promotes the generation of DA neurons and the enhancement of axon outgrowths of neurites in hES cells [16]. Conventionally, RA may function as a paracrine signal or an autocrine signal, entering the nucleus via retinaldehyde dehydrogenases (RALDHs) to bind RA receptors (RARs) or retinoid X receptors (RXRs) to activate the retinoic acid-response element (RARE) for gene transcription. Notably, the striatum expresses the highest endogenous levels of RA in the brain [17]. In our study, we found that RA efficiently induced differentiation of hTS cells into dopaminergic trophoblastic neural stem cells (tNSCs).

We then investigated whether tNSCs were effective in a PD rat model. We found that intracranial transplantation of tNSCs substantially regenerated the dopaminergic nigrostriatal pathway and functionally alleviated parkinsonian motor deficits in both acute and chronic PD rats. Together, these results suggest that hTS cells are a viable pluripotent cell source that circumvents socio-ethical concerns for the treatment of PD.

Materials and Methods

hTS Cell Culture and Differentiation

Tiny villous tissues (Fig. S1A) were well-minced in serum-free α-MEM (Sigma-Aldrich, St. Louis, MO) followed by trypsinization with 0.025% trypsin/EDTA (Sigma-Aldrich) (15 min) that was halted by adding α-MEM containing 10% FBS. Adherent cells were cultured in α-MEM, 20% FBS, and 1% penicillin-streptomycin at 37°C in 5% CO_2. Particularly, the doubling time of cell growth was 7.67 hr. Cell antibodies, primers, and differentiations used in this study are shown in Supporting Information (Table S1, Table S2).

Plasmids

Plasmid construct (F1B-GFP) used in this study is described in the Supporting Information (Text S1). hTS cells were co-transfected with a DNA mixture of F1B-GFP to yielded over 95% of transfection rate.

RT-PCR, Western Blots, Southern Blots, Immunoprecipitation, IP Assay, and Flow Cytometry

Methods performed in this work are described previously [18] and details are available in the Supporting Information (Text S1).

Chromatin Immunoprecipitation (ChIP)

Cells were serum-deprived for overnight and treated with RA (10 μM) for 30 min. The mixtures of protein G beads (Roche) and lysate were incubated with rocking at 4°C for 2 hr. After removing the beads by centrifugation, the lysate was added with IgG (Santa Cruz) and primary anti-Stat3 antibody (1:1,000, Santa Cruz) in accompany with new protein G beads. The mixtures were incubated with rocking overnight at 4°C. The immunoprecipitation complex was washed by RIPA lysis buffer and subjected for polymerase chain reaction (PCR) using Nanog promoter primer: forward, GACAGCCCCCACTTAACAAA and reverse, GCTTTTTCCCTCTGGCTCTT).

Microarray Data Analysis

Cells, with or without treatment of RA (10 μM), were cultured overnight and total RNAs were extracted using TRIzol reagent (Invitrogen) and subjected for Affymetrix microarray by using Affymetrix Human Genome U133 plus 2.0 GeneChip according to the manufacturer's guidance (Santa Clara, CA, http://www.affymetrix.com) performed at Genomic Medicine Center of National Taiwan University College of Medicine, Taipei, Taiwan). Data analysis was performed by using softwares of MetaCore, GeneSpring GX (version 7.3), and IPA Ingenuity System (version 8.8).

Immunofluorescence

For immunocytochemistry, after fixation with 4% paraformaldehyde in PBS (room temperature; 5 min) and washes, cells were permeabilized with 2% FBS/0.4% Triton X-100 in PBS (15 min), followed by 5% FBS blocking solution (2 hr) and rinsed three times. After incubation with specific primary antibody in PBS at 4°C overnight, appropriate FITC or PE or Texas Red or DyLight 488 or DyLight 594 conjugated secondary antibody was added for 1 hr at room temperature. By DAPI staining, nucleus (5 min) cells were subjected for microscopy.

For immunohistochemistry, brain sections were fixed in 4% paraformaldehyde (2 hr) and dehydrated in 30% sucrose in 0.02 M PBS (2 days) followed by processes in a freezing microtome (Leica). Coronal sections (30 μm) were treated with 0.3% H_2O_2 and Ready-To-Use blocking solution (IHC-101b, Bethyl Laboratories, Montgomery, TX) for 15 min followed by incubation with primary anti-TH antibody (1:1,000, Temecula, CA) at 4°C. Anti-rabbit second antibody (IHC-101d, Bethyl Laboratories) was used with DAB kit (Bethyl) for microscopic analysis.

Immunofluorescence Tissue Analysis

Coronal brain sections were immunostained with primary antibodies: TH (1:250; H-196, sc-14007; Santa Cruz), NeuN (1:100; MAB377; Millipore, Temecula, CA), and GFAP (1:200; BSB 5566, Declere; Bio SB, Santa Barbara, CA) in accompany with FITC-anti-TH, Cy-3-anti-NeuN, or Cy-3-anti-GFAP. Data were analyzed by TissueGnostics (TissueGnostics GmbH, Vienna, Austria). Evaluation was made by comparison between the lesioned (or the cell-implanted) side and the normal side. Cells with bizarre size or intensity of GFAP outside the normal (e.g., artifact or unusually heavy stain area or intensity in cell overlapping with Wilson's pencils) were excluded.

Animal Studies and Behavioral Assessments

Animal preparations were licensed and well-described previously [19], (Text S1). Before experiment, the apomorphine-induced rotation test after 6-OHDA injection was pre-evaluated weekly in PD rats to achieve a stable hemiparkinsonian status to avoid bias. In acute PD rats (body weight, 225–250 g), cells (3×10^6 cells/5 μl/5 min) were transplanted into the lesioned striatum at site (AP +1.0 mm, Lat +2.7 mm, Dep 6.4 mm) under anesthesia. The cell viability remained stable between 96% and 98% during the implantation procedures. In chronic PD rats, the body weight of rats was controlled between 560±65 g at pre-test and 548±46 g at post-test. Cells (1.5×10^6) were grafted at the same site. To obtain the brain sections, rats were anesthetized by sodium pentobarbitone (60 mg/kg i.p., Apoteksbolaget, Sweden) and trans-cardially perfused with saline (50 ml, 0.9% w/v) followed by ice-cold paraformaldehyde (200 ml, 10% w/v in 0.02 M PBS) were performed at 18- and 12-weeks in the acute and chronic PD rats, respectively. Brain sections were subjected for immunocytochemistry, and immunofluorescence tissue analysis as indicated. All behavioral assessments were performed as described previously [19], (Text S1).

Ethics Statement

Patient consent and sourcing of human tissue were approved by the Institutional Review Board on Human Subjects Research and Ethics Committees of Kaohsiung Medical University Hospital (KMUH-IRB-950140). The animal experiments were conducted according to the guideline and Approval of Animal Use Protocol of National Cheng Kung University (IACUC Approval No. 98010). The NCKU IACUC guidelines aim to reduce animal suffering by use of anesthesia, use of analgesics, and provide nutritional and fluid support.

Statistical Analysis

Data obtained from RT-PCR and flow cytometry were calculated by Student's t-test. For behavioral assessments, data were expressed as mean ± SEM. In acute PD rats, data of apomorphine-induced rotation tests were analyzed by using repeated measure analysis of variance (ANOVA) tests (SPSS Release 12.0 software) and applied least significant difference test (LSD) *post hoc* comparisons after repeated measure ANOVA tests between two groups. In chronic PD rats, paired Student's t-test was used to compare the two groups. p-value <0.05 was considered statistically significant.

Results

Ectopic Pregnancy Gives Rise to hTS Cells

A unique population of adherent cells was isolated from the early chorionic villi of embryos (gestational age: 5–7 weeks) in tubal ectopic pregnancies (Text S1). Neither feeder cells nor multiple inducers were used in the cell culture and the differentiation of hTS cells, as distinct from hES cells [3], [4], [5], circumventing potential side-effects and contaminations. After growth factor depletion, the main differentiated lineage of trophoblasts should be mature syncytial multinuclear trophoblasts. However, we only isolated mononuclear cytotrophoblast cells that adhered to the dish, as mature cells were removed in the isolation process. The population doubling time of hTS cells was 8 hr, which is distinct from the 48 hr population doubling time of the first trimester placenta-derived cytotrophoblast [20]. These cells expressed markers for hES cell-related specific stage embryonic antigen (SSEA)-1, −3 and −4 with immunolocalization identical to cytotrophoblasts in the villus (Fig. 1A). By immunocytochemistry, the cells expressed pluripotency transcription factors Oct4, Sox2, Nanog and Cdx2 in the nuclei at passage 9 and 17 (Fig. 1B, Fig. S1B).

The hTS cells expressed specific genes of the TE (i.e., Cdx2, Fgfr-2, Eomes, and BMP4), the ICM (i.e., Oct4, Nanog, FGF4, and Sox2) (Fig. 1C), as well as the three primary germ layers (Fig. 1D). We then evaluated for cellular homogeneity. With TissueGnostics analysis, we found that these cells expressed 100% and 97.3% Oct4-positive cells at passages 9 and 17, respectively (Fig. 1E). Flow cytometry with Oct4, Sox2, Nanog and Cdx2 at passage 15 demonstrated 97.9%, 95.0%, 98.7%, and 94.0% positive cells, respectively (Fig. 1F). Flow cytometry with surface markers CD44, CD73, CD105 and HLA-ABC at passages 3 and 9 also demonstrated over 97% positive (Fig. S1C). These results show that hTS cells are a highly homogenous population.

Flow cytometric analysis demonstrated that hTS cells expressed mesenchymal, but not hematopoietic, stem cell characteristics (Fig. S1D). Upon appropriate induction (Table S2), they gave rise to a variety of specific cell phenotypes such as osteoblasts, chondrocytes, myocytes, and adipocytes (Fig. S1E). Neither change in karyotype (46, XY) by chromosome analyses (Fig. S1F) nor in telomere length by Southern blots (Fig. 1G) was observed in a series of cell cultures. We defined these isolated cells as hTS cells.

An important question is whether hTS cells are distinct from PDMS cells. To investigate, we utilized the Affymetrix platform to compare global gene expressions between them. The results showed that the gene distribution differed not only with their histogram patterns (Fig. 1H), but also with their intensity values (Fig. 1I). Of the total 54,675 genes, 4,864 genes (8.9%) were distinct between them, including molecular and cellular functions and physiological system development (Table S3). Furthermore, both exhibited unique genes with respect to immune and mesenchymal characteristics (Fig. 1J and Table S4). Therefore, we concluded that hTS cells represented a unique TE-derived stem cell population distinct from PDMS cells obtained from the uterus.

LIF Maintains the Proliferative Capability of hTS Cells

Previous studies show that gradient levels of LIF exist in the fallopian tube [11] and that in the absence of LIF, ES cells differentiate into the TE fate [14]. To mimic the environment, we investigated the effect of LIF on hTS cells. We treated hTS cells with different concentrations of LIF (i.e., 500, 250, and 125 IU/ml) for 3-days each. Of the pluripotency-associated transcription factors, LIF (500 IU/ml) induced overexpressions of Oct4 and Sox2, but not Nanog and Cdx2 by RT-PCR (Fig. 2A). By flow cytometry, withdrawal of LIF (500 IU/ml) overexpressed Nanog and Cdx2; however, it suppressed Oct4 and Sox2 in a dose-dependent manner (Fig. 2B, Fig. 2C, and Fig. S2A). As the hTS cells travel from a high LIF environment to a low LIF environment along the fallopian tube, the relative Nanog/Cdx2 ratio increases (>2-fold), but the Oct4/Cdx2 ratio decreases (Fig. 2D). By using siRNAs and shRNAs, we also show that knockout of Nanog promoted expressions of Cdx2, and knockout of Cdx2 promoted Nanog (Fig. 2E, Fig. 2F, and Fig. S2B). Similar to the reciprocal relationship between Oct4 and Cdx2 in ES cells [21], we showed the existence of a reciprocal relationship between Nanog and Cdx2 in hTS cells. Expression of Cdx2 is required for maintaining the trophoblastic phenotype. These results suggested that Oct4 is involved in the maintenance of pluripotency at the early TE cells and Nanog enhanced the pluripotent capability in hTS cells at a later stage during transportation in the tube.

RA Generates Trophoblast Neural Stem Cells *in vitro*

We then attempted to generate neural cells by treating hTS cells with RA over time (i.e., 1-, 3-, 5- and 7-days). A population of neural stem cells (NSCs) was generated with phenotypes similar to those neural restricted precursor subtypes described in previous studies [22], including glial restricted precursors (GRP), neuronal restricted precursors (NRP), multipotent neural stem cells (MNS),

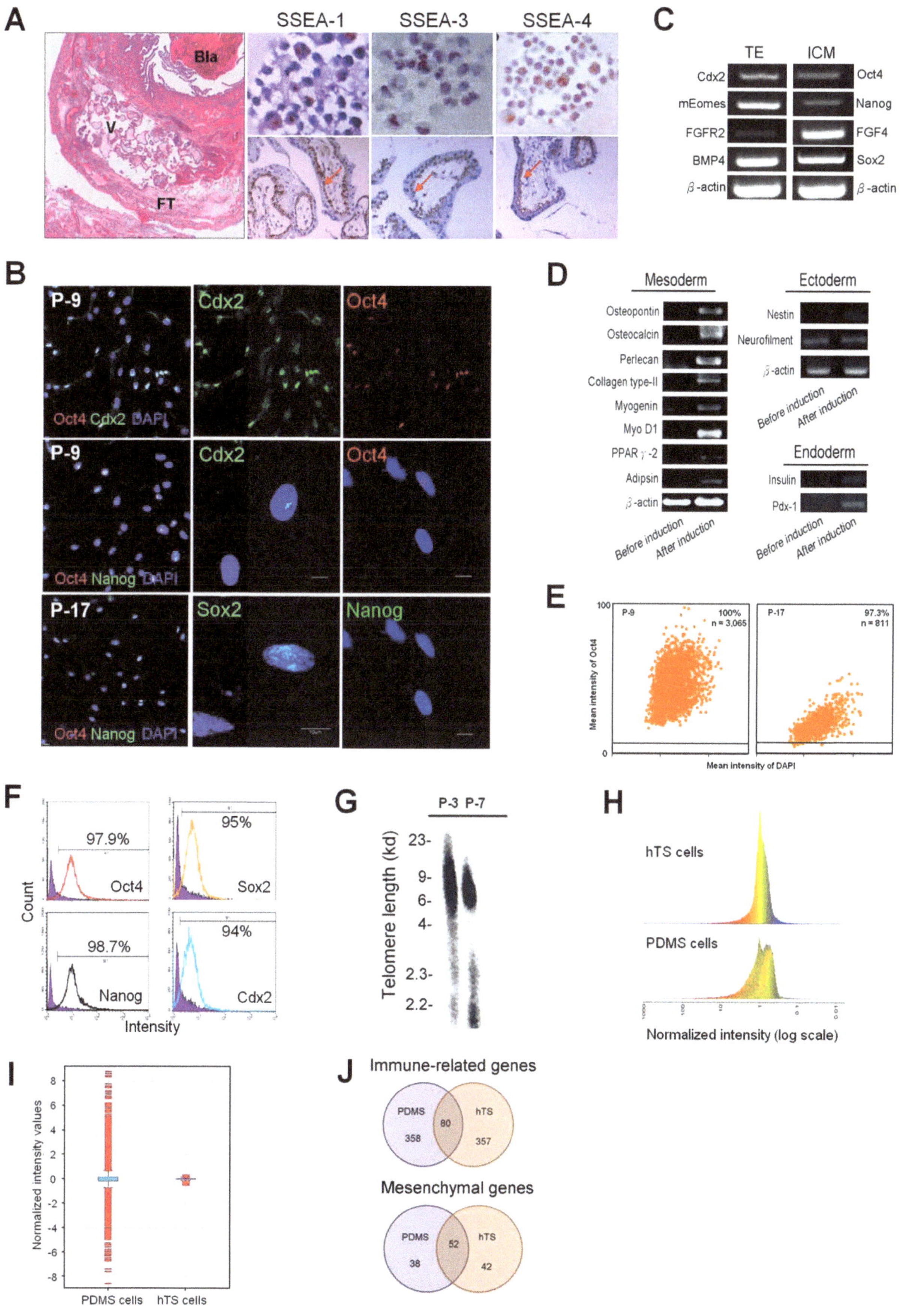
A
SSEA-1
SSEA-3
SSEA-4
Bla
V
FT
C
TE
ICM
Cdx2
Oct4
mEomes
Nanog
FGFR2
FGF4
BMP4
Sox2
β-actin
B
P-9
Cdx2
Oct4
Oct4 Cdx2 DAPI
Oct4 Nanog DAPI
P-17
Sox2
Nanog
D
Mesoderm
Ectoderm
Osteopontin
Osteocalcin
Perlecan
Collagen type-II
Myogenin
Myo D1
PPAR γ-2
Adipsin
Nestin
Neurofilment
Endoderm
Insulin
Pdx-1
Before induction
After induction
E
P-9
100%
n = 3,065
P-17
97.3%
n = 811
Mean intensity of Oct4
Mean intensity of DAPI
F
97.9%
Oct4
95%
Sox2
98.7%
Nanog
94%
Cdx2
Count
Intensity
G
P-3 P-7
Telomere length (kd)
23-
9-
6-
4-
2.3-
2.2-
H
hTS cells
PDMS cells
Normalized intensity (log scale)
I
Normalized intensity values
PDMS cells
hTS cells
J
Immune-related genes
PDMS
hTS
358
80
357
Mesenchymal genes
38
52
42

Figure 1. Characteristics of hTS Cells. (**A**) Histology of an ectopic pregnancy at fallopian tube by HE staining (left large panel). Bla: blastocyst, V: chorionic villi, FT: fallopian tube. Immunocytochemistry of SSEAs (red, arrow) showed SSEA-1 in the cytoplasm (left upper); SSEA-3 in the nucleus (middle upper); and SSEA-4 in both the cytoplasm and the membrane (right upper), corresponding to cytotrophoblasts in the ectopic chorionic villus (lower panel). (**B**) Immunofluorescent Oct4 and Cdx2 at passage 9 (P-9), Oct4 and Nanog expressed in the nuclei at passage 9 (P-9) and 17 (P-17) (left column). Scale bar: 20 μm. Expression of Cdx2, Oct4, Sox2, and Nanog in the nuclei in the amplified cells (middle and right columns). Scale bar: 10 μm. (**C**) Expression of specific genes of both trophectoderm (TE) and ICM by RT-PCR. (**D**) hTS cells expressed genes of three germ layers before induction (left column) and after appropriate induction (right column). (**E**) For cellular homogeneity, Oct4-positive cells in P-9 and P-17 with 100% and 97.3%, respectively, by TissueGnostics analysis. n = total cell number counted. (**F**) Flow cytometric analysis at passage 15, cells expressed positive Oct4, Nanog, Sox2, and Cdx2 were 97.9%, 98.7%, 95%, and 94%, respectively. (**G**) Southern blots showed no obvious change in telomere length at passages 3 (P-3) and 7 (P-7). (**H**) Distinction between hTS cells and PDMS cells in global gene expressions (total 54,675 genes; $p<0.05$ and fold change >2) with a homogenous histogram. (**I**) Different relative intensity values of genes (4,864 genes) between PDMS and hTS cells. Blue box indicating 50% of total genes and error bars indicating 25% and 75%. (**J**) Venn diagram illustrated the numbers of immue-related genes (upper panel) and mesenchymal genes (lower panel) in PDMS and hTS cells by microarray analysis. Overlap: common genes; Bilateral regions: unique genes.

astrocytes (AST), and undefined trophoblast giant cells (TGC) (Fig. 3A, left panel). Notably, the composition of these phenotypes remained relatively stable within 5 days of RA treatment. However, at 7 days, most of the NSCs became undefined TGC (Fig. 3A, right panel), compatible with previous reports [23]. The precursor cells expressed neurofilament protein, nestin, and astrocyte-specific glial fibrillary acidic protein (GFAP) measured at day-3 and day-5 after treatment (Fig. 3C). Immunocytochemically, we found that they expressed NSC-specific markers, for example: GFAP and fibroblast growth factor receptor-1 (FGFR1) for GRP; βIII-tubulin and microtubule-associated protein 2 (MAP2) for NRP; and GFAP and vimentin for MNS (Fig. 3C). We found consistency in cellular characteristics of the RA-induced NSCs, hereby defined as trophoblast neural stem cells (tNSCs) in the following studies.

In tNSCs, RA induction promoted expressions of NSC-related markers (e.g., nestin, neurofilament, Ngn3, MAP-2, NeuroD, CD133, and Oct4), of RA receptors (e.g., RARβ, RARγ, RXRα, and RXRβ), and most importantly, of RA-synthesizing enzymes RALDH-2 and -3 (Fig. 3D). The expression of RALDH-2 and -3 in hTS cells and tNSCs was unique because hES cells or even the retinol-treated ES cells do not express RALDH-2 [24]. By flow cytometry and immunocytochemistry, results showed that RA induced significant overexpression of Nanog and Oct4 but less so Cdx2 in hTS cells (Fig. 3E and Fig. 3F). To this end, we showed that RA efficiently induced hTS cells into tNSCs, maintaining a steady state within 5 days.

Withdrawal of LIF Promotes Nanog Expression for Pluripotency of tNSCs

Given that LIF can interplay with RA on neural differentiation of ES cells [25], we examined the effect of LIF on the RA-induced Nanog expression in tNSCs. To mimic the microenvironment of the fallopian tube, we first incubated hTS cells with LIF in variable concentrations (i.e., 500, 250, 125 IU/ml). On the 3rd day, we added RA (10 uM) for 1-day before analysis. Flow cytometry showed that a higher level of LIF significantly repressed the RA-induced Nanog (Fig. 3F and Fig. S3). These results suggested that as hTS cells move towards the isthmus, the better the RA's ability to maintain cellular pluripotency by Nanog expression.

RA Induces Nanog through Non-genomic Signaling Pathway

How does RA induce Nanog expression? We explored the molecular mechanisms of RA-induced Nanog in hTS cells. Previous studies report that eukaryotic Initiation factor 4B (eIF4B) is involved in the initiation phase of eukaryotic translation [26] through binding to internal ribosome entry site (IRES) [27]. c-Src mRNA has also been reported to contain an IRES element [28]. Activation of eIF4B might also be associated with c-Src mRNA activity for translation initiation [26], [29]. In the developing striatum and hippocampus, an increased Src kinase activity coincides not only with the peak period of neuronal differentiation and growth [30], but also the characteristics of self-renewal [31].

We found that RA promoted production of eIF4B in between 1–4 hr of incubation but faded away at 24 hr (Fig. 4A). This action was inhibited with eIF4B siRNA (Fig. 4B). We found that RA induced a rapid and transient expression of c-Src mRNA, peaking at 15 min (Fig. 4C), followed by production of c-Src protein at 1 hr (Fig. 4D). Notably, RA failed to induce c-Src mRNA elevation by qPCR at 1-day treatment (Fig. 4E).

Subsequently, we found that active c-Src bound directly to signal transducer and activator of transcription 3 (Stat3) (Fig. 4F) by phosphorylation at site Tyr705 to produce protein, which was inhibited by c-Src siRNA (Fig. 4G). This action was also inhibited by a selective c-Src inhibitor PP-1 analog (Fig. 4H). We identified a direct action of Stat3 on the Nanog gene promoter by chromatin immunoprecipitation (ChIP) assay (Fig. 4I). Nanog was produced in 4 hr, which was inhibited by PP1 analog (Fig. 4H) and by Stat3 siRNA (Fig. 4J).

For hTS cells, we propose the possibility of an RA-activated non-genomic eIF4B/c-Src/Stat3/Nanog signaling pathway through subcellular c-Src mRNA localization (Fig. 4K). The mechanism of RA-induced activation of eIF4B and the direct relationship between eIF4B and c-Src will require further study. How the RA-activated Nanog maintains hTS cell pluripotency will also require further study.

Cell Therapy Regenerates Nigrostriatal Pathway and Improves Behavioral Deficits in Acute and Chronic PD Rats

To determine the efficacy of cell therapy and the cell fate potential of tNSCs in animal studies, we grafted the cells into the brain of PD rats. We produced 6-hydroxydopamine (6-OHDA)-caused PD rats (Text S1). In acute PD rats, we performed the apomorphine-induced rotation test every 3-weeks up to 12-weeks post-cell therapy. Then we examined GFP-tagged hTS cells for cell fate at 18-weeks post-cell therapy. In chronic PD rats, we bred PD rats for over one year (average 12.3 months) in order to mimic the pathologically progressive nature of PD. After cell therapy, behavioral tests and validation of regeneration of dopaminergic nigrostriatal pathway were performed.

In Acute PD Rats

We transfected hTS cells with F1B(-540)-GFP plasmid construct (kindly provided by Dr. I. M. Chiu) to yield a success rate of over 95%. For cell therapy, acute PD rats were divided into three groups: group (a) received the GFP-tagged 1-day RA-induced tNSCs (n = 4); group (b) received GFP-tagged 5-day RA-induced tNSCs (n = 4); and group (c) received PBS solution as control

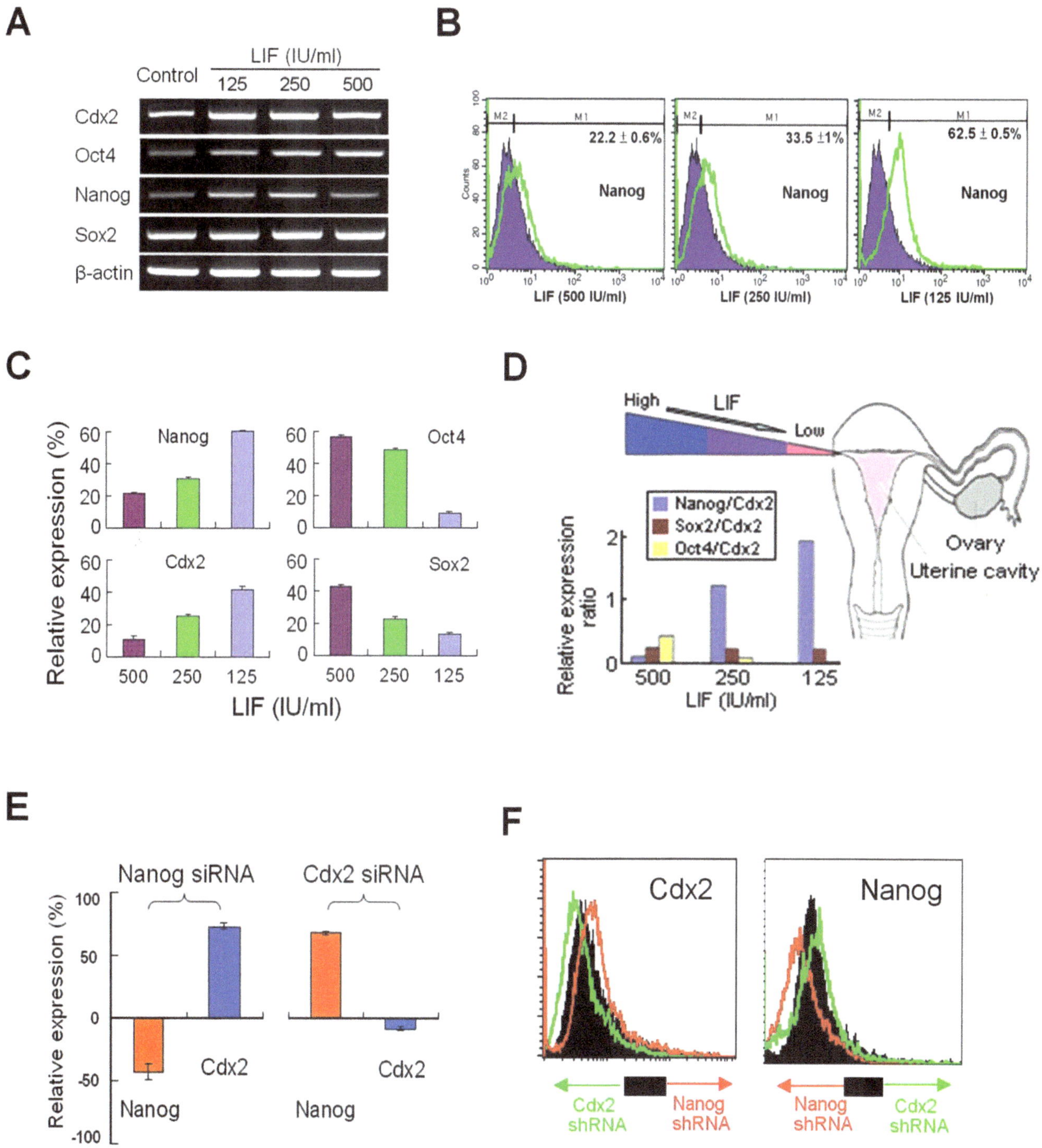

Figure 2. Regulation of Pluripotency Transcription Factors by LIF. (**A**) Expressions of Nanog, Cdx2, Sox2 and Oct4 mRNAs after treating different concentrations of LIF (**B**) By flow cytomery, withdrawal of LIF overexpressed Nanog and (**C**) enhanced expression of Nanog and Cdx2 but suppressed Oct4 and Sox2. (**D**) Physiological gradient of LIF levels from ampulla toward isthmus with an increased Nanog/Cdx2 ratio, but a decreased Oct4/Cdx2 ratio in a dose-dependent manner. (**E**) A reciprocal relationship between Nanog and Cdx2 evidenced by pretreatment with siRNAs (10^{-8} M, Sigma) and (**F**) with shRNAs (10^{-8} M, Sigma), respectively.

(n = 4). We transplanted the tNSCs into the lesioned striatum (STR) under anesthesia. We performed the apomorphine-induced rotation test every 3 weeks post-cell therapy (Fig. 5A). The results of group (a) showed a significant improvement in rotational behavior from 3-weeks to 12-weeks post-cell therapy. The results of group (b) showed significant improvement during the initial 6-weeks, but the improvement was gradually lost after the 6th week. No improvement was observed in control group (c).

Next we verified the recovery of the nigrostriatal pathway by immunocytochemical studies. In group (a), abundant newly generated TH-positive neurons appeared in the lesioned substantia nigra pars compacta (SNC) with multiple outgrowths

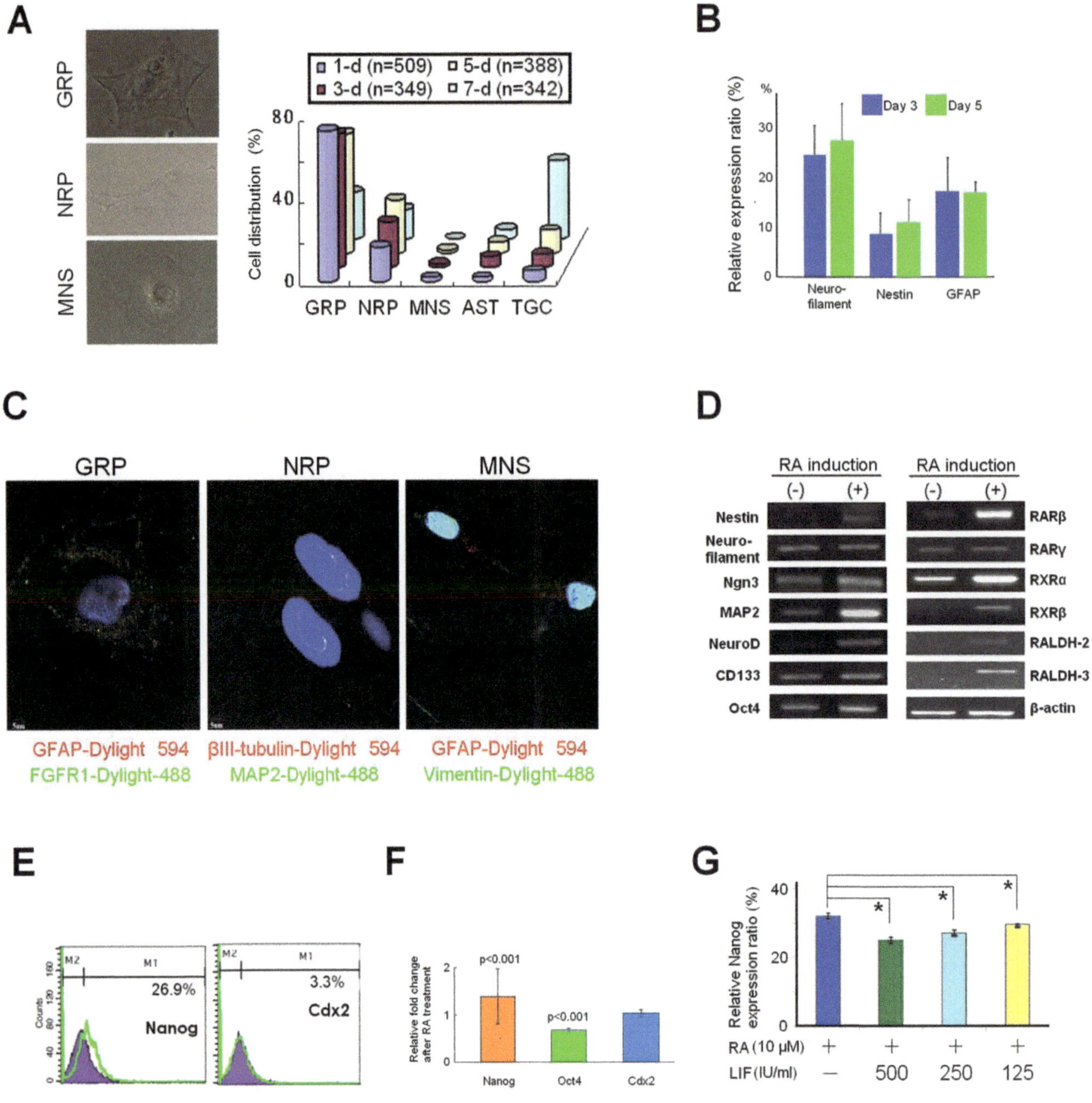

Figure 3. Biological Characteristics of tNSCs. (**A**) Frequency of NSC sub-phenotypes after RA (10 μM) induction over time: 1-, 3-, 5- and 7-days at first, second, third, and fourth row, respectively. n: total cells counted. GRP: glial restricted precursors; NRP: neuronal restricted precursors; MNS: multipotent neural stem cells; and TGC: undetermined trophoblast giant cells. (**B**) Flow cytometric comparison of RA-induced nestin, GFAP, and neurofilament expressions between 3-day and 5-day. Data represent mean ± SD (n = 3). (**C**) Immunocytochemical identification of GRP, NRP and MNS with specific markers. (**D**) Expression of RA-associated genes before and after RA induction for 1-day by RT-PCR. (**E**) RA-induced overexpression of Nanog by flow cytomertic and (**F**) immunocytochemical analysis (n = 60 for each item) by Student *t*-test. (**G**) Withdrawal of LIF significantly promoted expression of Nanog induced by RA (10 μM) by flow cytometry. *: p<0.01 compared to the control. All data indicate mean ± SD (n = 3).

to form neural circuitries with the surrounding host tissues (Fig. 5B and Fig. 5C). We did not observe such phenomena in group (b) (Fig. 5D) and in control group (c) (Fig. 5E). No immuno-suppressive agent was used and no teratoma formation was found.

To investigate the viability of those implanted GFP-tagged tNSCs in the brain, PD rats were sacrificed at 18-weeks post-implantation and their brain sections were examined by immunofluorescent analysis. The results revealed that the GFP-tagged tNSCs remained sporadically in clusters in the lesioned STR and SNC (Fig. 5F).

In Chronic PD Rats

During the breeding period, we performed the apomorphine-induced rotation test monthly to ascertain the rats' PD state until cell therapy. Behavioral assessments were performed every 3 weeks post-cell therapy, including the apomorphine-induced rotation test, the bar test for akinesia, the stepping test for rigidity, and the

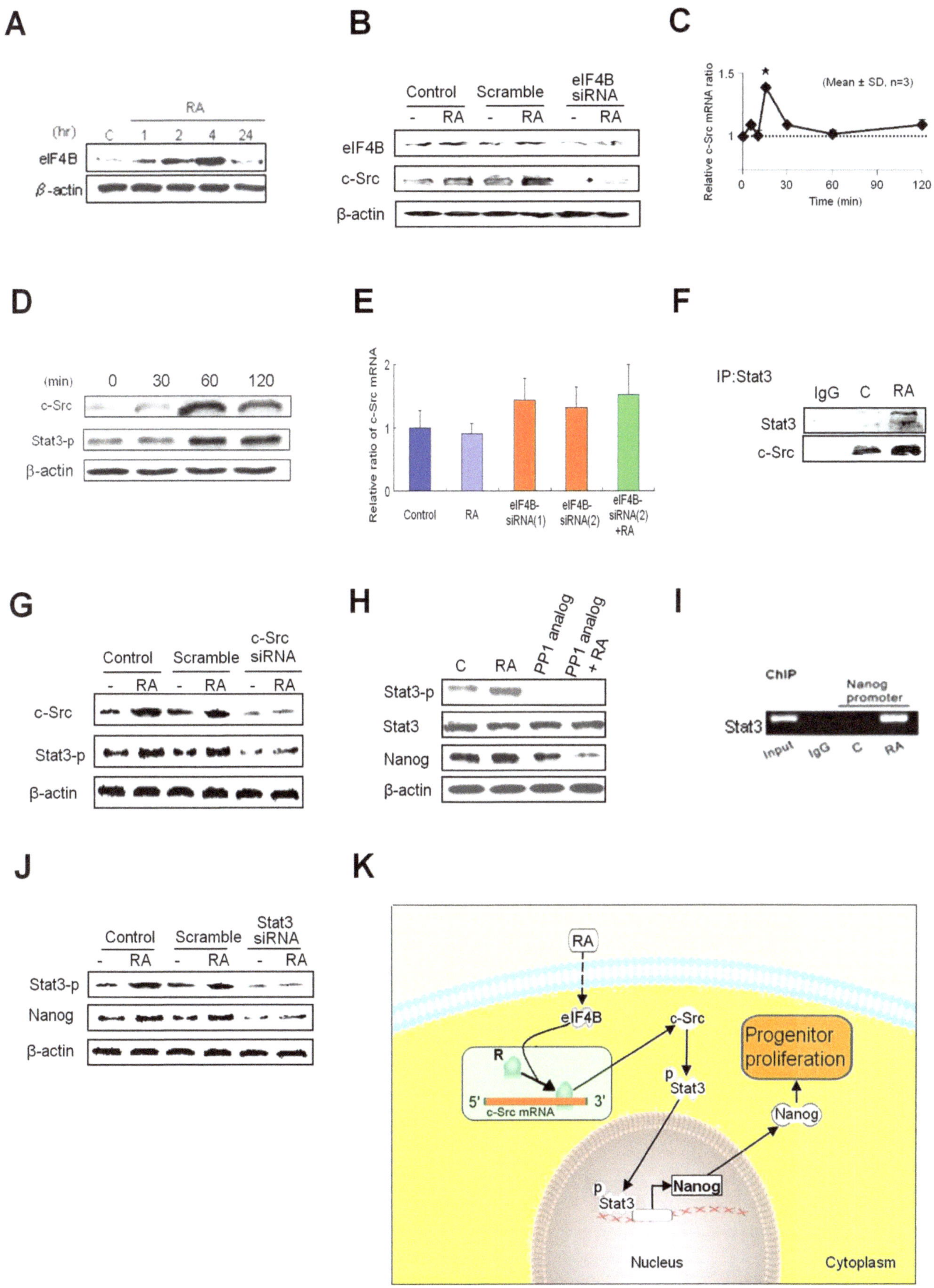
A
RA
(hr) C 1 2 4 24
eIF4B
β-actin
B
Control Scramble eIF4B siRNA
- RA - RA - RA
eIF4B
c-Src
β-actin
C
Relative c-Src mRNA ratio
(Mean ± SD, n=3)
Time (min)
D
(min) 0 30 60 120
c-Src
Stat3-p
β-actin
E
Relative ratio of c-Src mRNA
Control
RA
eIF4B-siRNA(1)
eIF4B-siRNA(2)
eIF4B-siRNA(2) +RA
F
IP:Stat3
IgG C RA
Stat3
c-Src
G
Control Scramble c-Src siRNA
- RA - RA - RA
c-Src
Stat3-p
β-actin
H
C RA PP1 analog PP1 analog + RA
Stat3-p
Stat3
Nanog
β-actin
I
ChIP
Nanog promoter
Stat3
Input IgG C RA
J
Control Scramble Stat3 siRNA
- RA - RA - RA
Stat3-p
Nanog
β-actin
K
RA
eIF4B
c-Src
R
5' 3'
c-Src mRNA
p Stat3
Progenitor proliferation
Nanog
p Stat3
Nanog
Nucleus
Cytoplasm

Figure 4. A Non-genomic RA Signaling Pathway. (**A**) Time course of RA induced production of eIF4B. (**B**) Activation of c-Src was inhibited by using eIF4B siRNA. (**C**) RA (10 μM) induced a rapidly transient elevation of c-Src mRNA peaking at 15 min in hTS cells. Data represent mean ± SD (n = 3); *:p<0.01 in Student's t test. (**D**) RA induced fast production of c-Src and phosphorylation of Stat3 in 1 hr by Western blots. β-actin: control. (**E**) qPCR assay showed that RA did not induce expression of c-Src mRNA at 1 day incubation and expression c-Src mRNA was not affected by using 2 different eIF4B siRNAs. (**F**) Stat3 directly interacted with c-Src by IP assay. (**G**) c-Src siRNA inhibited expression of Stat3. (**H**) c-Src inhibitor PP1 analog (4 μM) inhibited the RA-induced phosphorylation of Stat3 and RA induced overexpression of Nanog in hTS cells by Western blots. (**I**) RA induced binding interaction of Stat3 and Nanog promoter by ChIP assay. (**J**) Nanog expression was inhibited by Stat3 siRNA. (**K**) Schematic of the RA-induced c-Src/Stat3/Nanog pathway via subcellular c-Src mRNA localization in hTS cells. Dotted line indicates undetermined mechanism(s).

footprint analyses for postural imbalance and gait disorder as described previously [19]. Group I (n = 6) was the control, while Group II (n = 6) received 1-day RA-induced tNSCs. In group II, we observed a significant improvement of the apomorphine-induced rotations from 3 weeks to 12 weeks post-cell therapy (Fig. 6A), similar to the previous acute PD rats study. The bar test showed that the grasping time of the affected forelimb was significantly shortened at 3-weeks and continued to improve at 12-weeks (Fig. 6B). Assessments by step length (Fig. 6C), stride length (Fig. 6D), walking speed (Fig. 6E), and base of support (Fig. 6F) revealed significant improvement from 3-weeks to 12-weeks post-cell therapy. These studies were performed on a transparent walkway recorded by video camera (Fig. 6G). These results indicated that cell therapy with tNSCs was able to regenerate the dopaminergic nigrostriatal pathway and functionally improved the behavioral impairments in both acute and chronic PD rats.

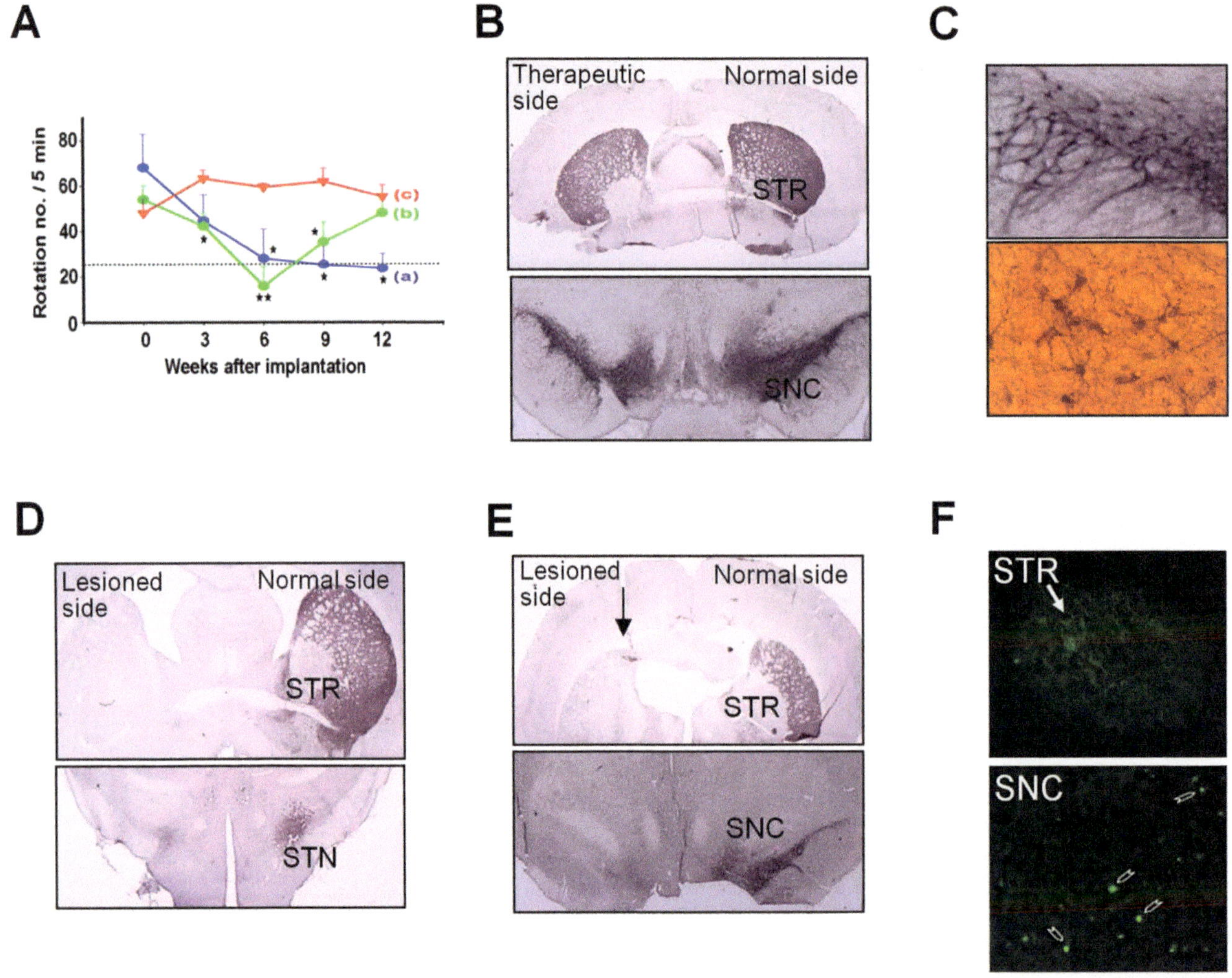

Figure 5. Transplantation of tNSCs in Acute PD Rats. (**A**) Apomorphine-induced rotation tests. Group (a) (blue line, n = 4) received 1-day RA-induced tNSCs. Group (b) (green line, n = 4) received 5-day RA-induced NSCs. Group (c) indicated control (red line, n = 4). Statistic analysis by repeated measure ANOVA: p-value = 0.001. LSD *post hoc* comparisons after repeated measure ANOVA in between two groups: *: p<0.05; **: p<0.01. (**B**) Regeneration of TH-positive cells at the STR (upper) and the SNC (lower) of lesioned side 18-week after cell therapy. Scale bar: 100 μm. (**C**). Amplified dopaminergic neuronal circuitry in the lesioned SNC (upper). A comparative normal feature shown in the lower panel. (**D**) No TH-positive immunoreactivity was found in the lesioned STR (upper panel) and subthalamic nucleus (STN, lower panel) of group (b) and (**E**) group (c). Arrow: needle track. (**F**) Presence of GFP-tagged tNSCs (arrow) at the STR (upper panel) and the SNC (lower panel) 18-weeks post-cell therapy.

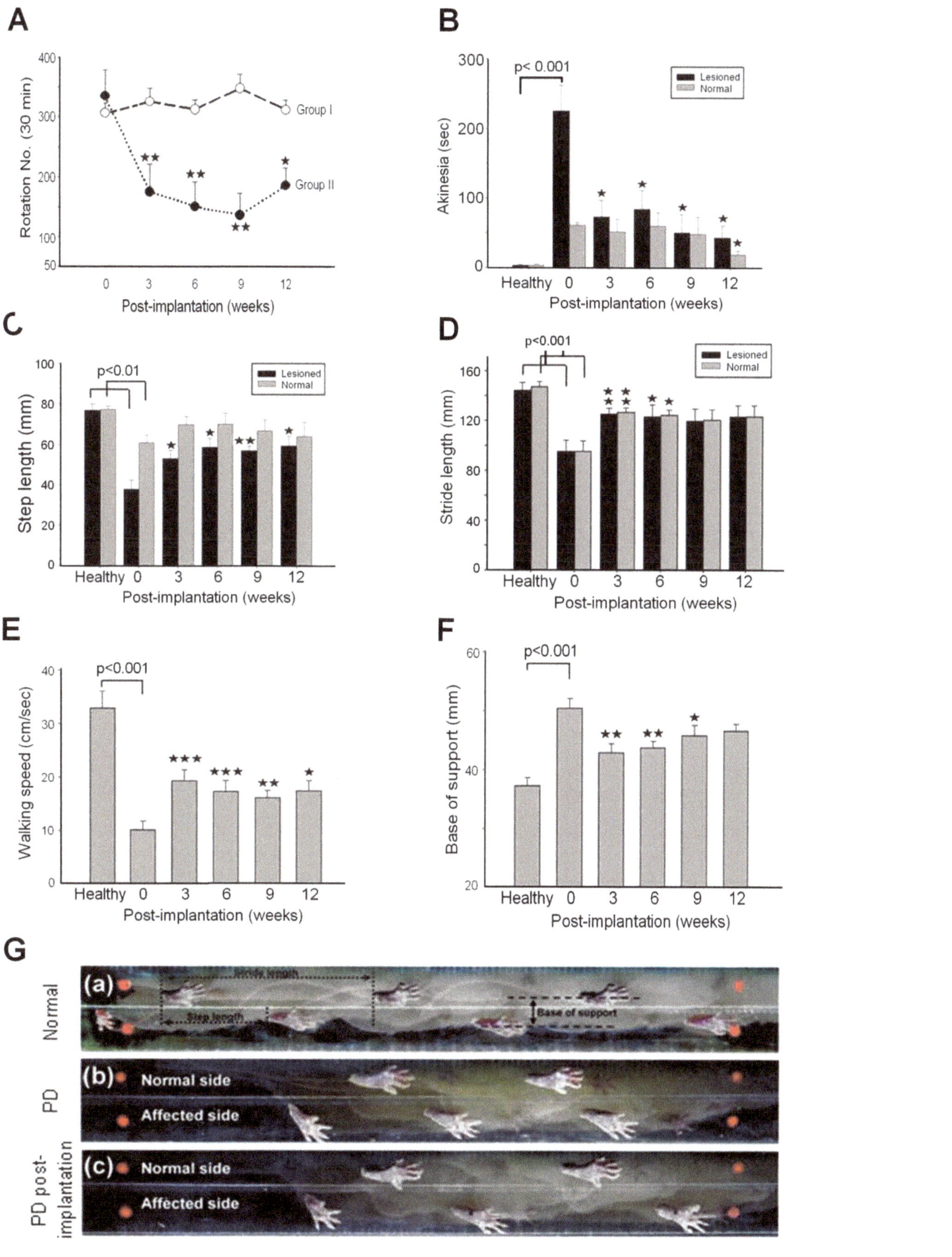
A
Rotation No. (30 min)
Group I
Group II
Post-implantation (weeks)
B
p< 0.001
Lesioned
Normal
Akinesia (sec)
Healthy
Post-implantation (weeks)
C
p<0.01
Step length (mm)
Post-implantation (weeks)
D
p<0.001
Stride length (mm)
Post-implantation (weeks)
E
p<0.001
Walking speed (cm/sec)
Post-implantation (weeks)
F
p<0.001
Base of support (mm)
Post-implantation (weeks)
G
Normal
PD
PD post-implantation
(a)
Stride length
Step length
Base of support
(b)
Normal side
Affected side
(c)
Normal side
Affected side

Figure 6. Behavioral Assessments in Chronic PD Rats. (**A**) Apomorphine-induced rotations analysis revealed a significant improvement in chronic PD rats that received tNSCs cell therapy (Group II, n = 6, black round line) compared to the control group (Group I, n = 6, blank round line). (**B**) Bar tests (sec) improved at 3-week post-cell therapy. (**C**) Shortened step length and (**D**) stride length were significantly improved after 3-week post-cell therapy in PD rats. (**E**) Significant improvement in walking speed (cm/sec) in PD rats at 3-week post-cell therapy. (**F**) Significant shortening in base of support (mm) was seen at 3-week post-cell therapy. (**G**) A well-designed cage with a video camera for footprint analyses. (a): in normal rats, (b): in PD rats pre-cell therapy, and (c): post-cell therapy. Asterisk: compared to the control. Student's t test: *:$p<0.05$; **: $p<0.01$; and ***: $p<0.001$.

tNSCs Regenerate the Dopaminergic Nigrostriatal Pathway *in vivo*

Next, we compared the regeneration of the dopaminergic nigrostriatal pathway after cell therapy with tNSCs by immunofluorescence tissue analysis between acute and chronic PD rats. We investigated brain sections of post-cell therapy acute PD rats (n = 2 at 1-week post-injury, n = 2 at 6-weeks post-injury, n = 2 as sham control for post-injury, n = 6 at 12-weeks post-cell therapy, and n = 2 as sham control for cell therapy) and post-cell therapy chronic PD rats (n = 2 at 12-weeks post-cell therapy and n = 2 as sham control for cell therapy).

In the post-injury PD rats at 6-weeks, the 6-OHDA caused progressive neural degeneration in the SNC, resulting in various cavities immunohistochemically (Fig. 7A). Quantitative analysis showed that the number of DA neurons in the SNC reduced to 48% at 1-week post-injury and to 13% at 6-weeks post-injury compared to the intact side (Fig. 7B and Fig. S4). In the STR, DA neurons were reduced to 78% at 1-week post-injury and to 4% at 6-weeks post-injury (Fig. 7B). In acute PD rats at 12-weeks post-cell therapy, tNSCs generated numerous DA neurons at the cavity wall with TH-positive nervous terminals projecting into the cavity (Fig. 7A, insert). We found that DA neurons had regenerated to 67% in the SNC and to 73% in the STR.

In chronic PD rats at 12-weeks post-cell therapy, immunohistochemical study revealed that DA neuronal circuitries were substantially regenerated in the therapeutic side of the SNC, mirroring the intact side (Fig. 7C), consistent with our previous studies. The recovery of DA neurons reached 78% in the SNC (Fig. 7D).

Glial cells are known to play as mediators in guiding the migration of neurons to their destinations or as sources of neural regeneration [32]. We found that in the STR, 6-OHDA in chronic PD rats caused not only degeneration of both GFAP(+) cells and DA neurons, but also disarrangement of the striatopallidal fibers (pencils of Wilson) and nigral axons. These phenomena were improved after tNSC therapy, showing numerous GFAP(+) cells embedded in the fine myelinated fibers (Fig. 7E). The GFAP(+) cells regenerated from 66% at 6-weeks post-injury to 94% at 12-weeks post-cell therapy in the lesioned STR (Fig. 7F). These results indicated that transplantation of tNSCs regenerated the dopaminergic nigrostriatal pathway in chronic PD rats. No teratoma formation was found in both the acute and the chronic PD rats. Longer term studies beyond 18-weeks post-transplantation may be useful.

tNSCs Exhibit Immune Advantages for Transplantation

Translation of stem cells into cell therapies requires the understanding of their antigenicity and immunogenicity for any clinical application to be useful. We first implanted the hTS cells into male severe combined immunodeficient (SCID) mice intramuscularly for 6–8 weeks. Histologically, no teratoma was found; but we observed myxoid-like bizarre cells between the muscle fibers (Fig. 7G). hTS cells appear to indicate immune advantages compared to hES cells with respect to teratoma formation [33], [34].

We then compared the expressions of immune-related genes among hTS cells, tNSCs, and hES cells. *In vitro*, HLA-ABC (MHC-I) was expressed highly in hTS cells (99.4%) and in tNSCs (99.7%), but lower in hES cells (12.9%); however, no surface HLA-DR (MHC-II) was expressed among them (Fig. 7F). No major difference was found among the three cells for innate immune system marker CD14, for effector-memory T-cells marker CD44, and for human endothelial cell proliferation marker CD105. Surprisingly, hES cells and tNSCs expressed the immature hematopoietic and endothelial cell marker CD34, linked to neurogenesis and angiogenesis [35], while hTS cells expressed significantly less. Both hTS and hES cells expressed a myeloid leukemia-associated marker CD33 and cancer stem cell marker CD133, while tNSCs expressed significantly less.

Discussion

The search for an alternative cell source for treatment of PD has been challenging [1], [2], [3], [4], [5], [6]. Here, we isolated hTS cells from preimplantation embryos in women with ectopic pregnancy. hTS cells are not only distinct from PDMS cells, but also exhibit hES cell-like qualities in pluripotency and viability in proliferation. Given that hTS cells originate from the TE and not from the ICM of the blastocyst, hTS cells circumvent the socio-ethical issues that surround the use of hES cells. Furthermore, ectopic pregnancies account for 1 to 2% of all pregnancies in industrialized countries and even higher in developing countries; thus, providing a locally abundant source of stem cells with relatively closer genetic make-up for patients in need of cell therapy.

It is interesting to note that hTS cells express both the markers of the ICM and TE. In murine embryos, Oct4 protein is detected in the ICM but not the TE. However, in porcine and bovine blastocysts, Oct4 protein is detected in both the ICM and the TE [36]. In human blastocysts, there is 31 times higher concentration of Oct4 mRNA in the ICM than that in the TE [37]. In hTS cells, we show co-expressions of pluripotent markers such as Oct4 and SSEA-4 along with TE-specific marker Cdx2. This may be specific to ectopic pregnancy-derived hTS cells and possibly due to the microenvironment of a tubal ectopic pregnancy. Further study of this microenvironment is required.

At the preimplantation stage, extrinsic factors like LIF [9] may influence the epigenetic gene regulation in embryos. Not surprisingly, LIF affected the fate choice of hTS cells similar to its action on ES cells [9], [25]. Mimicking the microenvironment in the fallopian tube, we showed that withdrawal of LIF in hTS cells downregulated Oct4, but upregulated Nanog and Cdx2. This fact suggested that Oct4 was initially responsible for the pluripotent state. However, hTS cell pluripotency became dependent of Nanog at a later stage, consistent with the notion that Nanog plays a crucial role in cell fate specification following the formation of the blastocyst [38]. The elevation of both Nanog and Nanog/Cdx2 ratio maybe responsible for the maintenance of pluripotency in hTS cells. Further study of this direct relationship is required. The overexpression of Cdx2 indicates hTS cell's trophoblastic phenotype as it enhances the formation of TS cells in ES cells [14].

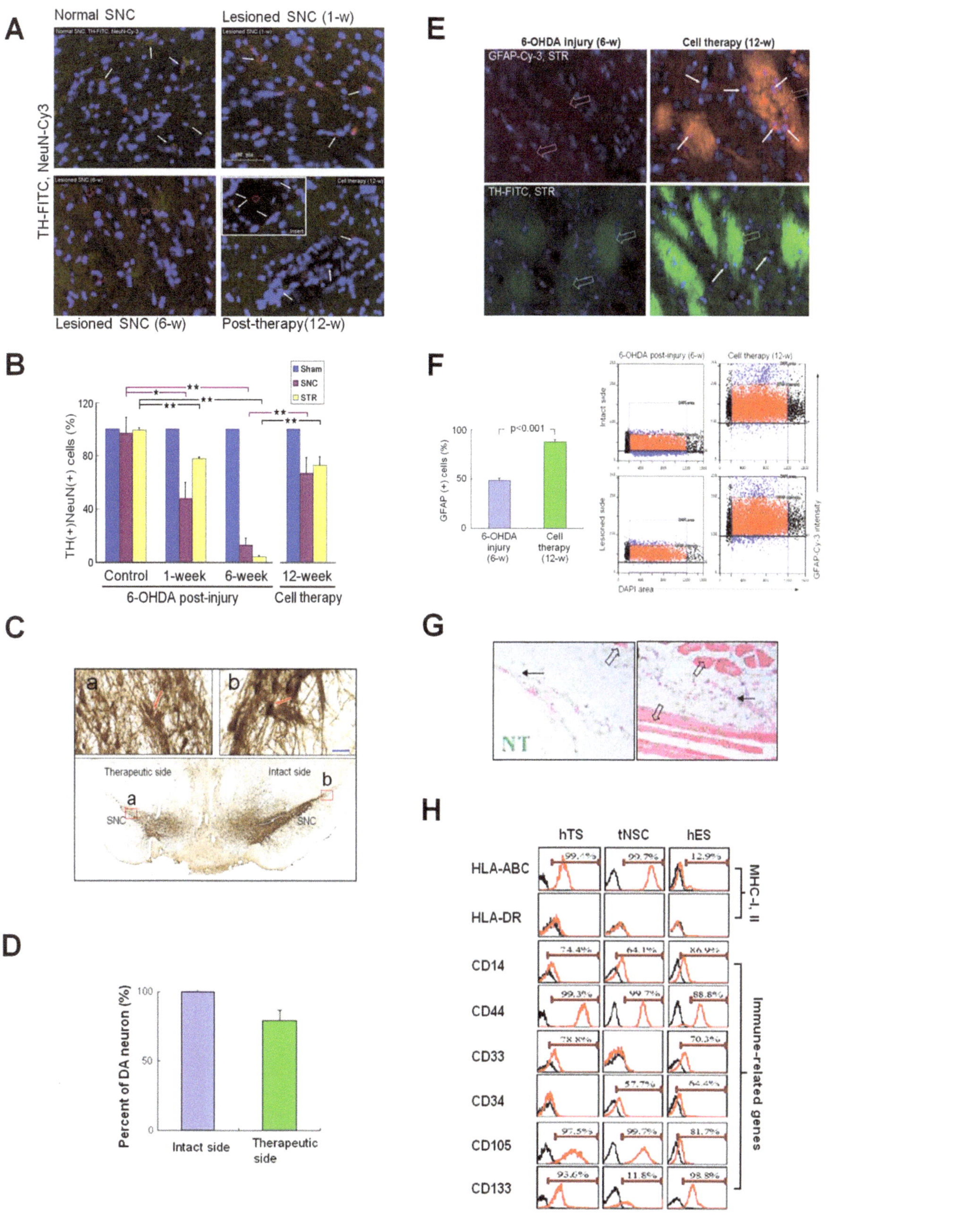

Figure 7. ***in vivo*** **Regeneration of TH(+) and GFAP(+) Cells with Less Immuno-responses.** (**A**) Immunohistofluoresence analysis. TH(+) and NeuN(+) motor neurons (arrow) in the SNC of control (left upper). Decreased TH(+) (arrow) at 1-week after 6-OHDA injury (right upper). Apparent reduction in TH(+) neurons with disarrangement of TH-positive neural terminals (green granules), and various degenerative cavity formation (red hollow circle) at 6-week post-injury (left lower). After transplantation at 12-weeks, TH(+) neurons (arrow) at wall of the degenerative cavity (red hollow

circle; insert) with TH(+) neural terminals (green color) projecting into the cavity (right lower). (**B**) Number of TH(+) cells at 1- and 6-weeks reduced to 48% and 13% in the lesioned SNC (red) and 78% and 4% in the lesioned STR (light blue), respectively, post-injury. After transplantation, TH(+) cells regrew up to 67% and 73% in the lesioned SNC and STR, respectively (right panel). Data analyzed by the software Tissuequest 2.0 (TissueGnostics Gmbh, Vienna, Austria). (**C**) Regeneration of dopaminergic neurons in the lesioned SNC (lower panel) with amplification (left upper, insert a) compared with the intact side (right upper, insert b). (**D**) Transplantation of tNSCs at 12-weeks yielded 78% of recovery rate in TH-positive neurons in the lesioned SNC compared to the intact side. Data represent mean ± SD (n = 3). (**E**) Degeneration of TH-FITC(+) and GFAP-Cy-3(+) Wilson's pencils (blank arrow) at 6-week post-injury in the lesioned STR (left column). At 12-weeks post-cell therapy (right column), several GFAP(+) cells (arrow) appeared inside the fine fibers of re-established Wilson's pencils (blank arrow). (**F**) For immunohistofluorescence imaging analysis, cells were counted in the gate (left scatter plots) determined by the location of cell size (8–10 μm in diameter) and its corresponding intensity of GFAP-Cy-3. Gate (red scatter plot): glial cells counted; black scatter plots: exclusive cells with bizarre size; blue scatter plots: cells with abnormal GFAP intensity. In the STR, the GFAP(+) cells were 65.5% in the lesioned side before treatment and became 93.9% after cell therapy compared to the intact side (right panel). Data represent mean ± SD (n = 3). Student *t*-test. (**G**) hTS cells implanted into the SCID mice (n = 6) raised only minor immunoreactions and without tumorigenesis observed. Myxoid-like bizarre cells (black arrow), muscle fibers (blank arrow), and needle track (NT). (**H**) Flow cytometric analysis of hTS cells, tNSCs, and hES cells. High expression of HLA-ABC in hTS cells and tNSCs, but less in hES cells. No HLA-DR expressed among them. Immune-related gene expressions, including CD14, CD44, CD33, CD34, CD105, and CD133, were compared among them.

RA was capable of determining the neuronal fate choice of hTS cells. hTS cells expressed RALDH-2 and -3, which enables the metabolism of retinol into RA [15], while ES cells did not [25]. We have found some evidence to suggest that RA-enabled hTS cell differentiation into stable tNSCs through a non-genomic eIF4B/c-Src/Stat3/Nanog signaling pathway via subcellular c-Src mRNA localization. The fast transient production of c-Src was possibly related to the RA-trigged machinery in local protein synthesis rather than conventional genetic processes. We suggest that the RA-activated eIF4B is possibly and indirectly associated with subcellular c-Src mRNA localization in hTS cells. How RA induced eIF4B activation remains to be further evaluated. Critical details of the regulatory mechanisms between eIF4B and c-Src will require further study. A more comprehensive knowledge of this pathway in tNSC proliferation may provide helpful insight to why Src-kinase activity increases at the peak period of neurogenesis in the developing striatum and hippocampus [30] and why the disruption of Stat3 signaling contributes significantly to neuronal death in PD [39].

A potential therapeutic advantage of tNSCs is its proliferation of a variety of restricted neural precursor subtypes. Of which, GRP and astrocytes may be attributed to the regeneration of both DA neurons and glial cells in the lesioned nigrostriatal pathway, substantially promoting dopaminergic neurogenesis [6]. Another advantage is that tNSCs facilitated more site-specific integration rather than regional incorporation, as we observed neural regeneration only in the lesioned sites. Application of tNSCs in other central nervous system diseases must be researched.

Previous studies report that hES cell-derived DA neurons do not survive in the grafted striatum and fail to improve the behavioral deficits in PD rats [40]. Cell therapy with tNSCs regenerated the dopaminergic nigrostriatal pathway and functionally improved the behavioral impairments in both acute and chronic PD rats. It is interesting to note from our animal studies that 1-day RA-induced tNSCs maintain long term effectiveness while the 5-day RA-induced tNSCs showed behavioral improvement in the beginning, but failed to be effective after 6-weeks. We believe that the transplanted 5-day RA-induced tNSCs were near the end of their effective window; that is, when most of the tNSCs transition into undefined giant cells. Contrastingly, the 1-day RA-induced tNSCs were extremely viable and once transplanted, maintained its regenerative effect. It may be helpful that tNSCs were implanted into the striatum of the brain, where cells would favorably meet an RA-enriched microenvironment [17], facilitating continuous proliferation *in vivo*. Moreover, the implanted tNSCs increased glial cells in the striatum, which is compatible with previous reports that RA induces the expression of GFAP [41] and that GFAP(+) progenitor cells may give rise to neurons and oligodendrocytes throughout the CNS [42]. The use of hTS cells in cell therapy may be optimal within an efficacy window: 1-day after RA-induction to tNSCs.

In chronic PD rats, tNSCs regenerated about two-thirds of the DA neurons in the nigrostriatal pathway at 12-weeks post-cell therapy. This positively suggests that tNSCs may also be effective for treatment of patients that have suffered PD symptoms over a long period of time. However, the appropriate dosage of tNSCs for treatment of PD remains to be evaluated.

Our investigation of hTS cell immune characteristics began with the intramuscular transplantation of hTS cells in SCID mice. Although we observed bizarre cells, we did not observe any teratoma formation. In subsequent animal studies, we neither used immunosuppressed mice nor immunosuppressants and no teratoma formation appeared. It has been reported that neural progenitor stem cells have immune privilege and have survived rejection in allografts [43]. It is interesting to note that by differentiation of hTS cells to tNSCs, RA-induced changes in expression of immune-related markers. For example, while the expression of CD34(+) increased, expressions of CD133(+) decreased. The biological significance is unclear. Nevertheless, it has been shown that autologous transplantation with CD34(+) immune-selected grafts is feasible in children with high-risk neuroblastoma [44], linking positive expression to graft feasibility. Also consistent with a recent report [45], lower CD133 expression in tNSCs may decreased the probability of tumorigenesis, as CD133(+) cells possess the capacity for unlimited self-renewal and can trigger brain tumor initiation [34]. Although it is promising that hTS cells were successfully employed for graft acceptance without teratoma formation in SCID mice and tNSCs in PD rats, further studies are warranted to investigate the possible immune responses of hTS cells or their derivatives after transplantation.

In conclusion, hTS cells are a promising cell source for regenerative medicine. The source from tubal ectopic pregnancies is abundant. The culture requires no feeder layer. The induction to tNSCs is not only simple (RA) but also fast (1-day). tNSCs also show early indications of immune-privilege. hTS cells and hTS cell-derived tNSCs may be remarkable candidates for shortening research timeframes, creating new disease models, and also for the treatment of patients with PD and other neurodegenerative diseases.

Supporting Information

Figure S1 (A) Microscopic feature of ectopic pregnancy-derived chorionic villi dissected. (B) At passage 9 and 17, hTS cells co-expressed Oct4 (red) and Nanog (green) in a homogeneous distribution. Scale bar: 20 μm. Nucleus: blue DAPI color. (C) Flow cytometry with Oct4, Sox2, Nanog and Cdx2 at passage 15 demonstrated 97.9%, 95.0%, 98.7%, and 94.0% positive cells, respectively. (D) Flow cytometric analyses revealed that hTS cells express markers of mesenchymal stem cells but not markers of hematopoietic stem cells. (E) Differentiation of hTS cells into specific phenotypes, including osteoblasts, chondrocytes, myocytes, and adipocytes by appropriate inductions. (F) Chromosome analyses revealed that normal karyotypes (44,XY) did not change at passages 3, 10, and 15 of cultivation in hTS cells.

Figure S2 (A) Withdrawal of LIF suppressed Oct4 and Sox2, but overexpressed Cdx2 by flow cytometry in hTS cells. (B) A reciprocal relationship between Nanog and Cdx2 evidenced by pretreatment with siRNAs (10^{-8} M, Sigma) by flow cytometry. Data represent mean ± SD, n = 3.

Figure S3 Flow cytometry showed that RA (10 μM) enhanced Nanog expression by withdrawal of LIF levels from 500, 250, 125, toward 0 IU/ml after 3-day incubation in hTS cells.

Figure S4 To qualify the TH(+)NeuN(+) neurons in the SNC pre- and post-cell therapy, coefficient of determination between TH-FITC and NeuN-Cy-3 was measured by immunohistofluorescent scatter plots in the chronic PD rats. Normal SNC: $R^2 = 0.72$ (left upper); 6-OHDA damaged SNC (1-week), $R^2 = 0.77$ (right upper); 6-OHDA damaged SNC (6-weeks), $R^2 = 0.25$ (left lower); SNC after tNSCs transplantation (12-weeks), $R^2 = 0.66$ (right lower). Result shown represents the average of 2 rats.

Table S1 All antibodies, primers, and reagents used in this study.

Table S2 Recipes used for cell differentiations.

Table S3 Significant distinction in biological functions in 4,864 distinguishable genes ($p<0.005$, fold change >2) between hTS cells and PDMS cells.

Table S4 Comparison of immune-related gene expression in relative intensity value between hTS cells and PDMS cells from Metasearch.

Text S1 Detailed experimental procedures and supporting references.

Acknowledgments

We thank Dr. Robert Silman and Mr. Yuta Lee for comments on the preparation of manuscript, Dr. I. M. Chiu for providing F1B(-540)-GFP plasmid construct, C. P. Chen for PDMS cells, Y. S. Liu for hES cells and Ms. H. T. Lee for technical assistance.

Author Contributions

Conceived and designed the experiments: JNL EMT. Performed the experiments: TTYL THH JJJC RMW CFT YCW MCK. Analyzed the data: JNL EMT THH SS. Contributed reagents/materials/analysis tools: JNL EMT THH SS. Wrote the paper: JNL EMT TTYL.

References

1. Freed CR, Greene PE, Breeze RE, Tsai WY, DuMouchel W, et al. (2001) Transplantation of embryonic dopamine neurons for severe Parkinson's disease. N Engl J Med 344: 710–719.
2. Olanow CW, Goetz CG, Kordower JH, Stoessl AJ, Sossi V, et al. (2003) A double-blind controlled trail of bilateral fetal nigral transplantation in Parkinson's disease. Ann Neurol 54: 403–414.
3. Ben-Hur T, Idelson M, Khaner H, Pera M, Reinhartz E, et al. (2004) Transplantation of human embryonic stem cell-derived neural progenitors improves behavioral deficit in Parkinsonian rats. Stem Cells 22: 1246–1255.
4. Cai J, Donaldson A, Yang M, German MS, Enikolopov G, et al. (2009) The role of Lmx1a in the differentiation of human embryonic stem cells into midbrain dopamine neurons in culture and after transplantation into a Parkinson's disease model. Stem Cells 27: 220–229.
5. Roy NS, Cleren C, Singh SK, Yang L, Beal SA, et al. (2006) Functional engraftment of human ES cell-derived dopaminergic neurons enriched by coculture with telomerase-immortalized midbrain astrocytes. Nat Med 12: 1259–1268.
6. Dunnett SB, Björklund A, Lindvall O (2001) Cell therapy in Parkinson's disease: stop or go? Nat Rev Neurosci 2: 365–369.
7. Parolini O, Alviano F, Bagnara GP, Bilic G, Bühring HJ, et al. (2008) Concise review: Isolation and characterization of cells from human term placenta: outcome of the first international workshop on placenta derived stem cells. Stem Cells 26: 300–311.
8. Yamanaka Y, Ralston A, Stephenson RO, Rossant J (2006) Cell and molecular regulation of the mouse blastocyst. Dev Dyn 235: 2301–2314.
9. Surani MA, Hayashi K, Hajkova P (2007) Genetic and epigenetic regulators of pluripotency. Cell 128: 747–762.
10. Chen HF, Chao KH, Shew JY, Yang YS, Ho HN (2004) Expression of leukemia inhibitory factor and its receptor is not altered in the decidua and chorionic villi of human anembryonic pregnancy. Hum Reprod 19: 1647–1654.
11. Wånggren K, Lalitkumar PG, Hambiliki F, Ståbi B, Gemzell-Danielsson K, et al. (2007) Leukaemia inhibitory factor receptor and gp130 in the human fallopian tube and endometrium before and after mifepristone treatment and in the human preimplantation embryo. Mol Hum Reprod 13: 391–397.
12. Williams RL, Hilton DJ, Pease S, Willson TA, Stewart CL, et al. (1988) Myeloid leukaemia inhibitory factor maintains the developmental potential of embryonic stem cells. Nature 336: 684–687.
13. Chambers I, Colby D, Robertson M, Nichols J, Lee S, et al. (2003) Functional expression cloning of Nanog, a pluripotency sustaining factor in embryonic stem cells. Cell 113: 643–655.
14. He S, Pant D, Schiffmacher A, Meece A, Keefer CL, et al. (2008) Lymphoid enhancer factor 1-mediated Wnt signaling promotes the initiation of trophoblast lineage differentiation in mouse embryonic stem cells. Stem Cells 26: 842–849.
15. Maden M (2007) Retinoic acid in the development, regeneration and maintenance of the nervous system. Nat Rev Neurosci 8: 755–765.
16. Li XJ, Du ZW, Zarnowska ED, Pankratz M, Hansen LO, et al. (2005) Specification of motoneurons from human embryonic stem cells. Nat Biotechnol 23: 215–221.
17. Kane MA, Chen N, Sparks S, Sparks S, Napoli JL (2005) Quantification of endogenous retinoic acid in limited biological samples by LC/MS/MS. Biochem J 388: 363–369.
18. Wang KH, Kao AP, Chang CC, Lee JN, Chai CY, et al. (2010) Modulation of tumorigenesis and oestrogen receptor-alpha expression by cell culture conditions in a stem cell-derived breast epithelial cell line. Biol Cell 102: 159–172.
19. Hsieh TH, Chen JJJ, Chen LH, Chiang PT, Liang JI, et al. (2011) Time course gait analysis of hemiparkinsonian rats following 6-hydroxydopamine lesion. Behav Brain Res 222: 1–9.
20. Rong-Hao L, Luo S, Zhuang LZ (1996) Establishment and characterization of a cytotrophoblast cell line from normal placenta of human origin. Hum Reprod 11: 1328–1333.
21. Niwa H, Toyooka Y, Shimosato D, Strumpf D, Takahashi K, et al. (2005) Interaction between Oct3/4 and Cdx2 determines trophectoderm differentiation. Cell 123: 917–929.
22. Panicker MM, Rao M (2001) Stem Cells and Neurogenesis. In: Marshak R, Gardner RL, and Gottlieb D, editors. Stem Cell Biology. New York: Cold Spring Harbor Laboratory Press, 399–438.
23. Yan J, Tanaka S, Oda M, Makino T, Ohgane J, et al. (2001) Retinoic acid promotes differentiation of trophoblast stem cells to a giant cell fate. Dev Biol 235: 422–432.
24. Chen L, Khillan JS (2008) Promotion of feeder-independent self-renewal of embryonic stem cells by retinol (vitamin A). Stem Cells 26: 1858–1864.
25. Martín-Ibáñez R, Urbán N, Sergent-Tanguy S, Pineda JR, Garrido-Clua N, et al. (2007) Interplay of leukemia inhibitory factor and retinoic acid on neural differentiation of mouse embryonic stem cells. J Neuron Res 85: 2686–2710.
26. Hellen CU (2009) IRES-induced conformational changes in the ribosome and the mechanism of translation initiation by internal ribosomal entry. Biochimica et Biophysica Acta (BBA) – Gene Regulatory Mechanisms 1789, 558–570.
27. Ochs K, Saleh L, Bassili G, Sonntag VH, Zeller A, et al. (2002) Interaction of translation initiation factor eIF4B with the poliovirus internal ribosome entry site. J Virol. 76: 2113–2122.

28. Allam H, Ali N (2009) Initiation factor eIF2-independent mode of c-Src mRNA translation occurs via an internal ribosome entry site. J Biol Chem 285: 5713–5725.
29. Gilbert WV (2010) Alternative ways to think about cellular internal ribosome entry. J Biol Chem. 285: 29033–29038.
30. Cartwright CA, Simantov R, Cowan WM, Hunter T, Eckhart W (1988) pp60c-src expression in the developing rat brain. Proc Natl Acad Sci USA 85: 3348–3352.
31. Annerén C, Cowan CA, Melton DA (2004) The Src family of tyrosine kinases is important for embryonic stem cell self-renewal. J Biol Chem 279: 31590–31598.
32. Barres BA (2008) The mystery and magic of glia: a perspective on their roles in health and disease. Neuron 60: 430–440.
33. Singh SK, Hawkins C, Clarke ID, Squire JA, Bayani J, et al. (2004) Identification of human brain tumour initiating cells. Nature 432: 396–401.
34. Hentze H, Soong PL, Wang ST, Phillips BW, Putti TC, et al. (2009) Teratoma formation by human embryonic stem cells: Evaluation of essential parameters for future safety studies. Stem Cell Res 2: 198–210.
35. Taguchi A, Soma T (2004) Administration of CD34+ cells after stroke enhances neurogenesis via angiogenesis in a mouse model. J Clin Invest 114: 330–338.
36. Kirchhof N, Carnwath JW, Lemme E, Anastassiadis K, Schöler H, et al. (2000) Expression pattern of Oct-4 in preimplantation embryos of different species. Biol Reprod 63, 1698–1705.
37. Hansis C, Grifo JA, Krey LC (2000) *Oct-4* expression in inner cell mass and trophectoderm of human blastocysts. Mol Hum Reprod 6: 999–1004.
38. Cavaleri F, Scholer HR (2003) Nanog: a new recruit to the embryonic stem cell orchestra. Cell 113: 551–552.
39. Wang L, Xu S, Xu X, Chan P (2009) (−)-Epigallocatechin-3-Gallate protects SH-SY5Y cells against 6-OHDA-induced cell death through STAT3 activation. J Alzheimers Dis 17: 295–304.
40. Park CH, Minn YK, Lee JY, Choi DH, Chang MY, et al. (2005) *In vitro* and *in vivo* analyses of human embryonic stem cell-derived dopamine neurons. J Neurochem 92: 1265–1276.
41. Ying M, Wang S, Sang Y, Sun P, Lal B, et al. (2011) Regulation of glioblastoma stem cells by retinoic acid: role for Notch pathway inhibition. Oncogene 30: 3454–3467.
42. Casper KB, McCarthy KD (2006) GFAP-positive progenitor cells produce neurons and oligodendrocytes throughout the CNS. Mol Cell Neurosci 31: 676–684.
43. Hori J, Ng TF, Shatos M, Klassen H, Streilein JW, et al. (2007) Neural progenitor cells lack immunogenicity and resist destruction as allografts. Ocul Immunol Inflamm 15: 261–273.
44. Marabelle A, Merlin E, Halle P, Paillard C, Berger M, et al. (2011) CD34+ immunoselection of autologous grafts for the treatment of high-risk neuroblastoma. Pediatr Blood Cancer 56: 134–142.
45. Li L, Baroja ML, Majumdar A, Chadwick K, Rouleau A, et al. (2004) Human embryonic stem cells possess immune-privileged properties. Stem Cells 22: 448–456.

Percutaneous Bone Marrow Transplantation Using Fractional Ablative Erbium:YAG Laser

Luis Rodriguez-Menocal[1,2], Marcela Salgado[1,2], Stephen Davis[1], Jill Waibel[1], Arsalan Shabbir[1], Audrey Cox[1], Evangelos V. Badiavas[1,2]*

1 Department of Dermatology and Cutaneous Surgery, University of Miami Miller School of Medicine, Miami, Florida, United States of America, **2** Interdisciplinary Stem Cell Institute, University of Miami Miller School of Medicine, Miami, Florida, United States of America

Abstract

Topical application of therapeutic agents has been a mainstay in Dermatology for the treatment of skin disorders but is not commonly used for systemic delivery. For a topically applied agent to reach distant body sites it must first overcome the barrier function of the skin and then penetrate into deeper structures before reaching the systemic circulation. This has limited the use of topically applied agents to those having specific charge, solubility and size restrictions. Pretreatment of the skin with ablative fractional laser appears to enhance the uptake of some topically applied drugs but the ability to effectively deliver agents to distant sites is largely unproven. In this report we used a fractional ablative Erb:YAG (Erbium/Yttrium Aluminum Garnet) laser to facilitate the transfer of bone marrow stem cells through the skin in a murine bone marrow transplant model. Chimerism could be detected in the peripheral blood of recipient C57BL/6 mice that were pretreated with ablative fractional laser and had topically applied enhanced green fluorescent protein (GFP) labeled bone marrow cells from syngeneic donor transgenic mice. This study indicates that fractional laser can be used to deliver stem cells through the skin and remain functionally intact.

Editor: Graca Almeida-Porada, Wake Forest Institute for Regenerative Medicine, United States of America

Funding: This work was supported by internal University of Miami funding sources. The funders had no role in study design, data collection and analysis, decision to publish, or preparation of the manuscript.

Competing Interests: The authors have declared that no competing interests exist.

* E-mail: ebadiavas@med.miami.edu

Introduction

Stem cell based therapies have the potential to address many disorders that currently have few if any therapeutic alternatives [1–3]. There are nevertheless special considerations that must be taken into account when considering cell-based therapeutics. Cells have several distinct properties that differ significantly from traditional agents such as their relatively large size and their rather fragile nature. The most common means of delivering cells systemically has been by intravenous infusion for purposes such as blood cell transfusion. For stem cell based therapies directed at specific tissue repair, a more localized delivery of cells might be preferred but how these locally administered stem cells will behave and where they will engraft is in many cases uncertain. Injection and surgical implantation techniques have been used when trying to administer cells locally to specific sites. With injections, the pressure gradient created during injection can cause significant damage to cells. The needle diameter chosen for injection also has to be carefully considered so cells are not damaged shearing forces. Alternatively, cells may also be delivered surgically by placement into the surgical field either alone or by incorporating them into a matrix material. Surgical techniques will however require compatibility with the wound healing process and embedding cells into matrices may impact their mobilization to targeted tissues. Developing unique methods of cell delivery will broaden the way in which cells can be used therapeutically. Delivering cells directly to broad areas at predetermined depths is also not possible with these techniques. We have examined the use of factional lasers to create micro-channels in the skin that may be used as conduits for cell delivery locally to tissues and also allow for systemic administration [4–6]. Fractional lasers can create microchannels and depending on the laser's parameters the size and depth of these channels can be controlled [7,8]. Nano and microparticles have been shown to be efficiently administered to depths down to 230 μm by using fractional lasers and these particles have been shown to stay in the dermis for longer than 1 month [9]. In this report we describe the delivery of lineage negative (Lin^-) bone marrow cells to the skin of irradiated mice using an ablative fractional laser with the goal of achieving functional bone marrow transplantation. Using fractional lasers, we have demonstrated that stem cells can be delivered to the skin, become systemically distributed, engraft into distant organs and remain functionally intact. The cells were given at predetermined depths and could cover broader areas in a more uniform fashion.

While this study may not suggest this model as an alternative for bone marrow transplantation, it does provides proof of concept that cell delivery by fractional lasers is a viable option for the delivery of intact functional cells, even to distant sites.

Materials and Methods

Animals

All animals and procedure were approved by the Institutional Animal Care and Use Committee (IACUC) at University of

Miami/Miller School of Medicine. Four week- old female recipient C57BL/6J or NOD/SCID mice were obtained from Jackson lab (Bar Harbor, ME, USA). Syngeneic male C57BL/6-Tg (UBC-GFP)30Scha/J mice were chosen as donors for bone marrow cells.

Irradiation

Recipient mice were placed in a ventilated acrylic container (28 cm diameter, 8 cm depth) between opposing gamma ray irradiation plates in a GAMMACELL50 irradiator. Total body irradiation (TBI) was perfomed before laser exposure and transplantation.

Mice were irradiated to at doses ranging from 200 Centigray (cGy) to 1,000 cGy to fully or partially ablate the recipient's bone marrow and create space for engraftment of donor cells. NOD/SSCID mice received a single dose of 400 cGy. C57BL/6 mice were treated with higher doses (400–1000 cGy) typically fractionated into two doses given 24 hours apart.

Fractional Laser Treatment

Following irradiation, mice were shaved using a motorized electric clipper and carefully inspected to be sure that only hair was shaved without injury to the underlying skin. Once mice were shaved, they were treated with a Sciton Er/YAG fractional laser. Laser parameters varied from 35 to 500 mj at 11–22% density with no coagulation. Depths of laser channels were determined to be between 35–500 microns depending on the settings. An articulated arm was used to deliver the laser beam onto the skin surface. The handpiece was able to create microscopic treatment zones. The dimension of this treatment area was 1 cm^2.

Isolation of Fresh Bone Marrow and Lineage Negative Cells (Lin^-)

Isolation of fresh bone marrow cells by grinding tibiae, femurs, spines and iliac crest from male C57BL/6-Tg (UBC-GFP) mice using a sterile mortar and pestle as previously described [10].

Lin^- cells were separated from whole bone marrow using a mouse cell enrichment kit (Myltenyc Biotechnology, USA). For depletion, cells are magnetically labeled with a cocktail of biotinylated antibodies against a panel of lineage antigens (CD5, CD45R [B220], CD11b, Anti-Gr-1 [Ly- 6G/C], 7–4, and Ter-119 antibodies) and Anti-Biotin MicroBeads that identify differentiated cells. After repeated wash to remove excess antibodies, the cells were incubated with magnetized microbeads that bind and eliminate the antigen-bound antibodies. Unbound lin^- cells were purified by magnetic separation using AutoMacs (Myltenyi Biotec Inc, Auburn CA) as a negative fraction under is depleted program. The lin^- cells were counted in a hematocytometer

Prior to delivery, the bone marrow cells were washed twice and suspended in PBS containing 2% heat inactivated FBS and 5 mM EDTA

Bone Marrow Cell Delivery

One day after the last irradiation the laser treatment was performed on the backs of recipient mice. Immediately after fractional laser treatment, a cylindrical chamber that was cut from a 1.5 ml polypropylene micro centrifuge tube was adhered to the laser treatment site (Figure 1) with Dermabond Topical Skin Adhesive (Ethicon, San Lorenzo, Puerto Rico). Approximately 150 µl of fluid containing 0.75×10^6 lineage negative GFP^+ bone marrow cells were placed within per chamber and the surface of the chamber sealed with tegaderm. Once sealed, the chambers were no longer manipulated. The chambers were noted to stay in

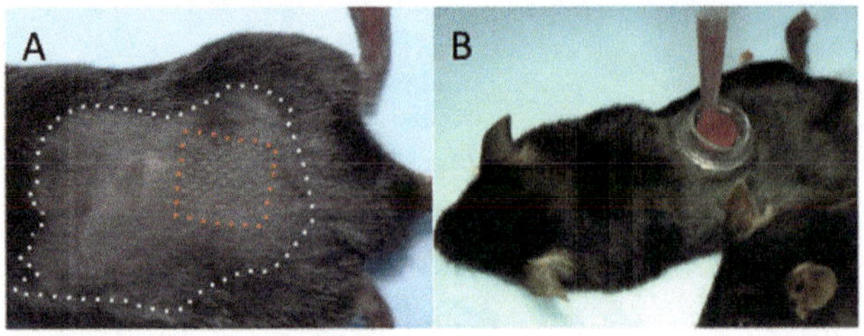

Figure 1. Application of cells following fractional ablative Erb:YAG laser. A) Irradiated C57/BL6 mouse. The white (outer) dashed line highlights the area that was shaved and the red (inner) dashed line highlights the laser treated area. B) Adhered chamber being filled with stem cell suspension.

place for at least 24 hours, after which they lost some adherence to the skin.

Control mice were; *i*) not irradiated and administered cells by laser, *ii*) irradiated and given an equivalent dose of Lin^- bone marrow cells by tail vein injection, *iii*) irradiated and not given bone marrow cells or *iv*) not irradiated and not given bone marrow cells.

Flow Cytometry Analysis

Three to four weeks after delivery of donor cells, blood was collected from the tail vein of treated mice to determine their level of chimerism in circulating blood cells. Heparinized mouse blood (0.3 mL) samples were centrifuged at 1,400 g for 5 min at 4°C, and each blood sample was incubated with 3 mL of NH4Cl, pH 7.2, for 10 min at room temperature to lyse the red blood cells. The remaining leukocytes were washed with phosphate-buffered saline (PBS), pH 7.4. As donor cells and their progeny are natively fluorescent (constitutively expressing GFP) labeling of cells with antibody prior to analysis was not required. Samples were analyzed on a BD LSR II Flow Cytometer (Becton, Dickenson and Company, Franklin Lakes, NJ).

DNA Isolation

DNA was isolated from blood samples using a QIAamp DNA Blood Mini Kit (QIAGEN, Valencia, CA) as per the manufacturers instructions. Liver, spleen and and skin tissues samples were snap frozen in liquid nitrogen and stored at −70°C. Samples were reduced to a fine powder using a stainless steel pulverizer (Biopulverizer; BioSpec Products,Inc., Bartlesville, OK) that had been prechilled in liquid nitrogen.

Pulverized material was immediately transferred to lysis buffer (RNeasy Protect; Qiagen, Valencia, CA) containing mercaptoethanol and further homogenized using a rotor (Pro 200; Pro Scientifc, Oxford, CT) at maximum speed for 1 minute, and stored at −70°C.

Nested-PCR Analysis

DNA was extracted from blood cells and tissue of recipient mice was prepared as described above and analyzed for the presence of donor specific sequences by Polymerase Chain Reaction (PCR). The reaction was carried out in a Bio-Rad C1000 thermal cycler (Bio-Rad, Hercules, CA) using the Platinum Blue PCR supermix protocol (Invitrogen, Carlsbad, CA) according to the manufacture's protocol. The GFP sequence was detected by a nested-PCR procedure using sequence specific primers (GFP primers: fwd: 5′-AAG TTC ATC TGC ACC ACC G- 3′; rev, 5′-TCC TTG AAG AAG ATG GTG CG-3′. Presence of the Y chromosome was demonstrated using a polymerase chain reaction (PCR) assay with Sry3-sequence (forward primer CGTGGTGAGAGGCACAA-GTT; reverse primer ATGGCATGTGGGTTCCTGTC according to primer-blast software NCBI. Similar concentrations of DNA were loaded on each sample tested.

As a control, GAPDH primers were used (FWD, 5′- AACT-TTGGCATTGTGGAAGG-3′; REV, 5′-ACACATTGGGGG-TAGGAACA-3′) with the annealing temperature at 57 °C for 28 cycles. The PCR products were separated in a 2.0% agarose gel and visualized by ethidium bromide staining.

Results

Application of Bone Marrow Cells

After irradiation, mice were treated with fractional Erb:YAG laser to create laser conduits (Figure 1) and chambers were applied to the backs of mice. Once loaded with cells, they chambers were sealed using an occlusive dressing (Tegaderm). All chambers remained intact overnight with most remaining adhered for more than 24 hours. No evidence of wounding related to applying the chambers was noted in any animal.

Generation of EGFP Chimeric NOD/SCID Mice.

Initial experiments were carried out in using female NOD/SCID mice as recipients of GFP labeled bone marrow cells taken from donor male transgenic C57BL/6-Tg (UBC-GFP) mice. Recipient mice were irradiated with a single dose of 200 or 400 cGy gamma radiation. Laser parameters were set, (35 Millijoule [mj] with 11% density and no coagulation) to create channels of approximately 35–50 μ in depth and 7.5×10^5 fresh total donor bone marrow cells were placed over the laser treated area. Control NOD/SCID irradiated and laser treated mice received no donor cells. Three weeks after treatment (post-irradiation) all animals that received 400 cGy irradiation but no donor cells were dead while those receiving 200 cGy irradiation but no donor cells survived. Animals receiving a lower dose of irradiation (200 cGy) and donor cells all survived (3 of 3 treated) but chimerism was low, averaging 5.46% as determined by circulating positive GFP cells (Figure 2). In animals that were treated with the higher dose of irradiation (400 cGy) and donor cells, only half (2 of 4) survived but their chimerism was significantly higher with an average of 29.25%. While these mice had significant levels of chimerism, none survived more than 3 months.

In an effort to improve on these results, we attempted to increase the depth of laser channels by increasing laser settings to 150 mJ with 22% density and no coagulation. These was done with the hope of delivering bone marrow cells deeper to better reach the circulation and have a greater chance for engraftment. We also treated some recipient NOD/SCID mice with donor cells that were depleted of lineage positive cells. Lineage negative (Lin^-) bone marrow cells are enriched for hematopoietic stem cells. All animals were treated with single dose of 400 cGy irradiation as this had yielded the highest chimerism in previous experiments. Irradiated and laser treated mice received 1×10^6 fresh bone marrow cells or 7.5×10^5 Lin^- cells. Despite these attempts, all irradiated recipient NOD/SCID mice receiving bone marrow cells (fresh or Lin^-) did not survive to three weeks. Positive control irradiated NOD/SCID mice receiving bone marrow cells by tail vein injection did however survive.

Generation of EGFP Chimeric C57/BL6 Mice.

We wished to further investigate the effects of radiation dose, an enriched population of hematoietic stem (Lin^-) cells and channel depth on chimerism. This proved difficult in NOD/SCID mice due to their relative fragility, particularly in terms of radiation dosing and tissue injury. We then chose to continue these experiments with syngeneic C57/BL6 mice. We have previously reported on the successful generation of long term chimeric animals (by tail vein injection) using syngeneic C57/BL6 mice [11]. Combined radiation doses (split into two administrations) of greater than 1000 cGy were generally regarded as lethal ablation and did not yield appreciable chimerism using this approach. We therefore set out to perform partial ablation with a single dose of 400 cGy.

We performed laser ablation to C57/BL6 mouse skin at various energy intensities with an initial goal to create channels into the skin at depths that would reach the subcutaneous vascular plexus in order for cells to reach the systemic circulation rapidly. We determined that laser conduits at depths of 350 to 500 microns (settings of 500 mJ, 22% density and no coagulation) would reach this plexus. Animals receiving this dosage however exhibited very little chimerism regardless of whether they were administered fresh bone marrow or lineage negative cells (Figure 3). In reducing the energy settings of the laser to the 50 or 150 mJ settings, there was only marginal improvement in chimerism. Adding an additional dose of (400 cGy) radiation also failed to produce a benefit in higher chimerism.

When we utilized the 35 mJ setting (as in our first experiments with NOD/SCID mice) we were able to achieve appreciable chimerism in mice given a single dose of 400 cGy radiation followed by the administration of lineage negative donor cells. Unlike the immune compromised mice however, these mice remained alive for more than a year and continued to remain chimeric (Figure 3). It was however perplexing that similar mice treated with the 50 mJ setting did not experience higher chimerism. In examining the histology of C57/BL6 mouse skin treated with these settings it did not seem that channel depth was appreciably different in terms of reaching specific tissue landmarks. The primary difference appeared to be in the amount of thermal damage or coagulation at the borders of the laser conduits. It is then possible that administered cells could have been inhibited to reach systemic circulation by a region of denatured collagen (Figure 4).

We next examined the possibility that there may be a greater benefit in attempting to deliver bone marrow cells through larger channels. Chambers containing lineage negative or fresh bone marrow cells were placed over 4 mm fresh full thickness wounds (made using a standard 4 mm punch biopsy tool). All mice had been previously irradiated with 400 cGy as was performed in laser treated animals. No animals survived at 21 days.

Analysis of Cell Distribution

In mice exhibiting low chimerism we attempted to determine if the low numbers of cells observed by FACS analysis was in part due to the loss of GFP expression in circulating donor cells. Quantitative PCR analysis of GFP and Sry DNA extracted from circulating cells demonstrated that this was not likely to be the case. Controls were peripheral blood from GFP and C57BL/6 mice.

To better investigate the fate of the transplanted GFP cells, necropsies were performed on the chimeric mice 2 months after transplantation. Skin spleen and liver were examined for the presence of GFP cells by PCR. All tissue had varying amounts of GFP cells (Figure 5) with the highest signal being found in spleen. This could indicate that donor cells may first traffic to and engraft in the spleen. It also was important for us to determine if cells detected in the marrow had truly engrafted or were just trafficking

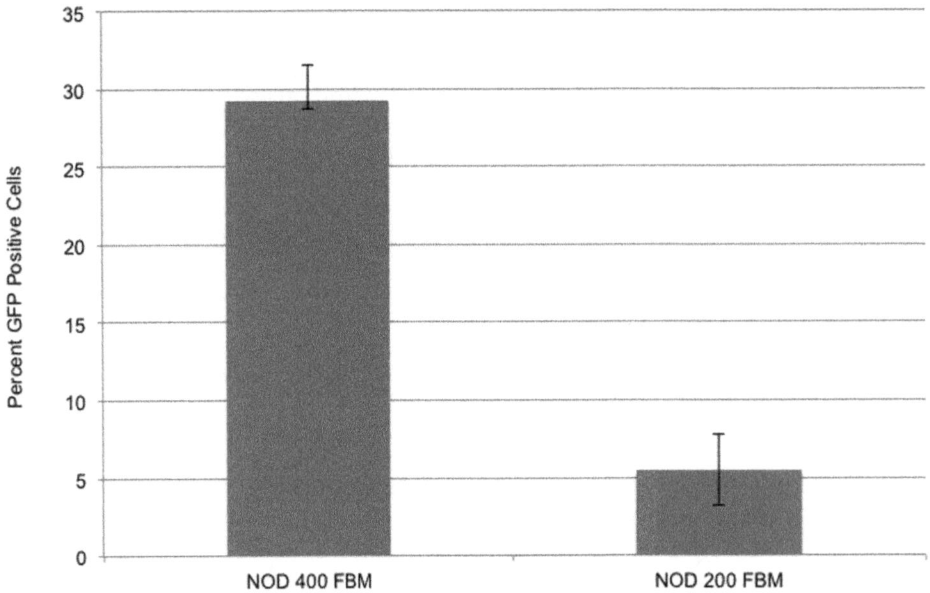

Figure 2. Percent chimerism as determined by the percentage of GFP positive nucleated cells within the blood. Bar graph on left represents the findings from NOD/SCID given 400 cGy gamma irradiation and treated with fractional laser and fresh bone marrow cells. Bar graph on the right represents the findings from NOD/SCID given 200 cGy gamma irradiation and treated with fractional laser and fresh bone marrow cells.

through. We then placed bone marrow from treated animals in culture for 2 weeks to eliminate any transiently trafficking or differentiated cells. PCR analysis of these cultures clearly indicated the presence of engrafted donor cells (Figure 6).

Discussion

In this report we have utilized a bone marrow transplantation model to demonstrate that fractional lasers can be used to assist in the percutaneous delivery of stem cells. The delivered cells were capable of reaching distant sites, remain functional and reconstitute organ function. While it is possible that fractional laser mostly provides only a breach in the barrier function of the skin, there are other points to consider. Efforts to deliver stem cells and create chimeric animals failed when lin^- cells were applied to larger, surgically excised areas created with a standard punch biopsy tool. This failure to achieve chimerism occurred despite the fact that punch biopsy wounds provided a much deeper channel which could hold a substantially greater volume of delivered cells in suspension than the (cumulative) channels created by fractional laser. This provided evidence that ablative fractional lasers might

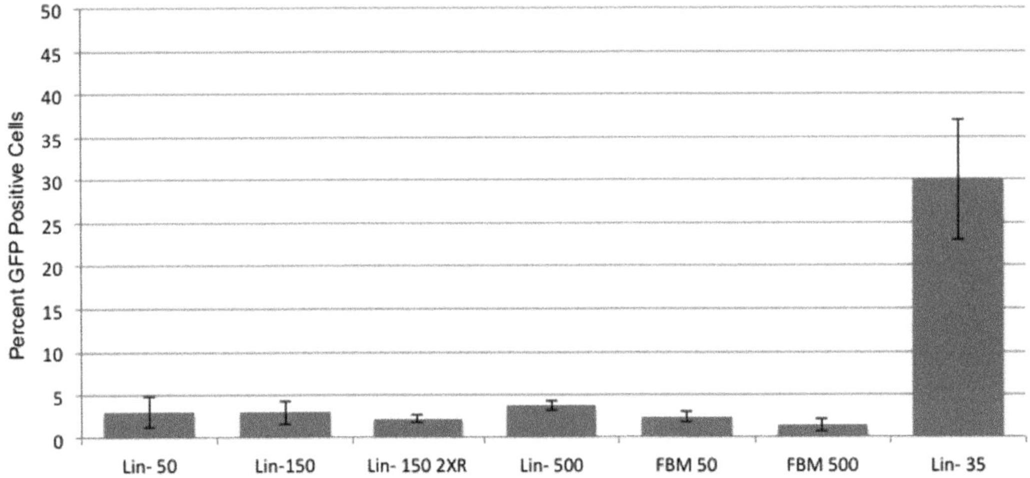

Figure 3. Percent chimerism as determined by the percentage of GFP positive nucleated cells within the blood of C57/BL6 mice gamma irradiated (400 cGy) after fractional laser treatment and administration of bone marrow cells. Lin- 50: Laser set at 50 mJ and Lin^- cells applied. Lin- 150: Laser set at 150 mJ and Lin^- cells applied. Lin- 150 2XR: An additional dose of 400 cGy gamma irradiation given (total of two doses), laser was set at 150 mJ, and Lin^- cells applied. FBM 50: Laser set at 50 mJ and fresh bone marrow cells applied. FBM 500: Laser set at 500 mJ and fresh bone marrow applied. Lin- 35: Laser set at 35 mJ and Lin^- cells applied. Of particular not is that these Lin- 35 treated mice remained alive and chimeric for more than one year.

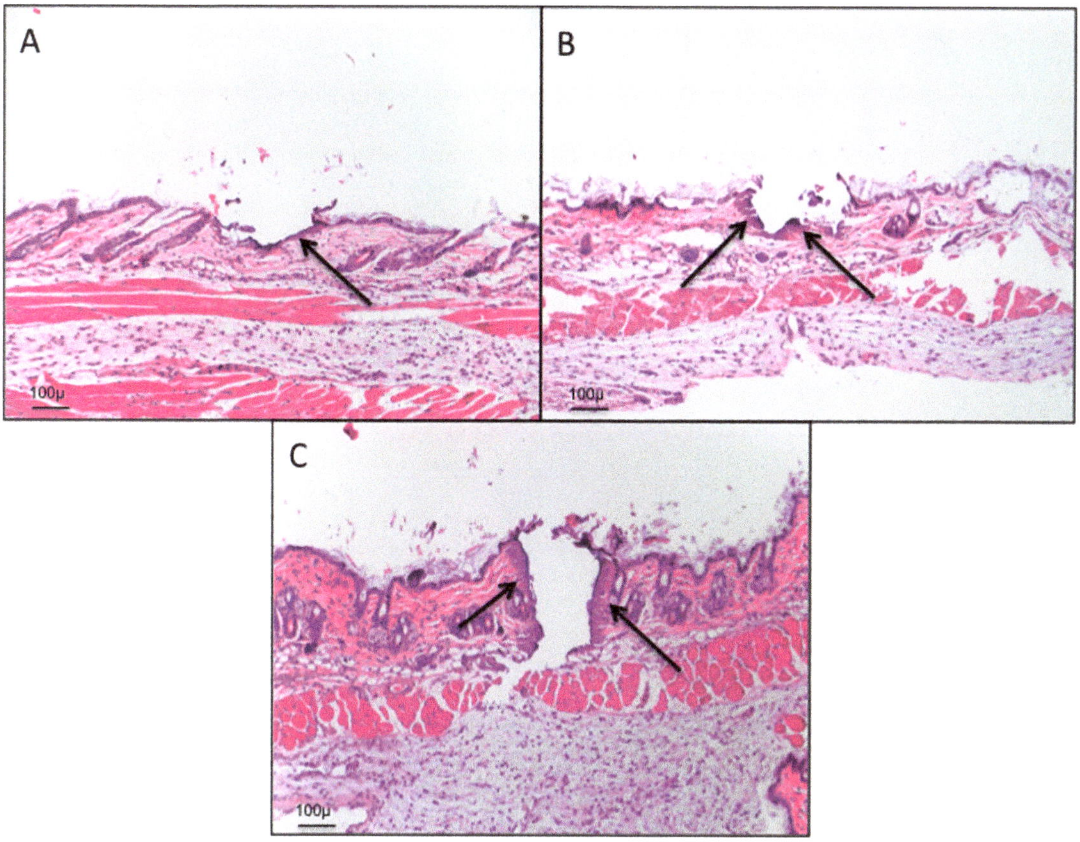

Figure 4. Histologic sections following laser treatment. A) C57/BL6 mouse skin treated with fractional Erb:YAG laser at the 35 μ setting (A), 50 μ setting (B) and 150 μ setting (C). There is increasing amounts of homogenization of the collagen (indicated by the arrows) at the edge of channels with greater applied laser energy and increased channel depth. The areas of collagen homogenization represent thermal damage.

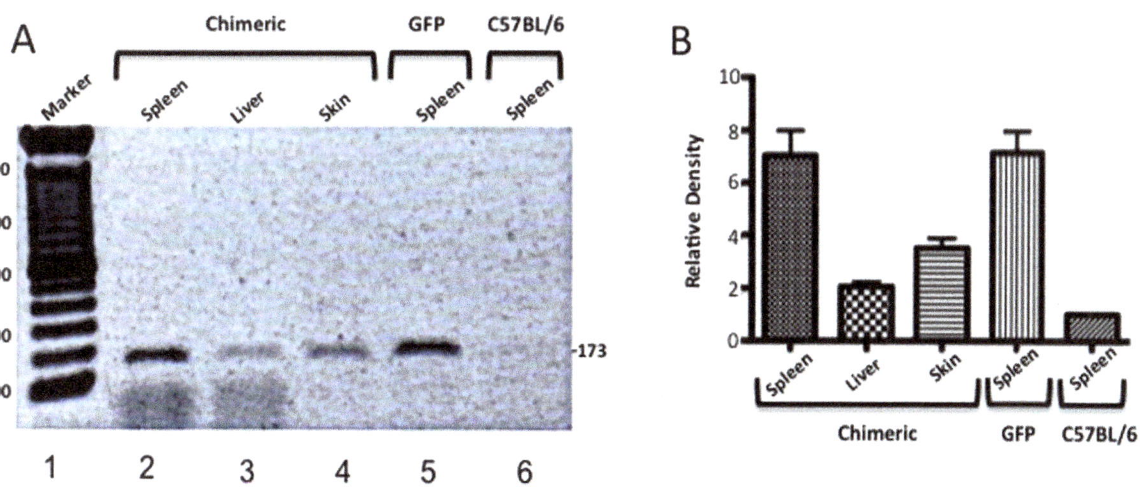

Figure 5. PCR of tissues derived from chimeric and control mice. PCR performed with GFP specific forward (5′-AAG TTC ATC TGC ACC ACC G-3′) and reverse (5′-TCC TTG AAG AAG ATG GTG CG-3′) primers. (A) Representative gel of GFP PCR products. Lane 1: MW Marker. Lanes 2, 3, & 4: PCR products derived from spleen the (Lane 2), liver (Lane 3) and skin (Lane 4) of a mouse administered Lin$^-$ bone marrow cells following fractional laser treatment. This animal exhibited 21.1% blood chimerism. Lane 5: PCR product from GFP transgenic mouse. Lane 6: Absence of PCR product in control C57/BL6 mouse. (B) Bar graph representation of relative expression of several (3 per group) GFP in chimeric (treated with fractional laser and Lin$^-$ bone marrow cells), GFP (transgenic) and control C57/BL6 mice. Chimeric mice exhibited 5–21.1% blood chimerism.

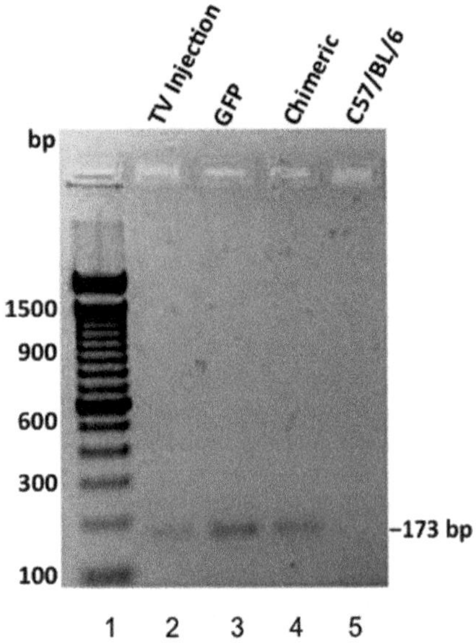

Figure 6. PCR for presence of GFP sequence in bone marrow cells in culture derived from chimeric and control mice. Lane 1: MW Marker. Lanes 3: Control irradiated mouse given Lin$^-$ bone marrow cells by tail vein injection. This mouse had 70% blood chimerism. Lane 3: Transgenic (donor) GFP mouse. Lane 4: Chimeric mouse treated with fractional laser and Lin$^-$ bone marrow cells. This mouse had 39.9% chimerism and survived more than one year. Lane 5: Control C57/BL6 mouse.

have other advantages for cell delivery beyond the ability to breach the barrier function of skin.

Given some of our observations, it is possible that laser created channels have an affinity to recruit particular cells. While the percentage of stem cells capable of reconstituting hematopoiesis is enhanced following lineage depletion of fresh bone marrow, their numbers are still quite rare [12]. Also, the amount of cells that could enter a laser created channel (measuring approximately 100 μ in diameter and 35 μ in depth) by simple diffusion would be expected to be at best restricted. Yet despite these limitations, we were able to create stable chimeras. Conduits created by ablative fractional lasers may then likely have distinct biologic features that can overcome these obstacles.

Treatment with ablative fractional lasers alone can induce a biologic response as has been widely demonstrated in reports where these have been used to reduce scars, photodamage and rhytides [13–15]. The conduits produced by fractional lasers are created by a unique, selective vaporization of tissue without the need for debridement; nor is there significant associated thermal damage. This type of minute tissue ablation thus differs significantly from injuries commonly occurring in nature. Studies of ablative fractional lasers have reported stimulation of tissue remodeling and the release of cytokines that appear to be distinct from other forms of treatment or injury [16,17]. Some of these released cytokines, such as Wnt5a, are active in several pathways involving cell adhesion and migration [16]. These factors are could have played important in recruiting stem cells to generate chimerics in our experiments.

Another surprising finding was that the increasing the depth of laser channels seemed to correlate with a reduced ability to create chimeric animals. This could have been due to the increased amount of thermal damage observed with increasing energies to create deeper channels. The larger area of thermal damage could have created a new barrier that would prevent the ability of cells within the laser channels to reach systemic distribution. Other possibilities include an altered cytokine response associated with the creation of deeper channels and/or administering higher energies to the skin. Further studies beyond the scope of this report will be required to answer these questions and determine the optimal energy parameters for fractional laser assisted delivery of cells and other materials.

The findings presented in this report do however support the supposition that a percutaneous means of cell and drug delivery can be accomplished through small ablative fractional laser channels where the delivered materials do not appear to be limited to materials having restricted size and solubility profiles. In our experiments this could not be accomplished by the direct application of cells after the surgical removal of skin. While the authors are not suggesting that delivery of hematopoietic stem cells via ablative fractional laser be used as a substitute for current bone marrow transplantation techniques, our findings do provide proof of principle that functional stem cell delivery at distant sites can be accomplished by these means. Ablative fractional lasers are however likely to application for the non-invasive delivery of functional cells to treat other local and systemic processes.

Author Contributions

Conceived and designed the experiments: EVB. Performed the experiments: LRM MS JW. Analyzed the data: EVB LRM MS SD JW AS AC. Contributed reagents/materials/analysis tools: EVB SD JW. Wrote the paper: LRM EVB MS SD JW AS AC.

References

1. Badiavas AR, Badiavas EV (2011) Potential benefits of allogeneic bone marrow mesenchymal stem cells for wound healing. Expert Opin Biol Ther 11: 1447–1454.
2. Badiavas EV, Ford D, Liu P, Kouttab N, Morgan J, et al. (2007) Long-term bone marrow culture and its clinical potential in chronic wound healing. Wound Repair Regen 15: 856–865.
3. Fathke C, Wilson L, Hutter J, Kapoor V, Smith A, et al. (2004) Contribution of bone marrow-derived cells to skin: collagen deposition and wound repair. Stem Cells 22: 812–822.
4. Waibel JS, Wulkan AJ, Shumaker PR (2013) Treatment of hypertrophic scars using laser and laser assisted corticosteroid delivery. Lasers Surg Med 45: 135–140.
5. Lee WR, Shen SC, Al-Suwayeh SA, Yang HH, Yuan CY, et al. (2011) Laser-assisted topical drug delivery by using a low-fluence fractional laser: imiquimod and macromolecules. J Control Release 153: 240–248.
6. Haak CS, Bhayana B, Farinelli WA, Anderson RR, Haedersdal M (2012) The impact of treatment density and molecular weight for fractional laser-assisted drug delivery. J Control Release 163: 335–341.
7. Zelickson BD, Walgrave SE, Al-Arashi MY, Altshuler GB, Yaroslavsky IV, et al. (2009) Semi-Automated method of analysis of horizontal histological sections of skin for objective evaluation of fractional devices. Lasers Surg Med 41: 634–642.
8. Laubach HJ, Tannous Z, Anderson RR, Manstein D (2006) Skin responses to fractional photothermolysis. Lasers Surg Med 38: 142–149.
9. Genina EA, Bashkatov AN, Dolotov LE, Maslyakova GN, Kochubey VI, et al. (2013) Transcutaneous delivery of micro- and nanoparticles with laser microporation. J Biomed Opt 18: 111406.
10. Quesenberry PJ, Dooner GJ, Tatto MD, Colvin GA, Johnson K, et al. (2010) Expression of cell cycle-related genes with cytokine-induced cell cycle progression of primitive hematopoietic stem cells. Stem Cells Dev 19: 453–460.
11. Badiavas EV, Abedi M, Butmarc J, Falanga V, Quesenberry P (2003) Participation of bone marrow derived cells in cutaneous wound healing. J Cell Physiol 196: 245–250.
12. Wognum AW, Eaves AC, Thomas TE (2003) Identification and isolation of hematopoietic stem cells. Arch Med Res 34: 461–475.
13. Hunzeker CM, Weiss ET, Geronemus RG (2009) Fractionated CO2 laser resurfacing: our experience with more than 2000 treatments. Aesthet Surg J 29: 317–322.

14. Waibel J, Beer K (2009) Ablative fractional laser resurfacing for the treatment of a third-degree burn. J Drugs Dermatol 8: 294–297.
15. Waibel J, Beer K (2008) Fractional laser resurfacing for thermal burns. J Drugs Dermatol 7: 59–61.
16. Kim JE, Won CH, Bak H, Kositratna G, Manstein D, et al. (2013) Gene profiling analysis of the early effects of ablative fractional carbon dioxide laser treatment on human skin. Dermatol Surg 39: 1033–1043.
17. Reilly MJ, Cohen M, Hokugo A, Keller GS (2010) Molecular effects of fractional carbon dioxide laser resurfacing on photodamaged human skin. Arch Facial Plast Surg 12: 321–325.

8

Stem Cell Therapy and Curcumin Synergistically Enhance Recovery from Spinal Cord Injury

D. Ryan Ormond[1], Craig Shannon[1], Julius Oppenheim[1], Richard Zeman[2], Kaushik Das[1], Raj Murali[1], Meena Jhanwar-Uniyal[1]*

1 Department of Neurosurgery, New York Medical College, Valhalla, New York, United States of America, **2** Department of Cell Biology and Anatomy, New York Medical College, Valhalla, New York, United States of America

Abstract

Acute traumatic spinal cord injury (SCI) is marked by the enhanced production of local cytokines and pro-inflammatory substances that induce gliosis and prevent reinnervation. The transplantation of stem cells is a promising treatment strategy for SCI. In order to facilitate functional recovery, we employed stem cell therapy alone or in combination with curcumin, a naturally-occurring anti-inflammatory component of turmeric (*Curcuma longa*), which potently inhibits NF-κB. Spinal cord contusion following laminectomy (T9–10) was performed using a weight drop apparatus (10 g over a 12.5 or 25 mm distance, representing moderate or severe SCI, respectively) in Sprague-Dawley rats. Neural stem cells (NSC) were isolated from subventricular zone (SVZ) and transplanted at the site of injury with or without curcumin treatment. Functional recovery was assessed by BBB score and body weight gain measured up to 6 weeks following SCI. At the conclusion of the study, the mass of soleus muscle was correlated with BBB score and body weight. Stem cell therapy improved recovery from moderate SCI, however, it had a limited effect on recovery after severe SCI. Curcumin stimulated NSC proliferation *in vitro*, and in combination with stem cell therapy, induced profound recovery from severe SCI as evidenced by improved functional locomotor recovery, increased body weight, and soleus muscle mass. These findings demonstrate that curcumin in conjunction with stem cell therapy synergistically improves recovery from severe SCI. Furthermore, our results indicate that the effect of curcumin extends beyond its known anti-inflammatory properties to the regulation of stem cell proliferation.

Editor: Cesar V. Borlongan, University of South Florida, United States of America

Funding: This work was funded from the American Association of Neurological Surgeons NREF/Medtronic Research Fellowship grant. The funders had no role in study design, data collection and analysis, decision to publish, or preparation of the manuscript.

Competing Interests: The authors have declared that no competing interests exist.

* E-mail: meena_jhanwar@nymc.edu

Introduction

Acute traumatic spinal cord injury (SCI) occurs at an estimated annual incidence of 15–40 cases per million [1]. The prevalence of SCI is estimated at 1.275 million people (0.4% of the population) in the United States alone [2]. Much of the morbidity associated with SCI occurs due to the limited intrinsic ability of the spinal cord to recover following transection or contusion. The pathophysiology of SCI is considered biphasic in nature. Primary injury results from mechanical force injuring the spinal cord, tearing axons, blood vessels, and causing cell membrane disruption. Secondary injury occurs via the subsequent edema, ischemia, inflammation, cytokine production, free radical damage, glial scar formation, apoptosis and necrosis that ensue. Perilesional astrocytes alter their phenotype and over the weeks following SCI form the glial scar composed of astrocytes and a dense extracellular matrix [3]. Therapeutic modalities that promote recovery could be better understood by elucidating the pathophysiology of SCI. Necrosis, apoptosis, and inflammation play critical roles in defining neuronal cell death and formation of the glial scar implicated in limiting SCI recovery. In addition, secreted inflammatory cytokines and growth factors generally up-regulate pro-survival molecules, such as NF-κB [4]. Thus, the major hurdles in complete recovery from SCI are mostly due to the upregulation of certain inflammatory molecules after injury that result in gliosis. Curcumin (diferuloylmethane), the active ingredient in turmeric (*Curcuma longa*), is a highly pleiotropic molecule with potent anti-inflammatory properties. The pleiotropic activities attributed to curcumin because of its chemical structure appear to influence multiple signaling pathways activated in SCI [5,6,7]. Some of these signaling pathways include phospholipase, lipooxygenase, cyclooxygenase 2, leukotrienes, thromboxane, prostaglandins, nitric oxide, collagenase, elastase, hyaluronidase, monocyte chemoattractant protein-1 (MCP-1), interferon-inducible protein, tumor necrosis factor (TNF), interleukin-12 (IL-12), and NF-κB [6,7,8]. It has also been shown to reduce the inflammatory cascade and may help prevent plaque deposition and protein oxidation in neurodegenerative conditions [6]. Curcumin's benefits in SCI have been recently reported in the literature, and this benefit may be due to a reduction in secondary injury to the spinal cord [9–16]. Furthermore, a study using curcumin with other dietary compounds showed enhancement of sensorimotor function, especially after exercise [15]. This suggests that curcumin added to the diet has a beneficial effect on recovery. In addition, the molecular mechanism underlying curcumin as a dietary supplement with Omega-3 fatty acids (DHA) has suggested that secondary injury can be prevented with such a treatment [16].

Stem cell therapies offer a promising way to replace neurons and oligodendrocytes lost to necrosis or apoptosis, help remyelinate axons, and increase the availability of neurotrophic factors to aid in local recovery [17–26]. Furthermore, stem cell therapy provides support to the surrounding milieu after injury to enhance axonal regeneration and/or neuroprotection. Transplantation of *in vitro* expanded NSC into rat spinal cord after contusion injury induced their differentiation into neurons or oligodendrocytes, and promoted functional recovery of limb movement [23,26]. Also, restoration of neurotrophic factors seen in stem cells may aid in supporting the surrounding milieu after injury in order to enhance axonal regeneration and neuroprotection [27]. Stem cell therapy has been explored in animal models as an aid to recovery following SCI. Curcumin, a potent anti-inflammatory compound, has been shown to improve functional recovery from SCI [5–14]. However the use of curcumin with stem cell therapy has not been well defined. The purpose of this study is to investigate stem cell therapy with or without curcumin in treatment of moderate as well as severe SCI. Herein we provide evidence that the combined use of stem cell therapy with curcumin aid in recovery after thoracic SCI in an animal model.

Materials and Methods

Animals

Animal work was performed with the approval of the New York Medical College IACUC under approval number: 137-1-1210.1. Adult female Sprague-Dawley rats (200–250 grams; Charles River, Troy, NY) were maintained on a routine 12:12 hour light/dark cycle with *ad libitum* access to food and water. The main purpose of using female rats in this study was that one of the common sequelae of SCI is neurogenic bladder. This results in overflow incontinence and loss of voluntary bladder control. Bladder distention and subsequent rupture is more common, and has a higher mortality rate, in males.

Neural Stem Cells

NSC were extracted from the subventricular zone (SVZ) of female Sprague-Dawley rat brains (Fig. 1A). Brain was removed following craniectomy from a euthanized animal and was placed in dissecting media. Brain was sectioned in multiple coronal sections approximately 3–4 microns thick rostral to the optic chiasm. The SVZ area was microdissected using the micropunch technique described previously [28] and areas micropunched as shown in Fig. 1A. The microdissected specimens were then digested using an enzyme digestion solution for 1–3 hrs at 37°C in a water bath. Following enzymatic digestion, the cells were resuspended in a specialized serum-free stem cell media containing growth factors such as epidermal growth factor, fibroblast growth factor, etc., and plated onto 12-well plates. The cells were incubated at 37°C in a 5% CO_2 chamber. Neurospheres formed in 5–6 days and the expansion was performed as previously described [29]. Neurospheres were stained for the stem cell marker nanog for authenticity, as shown in Fig. 1C.

Effect of Curcumin on Stem Cell Proliferation (*in Vitro*)

NSC were grown in stem cell media in the presence or absence of curcumin (500 nM or 1 uM). DMSO treated NSCs were used as controls. After 24 hours, neurosphere proliferation was determined by measuring the diameter of each neurosphere.

Spinal Cord Injury

Surgery was performed as described previously [14,30–32]. Briefly, aseptic T9–T10 laminectomy under anesthesia (pentobarbital sodium; dose 50–80 mg/kg IP; Hospira, Lake Forest, IL) was performed. Spinal cord contusion was induced using a weight-drop apparatus, where a guided 10 g rod was dropped 12.5 or 25 mm onto the exposed dura mater, representing moderate or severe SCI, respectively. Following contusion, the incision was closed with wound clips. The rats received 0.02% amoxicillin in the drinking water to prevent infection throughout the duration of the experiment. Body temperature of 37°C during anesthesia was maintained by a temperature-controlled heating pad and recorded with a rectal thermometer. During the recovery, urine was expressed manually twice daily to assist in urination until intrinsic function returned to normal.

Stem Cell Transplant

Rats in the stem cell treatment group underwent repeat surgery one week after injury. Under pentobarbital anesthesia (as described above) and sterile conditions, the original incision was reopened and the dura overlying the site of contusion was exposed. 1×10^6 neural stem cells in 20 ul of sterile PBS were then microinjected directly into the injured spinal cord. Stem cell injections were performed manually over ten minutes using a 30-guage needle. Injections were performed 1 mm into the spinal cord at the site of contusion injury. The incision was then closed in layers.

Treatments

Rats were randomized into the 3 (moderate SCI) or 4 (Severe SCI) treatment groups as shown in Table 1. The schemas of treatments/protocols for moderate or severe SCI used in this study are described in Fig. 1A.

Moderate SCI. Within 30 minutes following the contusion, rats were given either curcumin (60 mg/ml/kg BW in DMSO), or DMSO (1 ml/kg BW) via percutaneous, intramuscular injection near the site of injury. These injections were continued weekly for six weeks. Stem cell transplantation occurred one week following injury via direct injection into the spinal cord at the site of injury as described above. Rats were given either NSC or combinational treatment (NSC+curcumin) (Fig. 1A). Rats were given curcumin alone following this same protocol in a study previously reported by our laboratory [14].

Severe SCI. Within 30 minutes following the contusion, rats were given either curcumin (60 mg/ml/kg BW in DMSO), or DMSO (1 ml/kg BW) via percutaneous-intramuscular injection at the site of injury. Rats were given NSC, curcumin, or combinational treatment (NSC+curcumin). Rats given curcumin monotherapy continued to receive weekly injections at 60 mg/ml/kg BW in DMSO. Rats given stem cell transplant, following the initial dose of DMSO or high dose curcumin (60 mg/ml/kg in DMSO), then received daily IP injections of low dose curcumin (0.5 mmol/ml/kg BW), or an equivalent dose of DMSO alone, for two weeks following stem cell transplant. Stem cell transplant occurred one week following SCI.

Basso, Beattie and Bresnahan Score

Functional tests were performed before the injury and then weekly after the treatment. Locomotor activity was evaluated using the Basso, Beattie and Bresnahan (BBB) scores for hindlimb rat motor function and was recorded weekly [33]. Testing involved placing the rat on a large open field where a natural aversion to open spaces forced the rat to move towards the edges of the field.

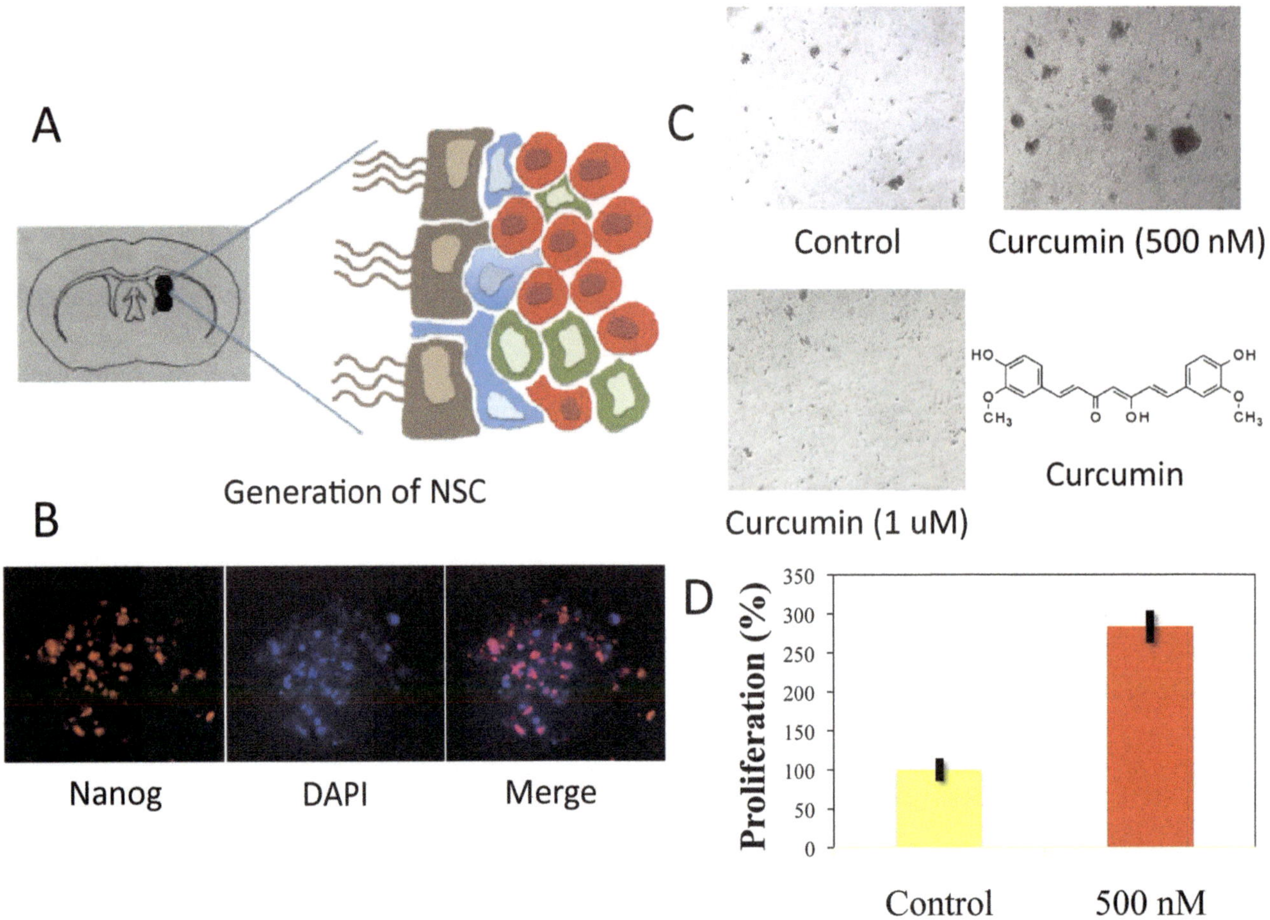

Figure 1. Curcumin enhances stem cell proliferation of NSC derived from SVZ in low doses, however it induces apoptosis at high doses. A: Schematic presentation of treatments timeline used in moderate and severe SCI. **B:** Micropunched area of the subventricular zone (SVZ), depicting the presence of ependymal cells (brown), astrocytes (blue), transit amplifying cells (green) and neuroblasts (red). **C:** Immunofluorescence analysis demonstrates the presence of stem cell marker nanog in neurospheres and DAPI staining for DNA. **D, E:** Effect of curcumin on neurosphere formation. At a low concentration of 500 nM, curcumin treatment stimulated proliferation of NSC in 24 hr. At higher concentrations (1 uM), curcumin caused autolysis and neurosphere fragmentation. The chemical structure of curcumin is also displayed.

Two independent blinded examiners observed hind-limb movements and assessed the rats' locomotor function [33]. Rats that demonstrated any functional recovery of limb movement within 72 hours of contusion injury were excluded from the study, allowing a more uniform cohort [33]. Rats falling more than two standard deviations above or below the mean were excluded from final statistical review. Data used for analysis included both left and right limb movements, which were recorded separately, but averaged in the final analysis. Body weight was also recorded weekly.

Table 1. Experimental cohorts: number of animals per cohort.

SCI severity	Control (DMSO)	Curcumin	NSC	NSC+Curcumin
Moderate	6	*	7	6
Severe	12	10	9	13

*Published [14].

Soleus Muscle Dissection

After 6 weeks, rats were sacrificed under pentobarbital anesthesia (100 mg/kg IP), and were then perfused with formalin prior to dissection. Perfusion fixation occurred in the following manner: a 5 cm incision rostral to caudal down the sternum was made followed by a 3 cm incision medial to lateral in order to cut the diaphragm. Afterwards, a needle attached to a vacuum-pump system was inserted through the apex of the left ventricle into the ascending aorta whereby formalin fixative was perfused into the rodent. Following fixation, a 1 cm incision was made in both posterior hindlimbs to retrieve soleus muscles. First, the Achilles' tendon was dissected out, and then the gastrocnemius muscle was elevated to expose the soleus muscle. This was identified, removed *en bloc*, and the weight of each soleus muscle was separately recorded, and averaged for final analysis.

Histopathological Analysis

Spinal cords at the region of injury were fixed in formalin and placed in paraffin blocks. Six µm axial sections were cut and placed onto the slides. The sections were then stained with Lox Blue staining. Spinal cord atrophy was evaluated by determining the epicenter using an Axiovert Zeiss microscope (Carl Zeiss

Microscopy, Thornwood, New York) with the AxioVision software. The spinal cord photographs were then measured utilizing ImageJ software to determine the spared tissue area as well as total spinal cord area. The peak contusion site (epicenter) was recorded as the area with the least amount of spared spinal cord tissue. Five different measurements were taken for each rat, which included one for the epicenter and two from either side of the epicenter (cranial and caudal). Total spinal cord values were also recorded for each section, and percentage of spared area in relation to total area was used for final analysis. The presented values are an average for each rat ± SEM (standard error of the mean).

Statistical Analysis

Data are presented as a mean ±SEM. BBB scores and body weight were analyzed by two-way analysis of variance (2-ANOVA) between groups over time followed by two-tailed t-test. Soleus muscle was analyzed using the two-tailed t-test. A correlation between the BBB scores and body weight and between BBB score and soleus muscle weight was established using the Pearson method. Statistical significance was determined at $p \leq 0.05$. For histopathological analysis, the statistical analysis was done using the two-tail T-test between DMSO treated versus other treatments individually.

Results

The Effect of Curcumin on NSC Proliferation

The effect of curcumin on NSC proliferation was examined in a cell culture model. NSC, as represented by neurospheres and staining positive with nanog, were grown in stem cell media in the presence or absence of curcumin. At a low concentration, curcumin (500 nM) caused a significant increase in neurosphere proliferation by over 180% in comparison to control. However, curcumin at higher doses (1 uM) caused widespread apoptosis and neurosphere fragmentation (Fig. 1 D, E).

BBB Score after Moderate SCI

Animals undergoing NSC transplant, or with combined treatment using NSC+curcumin, were tested after moderate SCI. Rats treated with NSC transplant demonstrated a significant improvement starting at week three. A significant difference in BBB score between cohorts was seen with the scores of 12.14±0.65 and 9.33±0.99 ($p = 0.028$) at week five and 12.86±0.93 and 9.67±1.02($p = 0.029$) at week six, for NSC treated and control rats, respectively. Combinational treatment with high dose curcumin and NSC transplant significantly improved the BBB score as compared to control. However, between combinational treatment and NSC alone, there was no noticeable difference in improvement based on BBB score (Fig. 2A). Curcumin monotherapy showed significant improvement in recovery from SCI in comparison to control, as we have previously reported [14].

BBB Score after Severe SCI

Curcumin and NSC treatment demonstrated similar results, in that they both initially demonstrated significant improvement at week two, but this significance was lost through the remainder of the experiment in comparison to controls (Fig. 2B). Rats undergoing combined NSC transplantation and curcumin showed consistent significant improvement in BBB score as compared to controls. This improvement began at week one, where BBB scores were 1.15±0.33 vs. 0.33±0.19 for combinational treatment and control, respectively ($p = 0.037$), and persisted through week six. While control rats continued to show some improvement over time, the combined NSC/curcumin cohort displayed more rapid and elevated BBB scores in comparison to controls. In fact, by week six, animals continued this significant difference with BBB scores of 8.81±0.82 and 4.75±0.78, for combinational treatment and control, respectively ($p = 0.0001$). Statistical analysis of combined NSC/curcumin treatment also demonstrated significant improvement in comparison to NSC or curcumin treatment alone.

Body Weight Gain after Moderate SCI

Weekly body weight is presented as a percent change from initial body weight recorded immediately prior to SCI. Many animals initially lost weight after SCI, but then began to increase in weight in a manner that often correlated with recovery. Body weight gain became statistically significant in NSC treated rats at week three and remained significant through week 5 (Fig. 3A). Combined treatment of NSC and curcumin in high doses after moderate SCI showed no statistically significant benefit in body weight gain in comparison to control. Correlation analysis once weight gain resumed (starting week two) demonstrated a linear relationship between BBB score and body weight after moderate SCI (Fig. 3B), suggesting a strong correlation between body weight gain and BBB score.

Body Weight Gain after Severe SCI

Weekly body weight changes are presented in Figure 3C, where the initial values are set equal to 100, as recorded at the time of contusion. All rats showed a decrease in body weight at week one, as expected, immediately following injury. Combinational treatment showed significance in comparison to control by week five ($p = 0.029$), curcumin alone showed a trend toward significance ($p = 0.20$), and NSC treatment showed no significance ($p = 0.55$). Correlation analysis demonstrated a linear relationship between body weight gain after week one in comparison to BBB score (Fig. 4D).

Soleus Muscle Mass after Moderate SCI

Soleus muscle weight (mg) was taken at the conclusion of the experiment. Values are reported as a percentage of the control = 100 (see Fig. 4). Rats treated after moderate SCI were not significantly different across cohorts (Fig. 4A). However, NSC treated rats showed higher soleus muscle weight in comparison to control or combined treatment. Specifically, rats had a soleus muscle weight of 100±6.01, 107.12±7.95, and 101.58±13.52 for control, NSC, and combined groups, respectively ($p = 0.48$ and $p = 0.92$ for NSC and combinational groups, respectively). Soleus muscle weight correlated with BBB score, implying that higher soleus muscle weight was associated with higher BBB score (Fig. 4B).

Soleus Muscle Mass after Severe SCI

After severe SCI, control animals had a soleus muscle weight of 100±7.09. Soleus muscle weight was higher in all groups in comparison to control, but only reached significance in the combined treatment cohort. Curcumin had a soleus muscle weight of 105.3±9.46. NSC transplant animals had a soleus weight of 115.54±8.51. Combinational therapy produced a soleus weight of 125.86±5.09. Importantly, soleus muscle weight was significantly better in combined treatment (NSC+curcumin) in comparison to control or in comparison to curcumin or NSC cohorts ($p = 0.005$, 0.03, and 0.046, respectively) (Fig. 4C). Correlation analysis demonstrated a linear relationship between BBB score and soleus

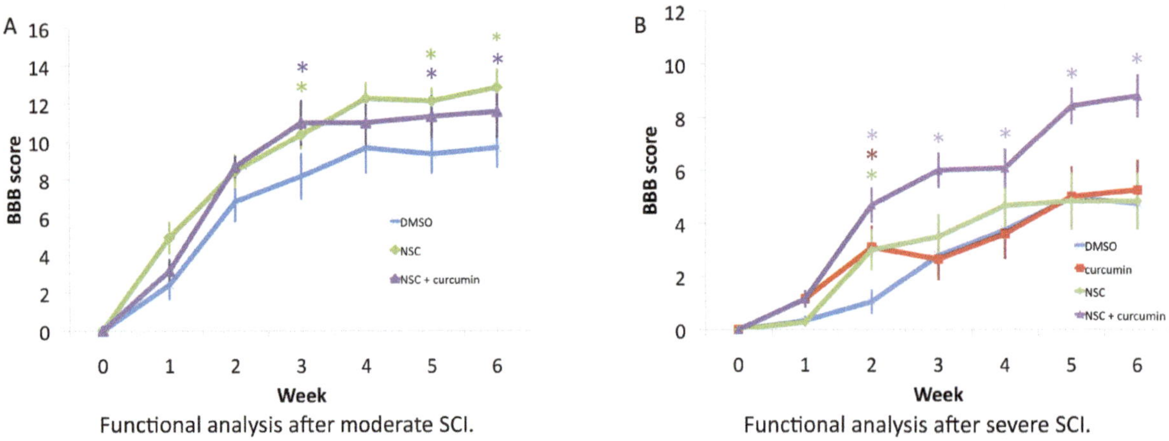

Figure 2. Stem cell therapy in conjunction with low-dose curcumin had a synergistic effect in recovery from severe SCI, as demonstrated by improved BBB score. A: Functional analysis (BBB) after moderate SCI showed that stem cell therapy with a high dose of curcumin rendered BBB scores, which were statistically significant as compared to control rats (DMSO), starting from week 3. **B:** In severe SCI, rats undergoing combined NSC transplantation and curcumin showed consistent improvement in BBB score as compared to DMSO treated controls. Statistical analysis of combined NSC+curcumin treatment also demonstrated significant improvement in comparison to NSC or curcumin treatment alone.

weight (Fig. 4D), demonstrating that the groups showing the most improved BBB score had the greatest soleus muscle mass.

Histopathological Analysis (Severe SCI)

Spared tissue in relation to the total area of NSC+Curcumin-, Curcumin-, and NSC-treated rats were presented along with control rats treated with DMSO (Fig. 5A). The rats treated with NSC+Curcumin and NSC monotherapy-treated rats showed better recovery as shown by the spared as compared to the total area.

The statistical analysis using the One Way ANOVA and post test analysis demonstrated no significant difference between the groups ($F(3,16) = 1.19$; NS), however, it should be noted that curcumin treated rats had the least spared area of the total area as shown in Fig. 5B.

Discussion

The data presented here demonstrate that NSC transplantation with curcumin synergistically enhanced recovery from severe SCI in a rat model. Recovery from SCI was demonstrated by improvement in parameters including BBB score, body weight, and soleus muscle weight. Stem cell therapies have been shown to be a promising way to regenerate neurons and oligodendrocytes lost during SCI, and have demonstrated efficacy in multiple animal studies [21–23,26,27]. This is an emerging field in spinal cord injury research, with several human studies currently being conducted worldwide [34]. Our findings differentiate the beneficial effects of NSC transplant by severity of injury. Although NSC treatment alone was sufficient to induce statistically significant recovery after moderate SCI, it could not induce significant recovery after severe injury except in combination with curcumin.

The rationale behind combinational therapy using stem cells and curcumin originates from the idea that curcumin, an anti-inflammatory compound, could reduce gliosis, while stem cells aid in reinervation by increasing the availability of neurotrophic factors at the site of injury. In addition, stem cells may also contribute oligodendrocytes, glia, and neurons to aid in recovery [21,22,23,26,27]. We have previously demonstrated the benefit of curcumin in moderate SCI where curcumin-treated rats showed marked improvement in functional recovery as early as the first week [14]. Consistently, this study demonstrated that after severe SCI there also was an initial benefit using curcumin. However, this effect was unsustained beyond week two (Fig. 3). Interestingly, combined stem cell treatment with low dose curcumin (0.5 mM/IP daily) significantly improved functional recovery starting from week 2, resulting in many animals recovering to weight bearing status even after severe SCI (Fig. 2B). It should be noted here that this effect of curcumin with stem cell transplant might not be attributable to its anti-inflammatory effect alone. A recent study demonstrated that low dose curcumin treatment increased the number of BrdU-labeled cells in mouse hippocampus, suggesting an enhanced proliferative effect *in vitro*, as well as *in vivo* [35]. Consistent with these findings, we observed curcumin in low doses induced stem cell proliferation *in vitro* (Fig. 1D). Curcumin, however, in higher doses (1 uM) caused apoptosis of stem cells. This proliferative effect may be one reason there is a synergistic effect of combinational treatment on SCI recovery. The lack of benefit of high dose curcumin in combination with stem cell transplant in the moderate injury cohort may also be explained by these findings, where the NSC+high dose curcumin cohort recovered less well than NSC treatment alone (Fig. 3A). Thus, it is possible that curcumin may not only aid the local cellular microenvironment, but also cause proliferation of transplanted stem cells in low doses.

Body weight and soleus muscle weight were used as surrogate markers of recovery following moderate or severe SCI. There was a decrease in body weight during the first week following SCI, however an increase in body weight began starting the second week in a manner that often correlated with functional recovery [14,30–33]. Combinational treatment with curcumin in high doses also demonstrated similar results after moderate injury, and did not significantly differ from controls. Combinational treatment improved body weight gain in a statistically significant manner following severe SCI. These data all correlate well with BBB score findings. Soleus muscle weight has also been used in previous studies as a surrogate marker of recovery of hindlimb motor function [14,30–33]. Following SCI, muscle atrophy is more pronounced in paralyzed muscles that contain a large proportion of slow fatigue-resistant muscle fibers, such as the soleus muscle,

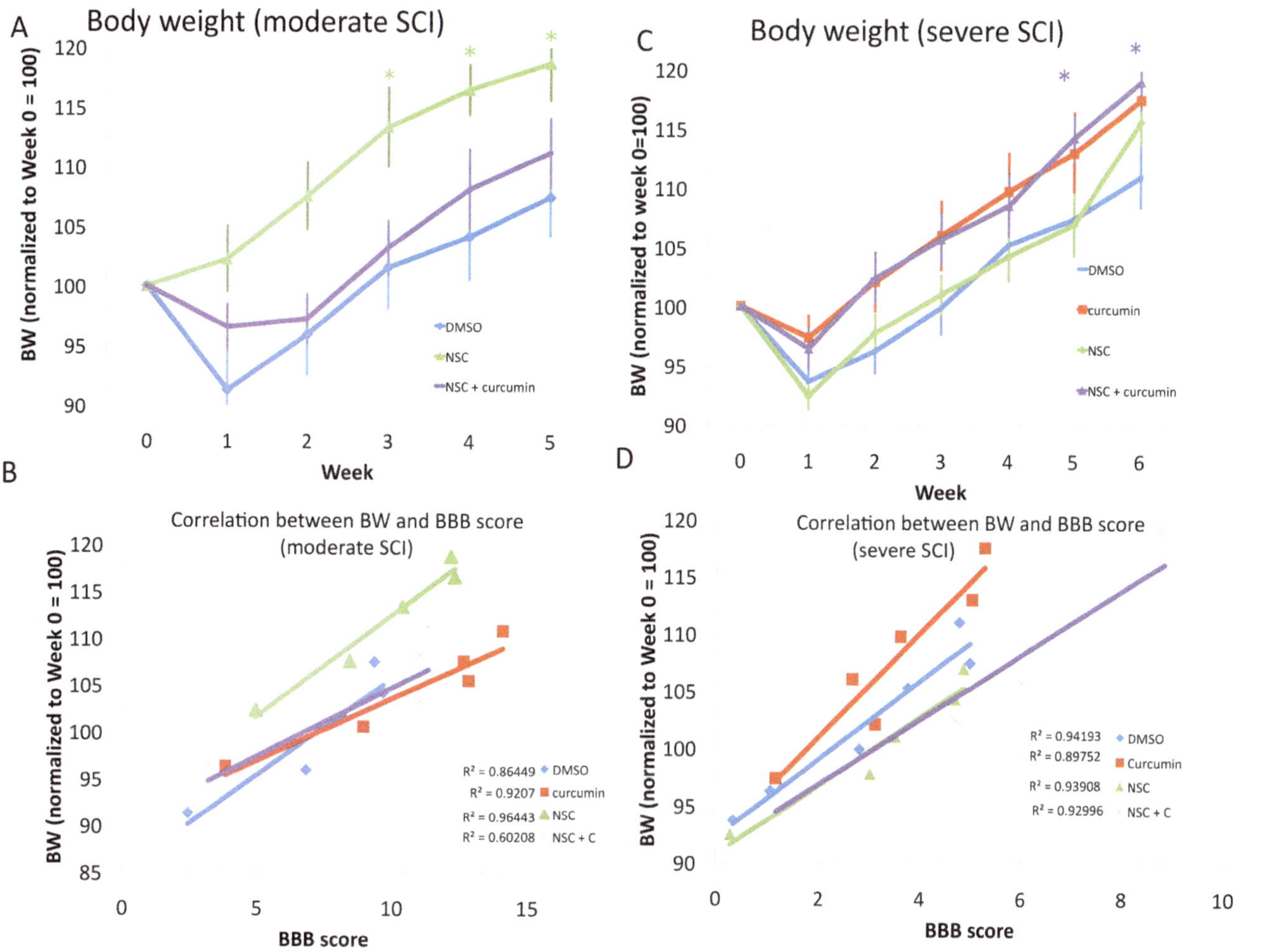

Figure 3. Stem cell therapy in conjunction with curcumin synergistically improves recovery from moderate or severe SCI, as demonstrated by improved body weight gain, which correlated with BBB score. A: Stem cell transplanted animals showed improved body weight gain in comparison to combinational treatment or DMSO treated controls. **B:** A positive correlation exists between body weight gain and improved BBB score in moderate SCI. R^2 values are listed next to their respective cohort. **C:** Combination of NSC therapy with curcumin treatment showed significant improvement in body weight gain in comparison to DMSO control. Curcumin treated rats showed a trend toward significance in BW gain. **D:** Correlation of BBB scores with body weight after severe SCI. R^2 values are listed next to their respective cohort.

that are largely responsible for maintaining posture and bearing weight [36]. Therefore, soleus muscle atrophies more quickly and recovers more slowly than other hindlimb muscles [37]. It also is anatomically simple to isolate and remove completely for assessing its mass [14,30–33,46]. Additionally, functional recovery following SCI, frequently measured by the BBB score, has been positively correlated to soleus muscle changes including improvement in muscle mass, MRI T2 changes, electrophysiological testing parameters, and a change in the proportion of slow Type 1 fibers [14,30–33,36–40]. This implies that enhanced locomotor activity leads to an improvement in muscle mass. In our study, soleus muscle mass also demonstrated improvement in stem cell transplanted animals after severe SCI in a statistically significant manner (Fig. 5). Correlation data also demonstrated significant correlation between BBB score and soleus muscle mass or body weight, adding credibility to the findings of significant synergistic improvement after severe SCI (Fig. 4–5).

Histopathology data somewhat correlated with functional outcomes. Combined treatment, with the best functional outcome, also had the largest spared tissue. NSC transplant monotherapy also fared better than control, although functional outcomes were not significantly different. Of note, curcumin monotherapy had the least spared tissue among the severe SCI cohorts, but functionally showed similar outcomes to control and NSC transplant alone. It should be emphasized that multiple previous studies have found no relation between functional outcome and the amount of spared tissue at the site of injury [42–44]. In general, these data serve to corroborate findings of BBB score, body weight, and soleus muscle weight, supporting the conclusion that improved functional outcomes have a physiologic correlation.

Various types of stem cells have been used for transplantation after SCI. These cell types most commonly include embryonic and adult NSC, fate-restricted neural-glial precursor cells, olfactory ensheathing glial cells, Schwann cells, mesenchymal stem cells, and bone-marrow stromal cells [20]. Adult NSC have been previously characterized in mammalian brain, and are abundant in the SVZ and dentate gyrus [27,45–48]. Endogenous NSC have even been found to exist in the spinal cord, although oligodendrocyte differentiation of these cells has not proven sufficient to promote remyelination after SCI [21–23,47]. This remains true even following infusion of exogenous growth factors [49–50]. More recently, however, studies have provided evidence of

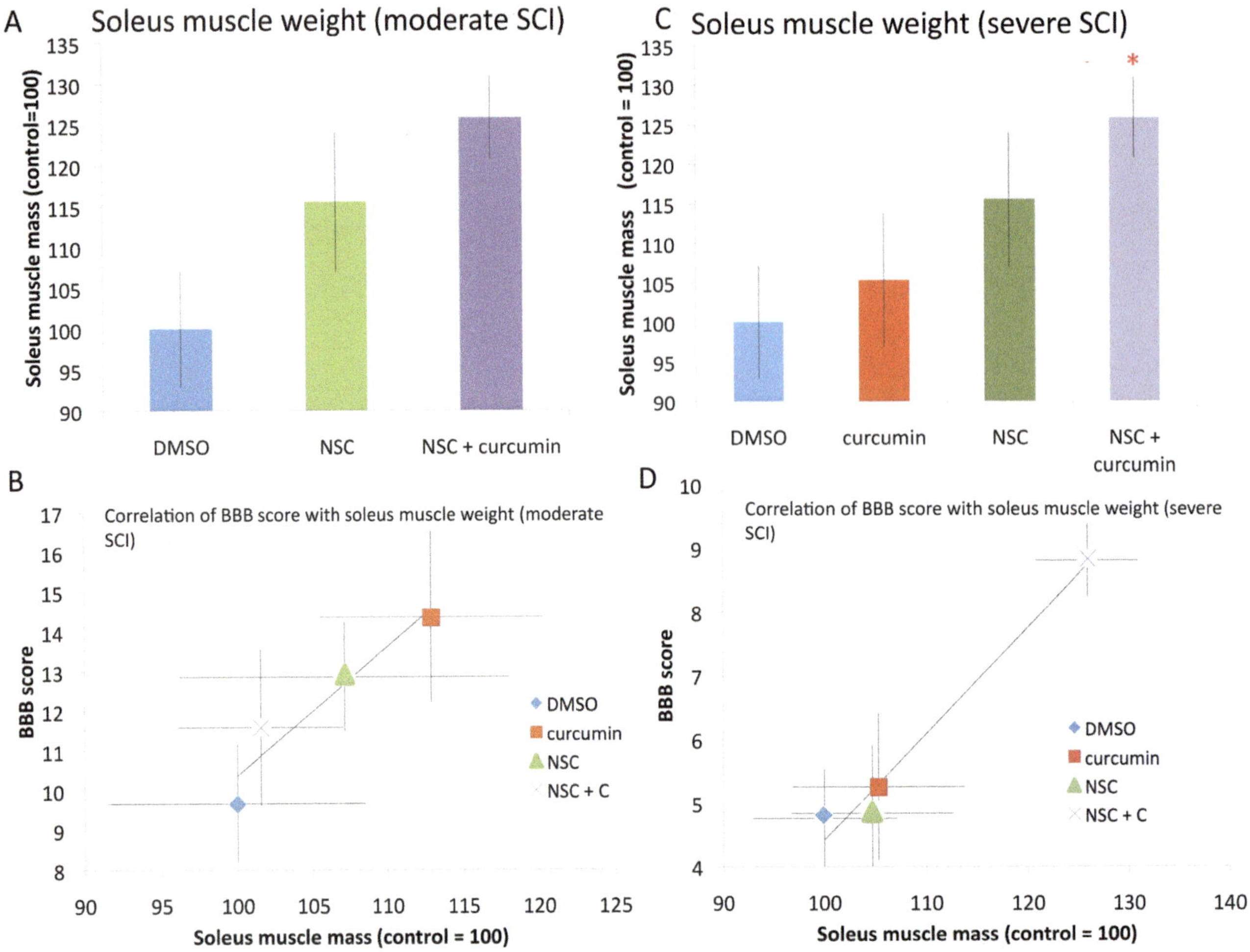

Figure 4. Soleus muscle weight as an indicator of recovery after moderate and severe SCI, and correlated with BBB score. A: Rats with greater recovery of hindlimb motor function have better soleus muscle weight at necropsy. While stem cell transplanted animals and combinational treatments had improved soleus muscle weight in comparison to DMSO controls, this did not reach statistical significance. **B:** Correlation of BBB score with soleus muscle weight after moderate SCI demonstrates strong correlation (0.97). Values presented are mean ± SEM. $*p<0.05$. **C:** Soleus muscle weight after severe SCI showed that all cohorts had larger soleus muscle mass compared to DMSO treated controls, however curcumin or stem cell monotherapy showed only a trend towards significance. NSC plus curcumin treatment improved soleus muscle weight significantly in comparison to DMSO controls. **D:** Correlation of BBB score with soleus muscle weight after severe SCI demonstrated a strong correlation ($r^2=0.91$). Values presented are mean ± SEM. $*p<0.05$.

functional motor recovery after transplantation of ependymal stem progenitor cells that were derived from adult rat spinal cord with a traumatic lesion [51]. In this study, cells were propagated and differentiated *in vitro* into oligodendrocyte precursor cells (OPCs) and engrafted after cell transplantation. Endogenous or transplanted NSCs can differentiate into oligodendrocytes and astrocytes, contributing to remyelination of axons and aiding in recovery [52]. While recruitment of these endogenous NSCs can occur as a part of spontaneous recovery, transplantation of NSCs provides an improved strategy for a faster and better recovery after SCI.

Adult neural stem cells are most commonly derived from the SVZ of the brain and have been found to differentiate to neurons, glia, and oligodendrocytes, although neuronal marker expression is generally rare [20]. This may be due to the local microenvironment at the site of injury, that may affect the differentiation of transplanted NSC. These adult NSC have been found to generate compact myelin in the injured spinal cord, although this has been somewhat inconsistently reported [20–23]. The choice of using adult NSC is due to their decreased risk of tumorigenesis [20–22], their ability to differentiate along a neural lineage with great consistency [18,20–23,26,53,54], and their repeated benefit to recovery using a thoracic contusion model with behavioral recovery evaluated with the open field BBB score [21–23,55–57]. All but one of these studies delivered stem cells directly to the injured spinal cord, similar to what was performed in this study [20,56].

Secondary damage to the spinal cord following SCI is induced primarily by inflammation and ischemia/reperfusion injury. White matter is especially susceptible to ischemic damage, especially by lipid peroxidation [58,59]. Oligodendrocytes are also vulnerable to superoxide produced during ischemia because these cells only contain low levels of superoxide antagonists like Mu-superoxide and catalase [58,60]. Apoptosis also occurs due, in part at least, to the inflammatory cascade, influx of calcium, and the activation of the caspase cascade [58,61,62]. A number of agents have been used to reduce ischemia/reperfusion injury to the spinal cord with varying degrees of success including statins, calcium channel

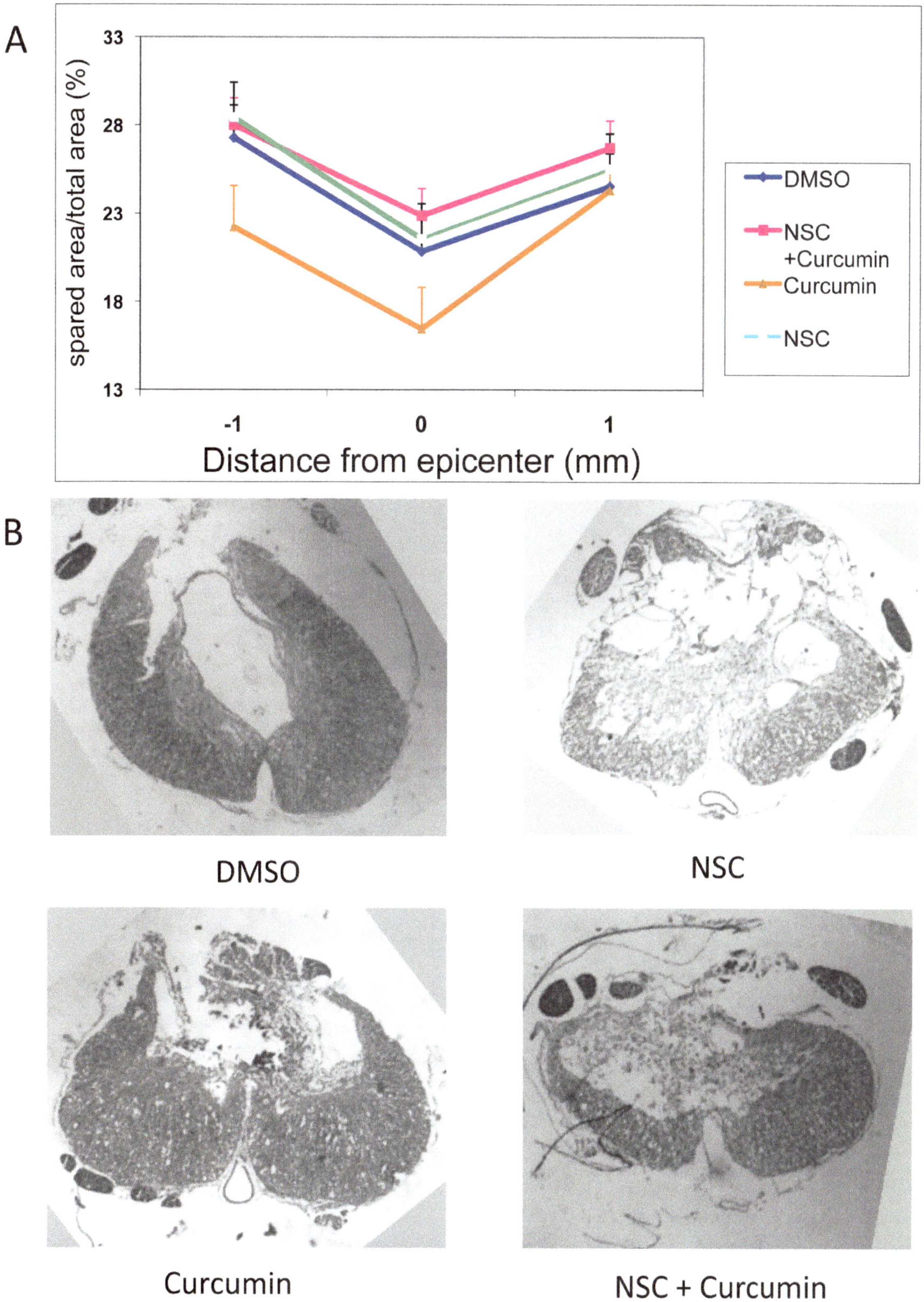

Figure 5. Histopathological examination after severe SCI. A: NSC plus curcumin and NSC all displayed enhanced spared area relative to the total area suggesting an improvement in recovery. DMSO treated rats also demonstrated increased spared area of the spinal cord, but somewhat less than the NSC plus curcumin, or NSC treatments. Curcumin-treated rats demonstrated less spared area than the other three treatments. **B:** Photomicrographs of representative samples of spinal cord sections across the area of contusion stained with Lox blue.

blockers, and caspase inhibitors, among others [58,61,62]. These studies demonstrate the risk of further SCI secondary to ischemia/reperfusion, and the possible benefits of using agents for their neuroprotective effects.

Curcumin is a potent antioxidant and free radical scavenger, and has been shown to improve ischemia-reperfusion injury in multiple studies [63–66]. Curcumin has a potent inhibitory effect on the NF-kB pathway in settings where it is activated [67–69]. NF-kB has been demonstrated to be highly upregulated in SCI, and is associated with glial scar formation [70–71]. Curcumin may exert its benefit in SCI recovery via inhibition of this pathway [9–12,14]. Absorbtion of curcumin in the GI tract has been debatable, and once absorbed into the blood stream, detectable levels are transient and require very high oral doses to achieve [72]. In order to alleviate the difficulties of poor bioavailability of curcumin, we dissolved curcumin in DMSO as a solvent to increase the rate of transfer across membranes, and administered it by IP injection or locally at the site of injury. Side effects of curcumin are limited, with no side effects noted in our animals [73]. It is noteworthy that despite the previously noted studies demonstrating curcumin's benefit after SCI, this study extends the findings by demonstrating that this benefit appears to be limited depending on the severity of injury [9–12,14]. For example, in severe injury, curcumin only benefitted recovery following combinational therapy with NSC. Limitations to recovery with existing therapies are not completely surprising, and have been noted in SCI studies in the past with other agents [74–76].

Our study demonstrates some of the limitations not only of curcumin, but also stem cell monotherapy, on recovery from SCI. While curcumin's bioavailability is limited, this may be useful where it is used in combinational therapy [35,73–74]. There is evidence, for example, that only very low concentrations of curcumin following IP injection cross the blood brain barrier [35,73–74]. However, it is precisely this low dose of curcumin that creates the synergistic effect seen in our study. A higher dose of curcumin in combination with NSC had a less significant effect than either therapy alone (Fig. 2A). These results emphasize that the synergistic effect of curcumin with stem cell therapy in severe SCI may occur via dual action of curcumin, i.e. it reduces the inflammation at the local SCI region induced by secondary injury and may also induce proliferation of stem cells required for faster regeneration. With the great interest in stem cell transplant after SCI, it is important to know what adjunctive agents may serve to enhance their effect. This is the first study to demonstrate significant functional recovery from severe SCI in animals treated with curcumin in combination with NSC transplant. Many of these animals regained independent ambulatory function even after severe SCI, a feat not accomplished with either therapy alone. Curcumin and NSC transplant appear to be a safe and effective treatment to improve recovery from SCI. Future studies would include the analysis of discrete biomarkers associated with inflammation and recovery after curcumin treatment.

Conclusions

The results of this study provide a strong basis for stem cell therapy in conjunction with curcumin in the treatment and management of SCI in humans.

Acknowledgments

We are thankful for Dr. Hong Peng for her help with histology.

Author Contributions

Conceived and designed the experiments: DRO MJ RZ. Performed the experiments: DRO CS JO. Analyzed the data: DRO CS JO MJ RZ RM KD. Contributed reagents/materials/analysis tools: DRO MJ RZ RM KD. Wrote the paper: DRO CS JO MJ RZ KD RM.

References

1. Sekhon LH, Fehlings MG (2001) Epidemiology, demographics, and pathophysiology of acute spinal cord injury. Spine 26(24 Suppl): S2–12.
2. Christopher & Dana Reeve Foundation (2009) One Degree of Separation: Paralysis and Spinal Cord Injury in the United States. Christopher & Dana Reeve Foundation.
3. Renault-Mihara F, Okada S, Shibata S, Nakamura M, Toyama Y, et al. (2008) Spinal cord injury: emerging beneficial role of reactive astrocytes' migration. Int J Biochem Cell Biol 40(9): 1649–53.
4. Ghosh S, Hayden MS (2008) New regulators of NF-kappaB in inflammation. Nat Rev Immunol 8(11): 837–48.
5. Aggarwal BB, Sung B (2009) Pharmacological basis for the role of curcumin in chronic diseases: an age-old spice with modern targets. Trends Pharmacol Sci 30(2): 85–94.
6. Begum AN, Jones MR, Lim GP, Morihara T, Kim P, et al. (2008) Curcumin structure-function, bioavailability, and efficacy in models of neuroinflammation and Alzheimer's disease. J Pharmacol Exp Ther 326(1): 196–208.
7. Chainani-Wu N (2003) Safety and anti-inflammatory activity of curcumin: a component of tumeric (Curcuma longa). J Altern Complement Med 9(1): 161–8.
8. Aggarwal BB, Harikumar KB (2009) Potential therapeutic effects of curcumin, the anti-inflammatory agent, against neurodegenerative, cardiovascular, pulmonary, metabolic, autoimmune and neoplastic diseases. Int J Biochem Cell Biol 41(1): 40–59.
9. Lin MS, Lee YH, Chiu WT, Hung KS (2011) Curcumin provides neuroprotection after spinal cord injury. J Surg Res 166: 280–289.
10. Lin MS, Sun YY, Chiu WT, Hung CC, Chang CY, et al. (2011) Curcumin attenuates the expression and secretion of RANTES after spinal cord injury in vivo and lipopolysaccharide-induced astrocyte reactivation in vitro. J Neurotrauma 28: 1259–1269.
11. Lee SG, Su ZZ, Emdad L, Sarkar D, Franke TF, et al. (2008) Astrocyte elevated gene-1 activates cell survival pathways through PI3K-Akt signaling. Oncogene 27(8): 1114–21.
12. Cemil B, Topuz K, Demircan MN, Kurt G, Tun K, et al. (2010) Curcumin improves early functional results after experimental spinal cord injury. Acta Neurochir (Wien) 152: 1583–1590.
13. Sahin Kavaklı H, Koca C, Alici O (2011) Antioxidant effects of curcumin in spinal cord injury in rats. Ulus Travma Acil Cerrahi Derg 17: 14–18.
14. Ormond DR, Peng H, Zeman R, Das K, Murali R, et al. (2012) Recovery from spinal cord injury using naturally occurring antiinflammatory compound curcumin. J Neurosurg Spine. 16(5): 497–503.
15. Joseph MS, Ying Z, Zhuang Y, Zhong H, Wu A, et al. (2012) Effects of Diet and/or Exercise in Enhancing Spinal Cord Sensorimotor Learning. PLoS ONE 7(7): 1–10.
16. Holly LT, Blaskiewicz D, Wu A, Feng C, Ying Z, et al. (2012) Dietary therapy to promote neuroprotetion in chronic spinal cord injury. J Neurosurg Spine 17(2): 134–140.
17. Teng YD, Lavik EB, Qu X, Park KI, Ourednik J, et al. (2002) Functional recovery following traumatic spinal cord injury mediated by a unique polymer scaffold seeded with neural stem cells. Proc Natl Acad Sci USA 99(5): 3024–9.
18. Vroemen M, Aigner L, Winkler J, Weidner N (2003) Adult neural progenitor cell grafts survive after acute spinal cord injury and integrate along axonal pathways. Eur J Neurosci 18(4): 743–51.
19. Ruff CA, Wilcox JT, Fehlings MG (2012) Cell-based transplantation strategies to promote plasticity following spinal cord injury. Exp Neurol. 235(1): 78–90.
20. Tetzlaff W, Okon EB, Karimi-Abdolrezaee S, Sparling JS, Plemel JR, et al. (2011) A Systematic Review of Cellular Transplantation Therapies for Spinal Cord Injury. J of Neurotrauma 28: 1611–1682.
21. Parr AM, Kulbatski I, Zahir T, Wang X, Yue C, et al. (2008) Transplanted adult spinal cord-derived neural stem/progenitor cells promote early functional recovery after rat spinal cord injury. Neuroscience 155(3): 760–70.
22. Parr AM, Kulbatski I, Wang XH, Keating A, Tator CH (2008) Fate of transplanted adult neural stem/progenitor cells and bone marrow-derived mesenchymal stromal cells in the injured adult rat spinal cord and impact on functional recovery. Surg Neurol 70(6): 600–7.
23. Karimi-Abdolrezaee S, Eftekharpour E, Wang J, Morshead CM, Fehlings MG (2006) Delayed transplantation of adult neural precursor cells promotes remyelination and functional neurological recovery after spinal cord injury. J Neurosci 26(13): 3377–89.
24. Yan J, Welsh AM, Bora SH, Snyder EY, Koliatsos VE (2004) Differentiation and tropic/trophic effects of exogenous neural precursors in the adult spinal cord. J Comp Neurol 480(1): 101–14.
25. Miller RH, Bai L (2012) Translating stem cell therapies to the clinic. Neurosci Lett. 519(2): 87–92.

26. Parr AM, Kulbatski I, Tator CH (2007) Transplantation of adult rat spinal cord stem/progenitor cells for spinal cord injury. J Neurotrauma 24(5): 835–45.
27. Temple S, Alvarez-Buylla A (1999) Stem cells in the adult mammalian central nervous system. Curr Opin Neurobiol 9(1): 135–41.
28. Jhanwar-Uniyal M, Beck B, Jhanwar YS, Burlet C, Leibowitz SF (1993) Neuropeptide Y projection from arcuate nucleus to parvocellular division of paraventricular nucleus: specific relation to the ingestion of carbohydrate. Brain Res 631(1): 97–106.
29. Marshall GP, Ross HH, Suslov O, Zheng T, Steindler DA, et al. (2008) Production of neurospheres from CNS tissue. Methods Mol Biol 438: 135–50.
30. Zeman RJ, Feng Y, Peng H, Etlinger JD (1999) Clenbuterol, a beta(2)-adrenoceptor agonist, improves locomotor and histological outcomes after spinal cord contusion in rats. Exp Neurol 159(1): 267–73.
31. Zeman RJ, Feng Y, Peng H, Visintainer PF, Moorthy CR, et al. (2001) X-irradiation of the contusion site improves locomotor and histological outcomes in spinal cord-injured rats. Exp Neurol 172(1): 228–34.
32. Zeman RJ, Peng H, Feng Y, Song H, Liu X, et al. (2006) Beta2-adrenoreceptor agonist-enhanced recovery of locomotor function after spinal cord injury is glutathione dependent. J Neurotrauma 23(2): 170–80.
33. Basso DM, Beattie MS, Bresnahan JC (1996) Graded histological and locomotor outcomes after spinal cord contusion using the NYU weight-drop device versus transection. Exp Neurol 139(2): 244–56.
34. Fehlings MG, Vawda R (2011) Cellular Treatments for Spinal Cord Injury: The Time is Right for Clinical Trials 8(4): 704–720.
35. Kim SJ, Son TG, Park HR, Park M, Kim MS, et al. (2008) Curcumin stimulates proliferation of embryonic neural progenitor cells and neurogenesis in the adult hippocampus. J Biol Chem 283(21): 14497–505.
36. Liu M, Stevens-Lapsley JE, Jayaraman A, Ye F, Conover C, et al. (2010) Impact of treadmill locomotor training on skeletal muscle IGF1 and myogenic regulatory factors in spinal cord injured rats. Eur J Appl Physiol 109(4): 709–20.
37. Liu M, Bose P, Walter GA, Anderson DK, Thompson FJ, et al. (2006) Changes in muscle T2 relaxation properties following spinal cord injury and locomotor training. Eur J of Appl Physiol 97(3): 355–61.
38. Lee Y, Lin C, Calozzo VJ, Robertson RT, Yu J, et al. (2007) Repair of spinal cord transection and its effects on muscle mass and myosin heavy chain isoform phenotype. J of Appl Physiol 103(5): 1808–1814.
39. Dupont-Versteegden EE, Houlé JD, Gurley CM, Peterson CA (1998) Early changes in muscle fiber size and gene expression in response to spinal cord transection and exercise. Am J Physiol. 275(4 Pt 1): C1124–33.
40. Landry E, Frenette J, Guertin PA (2004) Body weight, limb size, and muscular properties of early paraplegic mice. J Neurotrauma. 21(8): 1008–16.
41. Singh A, Balasubramanian S, Murray M, Lemay M, Houle J (2011) Role of spared pathways in locomotor recovery after body-weight-supported treadmill training in contused rats. J Neurotrauma. 28(12): 2405–16.
42. Rabchevsky AG, Streit WJ (1997) Grafting of cultured microglial cells into the lesioned spinal cord of adult rats enhances neurite outgrowth. J Neurosci Res 47(1): 34–48.
43. Guízar-Sahagún G, Ibarra A, Espitia A, Martínez A, Madrazo I, et al. (2005) Glutathione monoethyl ester improves functional recovery, enhances neuron survival, and stabilizes spinal cord blood flow after spinal cord injury in rats. Neuroscience. 130(3): 639–49.
44. Jakeman LB, Wei P, Guan Z, Stokes BT (1998) Brain-derived neurotrophic factor stimulates hindlimb stepping and sprouting of cholinergic fibers after spinal cord injury. Exp Neurol. 154(1): 170–84.
45. Weiss S, Dunne C, Hewson J, Wohl C, Wheatley M, et al. (1996) Multipotent CNS stem cells are present in the adult mammalian spinal cord and ventricular neuroaxis. J Neurosci. 16(23): 7599–609.
46. Johansson CB, Momma S, Clarke DL, Risling M, Lendahl U, et al. (1999) Identification of a neural stem cell in the adult mammalian central nervous system. Cell. 96(1): 25–34.
47. Namiki J, Tator CH (1999) Cell proliferation and nestin expression in the ependyma of the adult rat spinal cord after injury. J Neuropathol Exp Neurol. 58(5): 489–98.
48. Horner PJ, Power AE, Kempermann G, Kuhn HG, Palmer TD, et al. (2000) Proliferation and differentiation of progenitor cells throughout the intact adult rat spinal cord. J Neurosci. 20(6): 2218–28.
49. Kojima A, Tator CH (2002) Intrathecal administration of epidermal growth factor and fibroblast growth factor 2 promotes ependymal proliferation and functional recovery after spinal cord injury in adult rats. J Neurotrauma. 19(2): 223–38.
50. Martens DJ, Seaberg RM, van der Kooy D (2002) In vivo infusions of exogenous growth factors into the fourth ventricle of the adult mouse brain increase the proliferation of neural progenitors around the fourth ventricle and the central canal of the spinal cord. Eur J Neurosci. 16(6): 1045–57.
51. Moreno-Manzano V, Rodriguez-Jimenez FJ, Garcia-Rosello M, Lainez S, Erceq S, et al. (2009) Activated Spinal Cord Ependymal Stem Cells Rescue Neurological Function. Stem Cells 27: 733–743.
52. Barnabé-Heider F, Frisen J (2008) Stem cells for spinal cord repair. Cell Stem Cell 3: 16–24.
53. Cao Q, Benton RL, Whittemore SR (2002) Stem cell repair of central nervous system injury. J Neurosci Res. 68(5): 501–10.
54. Pfeifer K, Vroemen M, Blesch A, Weidner N (2004) Adult neural progenitor cells provide a permissive guiding substrate for corticospinal axon growth following spinal cord injury. Eur J Neurosci. 20(7): 1695–704.
55. Hofstetter CP, Holmström NA, Lilja JA, Schweinhardt P, Hao J, et al. (2005) Allodynia limits the usefulness of intraspinal neural stem cell grafts; directed differentiation improves outcome. Nat Neurosci. 8(3): 346–53.
56. Bottai D, Madaschi L, Di Giulio AM, Gorio A (2008) Viability-dependent promoting action of adult neural precursors in spinal cord injury. Mol Med. 14(9–10): 634–44.
57. Ziv Y, Avidan H, Pluchino S, Martino G, Schwartz M (2006) Synergy between immune cells and adult neural stem/progenitor cells promotes functional recovery from spinal cord injury. Proc Natl Acad Sci USA. 103(35): 13174–9.
58. Saito T, Tsuchida M, Umehara S, Kohno T, Yamamoto H, et al. (2011) Reduction of Spinal Cord Ischemia/Reperfusion Injury with Simvastatin in Rats. Neuroscience in Anesthesiology and Perioperative Medicine 113(3): 565–71.
59. Morell P, Quarles RH (1999) Myelin formation, structure and biochemistry: In: Siegel GJ, Agranoff BW, Fisher SK, Albers RW, Uhler MD, eds. Basic Neurochemistry: Molecular, Cellular and Medical Aspects. 6th edition. New York: Lippincott-Raven, 69–94.
60. Bernardo A, Greco A, Levi G, Minghetti L (2003) Differential lipid peroxidation, Mn superoxide, and bcl-2 expression contribute to the maturation-dependent vulnerability of oligodendrocytes to oxidative stress. J Neuropathol Exp Neurol 62: 509–19.
61. Fansa I, Altug ME, Melek I, Ucar E, Kontas T, et al. (2009) The Neuroprotective and Anti-inflammatory Effects of Diltiazem in Spinal Cord Ischaemia–Reperfusion Injury. The Journal of International Medical Research 37: 520–533.
62. Dolgun H, Sekerci Z, Turkoglu E, Kertmen H, Yilmaz ER, et al. (2010) Neuroprotective effect of mesna (2-mercaptoethane sulfonate) against spinal cord ischemia/reperfusion injury in rabbits. J Clin Neurosci. 17(4): 486–9.
63. Avci G, Kadioglu H, Sehirli AO, Bozkurt S, Guclu O, et al. (2012) Curcumin protects against ischemia/reperfusion injury in rat skeletal muscle. J Surg Res. 172(1): e39–46.
64. Pogue AI, Percy ME, Cui JG, Li YY, Bhattacharjee S, et al. (2011) Up-regulation of NF-kB-sensitive miRNA-125b and miRNA-146a in metal sulfate-stressed human astroglial (HAG) primary cell cultures. J Inorg Biochem. 105(11): 1434–7.
65. Barzegar A, Moosavi-Movahedi AA (2011) Intracellular ROS protection efficiency and free radical-scavenging activity of curcumin. PLoS One. 6(10): e26012.
66. Kuo CP, Lu CH, Wen LL, Cherng CH, Wong CS, et al. (2011) Neuroprotective effect of curcumin in an experimental rat model of subarachnoid hemorrhage. Anesthesiology. 115(6): 1229–38.
67. Ammon HP, Wahl MA (1991) Pharmacology of Curcuma longa. Planta Med 57: 1–7.
68. Satoskar RR, Shah SJ, Shenoy SG (1986) Evaluation of anti-inflammatory property of curcumin (diferuloyl methane) in patients with postoperative inflammation. Int J Clin Pharmacol Ther Toxicol 24: 651–654.
69. Shankar TN, Shantha NV, Ramesh HP, Murthy IA, Murthy VS (1980) Toxicity studies on turmeric (Curcuma longa): acute toxicity studies in rats, guineapigs & monkeys. Indian J Exp Biol 18: 73–75.
70. Brambilla R, Hurtado A, Persaud T, Esham K, Pearse DD, et al. (2009) Transgenic inhibition of astroglial NF-kappa B leads to increased axonal sparing and sprouting following spinal cord injury. J Neurochem 110: 765–778.
71. Rafati DS, Geissler K, Johnson K, Unabia G, Hulsebosch C, et al. (2008) Nuclear factor-kappaB decoy amelioration of spinal cord injury-induced inflammation and behavior outcomes. J Neurosci Res 86(3): 566–80.
72. Anand P, Kunnumakkara AB, Newman RA, Aggarwal BB (2007) Bioavailability of curcumin: problems and promises. Mol Pharm 4: 807–818.
73. Sharma RA, Steward WP, Gescher AJ (2007) Pharmacokinetics and pharmacodynamics of curcumin. Adv Exp Med Biol 595: 453–70.
74. Sharma S, Zhuang Y, Ying Z, Gomez-Pinilla F (2009) Dietary Curcumin Supplementation Counteracts reduction in levels of Molecules involved in Energy Homeostasis after Brain Trauma. Neuroscience 161(4): 1037–44.
75. Nesathurai S (1998) Steroids and spinal cord injury: revisiting the NASCIS 2 and NASCIS 3 trials. J Trauma 45(6): 1088–93.
76. Bracken MB, Shepard MJ, Holford TR, Leo-Summers L, Aldrich EF, et al. (1997) Administration of methylprednisolone for 24 or 48 hours or tirilazad mesylate for 48 hours in the treatment of acute spinal cord injury. Results of the Third National Acute Spinal Cord Injury Randomized Controlled Trial. National Acute Spinal Cord Injury Study. JAMA. 277(20): 1597–604.

9

Efficacy and Dose-Dependent Safety of Intra-Arterial Delivery of Mesenchymal Stem Cells in a Rodent Stroke Model

Dileep R. Yavagal[1,2,3]*, Baowan Lin[1], Ami P. Raval[1,2], Philip S. Garza[1], Chuanhui Dong[2], Weizhao Zhao[2], Erika B. Rangel[3], Ian McNiece[3], Tatjana Rundek[2], Ralph L. Sacco[2], Miguel Perez-Pinzon[1,2], Joshua M. Hare[3,4]

1 Cerebral Vascular Disease Research Laboratories, Leonard M. Miller School of Medicine, University of Miami, Miami, Florida, United States of America, **2** Department of Neurology, Leonard M. Miller School of Medicine, University of Miami, Miami, Florida, United States of America, **3** Interdisciplinary Stem Cell Institute, Leonard M. Miller School of Medicine, University of Miami, Miami, Florida, United States of America, **4** The Department of Medicine, Leonard M. Miller School of Medicine, University of Miami, Miami, Florida, United States of America

Abstract

Intra-arterial (IA) delivery of mesenchymal stem cells (MSCs) for acute ischemic stroke is attractive for clinical translation. However, studies using rat model of stroke have demonstrated that IA MSCs delivery can decrease middle cerebral artery (MCA) flow, which may limit its clinical translation. The goal of this study is to identify a dose of IA MSCs (maximum tolerated dose; MTD) that does not compromise MCA flow and evaluate its efficacy and optimal timing in a rat model of reversible middle cerebral artery occlusion (rMCAo). We sought to determine if there is a difference in efficacy of acute (1 h) versus sub-acute (24 h) IA MSCs treatment after rMCAo. Adult female Sprague-Dawley rats underwent rMCAo (90 min) and an hour later a single dose of MSCs (at de-escalating doses 1×10^6, 5×10^5, 2×10^5, 1×10^5 and 5×10^4) was given using IA route. MSCs were suspended in phosphate buffered saline (PBS) and PBS alone was used for control experiments. We measured the percent change in mean laser Doppler flow signal over the ipsilateral MCA in de-escalating doses groups to determine MTD. The results demonstrated that the lowering of IA MSC dose to 1×10^5 and below did not compromise MCA flow and hence an IA MSC dose of 1×10^5 considered as MTD. Subsequently, 1 h and 24 h after rMCAo, rats were treated with IA MSCs or PBS. The 24 h delivery of IA MSCs significantly improved neurodeficit score and reduced the mean infarct volume at one month as compared to control, but not the 1 h delivery. Overall, this study suggests that the IA delivery of MSCs can be performed safely and efficaciously at the MTD of 1×10^5 delivered at 24 hours in rodent model of stroke.

Editor: Toru Hosoda, Tokai University, Japan

Funding: This study was supported by Department of Neurology, University of Miami (DRY); James and Esther King Biomedical Research Program, Florida Department of Health Grant # 2KN09 (DRY); University of Miami intramural grant: Interdepartmental Research Development Initiative grant 2009–2010 (DRY) and the Anderson Family Gift 2010–2011, 2012–2014 (DRY). The funders had no role in study design, data collection and analysis, decision to publish, or preparation of the manuscript.

Competing Interests: The authors have declared that no competing interests exist.

* E-mail: dyavagal@med.miami.edu

Introduction

Stroke is the main cause of long-term disability and the third leading cause of death in the United States. The public health burden of stroke continues to be staggering with an estimated cost of 73.3 billion in 2010 [1]. Despite the approval of intravenous recombinant tissue plasminogen activator (rtPA) 18 years ago, and rapid growth in number of endovascular recanalization therapies for acute ischemic stroke (AIS), their impact on reducing stroke-related long-term disability is limited [2,3]. Hence, there continues to be a critical need for novel therapies for AIS. In this regard, over the last decade, various types of stem cell have been tested in several pre-clinical studies suggesting improved functional neurological outcomes after AIS [4–7]. The leading type of stem cell for clinical translation in stroke is the mesenchymal stem cell, which is an adult, non-hematopoietic progenitor cell with capacity to differentiate into a variety of cell lineages including osteoblasts, chondrocytes and neuron-like cells [8–11].

Mesenchymal stem cells (MSCs) are multipotent non-hematopoietic stem cells found mainly in the stromal fraction of the bone marrow as well as in the connective tissue of most organs [9–11]. MSCs can be easily isolated from adipose tissue, amniotic fluid, placenta and umbilical cord, though they are most commonly and efficiently derived from adult bone marrow. MSCs are an attractive cell source because they are relatively easy to obtain, expand and manipulate *in vitro* [12]. Moreover, adult MSCs do not confer the risk of tumorgenicity that pluripotent cells carry [13]. Importantly, MSCs are also immunoprivileged, with low MHC I and no MHC II antigen expression [12,13]. Their immunoprivileged nature obviates the need for immunosuppression in allogeneic administration of MSCs. Thus allogeneic MSCs, already produced from a healthy donor, can be given as an off-the-shelf product without delay if needed without the need of immunosuppression. This feature is especially attractive for future translation of MSCs into a treatment for ischemic stroke, which is presents most often without warning and may benefit from early administration of cell therapy [14–16].

There are multiple routes of stem cell delivery to the brain in AIS. Among these, the intra-arterial (IA) route of stem cell delivery has a high potential for clinical translation, in view of the increasing clinical application of endovascular therapy in the treatment of AIS [6,17–20]. Furthermore, IA delivery of stem cells after AIS is minimally invasive and results in a larger number and more diffuse distribution of stem cells in and around the infarcted area when compared to intra-parenchymal, intra-cerebroventricular and intravenous stem cell delivery [21]. IA delivery avoids the first pass trapping of stem cells in the lungs and liver seen with intravenous delivery [22]. Also, a recent study has shown superior functional and histological outcomes with IA delivery of stem cells as compared to intravenous administration [23]. However MSCs range in size from 5–50 microns and a major limitation to IA delivery of MSCs is the possibility of compromise of regional cerebral blood flow (rCBF) owing to those MSCs in the larger size range of 20–50 microns; thus a potentially novel therapeutic strategy can paradoxically worsen of cerebral ischemia outcome [10].

In the current study, first we hypothesized that compromise of MCA blood flow during IA delivery of MSCs was dose-dependent and therefore, lowering the dose of IA MSCs would mitigate rCBF compromise. The efficacy of IA MSC in focal cerebral ischemia, with doses lower than 1×10^6 has not been studied. Hence, we also hypothesized that the lower safe dose we identify, based on our first hypothesis, would be efficacious in focal cerebral ischemia. Furthermore, the efficacy of IA MSCs given immediately after reperfusion (within 60 minutes) as compared to the efficacy when given at 24-hours is unknown. Hence, we additionally hypothesized that the efficacy of the MTD of IA MSCs delivered at 24 hours (sub-acute) and 60 minutes (acute) in a rodent cerebral ischemia model would be the same.

Materials and Methods

All animal procedures were carried out in accordance with the Guide for the Care and Use of Laboratory Animals published by the U.S. National Institutes of Health and approved by the Animal Care and Use Committee of the University of Miami.

Female Sprague-Dawley rats (Charles River Laboratories, Inc., Wilmington, MA) weighing 260–310 g were used. The animals in the experimental groups were allocated in a randomized fashion. Investigators were blinded to dose and treatment group allocation during surgery and during outcome evaluations.

Surgery

The animals underwent Reversible Middle cerebral artery occlusion (rMCAo) following an overnight fast. Anesthesia was induced with 3% isofluorane and 70% nitrous oxide. Rats were intubated endotracheally and ventilated mechanically on a mixture of 1–0.5% isofluorane, 70% nitrous oxide and a balance of oxygen. The right femoral artery and vein were catheterized to permit to monitor blood pressure and to take arterial samples for blood gas and glucose assessments. Arterial PCO_2 and PO_2 were maintained in the normal range by ventilator adjustment. For immobilization, rats received pancuromium bromide 0.75 mg/kg i.v. Rectal temperature was measured continuously and maintained at 37–37.5°C by a heating pat under the rat's body. Cranial temperature was separately monitored by a 29-gauge thermocouple implanted into the right temporalis muscle and was maintained at 36–36.5°C by a warming lamp placed above the rat's head throughout the experiment. Our previous studies have shown that the cranial temperature 36–36.5°C corresponds to a brain temperature of 36.5–37°C [24]. Physiological variables (plasma glucose concentration, pH, PCO_2, PO_2 and mean arterial blood pressure (MABP)) were maintained normal before and after ischemia (Table S1 and S2; presented as supplementary data).

CBF monitoring

The left scalp was opened and the skull was exposed. A 2-mm bur hole was drilled on the left *sphenoid*, 3 mm below the up edge, 0.5 mm anterior and 6 mm lateral to the bregma, leaving the dura intact. A short Laser Doppler Flowmetry (LDF, Perimed Inc.) probe was placed above the dura and fixed on to the skull with glue and dental cement. From this point, the CBF inside the trunk of the ascending vessel from cortical branch of MCA was monitored. The LDF signal reading was recorded, starting at 30 min before the suture insertion and continuously throughout the experiment and 60 min after the cell infusion, using a specific designed monitoring system.

Reversible Middle cerebral artery occlusion (rMCAo)

The previously well-described suture technique for MCAo was utilized in this study [25–28]. The common carotid artery (CCA) was exposed through a midline incision on the ventral neck and carefully separated from the surrounding tissue including the adjacent vagus nerve by blunt dissection using microsurgery. Under an operation microscope, the branches of external carotid artery (ECA), superior thyroid and ascending pharyngeal arteries and further the terminal lingual and maxillary artery branches, were dissected and coagulated. Two long 5–0 silk sutures tied the ECA as distal to carotid bifurcation as possible. Two short (3 cm) 5–0 silk sutures were tied loosely around the segment of ECA closed to the bifurcation. The ICA was exposed, and the origin of the pterygopalatine artery (PPA) was visualized. An intraluminal 3.0 nylon suture, having a spherical and enlarged tip produced by heating near flame, introduced into the distal segment of ECA via a small incision on the vessel and advanced into the ECA lumen to reach the bifurcation. The 2 short silk sutures subsequently were tied to prevent bleeding. The ECA was cut at the distal segment between the 2 long-suture notes. The stump of the ECA was held toward the surgeon; thus the stump and the ICA were on a straight line. The suture was then turned into ICA with a special care not enter the PPA. The suture went advance in ICA until the sharp drop of regional cerebral blood flow confirmed by the LDF. The length of inserted suture inside the ICA from the bifurcation was 19–25 mm. A sudden and sharp decrease in the LDF signal was interpreted to indicate a successful MCA occlusion. The suture was gently withdrawn after the 90-min occlusion period. During the procedure and the 2-h recirculation, blood pressure and the blood gas were maintained in the normal ranges, cranial and rectal temperature were maintained at normal levels.

Inclusion and Exclusion Criteria

Rats that had less than 50% drop in LDF signal on suture insertion and premature mortality during rMCAo were excluded from the study prior to randomization and allocation to treatment groups.

Experimental Design

(1) Safety Study

Forty-three female rats underwent rMCAo. The suture was gently withdrawn after a 90-minute occlusion period (Figure 1A, B). At 60 minutes after recirculation, a 1 ml syringe was filled with the MSCs suspended in phosphate buffered saline (PBS) or PBS alone. The cell suspension in 0.5 mL PBS or PBS alone (0.5 mL) was delivered into the ipsilateral IA over 3 minutes. The different groups of animals received 1×10^6, 5×10^5, 2×10^5, 1×10^5 & 5×10^4 MSCs in PBS or only PBS. Following MSCs or PBS delivery, the catheter was carefully withdrawn and the ECA was tied. During the surgical procedure of rMCAo and IA MSCs deliver, we measure blood flow over ipsilateral MCA using laser Doppler flow signal (LDFS).

(2) Efficacy Study

In another cohort, 34 female rats underwent 90 minutes of rMCAo. Twenty-four hours after induction of rMCAo, rats were assigned to receive IA MSCs (1×10^5) or IA PBS or intravenous (IV) MSCs 1×10^6. In the IV MSCs group, the cells were given into the femoral vein at a 10-fold of the IA MSC dose in order to potentially increase the number of MSCs reaching the injured brain. In an additional experimental group, IA MSCs at 1×10^5 were given at 60 minutes after rMCAo to study the effect of acute timing on efficacy. The primary outcome measure was neurodeficit score and the secondary outcome was infarct volume at 4 weeks.

Donor rodent MSCs

MSCs used in this study were derived from green fluorescence protein (GFP) transgenic male Sprague-Dawley rats. At the Cell Manufacturing Program at the University of Miami, the bone marrow was harvested from rats and the mononuclear cells isolated by density gradient centrifugation. The cells were then cultured at 1 to 5×10^6 cells per ml in 25 ml of alpha MEM plus 20%FCS in T162 cm^2 culture flasks at 5%CO_2 and 37°C. MSCs exhibited spindle-shaped morphology and were characterized by (i) adherence to plastic, (ii) negativity for hematopoietic cell surface markers CD34 and CD45 (0.1±0.1% and 0.2±.0.1%, respectively) and positivity for CD73, CD90.2, and CD105 (88±5.4%, 99.2±0.7% and 95±3.7%, respectively). The cells were expanded to passage 6 and then the MSCs were suspended in 10% of cryopreservant dimethylsulfoxide (DMSO) and frozen in liquid nitrogen. Prior to injection, the cells were thawed rapidly and washed to remove (DMSO), and then re-suspended in PBS to the required cell dose. We evaluated cell viability prior to infusion using trypan blue technique. Only cell doses at and above 70% viability were administered in the study.

The presence of hematopoietic cell surface markers was investigated using fluorescence-activated cell sorting (FACS). Briefly, a total of $0.5-1\times10^6$ MSCs was used for FACS characterization. GFP positivity was detected in 79±5% of MSCs in culture. For surface markers, cells were incubated for 1 hour with FACS buffer (1% bovine serum albumin and 5% FBS diluted in distilled water) on ice, and subsequently 1 hour with the primary and secondary antibodies (washed 3 times for 5 minutes during centrifugation between the primary and secondary and after the secondary). Each analysis included at least 10,000 events and was performed on at least three separate cell preparations (BD FACSAria, University of Miami). All experiments for MSCs characterization were performed between passages 8 to 11 after isolation. Cells were incubated with anti-mouse PE-fluorochrome-conjugated antibodies against CD45, CD73, and CD90.2 (BD Biosciences, San Jose, CA), PE-fluorochrome-conjugated against CD34 (Santa Cruz Biotechnology, Santa Cruz, CA), PE/Cy7-fluorochrome-conjugated against CD105 (BioLegend, San Diego, CA), and their respective isotype controls (BD Biosciences, San Jose, CA).

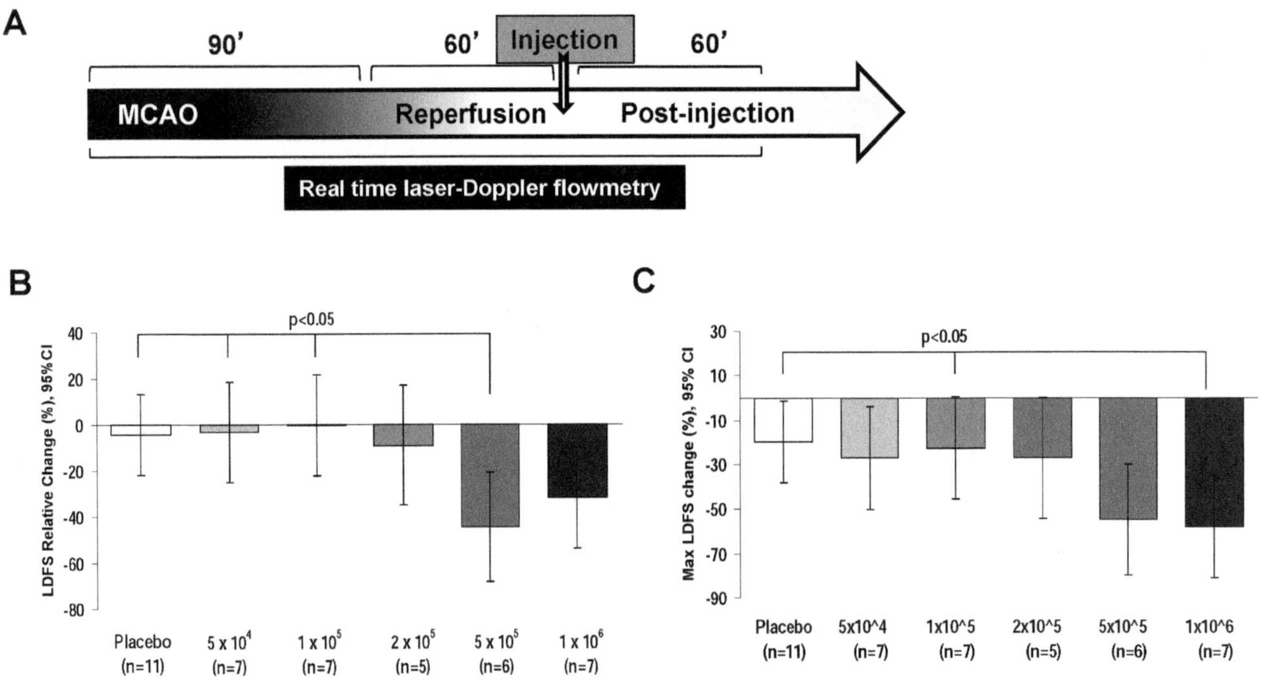

Figure 1. Lower doses of IA MSCs mitigate adverse effect IA injection on MCA blood flow. (A) Experimental timeline showing rMCAo for 90 minutes followed by withdrawal of the suture to allow reperfusion. At 60 minutes of reperfusion, IA MSC or vehicle only injection was given, followed by LDF monitoring for 60 min. (B) Comparison of relative LDFS worsening from baseline to final recording, among de-escalating dose groups (C) The 1×10^5 dose and placebo has significantly less maximum LDFS worsening as compared to the 1×10^6 dose. The comparisons in B and C were done using general linear modeling (GLM) to compare mean differences among groups. LDFS= Laser Doppler Flow Signal.

Intra-arterial MCA infusion

A PE-10 polyethylene catheter connected to a 30G needle and a 1-ml syringe was inserted into the stump of left ECA through an incision on the vessel under the operation microscope and was tightened to ECA by a 5–0 silk suture. The catheter went forward into ICA and avoided PPA, the extracranial branch of ICA, to approach the skull base [29]. The catheter was filled with heparinized physiological saline to prevent air bubble and coagulation. The syringe was filled with MSCs in PBS or PBS only. Using slow hand injection over 3 minutes, 0.5 ml volume of the cells in PBS or PBS only were then delivered into the ICA via the PE-10 catheter.

Histology and detection of MSCs in the brain

Under anesthesia rats were perfused via ascending aorta with FAM (a mixture of 40% formaldehyde, glacial acetic acid, and methanol, 1:1:8 by volume) for 20 min following a 2-min initial perfusion with physical saline. The rat heads were immersed in FAM for 1 day before the brains were removed. The brains were placed in FAM at 4°C for at least 1 additional day, and then coronal brain blocks were embedded in paraffin. All brains were cut into 10- µm thick sections from 5.5 mm to −7.5 mm from bregma at 9 standard levels to cover the whole infracted areas. Sections of the 9 levels were stained with hematoxylin and eosin to display the infracted areas and to obtain infarct volumes.

In a parallel study, adjacent sections were subjected to immunohistochemistry (IHC) studies with Vectastain ABC peroxidase method (Vector Labs, Burlingame, CA). Sections from levels 2–8 in the dose group 1×10^5 & 5×10^5 were tested by GFP-IHC to reveal the MSCs immigration inside the brain via injection through IA. To detect GFP positive MSCs, sections were incubated in the working dilution of 1:100 of anti-GFP (SC-101525, Santa Cruz Biotech, CA.) for 90 min at room temperature, and 3, 3″-diaminobenzidine (DAB) was used for visualization of the primary antibody binding. The dark-brown appearance was considered the specific labeling to indicate the MSCs. Hematoxylin counterstain was conducted to identify cells and brain regions. Blocked vessels were defined as: complete filling of the microvessel caliber with MSCs in axial section or segment of vessel in sagittal section of a vessel. The absolute total number of blocked microvessels per section was reported.

Both positive control, to confirm the negative labeling being truly negative, and negative control, to confirm the positive staining not false positive, were applied in the present study. The two negative controls including 1) mouse IgG 1 (X0931, Dako, Carpenteria, CA) replaced anti-GFP in the female brains of the present study and 2) anti-GFP applied to male rat brains having infarction from previous study were performed. Negative labeling represented in the 2 negative controls supported that the positive staining were true positive. Anti-glial fibrillary acidic protein (GFAP, Dako, Carpenteria) was employed in the IHC study as the positive control to ensure that the reduced sensitivity of reaction not due to improper technique because positive control slide, i.e. the known histological slide containing the specific antigen unavailable [26].

Neurodeficit Scoring

A standardized neurobehavioral test battery was conducted as described previously [27], which includes tests for postural reflex, sensorimotor integration and proprioception. Total neurological score ranged from a normal score of 0 to a maximal possible score of 12.

Statistical analyses

Descriptive statistics for physiological continuous variables were presented as means with standard deviations. For safety analysis, the primary outcome was the relative change in MCA blood flow from pre-injection to the final assessment after injection of MSCs or saline, whereas the secondary outcome was the maximum relative reduction in blood flow within the 60 minutes after injection. General linear model (GLM) was used to compare mean differences among groups in the relative change at the final LDFS measurement and in the maximum relative reduction in blood flow. Mixed model was used for analyzing for comparing LDFS over time between two groups. For efficacy analysis, mixed-effects model was used to assess differences among groups in neurodeficit scores based on the repeated measures. Due to skewed distribution, log transformation was first performed for infarct volume and ANOVA was used to compare the differences among groups at week 4 after treatment. All analyses were performed with SAS 9.3 (SAS Institute Inc., Cary, NC) or SPSS 16.0 (SPSS Inc., Chicago, IL).

Results

Lower IA MSC doses mitigate adverse impact on MCA blood flow

We first determined the impact of lowering the dose of IA MSC on MCA blood flow, starting with a dose of 1 million cells, which was previously shown to severely decrease MCA flow in one-thirds of animals [10]. In the animals receiving $1\times10^6$and 5×10^5 MSCs, there was a considerable decrease in the final post-injection MCA flow by 32% and 45% respectively, compared to the pre-IA MSC injection MCA flow (Figure 1B). Lowering the dose below 5×10^5 resulted in mitigation of the adverse effect of IA MSC injection on MCA flow. At the dose of 2×10^5, the MCA flow compromise was less as compared to the 5×10^5 dose group, but this was non-significant (LDFS decreased by 9% vs 45%, $p=0.06$). At the next lowered dose level of 1×10^5 MSCs, the MCA flow compromise was significantly less as compared to 5×10^5 dose group (LDFS decreased by 0.2% vs 45%, $p=0.01$). When we analyzed the maximum reduction in LDFS among the dose groups, the rats receiving $1\times10^6$and 5×10^5 MSCs had a maximum reduction of LDFS over 50%, whereas those receiving a dose at 1×10^5 MSCs had a maximum reduction in LDFS of 23%, similar to 20% in placebo group (Figure 1C). Furthermore, we found that as compared to the 5×10^5 dose, the 1×10^5 IA MSC dose injection led to less worsening in LDFS post injection over the course of 60 minutes post injection, becoming significantly improved at 45 minutes and 60 minutes post-injection (Figure 2A). Based on these results the 1×10^5 IA MSCs dose was utilized for subsequent studies and defined as maximum tolerated dose (MTD).

Mortality with MCAo and IA MSCs

The mortality rate of the MCAo procedure (premature mortality) in this study was 13.46%. In the part of the study evaluating safety of de-escalating doses, the mortality rate after IA MSC dose administration was: 8.3%, 28.6%, 12.5%, 42.9%, 42.9% and 42.9% in the placebo, 5×10^4, 1×10^5, 2×10^5, 5×10^5 and 1×10^6 IA MSC dose groups respectively.

Impact of MSC Dose on Microvascular Occlusion and MSC Distribution

Consistent with the effect on MCA flow, at 3–5 days post injection, we found a significantly higher number of brain microvessels showing complete occlusion with MSCs in rats

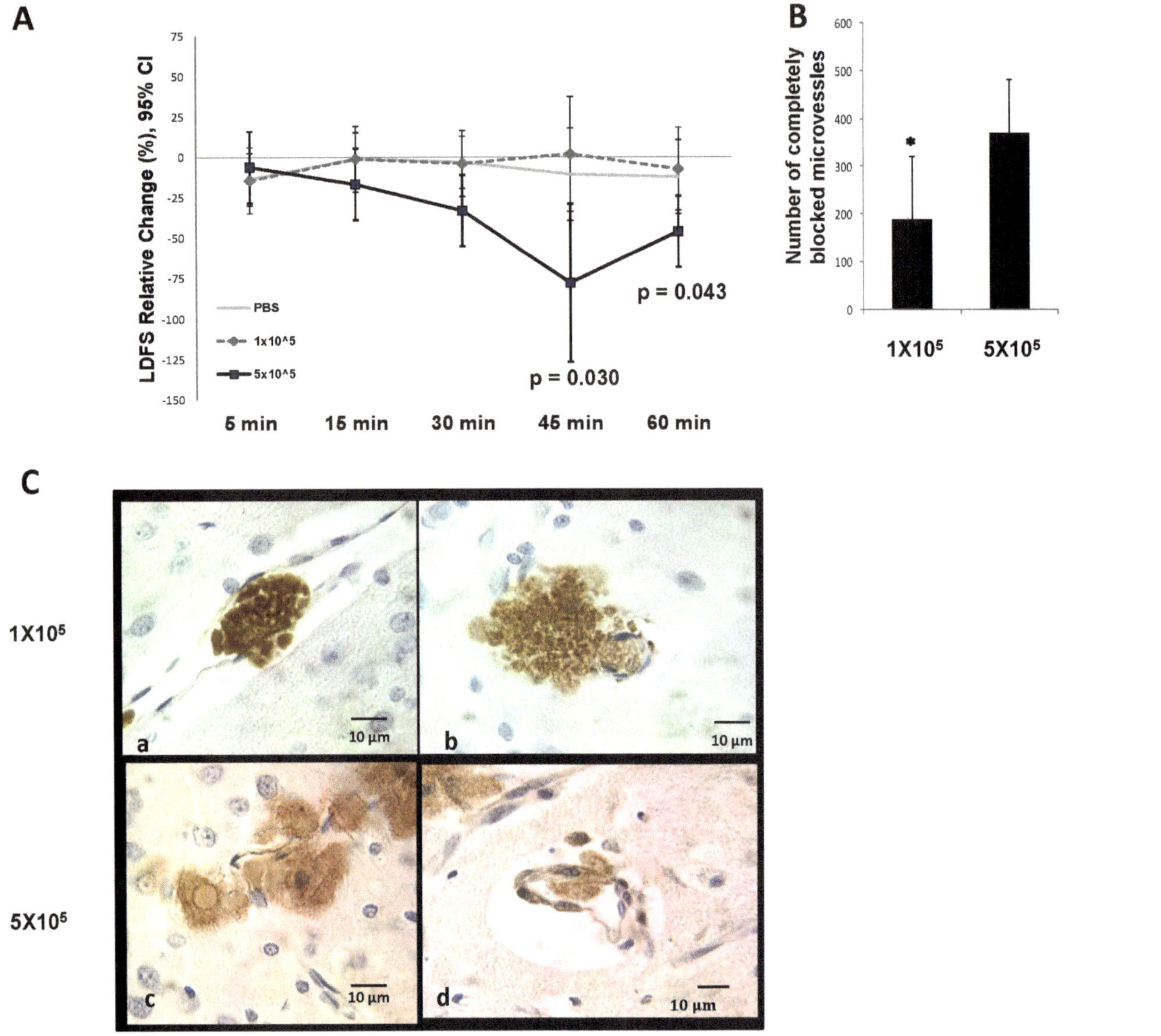

Figure 2. Comparison of LDFS changes over time and microvascular occlusion between dose-groups. (A) IA MSC dose of 1×10^5 and IA placebo have a transient and less MCA LDF worsening during 60 minutes post injection as compared to 5×10^5 dose using mixed model. LDFS = Laser Doppler Flow Signal. (B) On comparing the total number of microvessels with complete occlusion among the two dose groups, there were a significantly lower number of complete occlusions in the lower dose group. Mean ±SD, * $P<0.05$, ANOVA. (C) Representative brain sections from IA MSC 1×10^5 dose group showing GFP+ MSCs identified by 3, 3″-diaminobenzidine (DAB), showing localized complete filling of microvessels at 3–5 days post-injection, as well as MSCs just outside the vessel wall in brain parenchyma and (D) high power field of representative brain sections from IA 5×10^5 dose group showing single MSC partly inside and partly outside microvessel wall.

receiving IA MSCs at the higher dose of 5×10^5 as compared to the MTD of 1×10^5(366.5 vs. 185.75, $p=0.047$) (Figure 2B). However, in rats in both dose groups, MSCs were also seen in the brain parenchyma just outside the microvessels at the level of the blockage. (Fig. 2C) We noted two sections in one rat from the larger dose group in which individual MSCs were located across the vessel wall partly inside the vessel and rest in the adjacent brain parenchyma, consistent with them being in the process of diapedesis.

Efficacy Study of MTD of IA MSCs

We chose the 1×10^5 MSCs as the dose to use for our efficacy study of IA MSCs in rMCAo based on the above results showing that this was the highest dose group that had minimal adverse effect on MCA LDF post-injection. We compared the efficacy of IA MSCs given at 24 hours (IA MSC_24 h), at one hour (IA MSC_1 h), IV MSCs at a ten-fold dose given at 24 hours (IV MSC_24 h) and IC PBS given at 24 hours (IA PBS_24 h) post rMCAo. On mixed model analysis we found a significant interaction between treatment group and time point ($p=0.02$), with no significant difference among the groups at day 1 ($p=0.49$) and most significant difference at 28 days ($p<0.0001$) after treatment (Figure 3B). The IA MSC_24 h group had a significantly lower neurodeficit score (5.8, 95% CI: 4.4 to 7.2) at 28 days after treatment as compared to IA PBS_24 h (10.4, 8.5 to 12.3, $p<0.0001$), as well as to IV MSC_24 h at dose of 1×10^6 (10.3, 8.8 to 11.7, $p=0.0002$), and IA MSC_1 h (9.2, 7.3 to 11.1,

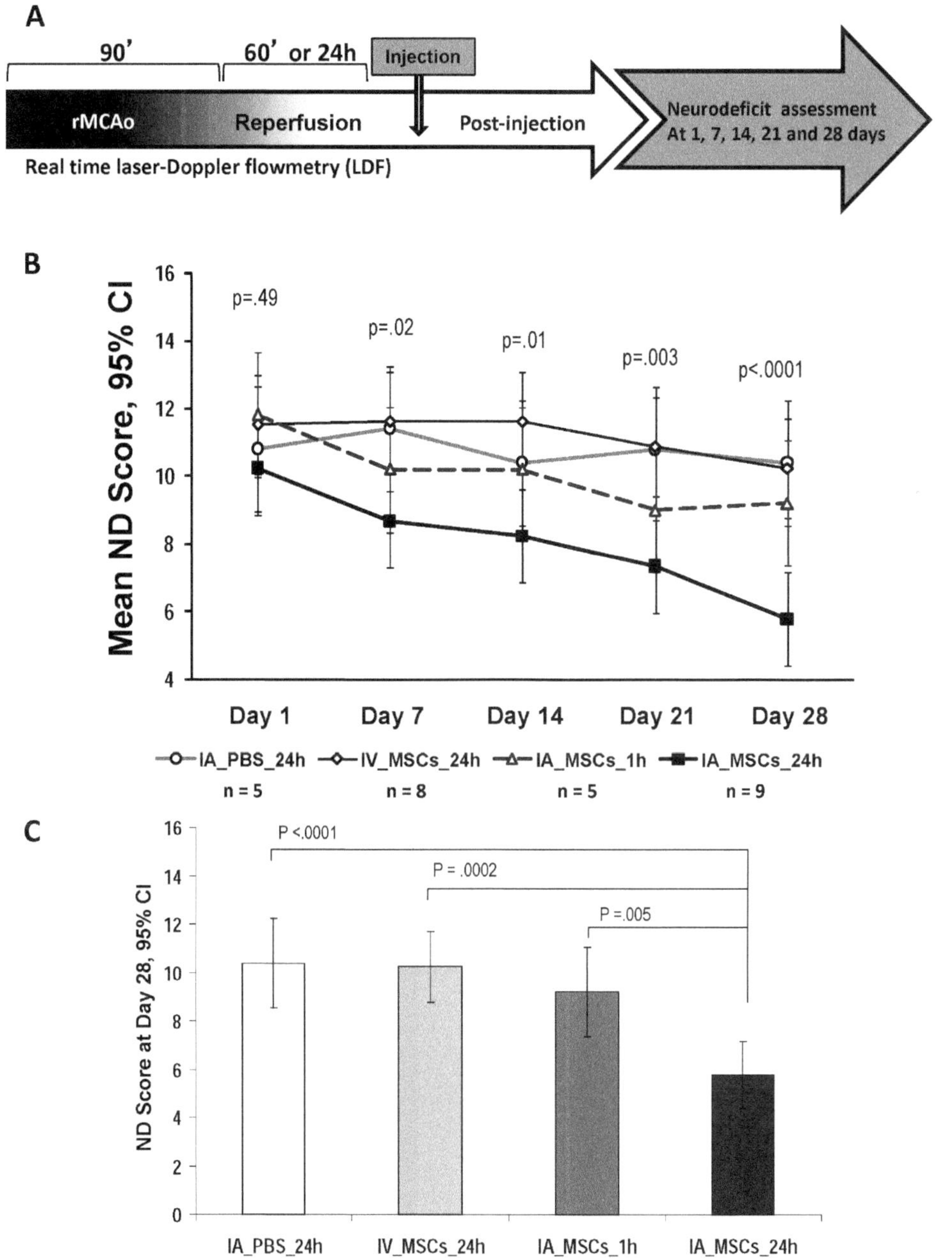

Figure 3. Functional neurologic outcomes are superior in the IA MSC_24 h group. (A) Experimental timeline of the efficacy study showing 90 min rMCAo followed by 24 hour reperfusion except in group receiving IA MSCs at 1 hour reperfusion followed by ND score assessment at 1,7,14,21 & 28 days. (B) On day 1 post rMCAo, the ND scores were not significantly different among groups. The ND score of the 1×10^5 group progressively decreased over time and at 28 days was significantly lower than the other groups. C, The day 28 ND score was significantly lower in the IA MSC_24 h group as compared to all the remaining groups. ND = Neurodeficit

$p = 0.005$) groups. The IA MSC_1 h group showed no significant difference in functional outcome compared to the IA PBS_24 h ($p = 0.37$) and IV MSC_24 h ($p = 0.38$) groups (Figure 3C).

Also, the mean infarct volume of the group treated with IA MSC_24 h group (4.6 mm3, 1.2 to 12.72) was significantly lower as compared to IA PBS_24 h (39.5, 11 to 136.1, $p = 0.01$) The groups treated with IV_24 h MSCs and IA MSC_24 h did not show a significant reduction in their infarct volume as compared to IA PBS_24 h (Fig. 4A). In order to compare the location of infarcted brain, we used an infarct topography frequency map, which showed a significantly decreased volume of infarction in the penumbral region in the IA MSC_24 h group as compared to the IA PBS_24 h group (Fig 4B).

Discussion

The major challenge to the IA method of delivery of MSCs is the possibility of vascular occlusion in small arterioles and capillaries owing to the large size of the cells ranging from 15–50 microns [10]. We report a major new finding of a dose response relationship in safety of IA delivery of allogeneic IA MSCs and

A

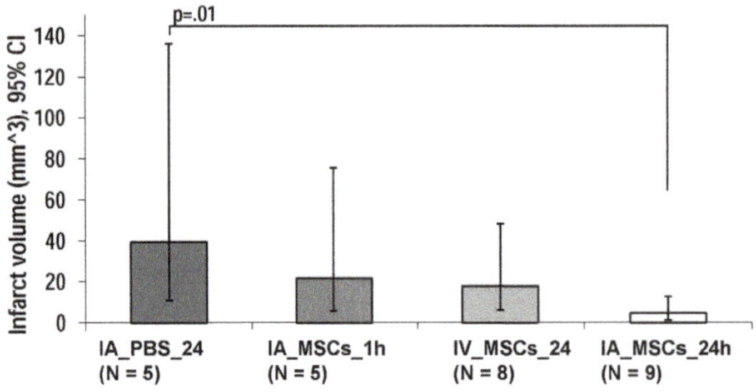

B

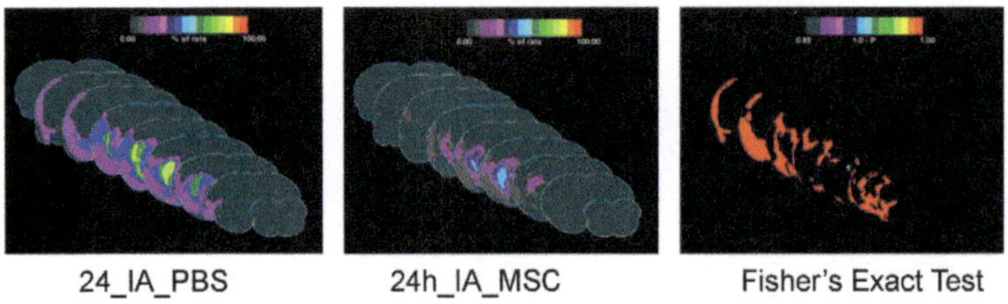

Figure 4. Comparison of infarct volume among treatment groups. (A) Geometric mean infarct volumes are compared among groups after log transformation to achieve normal distribution. Only the IA MSC_24 h group shows significantly reduced infarct volume as compared to the IA PBS_24 h. B, Frequency infarct map statistically comparing the location of the mean infarct volume in the IA MSC_24 h and IA PBS_24 h groups using color- coded representation of the percent of rats showing infarction in each brain region using "Fisher's exact test" with a color-coded representation of the "p-value", the color bar in "1-p" format. Displays are in "Coronal presentation"; middle sections are selected. The IA MSC_24 h group shows a much reduced infarction frequency, particularly surrounding the core as quantized by the Fisher test.

that lowering of the IA dose overcomes the decrease in blood flow in the middle cerebral artery after IA MSC injection in a rat model of acute ischemic stroke. Furthermore, the lower safe dose of 1×10^5 MSCs, when given IA at 24 hours results therapeutic effect to reduce infarct volume and improve functional status. Together these findings support further translational investigation to develop intra-arterial allogeneic MSCs as a novel therapy for AIS in the first 24 hours after stroke onset.

With the increasing number of catheter-based endovascular treatments for patients with AIS, intra-arterial delivery of stem cells has tremendous potential for clinical translation. Intra-arterial delivery of bone marrow mononuclear cells into the affected MCA was found to be safe within 3–7 days in a 20 patients trial of patients with MCA strokes [30]. The strategy of endovascular intra-arterial delivery of MSCs has been successfully implemented in early clinical studies of intra-coronary-delivered MSCs after acute myocardial infarction [31,32]. Given this tremendous promise for clinical application, pre-clinical studies addressing crucial translational hurdles in intra-arterial delivery of stem cells, including MSCs in stroke, are crucial in the efforts to bring this therapy to the bedside.

A previous study of IA MSC delivery reported severely reduced ipsilateral MCA blood flow (80–90%) in 35%, and moderately reduced (10–30%) in 47%, of animals after intra-carotid delivery of 1×10^6 MSCs given 30 minutes after rMCAo [10]. On dose de-escalation, starting at the dose of 1×10^6, we found a similar reduction in CBF signal on IA MSC delivery at the doses of 1×10^6 and 5×10^5. This adverse effect was somewhat ameliorated at the 2×10^5dose and resolved with further lowering of the dose to 1×10^5 consistent with a dose-response relationship in the safety of IA delivery of MSCs in rMCAo. In another recent study the safety of MSC delivery was tested using non-MCAo rats [33]. Lowering of MSCs dose from 2×10^6 to 1×10^6 dose decreased stroke lesions on MRI the rodent rMCAo model. Stroke lesions occurred frequently (12 out of 15 animals) when injecting 2×10^6 MSCs, but not after lowering the dose to 1×10^6 cells. While the decrease in dose of MCSs led to better safety, similar to our study, the safety of the 1×10^6 dose is likely due to the use of non-MCAo rats in that study. The twice bigger dose 2×10^6 produced frequent micro-strokes – mostly in the area of corpus callosum. With regards to mortality post cell injection, the groups receiving 2×10^5 and higher doses of MSCs had a mortality rate of 42.9% similar to that seen in other studies [10]. The 1×10^5 dose group and the PBS group had a numerically a lower mortality rate (12.5% and 8.3% respectively) as compared to the higher dose groups although not statistically significant. Similarly, mortality in 5×10^4 group is numerically higher than 1×10^5 however, there is no statistically significant difference between two groups. The surgical procedure of MCAo is highly invasive and observed mortality could be confounded by the mortality related to the MCAo in the safety phase of our study.

Despite approximately three times slower speed of injection as compared to a previous study we did not see mitigation of the adverse effect on blood flow in the MCA at the dose of 1×10^6 MSCs, the only dose tested in that study [10]. The speed of cell injection from our rodent study and others cannot be directly translated in humans. Apart from the large difference in scale, the size of microcatheters available for clinical use allow for excellent blood flow around the microcatheters in the ICA whereas, the PE 10 has a very snug fit in the rodent ICA. Testing of the IA injection speed into ICA in larger animal species may hence be needed.

The reduction of CBF signal over the MCA at the dose of 5×10^5 is seen shortly after intra-carotid delivery of MSCs at the higher doses and progressively worsens over 60 minutes (Figure 2A), suggesting a mechanical obstruction in the MCA, likely caused by microvascular "sludging" owing to stacking of cells in the microvasculature at higher doses. This is supported by our histology results that showed significantly increased microvascular occlusion at 3–5 days at the higher dose as compared to the lower dose and no evidence for other possible mechanisms for microvascular occlusion such as thrombosis or spasm. The occlusion of microvessels might be due to the higher ratio of the size of some MSCs (some being 15–50 microns in size) to the microvessels size. Along with intravascular MSCs, we found the vast majority of extravascular MSCs to be in the peri-vascular regions at this early time point. We found two separate instances in two separate animal brains of MSCs positioned across the endothelial lining partly inside a capillary and partly in the parenchyma, suggestive of transendothelial migration as the mechanism of transport of these cells into the brain, (Figure 2C, D) as seen in previous studies of intra-carotid stem cell infusion [34].

In the current study we observed the maximum reduction in LDFS was 20% following PBS delivery. Most importantly, the observed reduction in LDFS after PBS delivery occurred shortly after injection and was transient, lasting for short duration (~5 min) and reverted back to the baseline. In contrast, at the higher doses of 5×10^5 and 1×10^6 of IA MSCs, the reduction in LDFS lasted throughout the period of recording (60 min). Hence, the observed transient initial reduction with PBS injection is very unlikely to lead to worsening of ischemic injury

In our study, the animals receiving IA MSCs at the safe dose of 1×10^5 at 24 hours post rMCAo had a clearly superior functional neurologic outcome at one month as compared to controls receiving IA saline as well as compared to IV MSCs at the same time point. In parallel with the superior functional outcome, we also found the infarct volume to be significantly smaller only in the IA MSC_24 h group as compared to IA PBS_24 h. This decrease in infarct volume was found to be in the penumbral area (Figure 4B) in this group as compared to control suggesting that MSC-mediated neuroprotection as the likely mechanism of benefit. This mechanism has been confirmed in other studies that showed increased concentration of anti-apoptotic factors in the peri-infarct area in animals treated with MSCs.

While previous studies have shown efficacy of IA autologous bone marrow mononuclear cells (BMMCs) and fetal-derived neural stem cells [34,35], we found that allogeneic MSCs given via the IA route are efficacious in ameliorating neurological deficits in rodent cerebral ischemia without a need for immunosuppression. The application of allogeneic MSCs can facilitate rapid "off-the-shelf" therapy in patients with acute ischemic stroke. We found that not only of 1×10^5 IA MSCs lead to superior functional neurological recovery as compared to the control group given IA PBS, but also as compared to the group given IV MSCs at a 10-fold higher dose, indicating a robust relative efficacy of the IA route of delivery and its superiority to the IV route when MSCs are administered at 24 hours post rMCAo. Our results are consistent with the functional neurological benefit previously reported [15] with intra-arterial allogeneic MSCs administered at 24 hours post rMCAo. However, the dose of IA MSCs in our study was 10-fold lower as compared to the previous study. Our finding of efficacy at the MTD being safer as compared to higher doses, is novel and supports careful dose escalation studies in clinical translation of IA MSC therapy. In a previous study comparing the IA and IV routes of delivery of autologous BMMCs, efficacy was seen in the IA group but not the IV group [23]. Our results confirm a similar superiority of the IA route but add to this previous finding, as we also found a clear superiority of IA over the IV route on direct comparison of the effect of the two routes of delivery on functional recovery. Of note, in order to compensate for this systemic trapping of IV delivered MSCs, we chose an IV MSC dose that is 10 fold higher than the IA MSC dose so as to serve as an adequate control to IA MSCs that do not undergo systemic trapping. The choice of the IV dose was based on studies showing that IA administration of cells leads to a 7 to 12 times higher number of cells reaching the brain as compared to IV cell administration [34,36,37].

The optimal timing of IA stem cell delivery after AIS is unknown. We found that allogeneic IA MSCs given at 24 hours results in efficacy but not when given at 60 minutes after cerebral ischemia. The fact that there was no reduction in the infarct volume in the IA MSC_1 h group indicates that neuroprotection was not significant when MSCs are given at this hyperacute timing after rMCAo. This important finding may be a result of higher concentration of chemoattractant factors for transendothelial transport of MSCs at the more delayed time point, but we did not test this hypothesis. Another study has shown results consistent with the need to delay the administration of IA stem cells after stroke for efficacy [38]. In the prior study of IA neural stem cells in a mouse model of hypoxia-ischemia, authors found that increased phagocytosis of the cells in the post-ischemic brain when transplanted within 24 hours whereas transplantation at 3 days led to lower co-localization of MSCs with pan-monocytic marker Iba-1. These findings suggest that in the hyperacute phase the environment in the post-ischemic brain tissue may be detrimental to the administered IA cells and waiting for this inflammatory response to wane in the first few days is advantageous [38]. Therefore, while administration of IA stem cells via intra-carotid catheters immediately after endovascular reperfusion therapy for ischemic stroke in patients would avoid a second endovascular procedure to administer the cells a few days later, our data suggest that this hyperacute phase may not be an optimal treatment time point for IA cell therapy. Thus early clinical trial designs of IA allogeneic MSCs in ischemic stroke should consider administration of the cells in the first few days after stroke and not in conjunction or immediately following IA thrombolytic therapy.

One of the major limitations of the current study is that we did not test the possibility of slower IA injection to decrease cell sludging. However, the finding of efficacy at the MTD in our study offsets the concern that the higher doses than MTD may be necessary for benefit. Also, we did not comprehensively study the mechanism of action of MSCs, as our primary goal was safety and efficacy of the IA route of MSC delivery. We did not test for time points beyond 24 hours for IA allogeneic MSC injection. Although there is evidence for lack of efficacy of IA BMMC, as beyond 24 hours after rMCAo [39], the outer limit of the treatment time window remains to be established for allogeneic IA MSCs. Additionally, in the present study we monitored the LDF in the MCA for 60 minutes after IA MSCs delivery. However, effect of IA MSCs on LDF beyond 60 minutes is not known and remains to be investigated. Apart from LDF, the information such as presence of MSCs at perivascular locations and overall MSCs survival following IA MSCs needs investigation.

The present study used an experimental model of ischemic stroke in female rats to investigate efficacy of IA MSCs. The MSCs used for IA delivery were generated from male rats. Our approach of delivering MSCs derived from male rats in to the female experimental animals offer an added advantage of localizing MSCs in the brain by cytogenetic approach after IA or IV delivery in future. On the other hand, our use of female rats remains a limitation because it is well known that the circulating ovarian hormones influence ischemic pathology [40]. Future studies employing ovarian hormones-deprived (ovariectomized) female rats are needed to confirm the findings of current study.

In conclusion IA delivery of MSCs at low dose (1×10^5 cells) did not compromise MCA blood flow in a rat model of MCAo. Importantly, sub-acute delivery of MSCs at lowered dose protected brain from ischemic damage. Our study findings outline the core principles of MSC therapy for ischemic stroke using IA delivery. A clinical translational impact of these results needs to be determined in the future.

Supporting Information

Table S1 Physiological variables of protocol I. Values are mean ± SD (MABP, mean arterial blood pressure).

Table S2 Physiological variables of protocol II. Values are mean ± SD (MABP, mean arterial blood pressure).

Acknowledgments

We thank Drs. Myron Ginsburg, MD and Brant Watson, PhD for their critical review of the manuscript.

Author Contributions

Conceived and designed the experiments: DRY JH. Performed the experiments: BL. Analyzed the data: PSG AR CD WZ. Contributed reagents/materials/analysis tools: JH IM ER. Wrote the paper: DRY AR. Critically read manuscript: TR MAPP RLS JH.

References

1. Lloyd-Jones D, Adams RJ, Brown TM, Carnethon M, Dai S, et al. (2010) Executive summary: heart disease and stroke statistics—2010 update: a report from the American Heart Association. Circulation 121: 948–954.
2. Del Zoppo GJ, Saver JL, Jauch EC, Adams HP, Jr., American Heart Association Stroke C (2009) Expansion of the time window for treatment of acute ischemic stroke with intravenous tissue plasminogen activator: a science advisory from the American Heart Association/American Stroke Association. Stroke 40: 2945–2948.
3. Del Zoppo GJ, Moskowitz MA (2009) Translating interventions from ischemic stroke models to patients: the view in 2009. Front Neurol Neurosci 25: 34–38.
4. STEPS (2009) Stem Cell Therapies as an Emerging Paradigm in Stroke (STEPS): bridging basic and clinical science for cellular and neurogenic factor therapy in treating stroke. Stroke 40: 510–515.
5. Chen J, Li Y, Wang L, Zhang Z, Lu D, et al. (2001) Therapeutic benefit of intravenous administration of bone marrow stromal cells after cerebral ischemia in rats. Stroke 32: 1005–1011.
6. Mimura T, Dezawa M, Kanno H, Yamamoto I (2005) Behavioral and histological evaluation of a focal cerebral infarction rat model transplanted with neurons induced from bone marrow stromal cells. J Neuropathol Exp Neurol 64: 1108–1117.
7. Bliss T, Guzman R, Daadi M, Steinberg GK (2007) Cell transplantation therapy for stroke. Stroke 38: 817–826.
8. Williams BA, Keating A (2008) Cell therapy for age-related disorders: myocardial infarction and stroke—a mini-review. Gerontology 54: 300–311.
9. Pittenger MF, Martin BJ (2004) Mesenchymal stem cells and their potential as cardiac therapeutics. Circ Res 95: 9–20.
10. Walczak P, Zhang J, Gilad AA, Kedziorek DA, Ruiz-Cabello J, et al. (2008) Dual-modality monitoring of targeted intraarterial delivery of mesenchymal stem cells after transient ischemia. Stroke 39: 1569–1574.
11. Uccelli A, Moretta L, Pistoia V (2006) Immunoregulatory function of mesenchymal stem cells. Eur J Immunol 36: 2566–2573.
12. Bhatia R, Hare JM (2005) Mesenchymal stem cells: future source for reparative medicine. Congest Heart Fail 11: 87–91; quiz 92–83.
13. Williams AR, Hare JM (2011) Mesenchymal stem cells: biology, pathophysiology, translational findings, and therapeutic implications for cardiac disease. Circ Res 109: 923–940.
14. Liu X, Ye R, Yan T, Yu SP, Wei L, et al. (2013) Cell based therapies for ischemic stroke: From basic science to bedside. Prog Neurobiol pii: S03010082(13)00131-7.
15. Chen J, Li Y, Wang L, Lu M, Zhang X, et al. (2001) Therapeutic benefit of intracerebral transplantation of bone marrow stromal cells after cerebral ischemia in rats. J Neurol Sci 189: 49–57.
16. Blum A, Balkan W, Hare JM (2012) Advances in cell-based therapy for peripheral vascular disease. Atherosclerosis 223: 269–277.
17. Ikeda N, Nonoguchi N, Zhao MZ, Watanabe T, Kajimoto Y, et al. (2005) Bone marrow stromal cells that enhanced fibroblast growth factor-2 secretion by herpes simplex virus vector improve neurological outcome after transient focal cerebral ischemia in rats. Stroke 36: 2725–2730.
18. Li Y, Chen J, Wang L, Lu M, Chopp M (2001) Treatment of stroke in rat with intracarotid administration of marrow stromal cells. Neurology 56: 1666–1672.
19. Zhang L, Li Y, Romanko M, Kramer BC, Gosiewska A, et al. (2012) Different routes of administration of human umbilical tissue-derived cells improve functional recovery in the rat after focal cerebral ischemia. Brain Res 1489: 104–112.
20. Cloft HJ, Rabinstein A, Lanzino G, Kallmes DF (2009) Intra-arterial stroke therapy: an assessment of demand and available work force. AJNR Am J Neuroradiol 30: 453–458.
21. Li L, Jiang Q, Ding G, Zhang L, Zhang ZG, et al. (2010) Effects of administration route on migration and distribution of neural progenitor cells transplanted into rats with focal cerebral ischemia, an MRI study. J Cereb Blood Flow Metab 30: 653–662.
22. Fischer UM, Harting MT, Jimenez F, Monzon-Posadas WO, Xue H, et al. (2009) Pulmonary passage is a major obstacle for intravenous stem cell delivery: the pulmonary first-pass effect. Stem Cells Dev 18: 683–692.
23. Kamiya N, Ueda M, Igarashi H, Nishiyama Y, Suda S, et al. (2008) Intra-arterial transplantation of bone marrow mononuclear cells immediately after reperfusion decreases brain injury after focal ischemia in rats. Life Sci 83: 433–437.
24. Busto R, Dietrich WD, Globus MY, Valdes I, Scheinberg P, et al. (1987) Small differences in intraischemic brain temperature critically determine the extent of ischemic neuronal injury. J Cereb Blood Flow Metab 7: 729–738.
25. Longa EZ, Weinstein PR, Carlson S, Cummins R (1989) Reversible middle cerebral artery occlusion without craniectomy in rats. Stroke 20: 84–91.
26. Lin B, Ginsberg MD (2000) Quantitative assessment of the normal cerebral microvasculature by endothelial barrier antigen (EBA) immunohistochemistry: application to focal cerebral ischemia. Brain Res 865: 237–244.
27. Ley JJ, Vigdorchik A, Belayev L, Zhao W, Busto R, et al. (2005) Stilbazulenyl nitrone, a second-generation azulenyl nitrone antioxidant, confers enduring neuroprotection in experimental focal cerebral ischemia in the rat: neurobehavior, histopathology, and pharmacokinetics. J Pharmacol Exp Ther 313: 1090–1100.
28. Belayev L, Busto R, Zhao W, Fernandez G, Ginsberg MD (1999) Middle cerebral artery occlusion in the mouse by intraluminal suture coated with poly-L-lysine: neurological and histological validation. Brain Res 833: 181–190.
29. Woitzik J, Schilling L (2007) A new method for superselective middle cerebral artery infusion in the rat. J Neurosurg 106: 872–878.
30. Friedrich MA, Martins MP, Araujo MD, Klamt C, Vedolin L, et al. (2011) Intra-Arterial Infusion of Autologous Bone-Marrow Mononuclear Cells in Patients with Moderate to Severe Middle-Cerebral-Artery Acute Ischemic Stroke. Cell Transplant 21 Suppl 1:S13–21.
31. Strauer BE, Brehm M, Zeus T, Kostering M, Hernandez A, et al. (2002) Repair of infarcted myocardium by autologous intracoronary mononuclear bone marrow cell transplantation in humans. Circulation 106: 1913–1918.
32. Assmus B, Schachinger V, Teupe C, Britten M, Lehmann R, et al. (2002) Transplantation of Progenitor Cells and Regeneration Enhancement in Acute Myocardial Infarction (TOPCARE-AMI). Circulation 106: 3009–3017.
33. Janowski M, Lyczek A, Engels C, Xu J, Lukomska B, et al. (2013) Cell size and velocity of injection are major determinants of the safety of intracarotid stem cell transplantation. J Cereb Blood Flow Metab 33: 921–927.
34. Guzman R, De Los Angeles A, Cheshier S, Choi R, Hoang S, et al. (2008) Intracarotid injection of fluorescence activated cell-sorted CD49d-positive neural stem cells improves targeted cell delivery and behavior after stroke in a mouse stroke model. Stroke 39: 1300–1306.
35. Brenneman M, Sharma S, Harting M, Strong R, Cox CS, Jr., et al. (2010) Autologous bone marrow mononuclear cells enhance recovery after acute ischemic stroke in young and middle-aged rats. J Cereb Blood Flow Metab 30: 140–149.
36. Pendharkar AV, Chua JY, Andres RH, Wang N, Gaeta X, et al. (2010) Biodistribution of neural stem cells after intravascular therapy for hypoxic-ischemia. Stroke 41: 2064–2070.

37. Pluchino S, Zanotti L, Rossi B, Brambilla E, Ottoboni L, et al. (2005) Neurosphere-derived multipotent precursors promote neuroprotection by an immunomodulatory mechanism. Nature 436: 266–271.
38. Rosenblum S, Wang N, Smith TN, Pendharkar AV, Chua JY, et al. (2012) Timing of intra-arterial neural stem cell transplantation after hypoxia-ischemia influences cell engraftment, survival, and differentiation. Stroke 43: 1624–1631.
39. Minnerup J, Seeger FH, Kuhnert K, Diederich K, Schilling M, et al. (2010) Intracarotid administration of human bone marrow mononuclear cells in rat photothrombotic ischemia. Exp Transl Stroke Med 2: 3.
40. Raval AP, Saul I, Dave KR, DeFazio RA, Perez-Pinzon MA, et al. (2009) Pretreatment with a single estradiol-17beta bolus activates cyclic-AMP response element binding protein and protects CA1 neurons against global cerebral ischemia. Neuroscience 160: 307–318.

10

Attenuation of Hind-Limb Ischemia in Mice with Endothelial-Like Cells Derived from Different Sources of Human Stem Cells

Wing-Hon Lai[1], Jenny C. Y. Ho[1,2], Yau-Chi Chan[1], Joyce H. L. Ng[1], Ka-Wing Au[1], Lai-Yung Wong[1], Chung-Wah Siu[1,2], Hung-Fat Tse[1,2]*

1 Cardiology Division, Department of Medicine, Queen Mary Hospital, the University of Hong Kong, Hong Kong, HKSAR, China, **2** Research Center of Heart, Brain, Hormone and Healthy Aging, Li Ka Shing Faculty of Medicine, the University of Hong Kong, Hong Kong, HKSAR, China

Abstract

Functional endothelial-like cells (EC) have been successfully derived from different cell sources and potentially used for treatment of cardiovascular diseases; however, their relative therapeutic efficacy remains unclear. We differentiated functional EC from human bone marrow mononuclear cells (BM-EC), human embryonic stem cells (hESC-EC) and human induced pluripotent stem cells (hiPSC-EC), and compared their *in-vitro* tube formation, migration and cytokine expression profiles, and *in-vivo* capacity to attenuate hind-limb ischemia in mice. Successful differentiation of BM-EC was only achieved in 1/6 patient with severe coronary artery disease. Nevertheless, BM-EC, hESC-EC and hiPSC-EC exhibited typical cobblestone morphology, had the ability of uptaking Dil-labeled acetylated low-density-lipoprotein, and binding of *Ulex europaeus* lectin. *In-vitro* functional assay demonstrated that hiPSC-EC and hESC-EC had similar capacity for tube formation and migration as human umbilical cord endothelial cells (HUVEC) and BM-EC ($P>0.05$). While increased expression of major angiogenic factors including epidermal growth factor, hepatocyte growth factor, vascular endothelial growth factor, placental growth factor and stromal derived factor-1 were observed in all EC cultures during hypoxia compared with normoxia ($P<0.05$), the magnitudes of cytokine up-regulation upon hypoxic were more dramatic in hiPSC-EC and hESC-EC ($P<0.05$). Compared with medium, transplanting BM-EC (n = 6), HUVEC (n = 6), hESC-EC (n = 8) or hiPSC-EC (n = 8) significantly attenuated severe hind-limb ischemia in mice via enhancement of neovascularization. In conclusion, functional EC can be generated from hECS and hiPSC with similar therapeutic efficacy for attenuation of severe hind-limb ischemia. Differentiation of functional BM-EC was more difficult to achieve in patients with cardiovascular diseases, and hESC-EC or iPSC-EC are readily available as "off-the-shelf" format for the treatment of tissue ischemia.

Editor: Masuko Ushio-Fukai, University of Illinois at Chicago, United States of America

Funding: This work was supported by the Collaborative Research Fund of Hong Kong Research Grant Council (HKU 8/CRF/09, http://ugc.edu.hk/eng/rgc/crf/crf.htm); Theme-Based Research Scheme of Hong Kong University Grants Committee (T12-705/11, http://www.ugc.edu.hk/eng/rgc/theme/theme.htm); CRCG Small Project Funding of University of Hong Kong (Project No. 201007176290, JCH, http://www.rss.hku.hk/fund/small-project-funding); and Mr. Philip Wong Foundation Fund for Cardiac Stem Cell Research. The funders had no role in study design, data collection and analysis, decision to publish, or preparation of the manuscript.

Competing Interests: The authors have declared that no competing interests exist.

* E-mail: hftse@hkucc.hku.hk

These authors contribute equally to this work.

Introduction

In experimental studies [1] and clinical trials [2–4], human bone marrow (BM) derived stem cells have been used as potential cell therapy to enhance angiogenesis in ischemic tissue, however, their therapeutic application are limited by the ability to obtain sufficient number of stem cells for transplantation. Furthermore, aging and underlying cardiovascular risk factors [5], such as diabetes and heart failure [6] in patients with cardiovascular diseases affect the number and capacity of their circulating BM-derived progenitor cells [7,8] which further hamper the effectiveness of autologous BM cell therapy.

Prior studies [9] have demonstrated that functional endothelial-like cells (EC) can be derived from pluripotent human embryonic stem cells (hESC) for therapeutic angiogenesis. Nonetheless, the clinical application of hESC-EC is limited by the ethical issue related to the generation of hESC as well as the potential risk of immune rejection. Recent advances in the generation of human induced pluripotent stem cells (hiPSC) [10,11,12] may represent an alternative source to derive functional EC for treatment of cardiovascular diseases. As hiPSC can be generated in a patient-specific manner, which is genetically identical to individual patients, they can avoid the potential immune rejections after transplantation as well as the ethical concerns on generating hESC [11]. Indeed, functional EC have been successfully derived from hiPSC [13]. However, the therapeutic efficacy of hiPSC derived EC for angiogenesis as compared with hESC-EC or BM stem cells derived EC have not been studied. Here, we compared the *in-vitro* and *in-vivo* angiogenic effects of EC derived from hiPSC, hESC and human BM mononuclear cells.

Methods

Ethics Statement

The patient study was approved by the Institutional Review Board of the University of Hong Kong/Hospital Authority Hong Kong West Cluster (HKCTR-725, http://www.hkclinicaltrials.com), and all subjects provided written informed consent. The animal study protocol conforms to the Guide for the Care and Use of Laboratory Animals published by the United States National Institutes of Health and was approved by the ethics committee of the University of Hong Kong (1896–09).

Maintenance and Differentiation of Functional Endothelial-like Cells

Undifferentiated hiPSC: IMR90-iPSC (hiPSC-1, passage 15–25) and KS1-iPS (hiPSC-2, passage 15–25) [14] and hESC line: H1 (passage 40–50, WiCell Research Institute, Madison, WI) were maintained on Matrigel™ (BD Biosciences, MA)-coated dishes with mTeSR-1™ medium (Stem Cell Technologies, BC, Canada). Differentiation of EC from hiPSCs and hESCs were induced using formation of embryoid bodies (EB). Briefly, cell clusters were digested by 1 mg/ml dispase (Gibco, Gaithersburg, MD) and re-suspended in differentiation medium which consists of knockout-DMEM with 20% fetal calf serum (Hyclone, Logan, UT), 2 mM L-glutamine, 0.1 mM non-essential amino acids and 0.1 mM β-mercaptoethanol (Invitrogen, Carlsbad, CA) in non-coated dishes for 9 days. The resulting EB were plated on gelatin-coated 10-cm dish for 7 days. Then the central portions of the attached EBs were manually dissected out for further expansion using endothelial growth medium-2 (EGM-2, Lonza, Walkersville, MD) for 14 days. $CD45^-$ $CD31^+$ cells were then isolated by MoFlo XPD cell sorter (Beckman-Coulter, Fullerton, CA) and designated as hESC-EC and hiPSC-EC. For characterization, 4.8 μg/ml of DiI labeled acetylated low-density lipoprotein (DiI-AcLDL, Molecular Probes, Eugene, OR) was added and incubated at 37°C for 5 hour. Cells were washed by phosphate buffered saline, fixed in 2% paraformaldehyde for 10 minutes and then stained by 10 μg/ml of *Ulex europaeus* Lectin-FITC (Sigma Aldrich, St. Louis, MO) for 1 hour at room temperature [9]. CD31, von Willebrand factor (vWF) and intercellular adhesion molecule-1 (ICAM-1) immunofluorescence staining was performed by the endothelial cell characterization kit (Chemicon, Temecula, CA). Fluorescence-activated cell analysis (FACS) was performed with PE-labeled antibodies against CD31 (BD Bioscience, San Jose, CA), vWF (Beckman Coulter, Indianapolis, IN) and Kinase insert domain receptor (KDR; Sigma, St Louis, MO). Human umbilical cord endothelial cells (HUVEC) were cultured under standard condition in EGM-2 growth medium (Lonza, walkerville, MD) as a stable human endothelial cell control.

Differentiation of EC from Human Bone Marrow Cells

BM-MNCs were obtained from patients recruited into the placebo arm of our previous clinical trial on the use of direct endomyocardial transplantation of BM mononuclear cells for treatment of end-staged ischemic heart diseases [15] (Table 1). From each patient, 40 ml of BM blood was obtained from right iliac crest after local anesthesia. In brief, BM mononuclear cells were isolated by Ficoll (GE Healthcare, Amersham, UK) density gradient centrifugation, and plated to gelatin-coated plate at a density of 1×10^6 per ml in 6 wells plate with EGM2 medium. The viability of the cells at the time of harvest was greater than 95%. Attached cells were harvested at Day 14 with $CD31^+$ sorting as described above and designated as BM derived endothelial-like cells (BM-EC).

Angiogenic Tube Formation and Migration Assay

Tube formation of the hiPSC-1-EC, hiPSC-2-EC, hESC-EC, BM-EC and HUVEC were assessed with the *In-Vitro* Angiogenesis Assay Kit (Chemicon, Temecula, CA) with 1×10^4 cells as described with modifications [9]. Modified Boyden Chamber assay was preformed with 1×10^4 cells of cells (hiPSC-1-EC, hiPSC-2-EC, hESC-EC, BM-EC and HUVEC) placed in the upper chamber of the Transwell® pore Polycarbonate Membrane Insert (Corning, Lowell, MA) in EBM2 medium with 1% fetal bovine serum. The chamber was placed in a 24-well plate containing EGM2 medium with or without 50 ng/ml vascular endothelial growth factor (VEGF) as chemo-attractant and incubated at 37°C for 24 hours. Then the lower side of the filter was washed with phosphate buffered saline and fixed with 2% paraformaldehyde. Migrated cells were visualized by hematoxylin staining.

Cytokine Release

Conditioned medium was obtained by cultivating 1×10^5 hiPSC-1-EC, hiPSC-2-EC, hESC-EC, BM-EC and HUVEC with basal EBM2 medium (without supplement) for 5 days under normoxic (21% O_2, 5% CO_2) or hypoxic (0.5% O_2, 5% CO_2) condition using the BioSpherix OxyCycler C42 system (Bio-Spherix, Redfield, NY) and subjected to cytokine analysis: epidermal growth factor (EGF), basic fibroblast growth factor (FGF-B), hepatocyte growth factor (HGF), leptin, VEGF, placental growth factor (PIGF) and stromal derived factor-1 (SDF-1); adhesion molecules including, ICAM1 and vascular cell adhesion molecule (VCAM) using Procarta multiplex bead-based immuno-assay kit (Affymetrix, Fremont, CA) according to manufacturer's instructions. Data acquisition and analysis were done using Bio-Plex Manager software version 4.1.1. Cell numbers were counted after the 5 days incubation for normalization.

In-vivo Functional Effects of Cell Transplantation into Animal Model of Critical Hind-Limb Ischemia

Critical hind-limb ischemia was induced in male severe combined immunodeficient mice after anesthetized with xylazine (20 mg/kg) and ketamine (100 mg/kg) by intraperitoneal injection [16]. The femoral artery was excised from its proximal origin as a branch of the external iliac artery to the distal point where it bifurcates into the saphenous and popliteal arteries. After arterial ligation, mice were immediately assigned to the following experimental groups: hiPSC-1-EC group (n = 8), hiPSC-2-EC group (n = 8), hESC-EC group (n = 8), BM-EC group (n = 6), HUVEC group (n = 6) and the medium group (medium, n = 6). In each animal, a total of 3×10^6 cells (30 μl) or medium (EGM-2) was injected intramuscularly into 3 sites of the gracilis muscle at the medial thigh of the ischemic hind-limbs. Serial laser Doppler imaging analysis (Moor Instruments, Devon, UK) was performed to monitor the blood flow of hind-limbs immediately, at Day 7, 14, 21 and 28 after femoral artery ligation and cell-transplantation. The digital color-coded images were analyzed to quantify the blood flow in the region from the knee to the toe, the mean values of perfusion were calculated.

At Day 28, tissue from the ischemic limb was harvested and formalin fixed. Mouse monoclonal anti-vWF (Chemicon, Temecula, CA) was used to determine the capillary density on paraffin embedded tissues using standard immunostaining procedure. Arteries were stained with mouse monoclonal anti-α-smooth muscle actin antibody (Serotech, Raleigh, NC). Retention of transplanted cells was tracked by anti-human nuclear antigen antibody (HNA; Chemicon International, Temecula, CA) staining.

Table 1. Clinical characteristic of the patients' marrow MNC used in this study.

Patients Number	Sex	Age	CAD	CHF	HT	DM	PCI	CABG	Creatinine (umol/L)	Flow cytometry analysis %CD34^{+}	%CD3^{+}	%CD11b^{+}	%CD117
1	Male	61	+	+	-	-	+	-	106	1.26	37.34	21.34	2.92
2	Male	67	+	+	+	+	+	-	147	1.87	16.9	18.68	1.93
3	Male	70	+	+	+	+	+	+	104	1.32	15.66	14.41	1.29
4	Male	64	+	+	+	+	+	−	146	0.92	34.25	21	0.77
5	Male	60	+	+	+	+	+	−	162	1.49	20.81	21.67	2.01
6	Male	61	+	+	+	−	−	−	104	1.3	49.34	49.21	1.88

CAD = coronary arteries diseases; CHF = chronic heart failure; HT = hypertension; DM = diabetes mellitus; Lipid = hyperlipidemia; PCI = percutaneous coronary intervention; CABG = coronary artery bypass graft, '+' = present, '−' = absent.

To monitor the cell fate after transplantation of hiPSC-EC and hESC-EC, luciferase-expressing-hiPSC-1-EC and hESC-EC was generated, cultured *in-vitro* and expanded before transplantation. In brief, luciferase gene were transduced into hiPSC-1-EC and hESC-EC by lentivirus-mediated transfer and selected in the present of 1 ug/ml puromycin, respectively. After transplanting the luciferase-expressing-hiPSC-1-EC (n = 8) and hESC-EC (n = 8) into the ischemic limb, labeled cells were tracked with the IVIS 200 Bioluminescence Imaging System (Xenogen Biosciences, Cranbury, NJ) after intraperitoneal injection of 150 mg/kg D-luciferin (Xenogen Biosciences, Cranbury, NJ) at Day 7, 14 and 28 after cell transplantation. We failed to label BM-EC as they have limited passaging capacity. Signals were analyzed with the Living Image Software (Xenogen Biosciences, Cranbury, NJ).

Statistical Analysis

Data are expressed as mean ± SEM. Statistical comparisons were performed using Student's *t*-test or one-way ANOVA, as appropriate. The distribution pattern of cytokines levels were highly skewed, so these variables were log-transformed to normalize their distribution before analysis. Two-way ANOVA test was performed to compare different groups with time. Post-hoc Bonferroni's correction for multiple comparisons was used. All p values <0.05 were considered statistically significant.

Results

EC Differentiation

Differentiation of EC from hiPSC-1, hiPSC-2 and hESC were performed. After culturing for 3 weeks on EGM-2, EC outgrowth was observed from all hiPSC-1, hiPSC-2 and hESC (Figure 1A). BM mononuclear cells were obtained from a group of 6 patients and cultured in the EGM-2 for 14 days. However, functional EC could only be derived from 1/6 (17%) patient's BM mononuclear cells. Culture of BM mononuclear cells obtained in the other 5 patients failed to yield any CD31^{+} cells. During the differentiation, those BM mononuclear cells from these 5 patients failed to proliferate and the majority of the cells showed nuclear fragmentation, suggesting cellular apoptosis. This finding highlights the difficulty in obtaining functional EC from patient's autologous BM cells for transplantation.

Immunofluorescence staining showed that hiPSC-1-EC, hESC-EC and BM-EC display the characteristic phenotypes of endothelial cells, including uptake of DiI-Ac-LDL and expression of lectin as the control HUVEC (Figure 1A). Although double sorting for CD45 negative and CD31 positive was not performed in BM-EC due to the limited number of cells yielded, flow cytometry analysis revealed that only 5% of them were CD45 positive. Indeed, >90% of these differentiated EC were double positive for CD31 and KDR and >80% of them were CD31, vWF and KDR positive as determined by flow cytomety. There were no statistical significant difference in the expression of EC-surface markers between hiPSC-1-EC, hiPSC-2-EC, hESC-EC and BM-EC, and resemble to the expression in HUVEC (Figure 1B, all *P>0.05*).

Functional Assays of EC

The *in-vitro* functional capacity of EC was examined by their ability to form capillary like network on Matrigel in the tube-formation assay. CD31^{+} hiPSC-EC, hESC-EC, BM-EC, and HUVEC were plated onto Matrigel, lumen structure of vascular network were formed after 6 hours of plating (Figure 2A). There were no statistical differences in the tube formation ability between hiPSC-1-EC, hiPSC-2-EC, hESC-EC and BM-EC (Figure 2A, all *P>0.05*).

Migration of EC through the endothelial lining is a key step in angiogenesis and vascular repair. Therefore, the migration ability of the hiPSC-1-EC, hiPSC-2-EC, hESC-EC and BM-EC was compared using the modified Boyden chamber assay. As compared with HUVEC and their corresponding controls with medium alone, hiPSC-1-EC, hiPSC-2-EC, hESC-EC and BM-EC showed comparable enhancement in migration ability with VEGF as the endothelial specific chemo-attractant (*P<0.05*). However, there were no significant differences in their migration ability among the five groups (Figure 2B, all *P<0.05*).

Paracrine Secretion of Derived EC

The cytokine secretion profiles of EC derived from different sources were also investigated to provide insight into any potential difference in their paracrine effects after transplantation [17]. The cytokine profiles were quantified from the conditioned medium derived from hiPSC-1-EC, hiPSC-2-EC, hESC-EC, BM-EC and HUVEC using multiplex ELISA. During normoxic culture, similar pattern of cytokine expression (ICAM, VEGF, bFGF and leptin) were observed in the conditioned medium of hiPSC-1-EC, hiPSC-2-EC and hESC-EC, except that higher levels of HGF and PIGF were detected in hiPSC-1-EC and hiPSC-2-EC (Figure 3A). The conditioned medium of BM-EC had significantly higher levels of VEGF and VCAM and lower levels of ICAM1 than hESC-EC (all *P<0.05*, Figure 3A). Furthermore, the conditioned medium of BM-EC had significantly higher levels of VCAM and lower levels

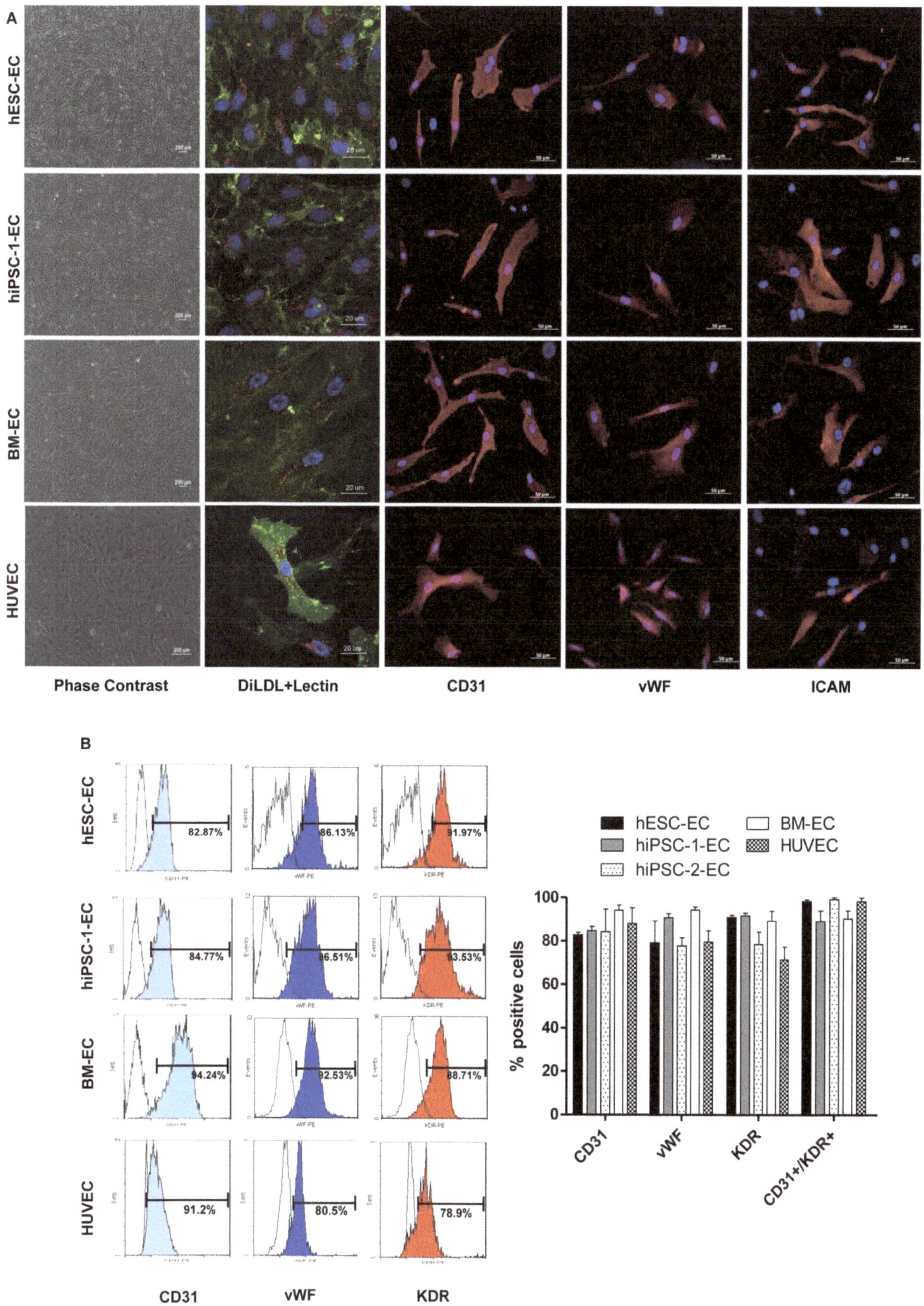
A
hESC-EC
hiPSC-1-EC
BM-EC
HUVEC
Phase Contrast
DiLDL+Lectin
CD31
vWF
ICAM
B
hESC-EC
hiPSC-1-EC
BM-EC
HUVEC
82.87%
86.13%
91.97%
84.77%
86.51%
93.53%
94.24%
92.53%
88.71%
91.2%
80.5%
78.9%
CD31
vWF
KDR
hESC-EC
hiPSC-1-EC
hiPSC-2-EC
BM-EC
HUVEC
% positive cells
100
80
60
40
20
0
CD31
vWF
KDR
CD31+/KDR+

Figure 1. Characterization of derived endothelial-like cells (EC). (A) Phase contrast photos (left panel) showing the morphology of hESC-EC, hiPSC-EC, bone marrow derived EC (BM-EC) and human umbilical cord endothelial cell (HUVEC) in a confluent culture. Immunofluorescence staining showed the expression of typical endothelial cells markers including *Ulex europaeus* lectin (green) and uptake of Di-Acetyl-LDL (red) (second panel) in hESC-EC, hiPSC-1-EC, BM-EC and HUVEC. Furthermore, expression of CD31, von Willebrand factor (vWF) and intercellular adhesion molecule 1 (ICAM) were also detected in hESC-EC, hiPSC-EC, BM-EC and HUVEC (B) Flow cytometry analysis revealed the expression of CD31, vWF, Kinase insert domain receptor (KDR) in hESC-EC, hiPSC-EC, BM-EC and HUVEC (left panel). There were no differences in the percentage of cells with expression of CD31, vWF, KDR and co-expression of CD31 and KDR between hESC-EC, hiPSC-EC and BM-EC (right panel).

of SDF1 than hiPSC-1-EC and hiPSC-2-EC (all $P<0.05$, Figure 3A). The conditioned medium of the HUVEC had significantly lower levels of leptin and SDF-1 (Figure 3A).

To mimic the micro-environment in the ischemic tissue, hiPSC-1-EC and hiPSC-2-EC, hESC-EC, BM-EC and HUVEC were subjected to hypoxic stress. Multiplex ELISA revealed that the expression of major angiogenic factors including EGF, HGF, VEGF, PIGF and SDF-1 were significantly up-regulated in hiPSC-1-EC, hiPSC-2-EC, hESC-EC, BM-EC and HUVEC under hypoxia as compared with normoxia (all $P<0.05$, Figure 3A and B). Multivariate radar chart analysis showed that the magnitude of cytokine up-regulation upon hypoxic induction is more dramatic in hiPSC-1-EC and hiPSC-2-EC and hESC-EC compared with BM-EC (Figure 3B).

Functional Improvement of Blood Flow in Ischemic Hind-limbs after Transplantation

The therapeutics efficacy of hiPSC-1-EC, hiPSC-2-EC, hESC-EC, BM-EC and HUVEC for therapeutic angiogenesis was compared in a mouse model of hind-limb ischemia. After the femoral artery ligation, the blood flow of the ischemic limb decreased to only 3.7±0.4% of that the non-ischemic control limbs, confirming the successful induction of acute hind-limb ischemia. From Day 7 onward, there was significant and progressive improvement of ischemic limb blood flow in all groups of mice receiving EC transplantation as compared with medium group (Figure 4A, all $P<0.05$). At Day 28, mice treated with hiPSC-1-EC (44.8±6.9%), hiPSC-2-EC (38.2%±2.6%), hESC-EC (37.4±2.5%), BM-EC (39.2±5.8%) and HUVEC (29.6%±4.9%) transplantation had significantly higher hind-limb blood flow than the medium group (14.4±6.2%, all $P<0.05$)(Figure 4A). Nonetheless, there were no significant differences in the blood flow of ischemic limb between those mice treated with hiPSC-1-EC, hiPSC-2-EC, hESC-EC and BM-EC throughout the study period (all p<0.05), and their rate of blood flow improvement were superior to the HUVEC group from Day 0 to 14 (Figure 4A).

Immunohistological staining for vWF in the ischemic limbs at Day 28 demonstrated increased in capillary density after hiPSC-1-EC, hiPSC-2-EC, hESC-EC and BM-EC transplantation as compared with HUVEC group and control media injection (Figure 4B, all $P<0.05$). On the other hand, immunostaining for α-smooth muscle actin failed to reveal any difference in the numbers of arteries among different groups (Figure 4B, $P<0.05$), suggesting the development of angiogenesis rather than arteriogenesis after EC transplantation.

Histological examination did not reveal any teratoma formation at the transplanted sites in any of the mice after EC transplantation. Cell retention of transplanted EC was also examined at Day 7 and Day 28 in a subgroup of animals by immunohistological staining for HNA in the ischemic limbs. Comparing the number of HNA positive cells at Day 7 after transplantation, minimal transplanted cell retention was observed in all three groups of mice after EC transplantation as indicated by the markedly reduced number of HNA positive staining in ischemic limb harvested on Day 28 (Figure 5A).

To further evaluate the cell retention level and the proliferative ability of hiPSC-EC, *in- vivo* live cells imaging was employed to detect and quantify the emitted bioluminescence from luciferase-labeled hESC-EC and hiPSC-EC progeny. According to bioluminescence emitted from labeled progeny, hESC-EC and hiPSC-EC could be detected at Day 7 after transplantation and then started to decline. Nevertheless, transplanted hESC-EC and hiPSC-EC were still detected at Day 28 after transplantation with a gradual drop in signal intensity (Figure 5B).

Discussion

Currently, BM-derived stem cells are the most common cell type investigated in clinical trials for treatment of cardiovascular diseases [3,4,18]. However, in order to overcome the scarcity and functional impairment of autologus BM stem cells, EC derived from pluripotent stem cells, such as hESC have been proposed as an alternative and unlimited source of cells for therapy. The latest advances in the generation of hiPSC [10] has shed light into the field of cell-based therapy by providing an alternative unlimited source for autologous transplantation which can potentially solve the problem of availability [19]. In the current study, we sought to evaluate the therapeutic efficacy of EC derived from different cell sources in treatment of ischemic cardiovascular diseases.

Here, we confirmed that one hESC line and two hiPSC lines could be differentiated into functional EC which express a comparable level of endothelial markers as BM-EC and HUVEC. *In-vitro* tube formation and cellular migration assays revealed that hESC-EC and hiPSC-EC have similar angiogenic function as BM-EC and HUVEC. Although there was some discrepancy in the angiogenic cytokine profile in the conditional medium between hiPSC-EC, hESC-EC, BM-EC and HUVEC at baseline, significant increased in the level of multiple major angiogenic cytokines, including EGF, HGF, VEGF, PIGF and SDF-1 were observed in the conditioned medium derived from all type of EC after hypoxia. Moreover, hiPSC-EC and hESC-EC had even greater response to hypoxic induction in their expression of angiogenic cytokines than the BM-EC and HUVEC. These angiogenic cytokines have been shown to enhance neovascularization in the ischemic sites [20].

Therefore, the increased expression of these angiogenic factors from the transplanted EC upon hypoxia in the ischemic condition of the hind-limb could provide paracrine effect to enhance neovascularization.

In concordance with these observations, our results also demonstrated a similar overall therapeutic efficacy of EC derived from hiPSC, hESC and BM-MNC as determined by laser Doppler blood flow up to Day 28 in the mouse model of hind-limb ischemia. Despite progressive improvement of blood flow over the ischemic limbs was observed in serial laser Doppler blood flow measurement, immunohistochemical staining and luciferase assay revealed only very small number of the EC remained and engrafted into the vasculature at the transplanted ischemic limbs. Therefore, majority of the beneficial effects exert by the

A

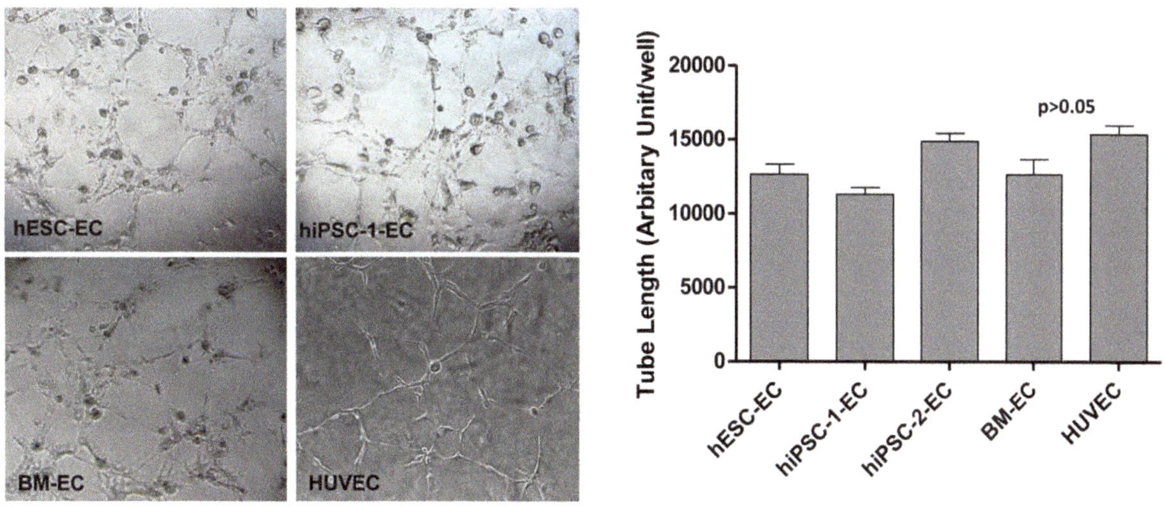

B

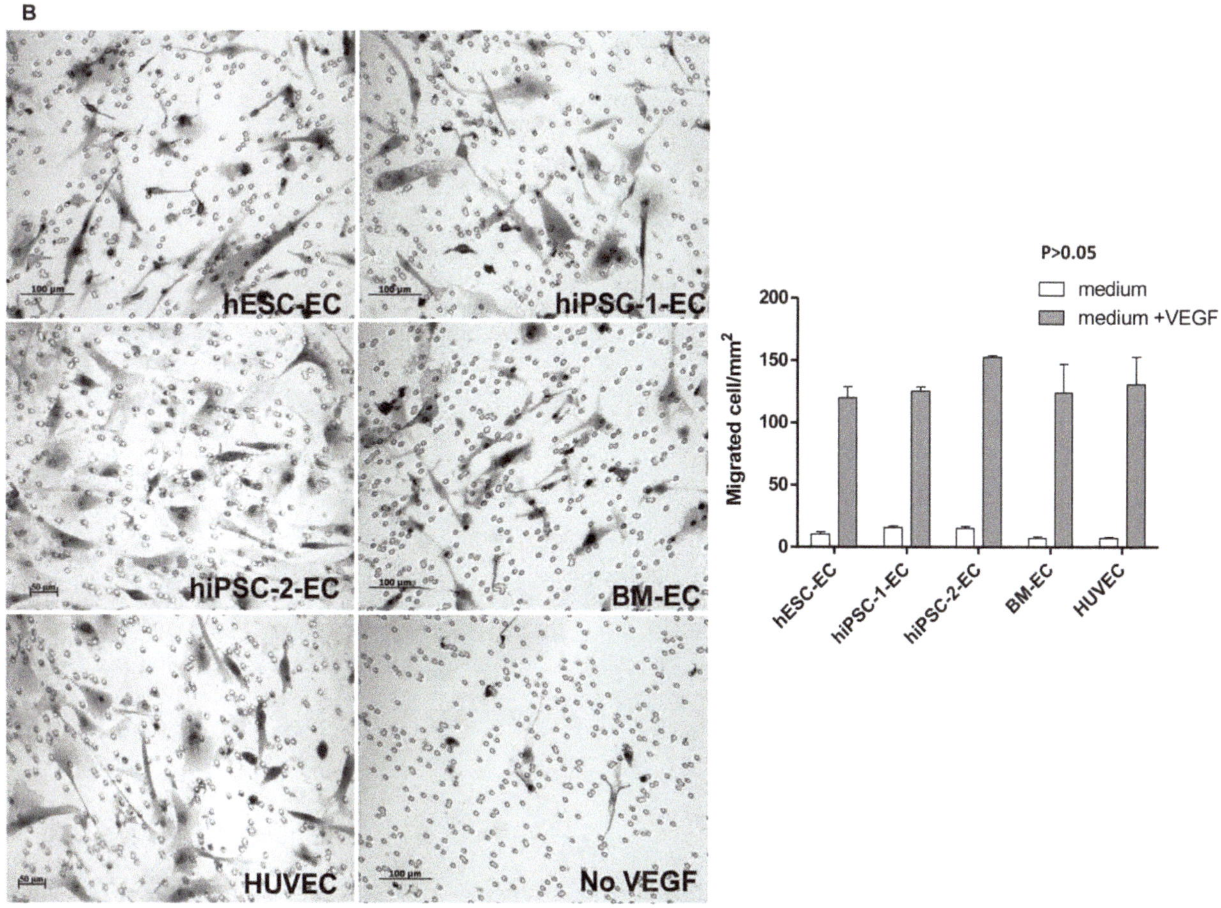

Figure 2. ***In- vitro* functional assessment of derived endothelial-like cells.** (A) Tube formation assay was performed with 1×10^4 cells seeded onto Matrigel and the presence of micro vessel formation was visualized by phase contrast microscope (100×) after 6 hours of incubation at 37°C. There was no significant difference in length of tube-like vessels formed by hESC-EC, hiPSC-EC, BM-EC and HUVEC ($p>0.05$). (**B**) Modified Boyden chamber assay was used to determine the migration ability of the hESC-EC, hiPSC-EC, BM-EC and HUVEC. These cells were placed in the inner chamber and allowed to migrate across the porous membrane toward the outer chamber with the presence of medium alone (control) or medium supplemented with vascular endothelial growth factor (VEGF). Then, the numbers of migrated cells per mm^2 present in the transwell were counted under light microscope (100×, upper panel). The numbers of migrated cells were significantly increased after supplement with VEGF in hESC-EC, hiPSC-EC, BM-EC and HUVEC as compared with their corresponding control (*$p<0.05$). However, there were no differences among those EC in the number of migrated cells with the presence of medium alone or medium supplemented with VEGF ($p<0.05$, lower panel).

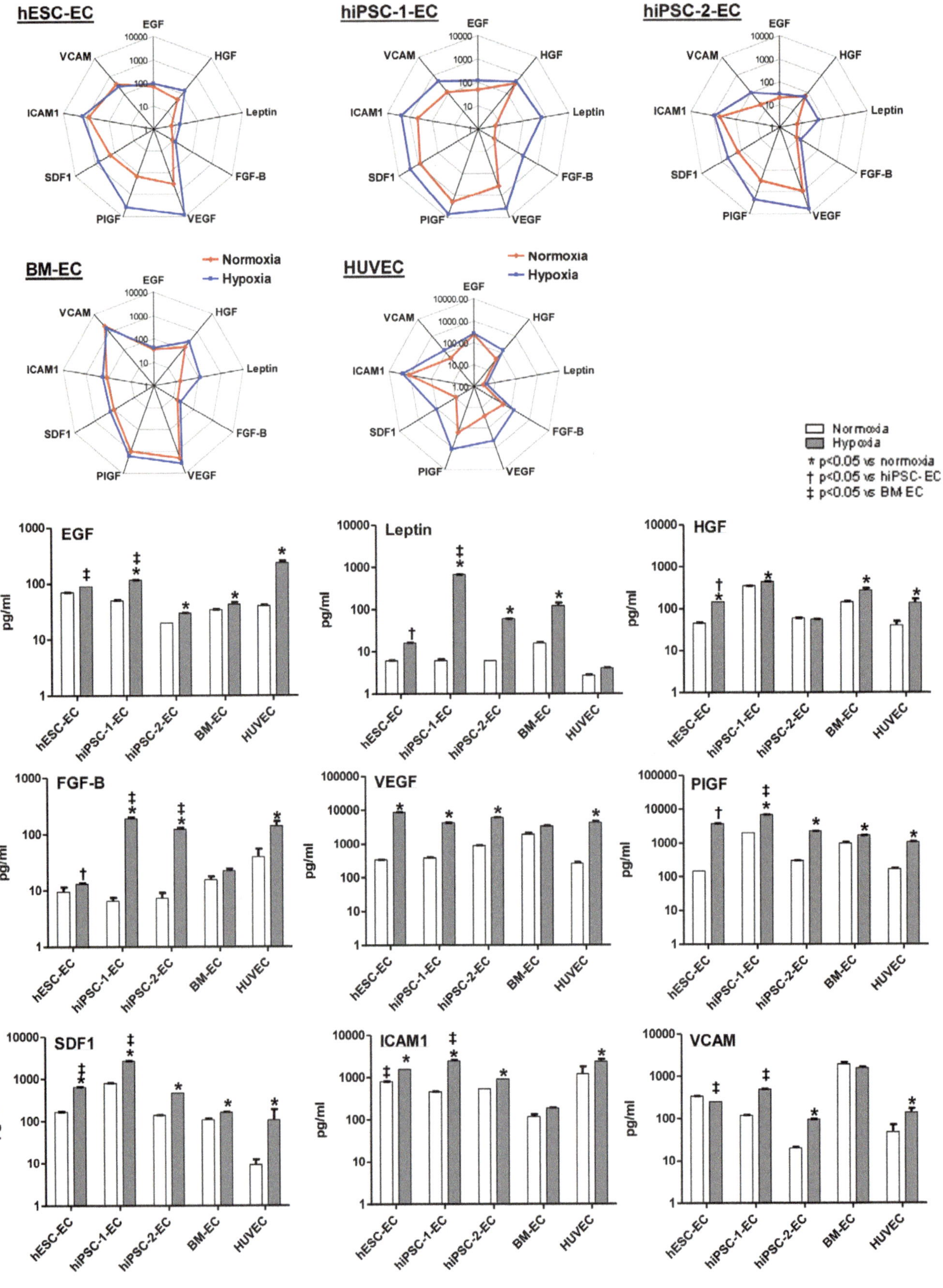

Figure 3. Angiogenic cytokine profiles of conditioned medium obtained from derived endothelial-like cells under normoxic and hypoxic condition. A total of 1×10^5 cells were plated into 6-well gelatin coated plate for overnight incubation, then the cells were washed and

cultivated in EBM2 basal medium without supplement for 5 days under normoxic (20% O_2) or hypoxic (0.1% O_2). Cytokine profiling of their conditioned mediums obtained were determined by Procarta multiplex bead-based immunoassay kit. (A) Bar charts and (B) multivariate radar chart showed the expression of epidermal growth factor (EGF), basic fibroblast growth factor (FGF-B), hepatocyte growth factor (HGF), leptin, vascular endothelial growth factor (VEGF), placental growth factor (PIGF) and stromal derived factor-1 (SDF-1); adhesion molecules including, ICAM1 and vascular cell adhesion molecule (VCAM) in the conditioned medium upon normoxic and hypoxic treatment.

transplanted EC is likely to be mediated via other mechanisms, such as paracrine secretion rather than *de-novo* formation of vasculature by the transplanted EC [1,21,22]. Furthermore, we only observed increased angiogenesis rather than arteriogenesis after EC transplantation.

Our results also confirmed previous findings [23] and showed that HUVEC cells have a lower capacity in improving blood flow and neovascularization in the hind-limb ischemic mouse model than those EC derived from BM, hESC or hiPSC. This finding suggests that the paracrine function of HUVEC is less effective than EC derived from other cell sources for enhancing neovascularization.

Since the initial description of successful generation of hiPSC from adult fibroblast, several clinical compliance protocols had been developed, such as the use of xeno-free and feeder-free system [14] and the generation of hiPSC from donor cells that does not require an invasive procedure, e.g. urine cells [24] and hair follicles [25]. Moreover, recent breakthrough of generating hiPSC using mRNA transfer [26] without the need of viral mediated transfer has brought the hiPSC technology closer to the clinic.

To our knowledge, this study is the first to compare the *in vitro* and *in vivo* therapeutic efficacy of EC derived from hiPSC, hESC, BM and normal human endothelial cells. Although hiPSC-EC only provide a similar therapeutic efficacy as BM-EC and hESC-EC, the use of patients-specific hiPSC-EC can circumvent issues related to the clinical use of hESCs, as well as the need for an invasive procedure for harvesting an adequate amount of BM cells

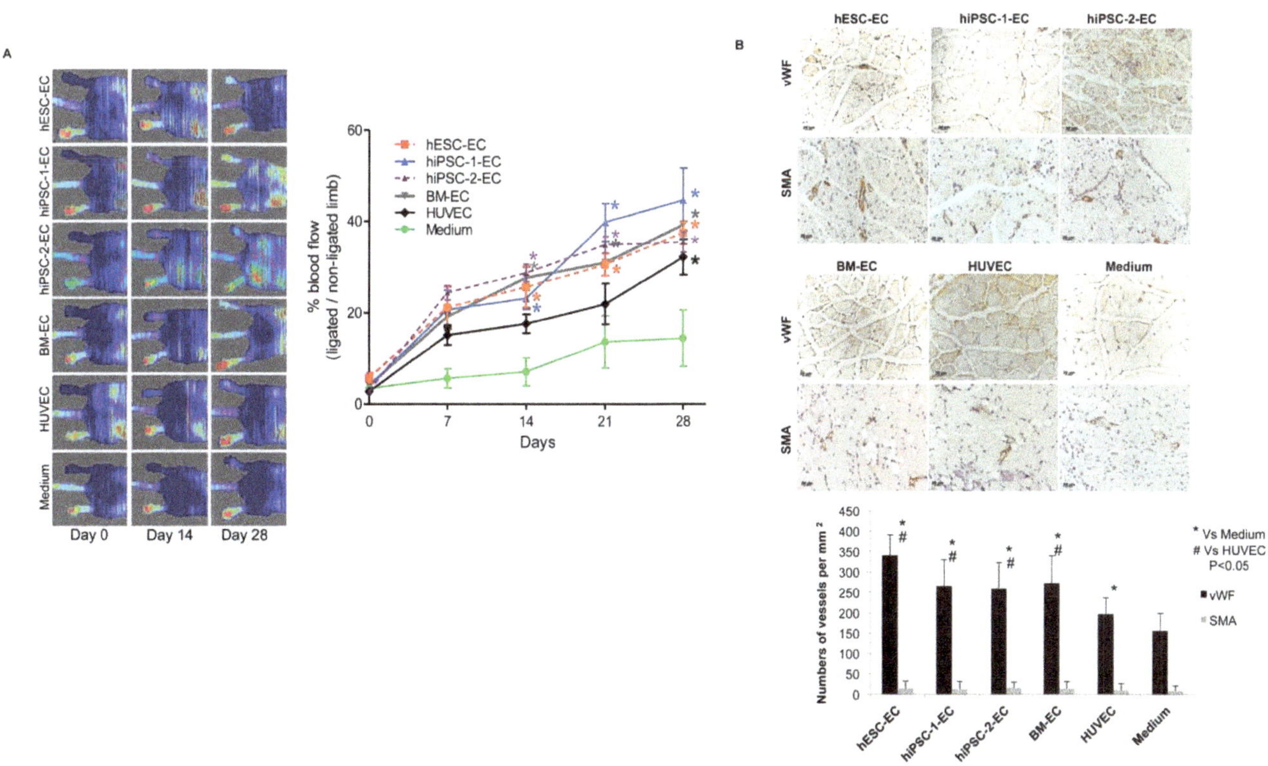

Figure 4. Improvement of blood perfusion in ischemic hind-limbs after transplantation of derived endothelial-like cells. (**A**) Laser Doppler scanning was preformed immediately after ligation of femoral artery to confirm the induction of hind-limb ischemia. Then repeated Doppler scanning was performed at Day 7, 14, 21 and 28 after transplantation of 3 million hESC-EC, two different hiPSC-ECs, BM-EC, HUVEC or medium alone into the ischemic muscle. Representative photos from the laser Doppler scanning at Day 0, Day 14 and Day 28 were shown. Blue color represents lower perfusion and red color represents higher perfusion (upper panel). In each animal, limb perfusion was quantified and the average of the perfusion ratio against the control limb was calculated in different time point (lower panel). As compared with medium injection, there was progressive and significant improvement of blood flow over the ischemic limbs after hESC-EC, hiPSC-ECs, BM-EC and HUVEC transplantation (n = 6 in each group, *$p < 0.05$ vs. medium control at different time points). However, there were no significant differences in the blood flow over the ischemic limbs among those mice after hESC-EC, hiPSC-EC and BM-EC transplantation ($p > 0.05$ at different time points). (**B**) Representative photos show the immunohistochemical staining for von Willebrand factor (vWF) and smooth muscle actin (SMA) of histological sections over the ischemic limbs at Day 28 after hESC-EC, hiPSC-EC, BM-EC or medium injection (upper panel). The number of vessels with positive staining (brown color) for vWF and SMA were determined under a light microscope with ×400 magnification with 20 random fields. As compared with medium injection, there were significant higher number of vessels with vWF positive but not SMA over the ischemic limbs after hESC-EC, hiPSC-ECs and BM-EC transplantation (*p vs. medium control and * vs HUVEC, $p < 0.05$). However, there were no significant differences in the number of vessels with vWF and SMA positive over the ischemic limbs among those mice after hESC-EC, hiPSC-EC and BM-EC transplantation ($p > 0.05$).

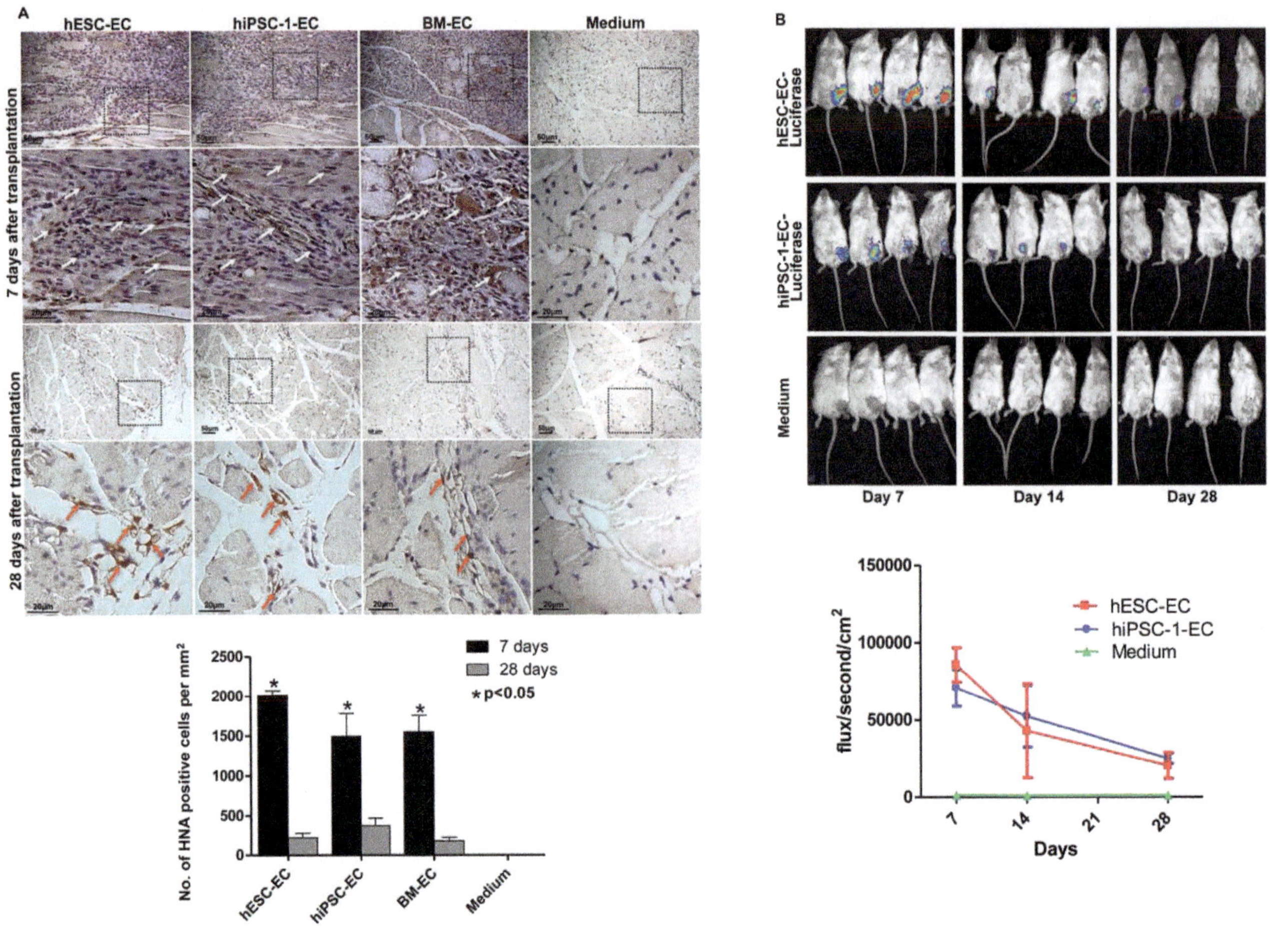

Figure 5. Retention of derived endothelial-like cells at Day 7, 14 and 28 after transplantation. (**A**) Representative photos show the immunohistochemical staining for human nuclear antigen (HNA) over the ischemic hind-limb at Day 7 and 28 after hESC-EC, hiPSC-1-EC, BM-EC or medium injection (upper panel). Red arrow indicated engrafted hESC-EC, hiPSC-EC, BM-EC in the ischemic limb with HNA positive (brown), indicating those cells were from human origin. At Day 7, clusters of HNA positive cells were observed at the injection sites after hESC-EC, hiPSC-1-EC, and BM-EC transplantation but not after medium injection. At Day 28, only a few HNA positive cells were found around the vasculature at the injection sites after hESC-EC, hiPSC-1-EC, and BM-EC transplantation. As compared with Day 28, the numbers of HNA positive cells detected at the injection sites were significantly higher at Day 7 after hESC-EC, hiPSC-1-EC, BM-EC transplantation (*$p<0.05$, lower panel). (**B**) Representative photos showing live cells tracking by quantifying luciferase activity using the Xenogen's *in vivo* optical imaging systems at Day 7, 14 and 28 after hESC-EC, hiPSC-1-EC, or medium injection (upper panel). Luciferase activity was plot as total flux measurement in photons/second/cm^2. Declined in intensity of luciferase activity over time after hESC-EC and hiPSC-1-EC transplantation was shown, suggesting progressive cell loss (n = 4 in each group, lower panel).

for transplantation. In this study, ECs with similar functional characteristics and therapeutic potential could be derived from two different hiPSC lines. Importantly, we did not observe any teratoma formation in mice after transplantation of both hESC-EC or hiPSC-EC, suggesting that the risk of uncontrolled proliferation or growth from those EC derived from pluripotent stem cells is low. Nevertheless, this finding needs to be further confirmed by future studies.

In this study, only 1 out of 6 specimens of BM mononuclear cells from patients with severe coronary artery disease could be able to differentiate into functional EC for transplantation due to their poor survival and proliferation capability as shown in previous studies [8]. Furthermore, the underlying concomitant conditions of these patients, such as heart failure, hypertension and diabetes [5–7] may also offset the function of their BM stem cells. The observation also highlight that autologous BM-EC transplantation might be feasible and effective in a small number of patients with cardiovascular diseases. The low successful rates to obtain functional EC from BM cells as well as their larger variability in characteristics are major limitations for their clinical use. Our results from EC derived from two hiPSCs and one hESC consistently demonstrated that they are not inferior to BM-EC for both *in-vivo* and *in-vitro* neovascularization. Importantly, functional EC could be derived from existing hESC and hiPSC lines despite multiple passages. Thus, these cell types are more stable sources of functional EC for potential therapeutic applications.

Taken together, our results showed that functional EC could be generated from hESC and iPSC with similar therapeutic efficacy for attenuation of severe hind-limb ischemia. Differentiation of functional BM-EC was more difficult to achieve in patients with cardiovascular diseases, and hESC-EC or iPSC-EC are readily available as "off-the-shelf" format for treatment of tissue ischemia.

This study has limitations. First, the precise mechanism of therapeutic benefit for each of the cell types remains unknown. In this study, we showed that only very small numbers of transplanted cells were detected at Day 28. This finding suggests that paracrine actions of the transplanted ECs are the likely mechanisms that contribute to increased angiogenesis. On the other hand, whether those transplanted ECs directly contribute to the paracrine effects is unknown, as other mechanisms such as recruitment of endogenous local or circulating progenitor cells into the ischemic tissues are not investigated in this study. Second, the precise role of different cytokines secreted by the transplanted cells remains unclear and only limited panel of cytokines were tested in this study.

Author Contributions

Conceived and designed the experiments: WH-L JC-H CW-S HF-T. Performed the experiments: WH-L JC-H JH-N YC-C LY-W KW-A. Analyzed the data: WH-L JC-H CW-S HF-T. Contributed reagents/ materials/analysis tools: CW-S HF-T. Wrote the paper: WH-L JC-H HF-T.

References

1. Tse HF, Siu CW, Zhu SG, Songyan L, Zhang QY, et al. (2007) Paracrine effects of direct intramyocardial implantation of bone marrow derived cells to enhance neovascularization in chronic ischemic myocardium. Eur J Heart Fail 9: 747–753.
2. Tse HF, Kwong YL, Chan JK, Lo G, Ho CL, et al. (2003) Angiogenesis in ischaemic myocardium by intramyocardial autologous bone marrow mononuclear cell implantation. Lancet 361: 47–49.
3. Kang S, Yang YJ, Li CJ, Gao RL (2008) Effects of intracoronary autologous bone marrow cells on left ventricular function in acute myocardial infarction: a systematic review and meta-analysis for randomized controlled trials. Coron Artery Dis 19: 327–335.
4. Martin-Rendon E, Brunskill SJ, Hyde CJ, Stanworth SJ, Mathur A, et al. (2008) Autologous bone marrow stem cells to treat acute myocardial infarction: a systematic review. Eur Heart J 29: 1807–1818.
5. Dimmeler S, Leri A (2008) Aging and disease as modifiers of efficacy of cell therapy. Circ Res 102: 1319–1330.
6. Fortini C, Toffoletto B, Fucili A, Puppato E, Olivares A, et al. (2010) Circulating stem cell vary with nyha stage in heart failure patients. J Cell Mol Med 15: 1726–1736.
7. Li M, Ho JC, Lai KW, Au KK, Xu A, et al. (2011) The decrement in circulating endothelial progenitor cells (EPCs) in type 2 diabetes is independent of the severity of the hypoadiponectemia. Diabetes Metab Res Rev 27: 185–194.
8. Vasa M, Fichtlscherer S, Aicher A, Adler K, Urbich C, et al. (2001) Number and migratory activity of circulating endothelial progenitor cells inversely correlate with risk factors for coronary artery disease. Circ Res 89: E1–7.
9. Cho SW, Moon SH, Lee SH, Kang SW, Kim J, et al. (2007) Improvement of postnatal neovascularization by human embryonic stem cell derived endothelial-like cell transplantation in a mouse model of hindlimb ischemia. Circulation 116: 2409–2419.
10. Takahashi K, Tanabe K, Ohnuki M, Narita M, Ichisaka T, et al. (2007) Induction of pluripotent stem cells from adult human fibroblasts by defined factors. Cell 131: 861–872.
11. Yamanaka S (2007) Strategies and new developments in the generation of patient-specific pluripotent stem cells. Cell Stem Cell 1: 39–49.
12. Yu J, Vodyanik MA, Smuga-Otto K, Antosiewicz-Bourget J, Frane JL, et al. (2007) Induced pluripotent stem cell lines derived from human somatic cells. Science 318: 1917–1920.
13. Li Z, Hu S, Ghosh Z, Han Z, Wu JC (2011) Functional Characterization and Expression Profiling of Human Induced Pluripotent Stem Cell- and Embryonic Stem Cell-Derived Endothelial Cells. Stem Cells Dev 20: 1701–1710.
14. Lai WH, Ho JC, Lee YK, Ng KM, Au KW, et al. (2010) ROCK inhibition facilitates the generation of human-induced pluripotent stem cells in a defined, feeder-, and serum-free system. Cell Reprogram 12: 641–653.
15. Tse HF, Thambar S, Kwong YL, Rowlings P, Bellamy G, et al. (2007) Prospective randomized trial of direct endomyocardial implantation of bone marrow cells for treatment of severe coronary artery diseases (PROTECT-CAD trial). Eur Heart J 28: 2998–3005.
16. Limbourg A, Korff T, Napp LC, Schaper W, Drexler H, et al. (2009) Evaluation of postnatal arteriogenesis and angiogenesis in a mouse model of hind-limb ischemia. Nat Protoc 4: 1737–1746.
17. Yoon CH, Hur J, Park KW, Kim JH, Lee CS, et al. (2005) Synergistic neovascularization by mixed transplantation of early endothelial progenitor cells and late outgrowth endothelial cells: the role of angiogenic cytokines and matrix metalloproteinases. Circulation 112: 1618–1627.
18. Siu CW, Tse HF (2012). Cardiac regeneration: messages from CADUCEUS. Lancet. 379: 870–1.
19. Seifinejad A, Tabebordbar M, Baharvand H, Boyer LA, Salekdeh GH (2010) Progress and promise towards safe induced pluripotent stem cells for therapy. Stem Cell Rev 6: 297–306.
20. Schultz A, Lavie L, Hochberg I, Beyar R, Stone T, et al. (1999) Interindividual heterogeneity in the hypoxic regulation of VEGF: significance for the development of the coronary artery collateral circulation. Circulation 100: 547–552.
21. Kim SW, Kim H, Cho HJ, Lee JU, Levit R, et al. (2010) Human peripheral blood-derived CD31+ cells have robust angiogenic and vasculogenic properties and are effective for treating ischemic vascular disease. J Am Coll Cardiol 56: 593–607.
22. Kim MH, Jin E, et al. (2011) Robust angiogenic properties of cultured human peripheral blood-derived CD31(+) cells. Int J Cardiol. 2011 Dec 19. [Epub ahead of print]
23. Bhang SH, Lee S, Lee TJ, La WG, Yang HS et al. (2012) Three-dimensional cell grafting enhances the angiogenic efficacy of human umbilical vein endothelial cells. Tissue Eng Part A. 18: 310–319.
24. Zhou T, Benda C, Duzinger S, Huang Y, Li X, et al. (2011) Generation of induced pluripotent stem cells from urine. J Am Soc Nephrol 22: 1221–1228.
25. Novak A, Shtrichman R, Germanguz I, Segev H, Zeevi-Levin N, et al. (2010) Enhanced reprogramming and cardiac differentiation of human keratinocytes derived from plucked hair follicles, using a single excisable lentivirus. Cell Reprogram 12: 665–678.
26. Warren L, Manos PD, Ahfeldt T, Loh YH, Li H, et al. (2010) Highly efficient reprogramming to pluripotency and directed differentiation of human cells with synthetic modified mRNA. Cell Stem Cell 7: 618–630.

Timing of Umbilical Cord Blood Derived Mesenchymal Stem Cells Transplantation Determines Therapeutic Efficacy in the Neonatal Hyperoxic Lung Injury

Yun Sil Chang[1,2]⁹, Soo Jin Choi[3]⁹, So Yoon Ahn[1], Dong Kyung Sung[2], Se In Sung[1], Hye Soo Yoo[1], Won Il Oh[3], Won Soon Park[1,2]*

1 Department of Pediatrics, Samsung Medical Center, Sungkyunkwan University School of Medicine, Seoul, Korea, **2** Samsung Biomedical Research Institute, Sungkyunkwan University School of Medicine, Seoul, Korea, **3** Biomedical Research Institute, MEDIPOST Co., Ltd., Seoul, Korea

Abstract

Intratracheal transplantation of human umbilical cord blood (UCB)-derived mesenchymal stem cells (MSCs) attenuates the hyperoxia-induced neonatal lung injury. The aim of this study was to optimize the timing of MSCs transplantation. Newborn Sprague-Dawley rats were randomly exposed to hyperoxia (90% for 2 weeks and 60% for 1 week) or normoxia after birth for 21 days. Human UCB-derived MSCs (5×10^5 cells) were delivered intratracheally early at postnatal day (P) 3 (HT3), late at P10 (HT10) or combined early+late at P3+10 (HT3+10). Hyperoxia-induced increase in mortality, TUNEL positive cells, ED1 positive alveolar macrophages, myeloperoxidase activity and collagen levels, retarded growth and reduced alveolarization as evidenced by increased mean linear intercept and mean alveolar volume were significantly better attenuated in both HT3 and HT3+10 than in HT10. Hyperoxia-induced up-regulation of both cytosolic and membrane $p47^{phox}$ indicative of oxidative stress, and increased inflammatory markers such as tumor necrosis factor-α, interleukin (IL) -1α, IL-1β, IL-6, and transforming growth factor-β measured by ELISA, and tissue inhibitor of metalloproteinase-1, CXCL7, RANTES, L-selectin and soluble intercellular adhesion molecule-1 measured by protein array were consistently more attenuated in both HT3 and HT3+10 than in HT10. Hyperoxia-induced decrease in hepatocyte growth factor and vascular endothelial growth factor was significantly up-regulated in both HT3 and HT3+10, but not in HT10. In summary, intratracheal transplantation of human UCB derived MSCs time-dependently attenuated hyperoxia-induced lung injury in neonatal rats, showing significant protection only in the early but not in the late phase of inflammation. There were no synergies with combined early+late MSCs transplantation.

Editor: Dimas Tadeu Covas, University of Sao Paulo – USP, Brazil

Funding: This work was supported by a grant from the Korea Healthcare technology R&D Project, Ministry for Health, Welfare & Family Affairs, Republic of Korea (A102136), by the Basic Science Research Program through the National Research Foundation of Korea (NRF) funded by the Ministry of Education, Science, and Technology (S-2011-0317-000), and by a Samsung Biomedical Research Institute Grant (#SBRI CB11271). The funders had no role in study design, data collection and analysis, decision to publish, or preparation of the manuscript.

Competing Interests: The authors have the following interests: co-authors Soo Jin Choi and Won-il Oh are employed by MEDIPOST Co., Ltd. Co-author Won-il Oh is a board member and holds stocks in MEDIPOST Co., Ltd. Samsung Medical Center and MEDIPOST Co., Ltd. have issued or filed patents for "Method of treating lung diseases using cells separated or proliferated from umbilical cord blood" under the name of Yun Sil Chang, Won Soon Park and Yoon Sun Yang. The relevent application number is: PCT/KR2007/000535. There are no further patents, products in development or marketed products to declare.

* E-mail: wonspark@skku.edu

⁹ These authors contributed equally to this work.

Introduction

Recent improvements in neonatal intensive care medicine have resulted in marked improvements in the survival of the premature infants [1]. However, bronchopulmonary dysplasia (BPD), a chronic lung disease that follows ventilator and oxygen therapy in the premature infants, still remains a major cause of mortality and morbidity with few effective treatments [2,3].

Although the pathogenesis of BPD has not been clearly elucidate yet, oxidative stress and the ensuing inflammation mediated by neutrophils [4] and pro-inflammatory cytokines [5] is believed to play a seminal role in the lung injury process leading to the development of BPD [6]. Recently, we have shown that local intratracheal but not systemic intraperitoneal xenotransplantation of human umbilical cord blood (UCB)-derived mesenchymal stem cells (MSCs) attenuates hyperoxia induced lung injuries such as impaired alveolarization, increased apoptosis and fibrosis in the immunocompetent neonatal rats [7]. Furthermore, these protective effects of stem cell transplantation were dose dependent [8]. Overall, these findings suggest that human UCB derived MSCs transplantation could be a novel therapeutic modality for BPD. However, while the administration of human UCB-derived MSCs at postnatal day (P) 5 was effective in our previous studies [7,8], the optimal timing for their administration has not been determined yet.

Previously, we have shown that the protective effects of human UCB-derived MSCs transplantation are primarily mediated by their anti-inflammatory effects rather than by their regenerative capabilities [7,8]. These findings suggest that the therapeutic time window of stem cell transplantation could be narrow, i.e., only during the early but not the late phase of inflammatory responses. In the present study, we thus tried to determine the optimal timing at which intratracheally delivered human UCB-derived MSCs could attenuate the hyperoxia-induced lung injuries in the newborn rat pups. We firstly conducted time course experiments of inflammatory responses by measuring inflammatory cytokines such as tumor necrosis factor (TNF)-α, interleukin (IL)-1α, 1β and 6 levels at P 0, 3, 5, 7, 10 and 14 in the hyperoxia-induced neonatal lung tissue. After then, we tried to determine the optimal timing by comparing the therapeutic efficacy of early (P3) versus late (P10) intratracheal administration of human UCB derived MSCs in attenuating the hyperoxia-induced lung injuries in the newborn rat pups. We also tried to determine whether combined early (P3)+late (P10) stem cell transplantation has any synergistic effects.

Materials and Methods

Cell Preparation

This study was approved by Institutional Review Board of Samsung Medical Center and by Medipost, Co., Ltd, Seoul, Korea. As previously reported, UCB was collected from umbilical veins after neonatal delivery with informed consent from pregnant mothers, and MSCs were isolated and cultivated from human UCB [9,10]. The cells expressed CD105 (99.6%) and CD73 (96.3%), but not CD34 (0.1%), CD45 (0.2%) and CD14 (0.1%) [7]. They were positive for HLA-AB (96.8%), but generally not for HLA-DR (0.1%). The cells also expressed pluripotency markers such as octamer-binding transcription factor 4 (Oct 4; 30.5%) [11] and stage-specific embryonic antigen 4 (SSEA-4; 67.7%) [12]. Human UCB-derived MSCs differentiated into various cell types such as respiratory epithelium, osteoblasts, chondrocytes and adipocytes with specific *in vitro* induction stimuli [7,10,12,13]. We confirmed the differentiation potential and karyotypic stability of the human UCB-derived MSCs up to the 11th passage.

Animal model

The experimental protocols described herein were reviewed and approved by the Animal Care and Use Committee of Samsung Biomedical Research Institute, Seoul, Korea. This study was also performed in accordance with the institutional and National Institutes of Health guidelines for laboratory animal care. Timed pregnant Sprague-Dawley rats (Orient Co., Seoul, Korea) were housed in individual cages with free access to water and laboratory chow. The rat pups were delivered spontaneously and reared with their dams. The experiment began within 10 h after birth, and continued through P21. Rat pups were randomly divided into four experimental groups; normoxia control group (NC), hyperoxia control group (HC), hyperoxia with early at P3 (HT3), late at P10 (HT10), or combined early+late at P3+10 (HT3+10) human UCB-derived MSCs transplantation group. Rat pups of NC were kept with a nursing mother rat in the standard cage at room air throughout the experiment. Rat pups of hyperoxia groups were maintained with a nursing mother in the standard cage within a 50 liter Plexiglas chambers in which the hyperoxia (oxygen concentration of 90%) was maintained until P14, and after then oxygen concentration was reduced to 60% until P21. Humidity and environmental temperature were maintained at 50% and 24°C, respectively. Nursing mother rats were rotated daily between litters in the normoxia and hyperxoxia groups to avoid oxygen toxicity. Survival and body weight of rat pups in each group were checked daily throughout the experiment. The rat pups of NC and HC were sacrificed at P 1, 3, 5, 7, 10 and 14 for time course experiments and at P21 for group comparison under deep pentobarbital anesthesia (60 mg/kg, intraperitoneal), and the whole lung tissue was obtained for morphometric and biochemical analyses. Six to eight animals were used in each subgroup of analysis.

Transplantation of human UCB-derived MSCs

The human UCB-derived MSCs from the 5th passage from a single donor were labeled using a PKH26GL Red Fluorescent Cell Membrane Labeling Kit (Sigma-Aldrich, St. Louis, MO, USA) for transplantation according to the manufacturer's protocol in the present study, as previously reported [7,8]. For donor cell transplantation, 5×10^5 cells in 0.05 ml phosphate buffered saline (PBS, pH 7.4) were administered intratracheally at P 3, P 10 or P 3+10. For NC and HC, equal volume of PBS was given intratracheally at P3 and P10. For intratracheal transplantation, the rats were anesthetized with an intraperitoneal injection of ketamine and xylazine mixture (45 mg/kg and 8 mg/kg, respectively), and restricted on a board at a fixed angle. MSCs were administered into the trachea through a 30-gauge needle syringe. After the procedure, the animals were allowed to recover from anesthesia, and were returned to their dams. There was no mortality associated with the transplantation procedure.

Tissue preparation

The lungs were resected after transcardiac perfusion with ice-cold phosphate buffered saline (PBS), snap-frozen in liquid nitrogen, and stored at −80°C for later biochemical analyses.

For morphometric analyses, lungs were fixed in situ by tracheal instillation of 10% buffered formalin at a constant inflation pressure of 20 cm H_2O, and then fixed overnight at room temperature in the same fixative. The fixed right lungs were embedded in paraffin, and the left lungs were embedded in an optimal cutting temperature (OCT) compound (SAKURA 4583, Sakura, Torrance, CA, USA). Blocks of the OCT compound were sectioned at 10 μm on a cryostat (Shandon Cryotome, Thermo Electron Co., Waltham, MA, USA) and stored in a deep freezer until analyzed by immunohistochemistry. Four-micrometer-thick sections were cut from the paraffin blocks, and stained with hematoxylin and eosin. Images of each section were captured with a magnifier digital camera through an Olympus BX40 microscope (Olympus optical Co. Ltd., Tokyo, Japan), and were saved as JPEG files.

Morphometry

The level of alveolarization was determined by measuring the MLI and mean alveolar volume. The mean inter-alveolar distance was measured as MLI, by dividing the total length of the lines drawn across the lung section by the number of intercepts encountered, as described by Thurlbeck [14]. The mean alveolar volume was calculated using the method reported by Snyder et al. [15,16]. Briefly, a grid containing equally spaced crosses was placed on a uniformly enlarged photomicrograph of each lung field. The diameters (ℓ) of the alveoli containing a cross were measured along the horizontal axis of the cross. The cube of the alveolar diameter times π and divided by 3 ($\ell^3\pi/3$) was used to estimate the mean alveolar volume. A minimum of two sections per rat and six fields per each section were examined randomly for each analysis.

TUNEL assay

The immunofluorescent TUNEL staining with an in situ cell death detection kit (S7110 ApopTag, Chemicon, Temecula, CA,

USA) was done to measure the extent of apoptosis in the lung. Paraffin section slides were deparaffinized, rehydrated, and digested with Proteinase K (20 μg/ml in PBS) (Sigma Co., St. Louis, MO, USA) at room temperature for 15 minutes and then washed in PBS for 10 minutes. Sections were then incubated with equilibration buffer for 1 minute and immediately incubated with working strength TdT enzyme in a humidified chamber at 37°C for 1 hour. Each section was immersed in a stop/wash buffer and gently rinsed with PBS. Fluorescein isothiocyanate (FITC)-labeled anti-digoxigenin conjugate was applied to the sections which were then incubated at room temperature for 30 minutes in the dark. Nuclear counterstaining was performed with propidium iodide (0.5 μg/ml, Sigma Co., St. Louis, MO, USA). Slides were washed again in PBS, mounted with Vectasheild mounting solution (Vector laboratories, Burlingame, CA, USA), and visualized with a fluorescent microscope (Nikon E600 fluorescence microscope, Tokyo, Japan) using an excitation wavelength of 460–490 nM. Ten non-overlapping fields with a magnifying power of ×200 were examined to count TUNEL positive cells.

Quantification of the PKH26 positive cells

Ten-μm-thick cryosections were mounted with a Vector shield mounting solution containing DAPI (H-1200, Vector, Burlingame, CA, USA). The cell counts for the transplanted or donor-derived cells were measured using PKH26 red fluorescence, as described above after combining the ×20 objective images of the DAPI-stained nuclei signals. Five fields per section were selected randomly, focused, and counted with the naked eye under a fluorescence microscope (Nikon E600, Nikon, Tokyo, Japan) using a filter to detect the PKH26 red fluorescence. The PKH26 red fluorescence was counted manually and averaged per high power field (HPF) in a single animal. Two random sections per animal were evaluated in a blinded manner.

Western blot for $p47^{phox}$

The upregulation of $p47^{phox}$, a subunit of NADPH oxidase, both in the cytoplasmic and plasma membrane portion serves as an indicator of NADPH oxidase activation that is responsible for generating reactive oxygen species [17,18]. Tissue slides were incubated for 40 min in PBS containing 0.1% Triton X-100. After blocking with 0.5% BSA, slides were incubated overnight at 4°C with anti-$p47^{phox}$ antibody (1:200) and then exposed for 2 hr at room temperature to FITC-conjugated goat anti mouse immunoglobulin-G (1:200; BD Biosciences, USA). Vectasheld mount medium with DAPI (Vector Laboratories) was used to preserve. Confocal microscopic examination was carried out at 400× magnification using Bio-Rad Radiance 2100 (Bio-Rad Laboratories Inc. Hercules, CA, USA) with krypton/argon laser, and images were achieved using the Laser shop 2000 software (Bio-Rad Laboratories, Inc.). For this purpose, tissue sample from each animal were separated into membrane and cytosolic components for western blot examination. Tissues were homogenized in ice-cold hypertonic solution and centrifuged at 600 g for 10 minutes. The supernatant was ultra-centrifuged at 100,000 g for 1.5 hours. The supernatant contained the cytosolic fraction and the membrane-particulate pellet was resuspended in hypotonic solution containing 1% Triton X-100. Samples were analyzed by western blotting using antibodies against the NADPH oxidase cytosolic subunit, $p47^{phox}$ (1:500, BD Biosciences, San Diego, CA, USA). The bands were recognized by horseradish peroxidase-conjugated anti-mouse secondary antibody (1:1,000, DAKO, Glostrup, Denmark), and then western blots were developed with enhanced chemiluminescence detection reagents (Amersham Pharmacia, Uppsala, Sweden), and exposed to X-ray film (Fuji Photo Film, Tokyo, Japan). The blots were re-probed with antibodies against GAPDH (1:1,000, Santa Cruz Biotechnology Inc., Santa Cruz, CA, USA). To determine the relative degree of membrane purification, the membrane fraction was subjected to immunoblotting for calnexin (1:500 Santa Cruz Biotechnology Inc), a membrane marker.

Protein macroarray

Each lung lysate was analyzed using a rat cytokine array kit (Proteome Profiler™; R&D Systems, Minneapolis, MN, USA). A total of 250 μg of lysate was incubated in the nitrocellulose membrane array overnight at 4°C. After washing away the unbound protein, the array was incubated with a cocktail of phospho-site-specific biotinylated antibodies for 2 h at room temperature, followed by streptavidin–HRP for 30 min. Signals were visualized with chemiluminescent reagents (Amersham Biosciences, Pittsburgh, PA, USA), and recorded on X-ray film. The arrays were scanned, and optical densities were measured using Image J software (NIH) and compared among the experimental groups. The protein macroarray analysis included inflammatory cytokines of interest, including tissue inhibitor of metalloproteinase (TIMP)-1, Chemokine (C-X-C motif) ligand 7 (CXCL7), regulated upon activation normal T-expressed and presumably secreted (RANTES), L-selectin and the soluble form of intercellular adhesion molecule (sICAM)-1.

Myeloperoxidase activity

The activity of MPO, an indicator of neutrophil accumulation, was determined by modification of the method by Gray et al. [19]. The lung tissues were homogenized in a phosphate buffer (pH 7.4) and centrifuged at 30,000 g for 30 min. The pellet was resuspended in another phosphate buffer (50 mM, pH 6.0) containing 0.5% hexadecyltrimethyl ammonium bromide. MPO activity in the resuspended pellet was assayed by measuring absorbance changes spectrophotometrically at 460 nm, using 0.167 mg/mL of O-dianisidine ihydrochloride and 0.0005% hydrogen peroxide. One unit of MPO activity was defined as the quantity of enzyme degrading 1 μM of peroxide/min.

Hepatocyte growth factor (HGF)

Total RNA in the sample was extracted using RNA Trizol according to the manufacturer's protocol (Invitrogen Corporation, Carlsbad, CA, USA). Total RNA concentration was measured by spectrophotometry (Nanodrop Wilmington, DE, USA) at 260 nm. One microgram of RNA was used to produce cDNA with a Protoscript® II RT-PCR kit (New England Biolabs, Ipswich, MA, USA). PCR primers for rat hepatocyte growth factor and rat glyceraldehyde-3-phosphate dehydrogenase (GAPDH) were designed with Primer3 (Whitehead Institute, Cambridge, MA, USA) and synthesized by Bioneer Inc. (Bioneer, Daejeon, Korea). The sequence of primers used was as follows: rat HGF (sense-accctggtgtttcacaagca- antisense-aggggtgtcagggtcaagag-), rat GAPDH (sense- ggccaaaagggtcatcatct-, antisense-gtgatggcatggactgtggt-). PCR products were run on a 1.2% agarose gel electrophoresis, visualized by ethidium bromide and scanned by a Gel Doc 2000 analyzer (Bio-Rad, Hercules, CA, USA). The expression levels for each gene were semi-quantified by densitometric analysis using software (Quantity One, Bio-Rad, Hercules, CA, USA). Relative expression levels were estimated by the density ratio of rat GAPDH to rat HGF.

Statistical Analysis

The data are expressed as the mean ± SEM. Survival rates were compared using the Kaplan-Meier analysis followed by a log rank test. For continuous variables with a normal distribution, the groups were compared using a *t*-test with a Bonferroni correction. Continuous variables that were not normally distributed were analyzed using the Wilcoxon rank test with a Bonferroni correction. All data were analyzed using Stata software (ver. 11.0, StataCorp LP, College Station, TX, USA). Values of $p<0.05$ were considered statistically significant.

Results

Temporal profile of inflammatory responses

In time course experiments of inflammatory responses by measuring TNF-α, IL-1α, IL-1β and IL-6 levels with ELISA at P 1, 3, 5, 7, 10 and 14 in the lung tissue, IL-6 levels after P5 and TNF-α, IL-1α and IL-1β levels after P7 in HC became significantly increased compared to NC up to P 14 (Fig. 1).

Survival rate and body weight gain

Exposure to oxygen (HC) significantly reduced the survival rate to 70.8% ($P<0.05$ vs. NC) at the end of experiment (P21) compared to the 100% survival rate of NC. On the contrary, survival rates of HT3 (91.7%, $P>0.05$ vs. NC) and HT3+10 (87.5%, $P>0.05$ vs. NC) were not different when it compared to NC. However, survival rate of HT10 (75.0%, $P<0.05$ vs. NC) was significantly lower than that of NC (Fig. 2).

Although birth weight was not significantly different between the five experimental groups (7.0±0.13 g, 7.1±0.04 g, 7.1±0.04 g, and 7.1±0.03 in NC, HC, HT3, HT10 and HT3+10, respectively), body weight at P21 in HC (34.3±4.6 g, $P<0.01$) was significantly lower compared to NC (41.7±2.3 g), and this retarded body weight gain observed in HC was significantly improved in HT3 (39.5±4.7 g, $P<0.01$ vs. HC) and HT3+10 (38.8±4.4 g, $P<0.01$ vs. HC), but not in HT10 (35.7±4.2 g, $P>0.05$ vs. HC, $P<0.05$ vs. HT3, $P<0.05$ vs. HT3+10).

Lung histopathology

Fig. 3A presents typical photomicrographs showing the histopathological differences observed by optical microscopy in each experimental group at P 21. While uniform and small alveoli were observed in NC, impaired alveolar growth, as evidenced by fewer and larger alveoli, focal airspace enlargement and heterogeneous alveolar size were observed in HC compared to NC. After MSCs transplantation, the hyperoxia-induced impairments in alveolar growth and morphological changes were attenuated, particularly in both HT3 and HT3+10 compared to HT10.

In morphometric analyses, the MLI and mean alveolar volume, indicating the size and volume of the alveoli respectively, were significantly higher in HC (67.2±1.2 μm in MLI, $P<0.001$ vs. NC; $19.4\pm1.2\times10^4$ μm^3 in mean alveolar volume, $P<0.001$ vs. NC) than in NC (40.8±1.0 μm in MLI and $3.1\pm0.04\times10^4$ μm^3 in mean alveolar volume). (Fig. 3B). After MSCs transplantation, these hyperoxia-induced morphometric abnormalities were better attenuated in HT3 (54.0±1.4 μm in MLI, $P<0.001$ vs. HC; $6.7\pm0.9\times10^4$ μm^3 in mean alveolar volume $P<0.001$ vs. HC) and HT3+10 (55.3±0.8 μm vs. HC, $P<0.001$; $6.2\pm0.5\times10^4$ μm^3 in mean alveolar volume $P<0.001$ vs. HC) compared to those in HT10 (59.4±1.7 μm in MLI, $P<0.001$ vs. HC, $P<0.01$ vs. HT3, , $P<0.05$ vs. HT3+10 ; $10.4\pm1.0\times10^4$ μm^3 in mean alveolar volume $P<0.001$ vs. HC, $P<0.01$ vs. HT3, $P<0.01$ vs. HT3+10) (Fig. 3B).

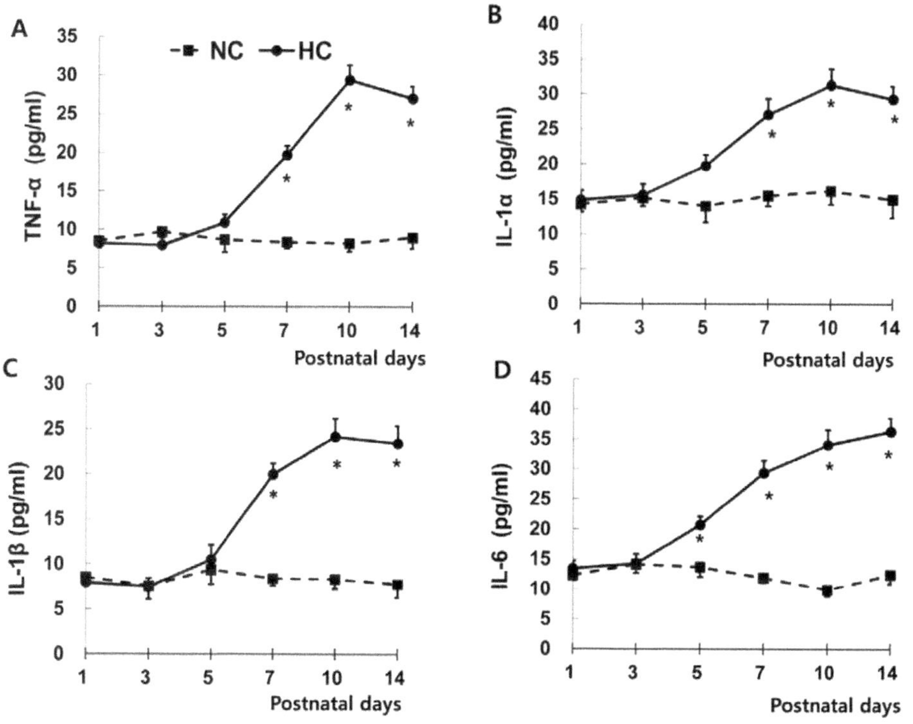

Figure 1. Temporal profiles of inflammatory cytokines. Tumor necrosis factor (TNF)-α, interleukin (IL)-1α, IL-1β, and IL-6 levels measured with ELISA at P 1, 3, 5, 7, 10 and 14 in the rat lung tissue. NC, Normoxia control group; HC, hyperoxia control group. Data; mean±SEM. *$P<0.05$ compared to NC.

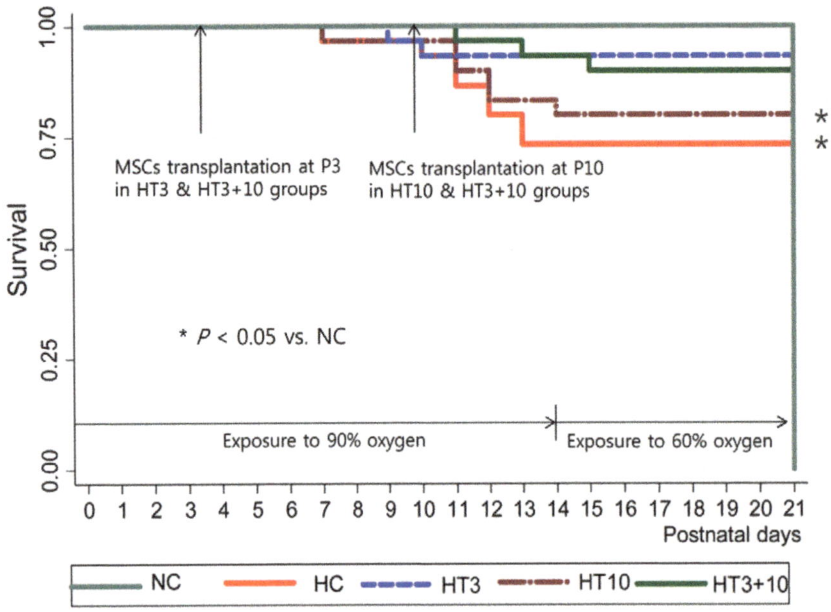

Figure 2. Survival curve. Kaplan-Meier survival curve up to P 21 showing decreased survival. HC compared to NC, and improved survival in both HT3 and HT3+10, but not in HT10. NC, Normoxia control group; HC, hyperoxia control group; HT3, hyperoxia with stem cell transplantation group at P3; HT10, hyperoxia with stem cell treatment group at P10; HT3+10, hyperoxia with stem cell treatment group at P3 and P10. $^{*}P<0.05$ compared to NC.

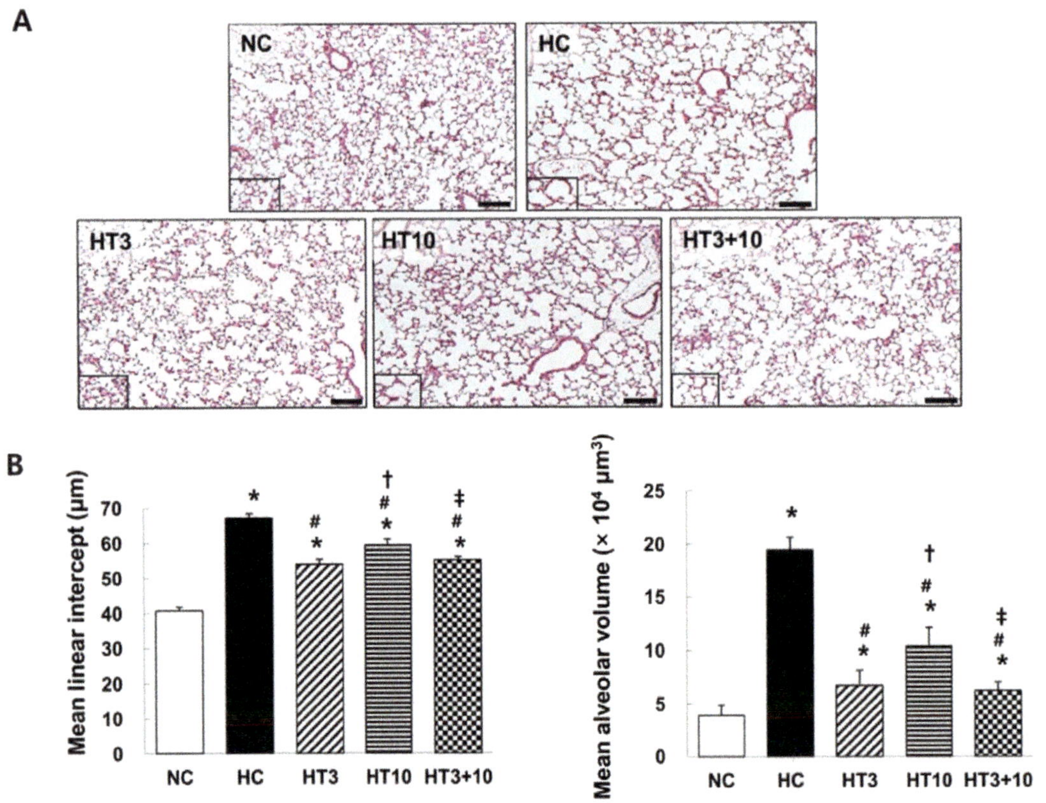

Figure 3. Histology and morphometric analysis of the surviving P21 rat lung. (A): Representative optical microscopy photomicrographs of the lungs stained with hematoxylin and eosin (scale bar = 100 μm). (B): Degree of alveolarization measured by the mean linear intercept (*left*) and mean alveolar volume (*right*). NC, Normoxia control group; HC, hyperoxia control group; HT3, hyperoxia with stem cell transplantation group at P3; HT10, hyperoxia with stem cell treatment group at P10; HT3+10, hyperoxia with stem cell treatment group at P3 and P10. Data; mean ± SEM. $^{*}P<0.05$ compared to NC, $^{\#}P<0.05$ compared to HC, $^{\dagger}P<0.05$ compared to HT3, $^{\ddagger}P<0.05$ compared to HT10.

The number of TUNEL positive cells in the lung of P21 rats per high power field was significantly increased in HC (15.2±1.1, $P<0.001$) compared to NC (1.1±0.2). This hyperoxia-induced increase in the number of TUNEL positive cells was significantly attenuated in both HT3 (7.6±0.8, $P<0.001$ vs. HC) and HT3+10 (6.6±0.3, $P<0.001$ vs. HC), but not in HT10 (17.4±0.6, $P>0.05$ vs. HC, $P<0.001$ vs. HT3, $P<0.001$ vs. HT3+10) (Fig. 4).

The deposition of PKH26 red fluorescence positive donor cells was observed only in the MSCs transplantation groups, but not in NC and HC (Fig. 5A). The number of donor cells identified per lung field was significantly larger in HT10 (21.5±2.9, $P<0.001$ vs. HT3) and HT3+10 (25.4±1.7, $P<0.001$ vs. HT3) than in HT3 (10.6±1.6). However, there were no significant differences in the donor cells between HT10 and HT3+10 (Fig. 5B).

Cytosolic and membrane expressions of $p47^{phox}$

Since NADPH oxidase produces oxygen free radicals in both phagocytic [20] and nonphagocytic cells [21,22], hyperoxia-induced production of ROS was evaluated by NADPH oxidase activation, as evidenced by increased cytosolic and membrane expression of a cytosolic subunit of NADPH oxidase $p47^{phox}$. In fluorescent microscopy, increased $p47^{phox}$ was observed in HC compared to NC, indicating the activation of NADPH oxidase. After MSCs transplantation, the hyperoxia-induced increase in $p47^{phox}$ was attenuated, particularly in both HT3 and HT3+10 compared to HT10 (Fig. 6A). In western blot analyses, significantly higher levels of $p47^{phox}$ were observed in HC both in the cytosolic ($P<0.01$) and membrane ($P<0.05$) fractions than in NC. While the hyperoxia-induced increase in the cytosolic expression of $p47^{phox}$ was significantly attenuated in the MSCs transplantation groups ($P<0.05$ vs. HC) the increase in the membrane fraction was attenuated in both HT3 ($P>0.05$ vs. NC) and HT3+10 ($P>0.05$ vs. NC), but not in HT10 ($P<0.05$ vs. NC).

ELISA and Protein array of cytokines

In HC, significantly increased levels of TNF-α ($P<0.001$ vs. NC), IL-1α ($P<0.001$ vs. NC), IL-1β ($P<0.01$ vs. NC), IL-6 ($P<0.001$ vs. NC) and TGF-β ($P<0.01$ vs. NC) measured by ELISA and TIMP-1 ($P<0.001$ vs. NC), CXCL7 ($P<0.001$ vs. NC), RANTES ($P<0.001$ vs. NC), L-selectin ($P<0.001$ vs. NC) and sICAM-1 ($P<0.05$ vs. NC) measured by protein array were observed compared to NC (Fig. 7). The hyperoxia-induced increase in these cytokine levels was significantly attenuated in both HT3 and HT3+10, but not in HT10, and the attenuation of IL-1α and IL-6 was more profound in HT3 (IL-1α, $P>0.05$ vs. NC, $P<0.01$ vs. HC; IL-6, $P>0.05$ vs. NC, $P<0.001$ vs. HC) and HT3+10 (IL-1α, $P>0.05$ vs. NC, $P<0.01$ vs. HC; IL-6, $P>0.05$ vs. NC, $P<0.001$ vs. HC) than in HT10 (IL-1α, $P<0.05$ vs. NC, $P<0.05$ vs. HC; IL-6, $P<0.01$ vs. NC, $P<0.01$ vs. HC, $P<0.01$ vs. HT3, $P<0.01$ vs. HT3+10).

ED1 positive cells, Myeloperoxidase activity and Collagen levels

The ED1 positive alveolar macrophages were significantly higher in HC (13.6±1.8, $P<0.001$) than in NC (1.0±0.1). This hyperoxia- induced increase in ED1 positive cells was significantly attenuated with MSCs transplantation, and this attenuation was more profound in HT3 (4.9±0.8, $P<0.001$ vs. HC) and HT3+10 (4.9±0.2, $P<0.001$ vs. HC) than in HT10 (7.9±1.1, $P<0.01$ vs. HC, $P<0.05$ vs. HT3, $P<0.05$ vs. HT3+10) (Fig. 8A).

The MPO activity in HC (8.2±0.5 U, $P<0.001$) was significantly higher than in NC (1.5±0.2 U). The hyperoxia-induced increase in MPO activity was significantly attenuated in both HT3 (6.0±0.2 U, $P<0.001$ vs. HC) and HT3+10 (6.0±0.3 U, $P<0.001$ vs. HC), but not in HT10 (8.6±0.8 U, $P>0.05$ vs. HC, $P<0.01$ vs. HT3, $P<0.01$ vs. HT3+10) (Fig. 8B).

The lung collagen levels at P21 were significantly higher in HC (149±5 μg/mg protein, $P<0.001$) than in NC (81±5 μg/mg protein). This hyperoxia-induced increase in the lung collagen levels was significantly attenuated in both HT3 (124±3 μg/mg protein, $P<0.01$ vs. HC) and HT3+10 (126±4 μg/mg protein, $P<0.01$ vs. HC), but not in HT10 (142±4 μg/mg protein, $P>0.05$ vs. HC, $P<0.05$ vs. HT3, $P<0.05$ vs. HT3+10) (Fig. 8C).

HGF and VEGF

The HGF in the rat lung measured by RT-PCR at P21 were significantly lower in HC ($P<0.05$) than in NC. The hyperoxia-induced decrease in lung HGF was significantly up-regulated in both HT3 ($P<0.05$ vs. HC) and HT3+10 ($P<0.01$ vs. HC), but not in HT10 ($P>0.05$ vs. HC, $P<0.05$ vs. HT3, $P<0.01$ vs. HT3+10) (Fig. 9A). The VEGF levels in the rat lungs measured by ELISA at P21 were significantly lower in HC (23.5±1.9 pg/ml, $P<0.001$) than in NC (39.5±4.3 pg/ml). This hyperoxia-induced

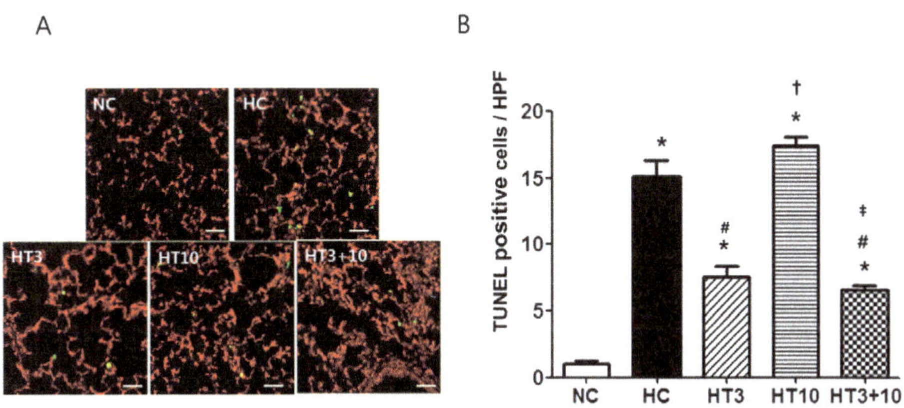

Figure 4. TUNEL positive cells in the distal lungs of the P 21 rat pups. (A): TUNEL positive cells were labeled with FITC (*green*) and the cell nuclei were labeled with propidium iodide (*red*) (Scale bar; 25 μm). (B): Number of observed TUNEL positive cells per high power field. NC, Normoxia control group; HC, hyperoxia control group; HT3, hyperoxia with stem cell transplantation group at P3; HT10, hyperoxia with stem cell treatment group at P10; HT3+10, hyperoxia with stem cell treatment group at P3 and P10. Data; mean±SEM. *$P<0.05$ compared to NC, # $P<0.05$ compared to HC,† $P<0.05$ compared to HT3, ‡ $P<0.05$ compared to HT10.

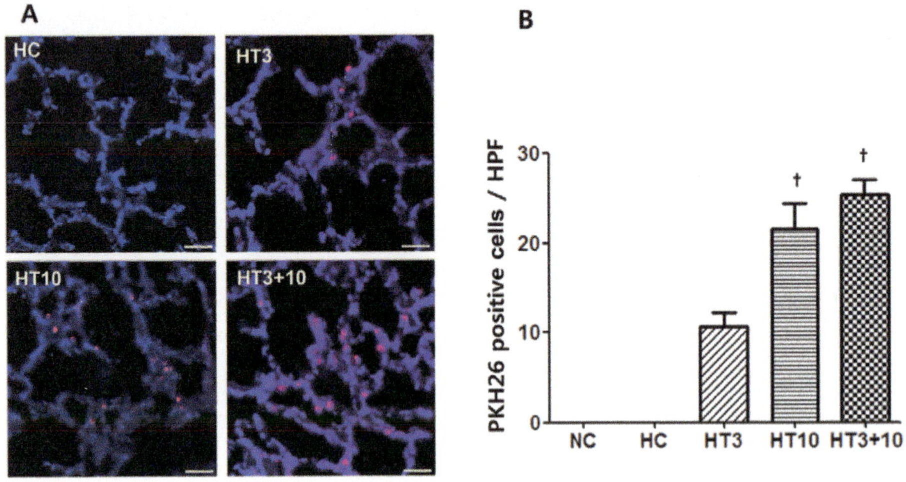

Figure 5. Donor cell localization in the lung of the P 21 rats. (A): Fluorescent microscopic observation of the PKH26 labeled human UCB-derived MSCs (donor cells, *red*) localized in the lungs of the P 21 newborn rats and the nuclei labeled with DAPI (*blue*) (Scale bar; 25 μm). (B): Number of PKH26 positive cells in the lung per high power field. NC, Normoxia control group; HC, hyperoxia control group; HT3, hyperoxia with stem cell transplantation group at P3; HT10, hyperoxia with stem cell treatment group at P10; HT3+10, hyperoxia with stem cell treatment group at P3 and P10. Data; mean±SEM. † $P<0.05$ compared to HT3.

decrease in the lung VEGF level was attenuated in both HT3 (30.6±2.0 pg/ml, $P<0.05$ vs. HC) and HT10 (33.4±1.6 pg/ml, $P<0.01$ vs. HC), but not in HT10 (28.1±1.7 pg/ml, $P>0.05$ vs. HC) (Fig. 9B).

Discussion

In the present study, prolonged exposure of newborn rat pups to hyperoxia for 3 weeks increased mortality, retarded growth, and developed lung injuries similar to those seen in the premature human infants with BPD [23,24], exhibiting decreased alveolarization as evidenced by increased MLI and alveolar volume [25], and significantly increased TUNEL positive cells [26]. Previously, we have shown that the neonatal hyperoxic lung injuries induced by ≥90% oxygen for 2 weeks were significantly attenuated with intratracheal human UCB-derived MSCs transplantation at P5 [7], and these beneficial effects were dose dependent [8]. In the present study, hyperoxic exposure was extended to 3 weeks to guarantee the comparable time span after MSCs transplantation at P10, and the oxygen concentration during the third week was reduced from 90% to 60% because of the concern about increased mortality due to prolonged high oxygen exposure. Only MSCs given at P3 but not at P10 showed protective effects against the hyperoxia-induced lung injuries, and no synergistic effects were observed with combined early+late at P3+10 MSCs transplantation in this study. Overall, these findings suggest that the therapeutic efficacy of human UCB-derived MSCs transplantation is time-dependent, showing protection only during the early but not late phase of inflammation.

In time course experiments of the present study, lung inflammatory cytokine levels such as TNF-α, IL-1α, IL-1β and IL-6 in HC became significantly higher after P5~7 compared to NC. Intratracheal MSCs given at P3 or P5, but not at P10 protected against neonatal hyperoxic lung injuries in our previous [7,8] and present studies. These findings suggest that increased secretion of proinflammatory cytokines and the ensuing full blown host inflammatory milieu might be a strong inhibitor of MSCs protection against neonatal hyperoxic lung injury, and thus the therapeutic time window of stem cell transplantation in BPD might be limited only to the early phase of inflammation.

In the present study, the number of PKH26 positive cells at P21, indicative of donor cell localization in the lung tissue, was significantly higher in HT10 and HT3+10 than in HT3. These findings suggest that substantial donor cell losses occur early after intratracheal MSCs transplantation [27]. However, significant protective effects against neonatal hyperoxic lung injuries were observed in both HT3 and HT3+10, but not in HT10. Similar results of beneficial effects of MSCs despite low engraftment have also been reported in other disease models such as acute renal failure [28], and cardiac injury [29]. In our previous study [7,8], we also observed only very few donor cells differentiate into type II pneumocytes, Overall, these findings suggest that even a small number of MSCs survived could mediate their therapeutic effects against neonatal hyperoxic lung injuries primarily by secreting key bioactive mediators rather than by direct tissue repair [30], and early timing of MSCs transplantation is critical for their best protective effects. Moreover, although donor cells rapidly faded away after transplantation, the favorable effects of MSCs were persistent up to P21. These findings suggest that the paracrine effects initially induced by stem cells might play a pivotal role in tissue repair, and are sustained later by the intact host tissue protected by MSCs transplantation [31]. However, we cannot exclude the possibility that rejection might have occurred in our study because transplant was performed in a xenograft model; therefore, further studies are needed to clarify this.

Oxidative stress to the immature lung is a well-known risk factor for the development of BPD [32]. NADPH oxidase is a multi-component enzyme complex responsible for production of superoxide anion (O_2^-), which generates other reactive oxygen species such as hydrogen peroxide, hydroxyl radical, and hypochlorous acid [20]. Upon its activation, the cytoplasmic subunits p47^{phox}, p67^{phox}, p40^{phox} and Rac translocate to membrane bound cytochrome [33]. The membrane translocation of the p47^{phox} can thus be served as an *in vivo* indicator of NADPH oxidase activation in the lung tissue. In our previous study [8], we have observed the dose-dependent anti-oxidative effects of intratracheal MSCs transplantation. In the present study,

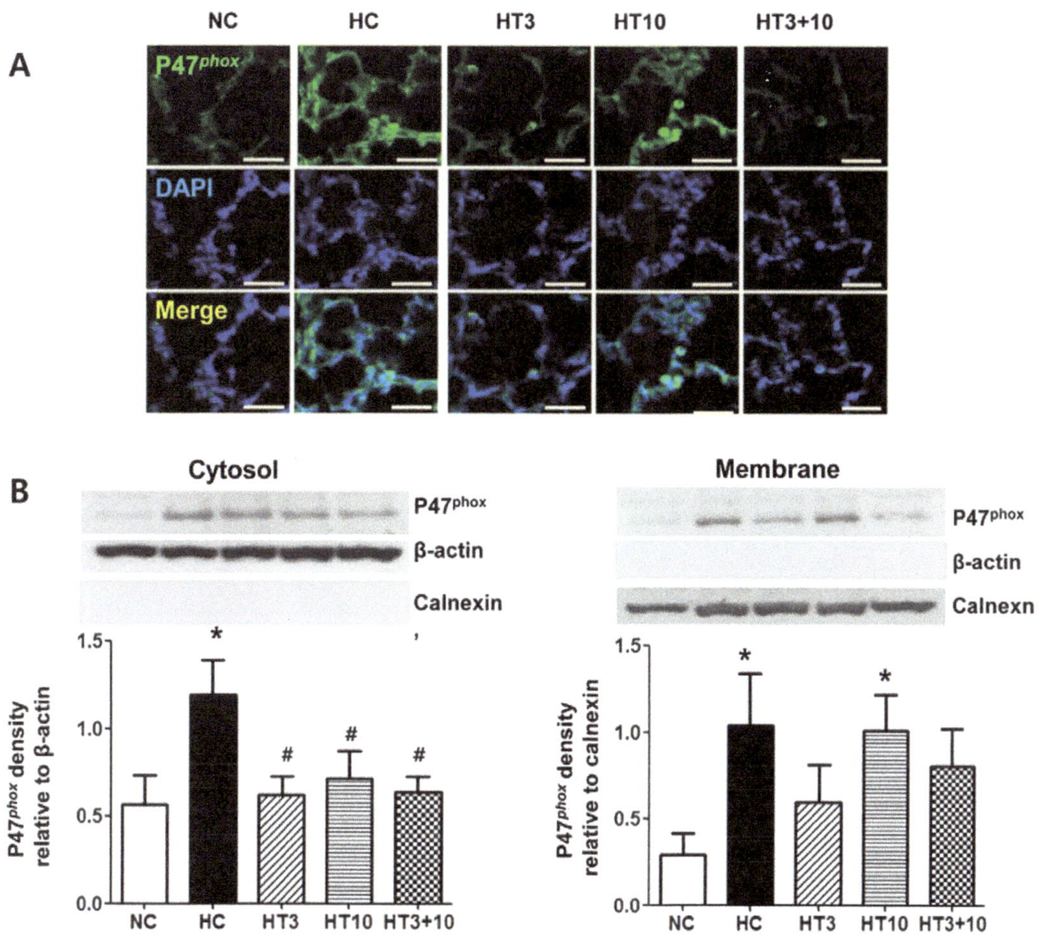

Figure 6. p47^{phox}, cytosolic subunit of nicotinamide adenine dinucleotide phosphate oxidase in the P21 rat lung tissue. (A): Fluorescent microscopic observation of p47^{phox} (*green*) localized in the lungs of the P21 rats and the nuclei labeled with DAPI (*blue*) (Scale bar; 25 μm). (B): Representative western blots (*top*) and densitometric histograms (*bottom*) in the cytosol (*left*) and membrane (*right*) fractions of P21 rat lung homogenates. NC, Normoxia control group; HC, hyperoxia control group; HT3, hyperoxia with stem cell transplantation group at P3; HT10, hyperoxia with stem cell treatment group at P10; HT3+10, hyperoxia with stem cell treatment group at P3 and P10. Data; mean±SEM. $^{*}P<0.05$ compared to NC, $^{\#}$ $P<0.05$ compared to HC.

increased expression of the p47^{phox} protein was observed both in the cytosolic and membrane fraction in HC. Although significant attenuation of cytosolic expression of the p47^{phox} was observed after MSCs transplantation, a decrease in membrane translocation of the p47^{phox} was observed in both HT3 and HT3+10, but not in HT10. These findings suggest that MSCs transplantation at the early rather than late phase of inflammation will be the optimal timing for their best anti-oxidative effects.

In our previous studies [7,8], we have shown that inflammatory responses mediated by neutrophils [4] and proinflammatory cytokines [5] play a pivotal role in the development of BPD [6], and that the protective effects of MSCs therapy against hyperoxia-induced lung injuries are mediated primarily by their anti-inflammatory effects rather than by their regenerating capacity. In the present study, hyperoxia-induced increase in ED1 positive alveolar macrophages, lung myeloperoxidase activity and cytokines such as IL-1α, IL-1β, IL-6, TNF-α measured by ELISA, other inflammatory markers such as TIMP-1, CXCL7, RANTES, L-selectin and sICAM-1 measured by protein array were consistently attenuated by MSCs administration at both early (HT3) and combined early+ late (HT3+10), but not at the late (HT10) phase of inflammation. These findings suggest a full-blown host inflammatory micro-environment might hinder the anti-inflammatory effects of the transplanted MSCs. Further studies will be necessary to elucidate the inhibitory mechanism of endogenous inflammatory factors on transplanted stem cells function.

In our previous studies [7,8], we have shown that the anti-fibrotic effect of transplanted MSCs was associated and probably mediated by their down-modulation of hyperoxia-induced pulmonary inflammatory responses. Significant attenuation of hyperoxia-induced increase in both fibrogenic cytokines such as TGF-β and TIMP-1 [34] along with other inflammatory cytokines, and collagen levels in both HT3 and HT3+10, but not in HT10 support the assumption that the early timing of MSCs transplantation is critical for their best anti-inflammatory and the ensuing anti-fibrotic effects.

In the present study, hyperoxia-induced decrease in growth factors such as VEGF and HGF was significantly up-regulated in both HT3 and HT3+10, but not in HT10 despite higher donor cell localization in the lung tissue at P21 in HT10 than in HT3. These findings suggest that the protective effects of MSCs transplantation such as promotion of angiogenesis, anti-apoptotic effects and reduced inflammation are strongly associated or

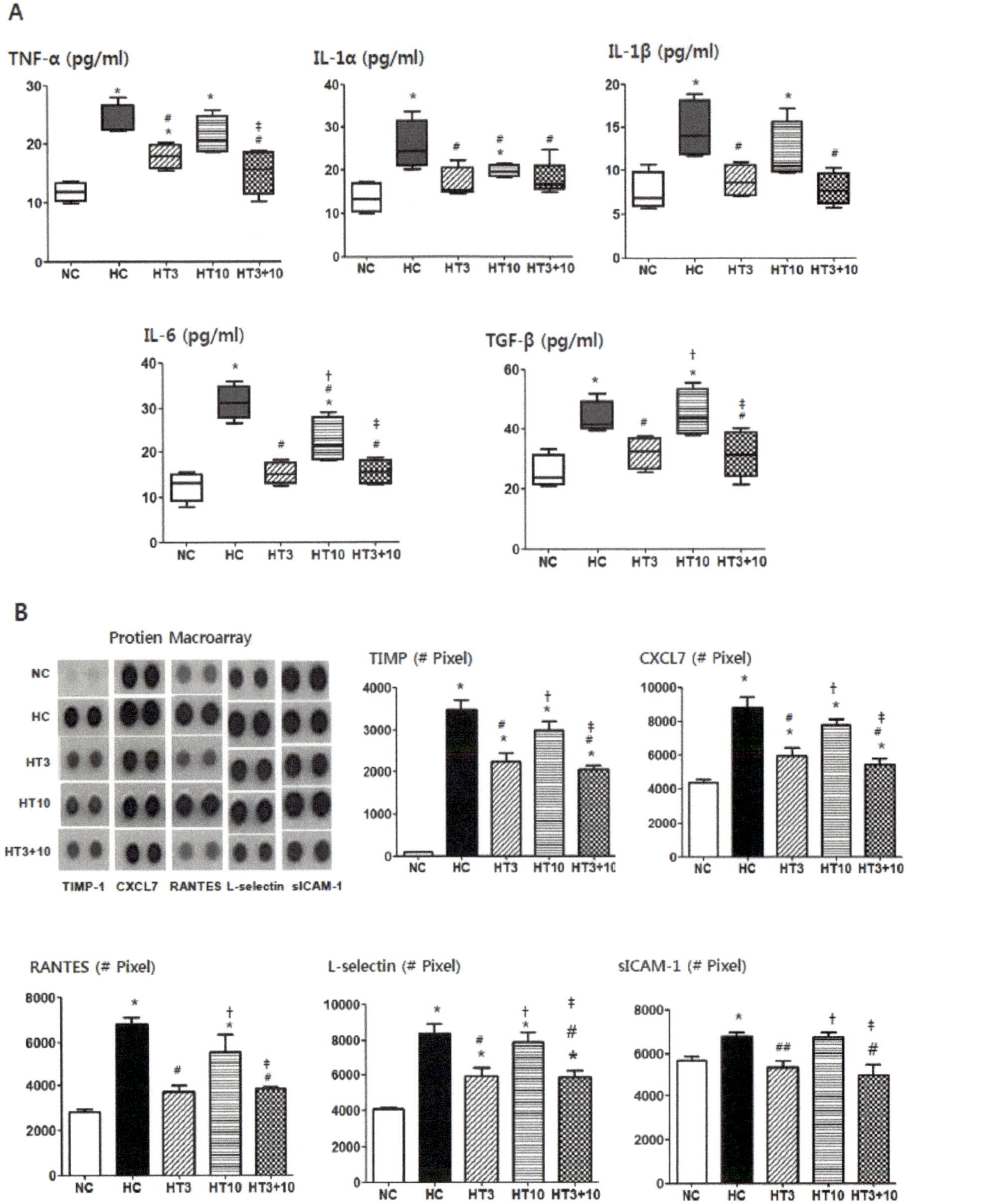

Figure 7. Histograms of inflammatory cytokines and chemokines in the hyperoxic lung injury after MSCs transplantation. TNF-α, IL-1α, IL-1β, IL-6, and TGF-β levels measured by ELISA (A) and TIMP-1, CXCL7, RANTES, L-selectin and sICAM-1 levels measured by protein array at P21 in the rat lung tissue (B,C). NC, Normoxia control group; HC, hyperoxia control group; HT3, hyperoxia with stem cell transplantation group at P3; HT10, hyperoxia with stem cell treatment group at P10; HT3+10, hyperoxia with stem cell treatment group at P3 and P10. Data; mean±SEM. $^{*}P<0.05$ compared to NC, $^{\#}$ $P<0.05$ compared to HC,† $P<0.05$ compared to HT3, ‡ $P<0.05$ compared to HT10.

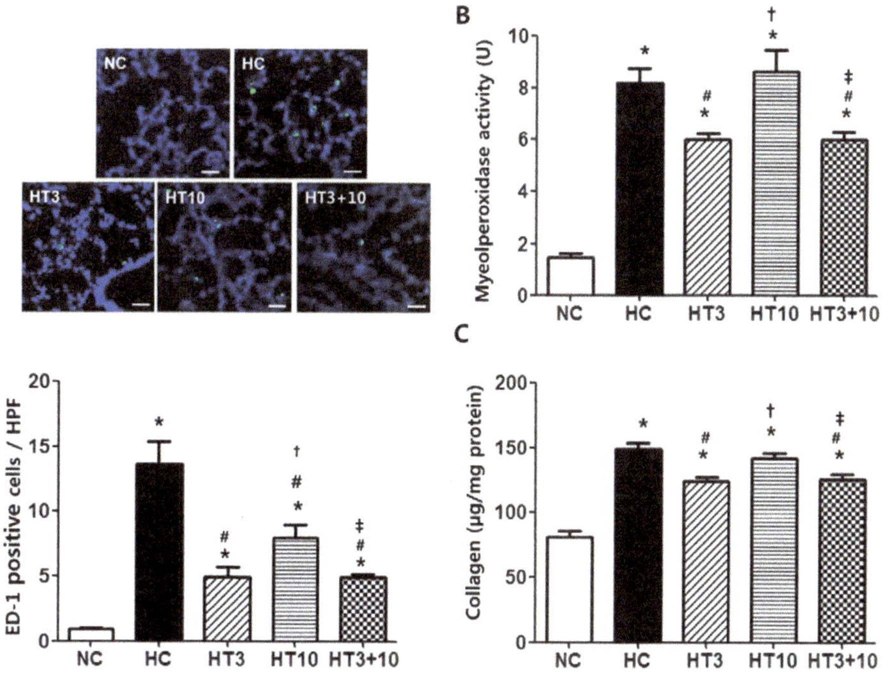

Figure 8. Inflammation with fibrosis in the hyperoxic lung injury after MSCs transplantation. ED1 positive cells indicative of alveolar macrophage were labeled with FITC (*green*) and the nuclei were labeled with DAPI (*blue*) (Scale bar; 25 µm) (*at top*) and number of observed ED1 positive cells per high power field (*below*) (A), myeloperoxidase activity (B), and collagen levels (C) of the rat P21 lung tissues. NC, Normoxia control group; HC, hyperoxia control group; HT3, hyperoxia with stem cell transplantation group at P3; HT10, hyperoxia with stem cell treatment group at P10; HT3+10, hyperoxia with stem cell treatment group at P3 and P10. Data; mean±SEM. $^{*}P<0.05$ compared to NC, $^{\#}$ $P<0.05$ compared to HC,† $P<0.05$ compared to HT3, ‡ $P<0.05$ compared to HT10.

probably mediated by enhanced secretion of these growth factors [35]. Moreover, as the full blown host inflammatory milieu might hinder the secretion of these growth factors by MSCs [36] the timing of MSCs transplantation early in the inflammation is essential for enhanced expression of these growth factors. Further studies will be necessary to clarify this.

In summary, intratracheal transplantation of human UCB derived MSCs time-dependently attenuated hyperoxia-induced lung pathology such as decreased alveolarization and increased

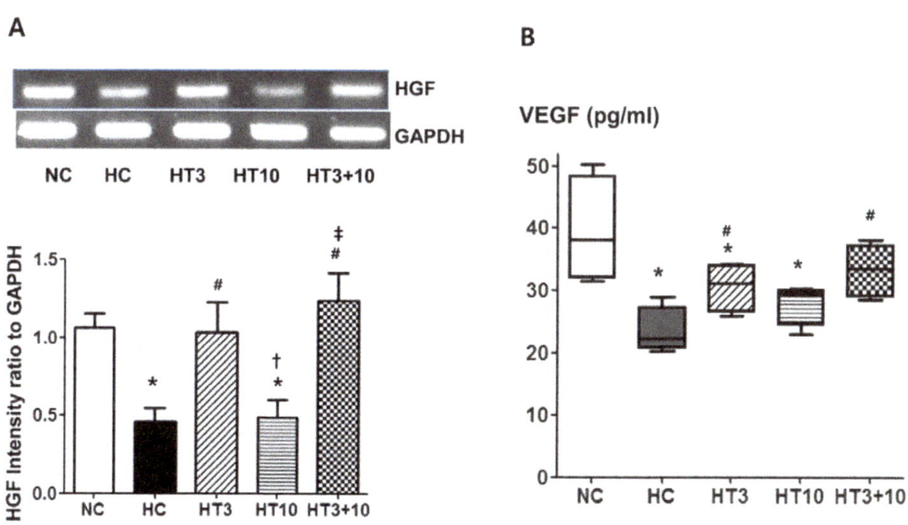

Figure 9. VEGF and HGF in the hyperoxic lung injury after MSCs transplantation. Representative RT-PCR blots (*at top*) and densitometric histograms (*below*) for HGF (A) and VEGF measured with ELISA (B) in the P21 rat lungs. NC, Normoxia control group; HC, hyperoxia control group; HT3, hyperoxia with stem cell transplantation group at P3; HT10, hyperoxia with stem cell treatment group at P10; HT3+10, hyperoxia with stem cell treatment group at P3 and P10. Data; mean±SEM. $^{*}P<0.05$ compared to NC, $^{\#}$ $P<0.05$ compared to HC,† $P<0.05$ compared to HT3, ‡ $P<0.05$ compared to HT10.

apoptotic cells, oxidative stress, inflammatory responses and the ensuing fibrosis, and up-regulated growth factors such as HGF and VEGF, showing significant protection only in the early at P3 but not in the late at P10 phase of inflammation. There were no synergies with combined early+late MSCs transplantation. These findings are expected to have important implications for future clinical translation to determine the optimal timing of MSCs transplantation for the currently untreatable neonatal hyperoxic lung disease i.e., BPD in premature infants.

Author Contributions

Conceived and designed the experiments: WSP. Performed the experiments: SYA DKS HSY. Analyzed the data: YSC SIS. Contributed reagents/materials/analysis tools: WIO. Wrote the paper: YSC SJC.

References

1. Saigal S, Doyle LW (2008) An overview of mortality and sequelae of preterm birth from infancy to adulthood. Lancet 371: 261–269.
2. Avery ME, Tooley WH, Keller JB, Hurd SS, Bryan MH, et al. (1987) Is chronic lung disease in low birth weight infants preventable? A survey of eight centers. Pediatrics 79: 26–30.
3. Bregman J, Farrell EE (1992) Neurodevelopmental outcome in infants with bronchopulmonary dysplasia. Clin Perinatol 19: 673–694.
4. Fahy JV (2009) Eosinophilic and neutrophilic inflammation in asthma: insights from clinical studies. Proc Am Thorac Soc 6: 256–259.
5. Choo-Wing R, Nedrelow JH, Homer RJ, Elias JA, Bhandari V (2007) Developmental differences in the responses of IL-6 and IL-13 transgenic mice exposed to hyperoxia. Am J Physiol Lung Cell Mol Physiol 293: L142–150.
6. Warner BB, Stuart LA, Papes RA, Wispe JR (1998) Functional and pathological effects of prolonged hyperoxia in neonatal mice. Am J Physiol 275: L110–117.
7. Chang YS, Oh W, Choi SJ, Sung DK, Kim SY, et al. (2009) Human umbilical cord blood-derived mesenchymal stem cells attenuate hyperoxia-induced lung injury in neonatal rats. Cell Transplant 18: 869–886.
8. Chang YS, Choi SJ, Sung DK, Kim SY, Oh W, et al. (2011) Intratracheal transplantation of human umbilical cord blood derived mesenchymal stem cells dose-dependently attenuates hyperoxia-induced lung injury in neonatal rats. Cell Transplant.
9. Jang YK, Jung DH, Jung MH, Kim DH, Yoo KH, et al. (2006) Mesenchymal stem cells feeder layer from human umbilical cord blood for ex vivo expanded growth and proliferation of hematopoietic progenitor cells. Ann Hematol 85: 212–225.
10. Yang SE, Ha CW, Jung M, Jin HJ, Lee M, et al. (2004) Mesenchymal stem/progenitor cells developed in cultures from UC blood. Cytotherapy 6: 476–486.
11. Boiani M, Scholer HR (2005) Regulatory networks in embryo-derived pluripotent stem cells. Nat Rev Mol Cell Biol 6: 872–884.
12. Gang EJ, Bosnakovski D, Figueiredo CA, Visser JW, Perlingeiro RC (2007) SSEA-4 identifies mesenchymal stem cells from bone marrow. Blood 109: 1743–1751.
13. Lee JK, Lee MK, Jin HJ, Kim DS, Yang YS, et al. (2007) Efficient intracytoplasmic labeling of human umbilical cord blood mesenchymal stromal cells with ferumoxides. Cell Transplant 16: 849–857.
14. Thurlbeck WM (1982) Postnatal human lung growth. Thorax 37: 564–571.
15. Cho SJ, George CL, Snyder JM, Acarregui MJ (2005) Retinoic acid and erythropoietin maintain alveolar development in mice treated with an angiogenesis inhibitor. Am J Respir Cell Mol Biol 33: 622–628.
16. Snyder JM, Jenkins-Moore M, Jackson SK, Goss KL, Dai HH, et al. (2005) Alveolarization in retinoic acid receptor-beta-deficient mice. Pediatr Res 57: 384–391.
17. Babior BM (1999) NADPH oxidase: an update. Blood 93: 1464–1476.
18. Cross AR, Segal AW (2004) The NADPH oxidase of professional phagocytes–prototype of the NOX electron transport chain systems. Biochim Biophys Acta 1657: 1–22.
19. Gray KD, Simovic MO, Chapman WC, Blackwell TS, Christman JW, et al. (2003) Endotoxin potentiates lung injury in cerulein-induced pancreatitis. Am J Surg 186: 526–530.
20. Robinson JM, Badwey JA (1995) The NADPH oxidase complex of phagocytic leukocytes: a biochemical and cytochemical view. Histochem Cell Biol 103: 163–180.
21. Hohler B, Holzapfel B, Kummer W (2000) NADPH oxidase subunits and superoxide production in porcine pulmonary artery endothelial cells. Histochem Cell Biol 114: 29–37.
22. Li JM, Shah AM (2002) Intracellular localization and preassembly of the NADPH oxidase complex in cultured endothelial cells. J Biol Chem 277: 19952–19960.
23. Coalson JJ (2003) Pathology of new bronchopulmonary dysplasia. Semin Neonatol 8: 73–81.
24. Jobe AJ (1999) The new BPD: an arrest of lung development. Pediatr Res 46: 641–643.
25. Kunig AM, Balasubramaniam V, Markham NE, Morgan D, Montgomery G, et al. (2005) Recombinant human VEGF treatment enhances alveolarization after hyperoxic lung injury in neonatal rats. Am J Physiol Lung Cell Mol Physiol 289: L529–535.
26. McGrath-Morrow SA, Stahl J (2001) Apoptosis in neonatal murine lung exposed to hyperoxia. Am J Respir Cell Mol Biol 25: 150–155.
27. Reinecke H, Murry CE (2002) Taking the death toll after cardiomyocyte grafting: a reminder of the importance of quantitative biology. J Mol Cell Cardiol 34: 251–253.
28. Togel F, Hu Z, Weiss K, Isaac J, Lange C, et al. (2005) Administered mesenchymal stem cells protect against ischemic acute renal failure through differentiation-independent mechanisms. Am J Physiol Renal Physiol 289: F31–42.
29. Gnecchi M, He H, Noiseux N, Liang OD, Zhang L, et al. (2006) Evidence supporting paracrine hypothesis for Akt-modified mesenchymal stem cell-mediated cardiac protection and functional improvement. FASEB J 20: 661–669.
30. Aslam M, Baveja R, Liang OD, Fernandez-Gonzalez A, Lee C, et al. (2009) Bone marrow stromal cells attenuate lung injury in a murine model of neonatal chronic lung disease. Am J Respir Crit Care Med 180: 1122–1130.
31. Cho HJ, Lee N, Lee JY, Choi YJ, Ii M, et al. (2007) Role of host tissues for sustained humoral effects after endothelial progenitor cell transplantation into the ischemic heart. J Exp Med 204: 3257–3269.
32. Saugstad OD (2003) Bronchopulmonary dysplasia-oxidative stress and antioxidants. Semin Neonatol 8: 39–49.
33. Bastian NR, Hibbs JB, Jr. (1994) Assembly and regulation of NADPH oxidase and nitric oxide synthase. Curr Opin Immunol 6: 131–139.
34. Moodley Y, Atienza D, Manuelpillai U, Samuel CS, Tchongue J, et al. (2009) Human umbilical cord mesenchymal stem cells reduce fibrosis of bleomycin-induced lung injury. Am J Pathol 175: 303–313.
35. Deuse T, Peter C, Fedak PW, Doyle T, Reichenspurner H, et al. (2009) Hepatocyte growth factor or vascular endothelial growth factor gene transfer maximizes mesenchymal stem cell-based myocardial salvage after acute myocardial infarction. Circulation 120: S247–254.
36. Kokaia Z, Martino G, Schwartz M, Lindvall O (2012) Cross-talk between neural stem cells and immune cells: the key to better brain repair? Nat Neurosci 15: 1078–1087.

12

Optimization of Multimodal Imaging of Mesenchymal Stem Cells Using the Human Sodium Iodide Symporter for PET and Cerenkov Luminescence Imaging

Esther Wolfs[1], Bryan Holvoet[1], Rik Gijsbers[2], Cindy Casteels[1], Scott J. Roberts[3], Tom Struys[4], Michael Maris[2], Abdelilah Ibrahimi[2], Zeger Debyser[2], Koen Van Laere[1], Catherine M. Verfaillie[5], Christophe M. Deroose[1]*

1 Nuclear Medicine and Molecular Imaging, Department of Imaging and Pathology, KU Leuven, Leuven, Belgium, **2** Laboratory of Molecular Virology and Gene Therapy, Department of Pharmaceutical and Pharmacological Sciences, KU Leuven, Leuven, Belgium, and Leuven Viral Vector Core, KU Leuven, Leuven, Belgium, **3** Skeletal Biology and Engineering Research Center, Department of development and regeneration, KU Leuven, Leuven, Belgium, **4** Lab of Histology, Department of morphology, Biomedical Research Institute, Universiteit Hasselt, Hasselt, Belgium, **5** Stem Cell Institute Leuven, Department of development and regeneration, KU Leuven, Leuven, Belgium

Abstract

Purpose: The use of stably integrated reporter gene imaging provides a manner to monitor the *in vivo* fate of engrafted cells over time in a non-invasive manner. Here, we optimized multimodal imaging (small-animal PET, Cerenkov luminescence imaging (CLI) and bioluminescence imaging (BLI)) of mesenchymal stem cells (MSCs), by means of the human sodium iodide symporter (hNIS) and firefly luciferase (Fluc) as reporters.

Methods: First, two multicistronic lentiviral vectors (LV) were generated for multimodal imaging: BLI, ^{124}I PET/SPECT and CLI. Expression of the imaging reporter genes was validated *in vitro* using $^{99m}TcO_4^-$ radioligand uptake experiments and BLI. Uptake kinetics, specificity and tracer elution were determined as well as the effect of the transduction process on the cell's differentiation capacity. MSCs expressing the LV were injected intravenously or subcutaneously and imaged using small-animal PET, CLI and BLI.

Results: The expression of both imaging reporter genes was functional and specific. An elution of $^{99m}TcO_4^-$ from the cells was observed, with 31% retention after 3 h. After labeling cells with ^{124}I *in vitro*, a significantly higher CLI signal was noted in hNIS expressing murine MSCs. Furthermore, it was possible to visualize cells injected intravenously using BLI or subcutaneously in mice, using ^{124}I small-animal PET, CLI and BLI.

Conclusions: This study identifies hNIS as a suitable reporter gene for molecular imaging with PET and CLI, as confirmed with BLI through the expression of Fluc. It supports the potential for a wider application of hNIS reporter gene imaging and future clinical applications.

Editor: Pranela Rameshwar, Rutgers - New Jersey Medical School, United States of America

Funding: CMD is a post-doctoral fellow of the Clinical Research Fund of the University hospitals Leuven (www.fwo.be). The funders had no role in study design, data collection and analysis, decision to publish, or preparation of the manuscript.

Competing Interests: The authors have declared that no competing interests exist.

* E-mail: christophe.deroose@uzleuven.be

Introduction

The development of imaging reporter genes is of great importance to the field of molecular imaging, as their encoded proteins can permit longitudinal non-invasive imaging of biological processes at the cellular and subcellular level.

One of these imaging reporter genes encodes for the well-characterized human sodium iodide symporter (hNIS) [1]. hNIS is situated in the basolateral membrane of thyroid follicular cells, as well as other tissues such as the stomach mucosa, salivary glands and the lactating mammary gland. hNIS is an intrinsic plasma membrane glycoprotein with 13 transmembrane domains that mediates the first step in the formation of thyroid hormones by the transport of iodide into the thyroid follicular cells against a concentration gradient [1]. Furthermore, hNIS is able to transport radioactive forms of iodide, as well as other anions such as technetium pertechnetate ($^{99m}TcO_4^-$) [2]. Due to its relatively low endogenous expression in extrathyroidal tissues, the wide availability of FDA-approved radioactive probes and the experience with hNIS imaging, hNIS satisfies most of the criteria to be a suitable reporter gene for imaging purposes [3]. Furthermore, hNIS is a human protein and its use as an imaging reporter gene will therefore not evoke immunological responses [4].

Exogenous expression of hNIS can be applied for non-invasive and nuclear imaging of grafted cells. This can lead to a better

Table 1. Vector constructs used for gene transfer.

Vector type	Expression cassette	Cell type
LV	LV_hNIS	HEK293T
	LV_eGFP	
Bicistronic LV	LV_hNIS-T2A-Fluc	HEK293T
	LV_hNIS-T2A-eGFP	
	LV_Fluc-T2A-eGFP	
Multicistronic LV	LV_hEF1α-3flagFluc-T2A-hNIS-IRES-PuroR	MSCs
	LV_EF1α-3flagFluc-IRES-PuroR	

understanding of stem cell homing, and assist in further applying and developing stem cell-based therapies.

Mesenchymal stem cells (MSCs) are non-hematopoietic multipotent stem cells with an innate ability to differentiate towards multiple mesenchymal lineages [5,6]. These cells appear to be good candidates for clinical use; they can be expanded easily *in vitro* and lack immunogenicity [7]. MSCs also possess immune modulating properties, through the inhibition of immune cell function and proliferation, and their use as immunomodulators is being explored clinically [8]. Besides their role in tissue regeneration, MSCs have significant trophic effects on endogenous (stem) cells [9]. Furthermore, they have also been proven to migrate towards multiple tumors *in vivo*, which renders them good candidates as delivery vehicles for antitumor therapy [10–12].

Because of these characteristics, the interest in MSCs for clinical applications has increased over the past years. As such, MSCs have been used in several clinical phase I and phase II studies concerning acute myocardial infarction [13], but the most promising results have been obtained in human graft-versus-host disease and allograft rejection studies [14–18].

To further improve stem cell-based treatments, cellular homing as well as survival after engraftment needs to be studied in greater detail. Imaging reporter genes can play an important role in the non-invasive longitudinal follow-up of grafted cells. In the present work, we optimized lentiviral vector (LV) transduction of murine MSCs, inducing the expression of the imaging reporter genes firefly luciferase (Fluc) for bioluminescence imaging and hNIS for emission tomography (PET/SPECT) and Cerenkov luminescence imaging (CLI).

CLI uses Cerenkov radiation for molecular imaging. Cerenkov radiation is an electromagnetic radiation emitted when a charged particle travels at a speed beyond the speed of light in a dielectric medium. The upper limit on speed (c), is the speed of light in a vacuum, but in a particular medium the speed of light is a fraction of c (in water ~0.75), and positrons from PET isotopes can be emitted at speeds higher than these values. The charged particle travels through the medium and thereby temporarily displaces the electrons in the medium. While returning to their ground state, the electrons will emit visible light photons that can be detected using a BLI system [19,20]. It is thus possible to image certain PET

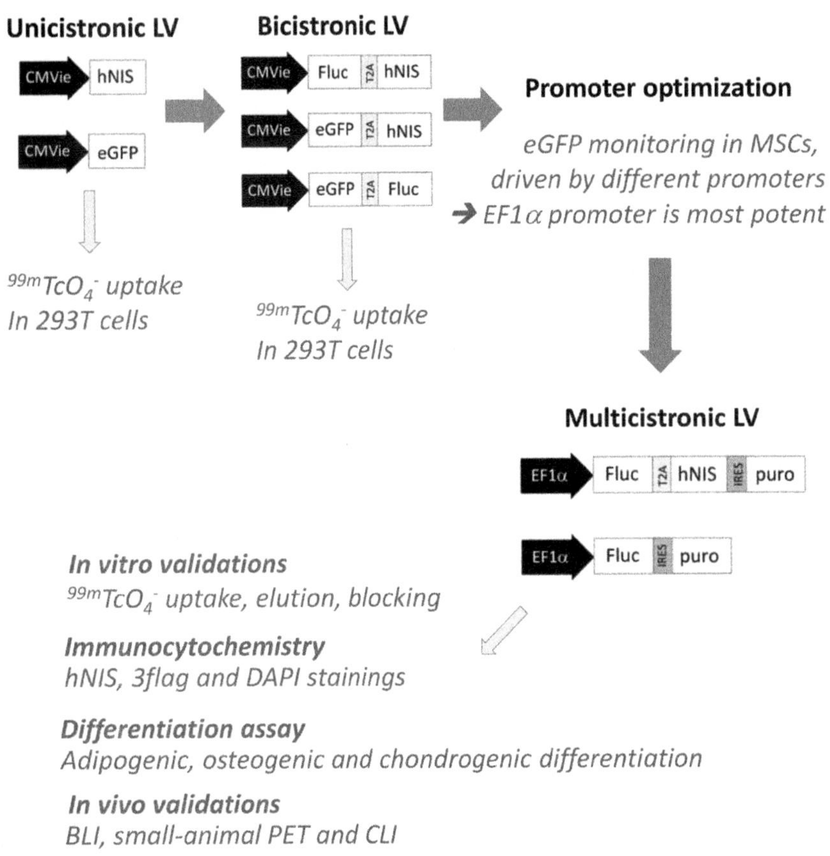

Figure 1. Diagram illustrating the steps involved in this study. Overview of the lentiviral vector constructs used to induce reporter gene expression and the experiments performed with these respective vectors.

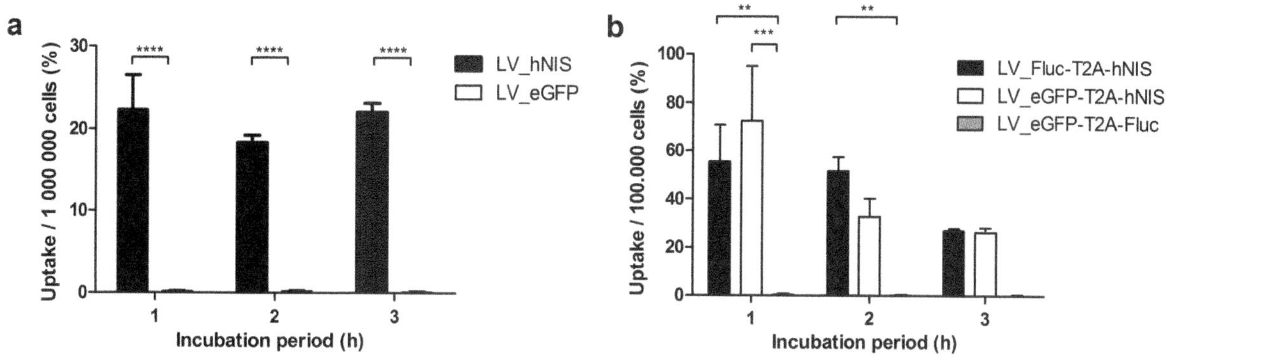

Figure 2. Results of uptake experiments with $^{99m}TcO_4^-$ in HEK293T cells after the induction of hNIS expression. Uptake ratios resulting from the transduction with a unicistronic LV (b) and a bicistronic LV (c). Significantly higher uptakes were observed in cells expressing hNIS. **: $p<0.01$; ***: $p<0.001$; ****:$p<0.0001$.

radionuclides via an optical system with high sensitivity and short scanning times.

Here, the longitudinal expression of these imaging reporter genes was evaluated both *in vitro* and *in vivo*.

Materials and Methods

Cell culture

HEK293T cells were used for testing the functionality of the different vectors described below. Cells were cultured under normal oxygen conditions in a 5% CO_2 humidified incubator at 37°C. Growth medium consisted of Dulbecco's modified eagle medium (DMEM, Gibco, Invitrogen, Carlsbad, CA, USA), 10% fetal bovine serum (FBS; Lonza BioWhittaker, Basel, Switzerland) and 1% Penicillin/Streptomycin (Gibco).

Murine MSCs from C57Bl/6 mice were obtained from the lab of Prof. dr. D Prockop, Tulane University, USA [21]. Cells were cultured in a 5% CO_2 humidified incubator under normoxic conditions at 37°C in growth medium containing Iscove's Modified Dulbecco's Medium (IMDM, Gibco), 10% FBS, 10% horse serum (Biochrom, Berlin, Germany), 1% L-glutamine (Gibco) and 1% penicillin/streptomycin.

Lentiviral vector optimization and transduction

For the optimization of the final multicistronic LV construct, different vector types and constructs were generated and tested, (for an overview see Table 1 and Fig. 1). LV constructs carrying a cytomegalovirus immediate early (CMVie) promoter driving hNIS or enhanced green fluorescent protein (eGFP) as a control (LV_hNIS and LV_eGFP, respectively) were used at first in HEK293T cells as a proof-of-principle to demonstrate hNIS functionality. Subsequently, bicistronic LV were generated, combining hNIS with alternative imaging reporter modules where a CMVie promoter drives hNIS and Firefly luciferase (Fluc, for bioluminescence imaging) or hNIS and eGFP (fluorescence) coupled by a *Thosea asigna* virus 2A (T2A) sequence, LV_Fluc-T2A-hNIS and LV_eGFP-T2A-hNIS, respectively. As a control, a LV was included encoding both eGFP and Fluc coupled by a T2A sequence, LV_eGFP-T2A-Fluc [22]. In parallel, we determined the ideal promoter for efficient MSC transcription, employing a LV that drives eGFP from different promoters, such as human elongation factor 1α (EF1α), human Cyclophilin A (CypA), or viral Spleen focus forming virus LTR (SFFV), and CMVie. These LV were referred to as LV_hEF1α-eGFP, LV_CypA-eGFP, LV_SFFV-eGFP, LV_CMVie-eGFP, respectively. MSCs were transduced with the respective vectors using a protocol as reported earlier [23]. eGFP fluorescence was monitored using fluorescence activated sorting (FACS) and on day 37, the 5% brightest population was isolated from each condition to overcome the lack of puromycin resistance in the expression cassettes. Long-term eGFP expression was monitored using FACS until 60 days. Results are given as total fluorescence (fraction of total cells that are fluorescent×mean fluorescence intensity).

Finally, a multicistronic LV carrying the hEF1α promoter to drive triple flag tagged firefly luciferase (3flagFluc), a T2A sequence, the human sodium iodide symporter (hNIS), an internal ribosomal entry site (IRES), and a puromycin resistance gene (PuroR), LV_hEF1α-3flagFluc-T2A-hNIS-IRES-PuroR was tested. As a control, the same vector lacking hNIS was used, LV_EF1α-3flagFluc-IRES-PuroR. For the sake of clarity, MSCs transduced with the LV_EF1α-3flagFluc-T2A-hNIS-IRES-PuroR will be referred to as Fluc-hNIS expressing MSCs. MSCs transduced with the EF1α-3flagFluc-IRES-Puro LV, will be referred to as Fluc expressing MSCs.

To create stable cell lines, cells were seeded into medium containing serial dilutions of LV and cells were incubated for 48 hours. For the selection of transduced cells, 2 μg/mL puromycin (Merck Millipore, Darmstadt, Germany) was added to the growth medium, and cells were maintained under this condition.

In vitro radiotracer uptake experiments

Cells were plated in triplicate in 24-well plates at a density of 10^5 cells per well in normal growth medium, and kept under standard incubation conditions. After 24 hours, cells were washed with PBS and incubated with 250 μL of pertechnetate ($^{99m}TcO_4^-$) tracer solution (0.74 MBq/mL in DMEM; Gibco) for different periods ($n = 3$). All *in vitro* data are shown as decay-corrected values.

After incubation, cells were washed 3 times with phosphate buffered saline (PBS; Gibco), and tracer concentration in the cell fraction was measured using a gamma counter (Perkin Elmer, Waltham, MA, USA). Uptake values were corrected for the cell number in the according samples, as measured using a nucleocounter system (Chemometec, Allerød, Denmark).

The elution of $^{99m}TcO_4^-$ initially taken up by the cells was measured by incubating the cells for one hour with $^{99m}TcO_4^-$ (0.74 MBq/mL), washing the cells and incubating them on tracer-free DMEM for varying periods. The activity in the cells, the supernatant and the elution medium was measured and elution rates were calculated ($n = 3$).

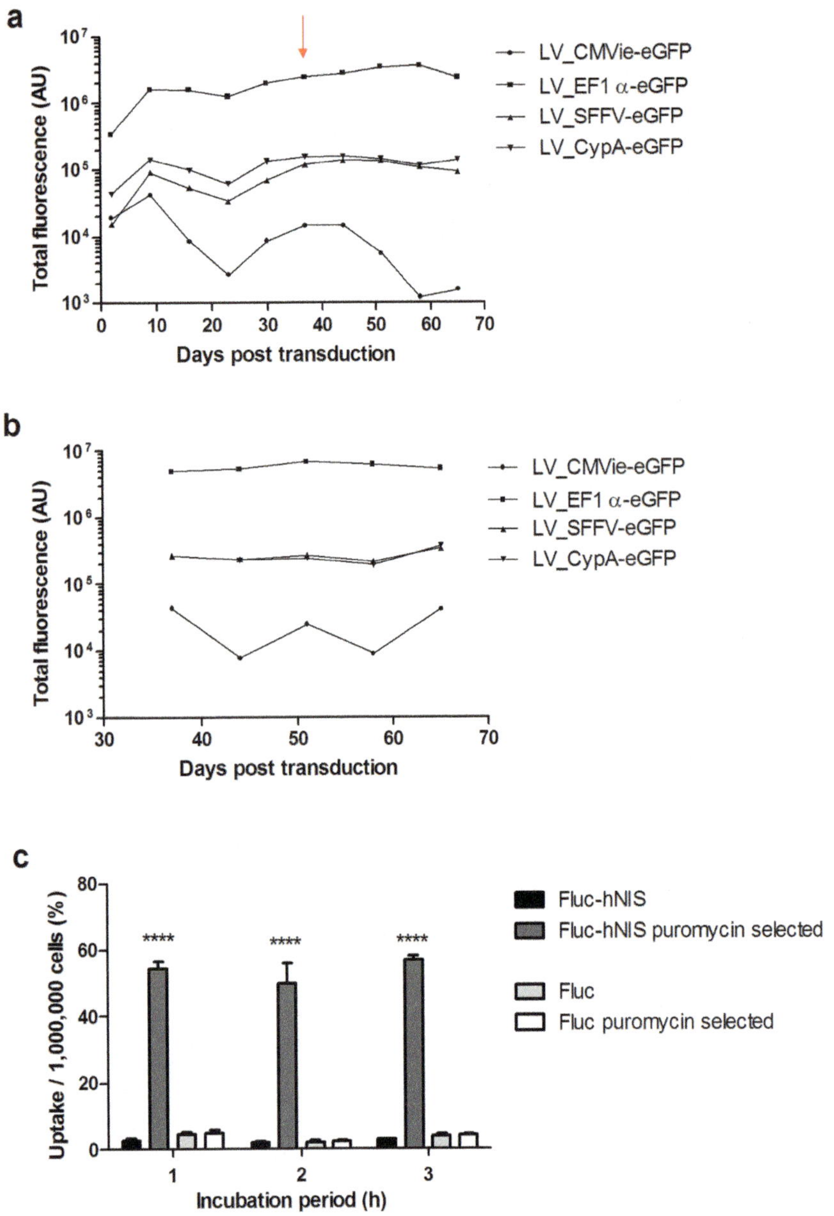

Figure 3. Optimization of LV constructs and selection. Different promoters were tested with FACS (AU: arbitrary units) for eGFP expression in MSCs transduced with different LV driven by different human or viral promoters (a). On day 38, FACS selection of the highest expressors was done (red arrow on panel (a)), and this population was also monitored (b). $^{99m}TcO_4^-$ uptake experiment in Fluc-hNIS or Fluc expressing MSCs with or without puromycin selection (c). ****:$p<0.0001$.

A blocking experiment was performed with sodium perchlorate ($NaClO_4$) using three different concentrations solved in DMEM: 10, 25 and 50 μM. The cells were incubated with the blocking solutions containing the tracer (0.74 MBq/mL $^{99m}TcO_4^-$ in DMEM+ $NaClO_4$) and activity in supernatant and cells was measured using a gamma counter (n = 3).

Immunocytochemistry

To further confirm hNIS and 3flag expression in MSCs transduced with the LV_EF1a-3flagFluc-T2A-hNIS-IRES-PuroR and the LV_EF1a-3flagFluc-IRES-PuroR, immunofluorescent stainings were performed. Cells were fixed using unifix for 20 min at 4°C, permeabilized with 0.05% Triton for 30 min at room temperature (only in case of 3flag staining) and blocked with 10% normal donkey serum (Millipore) for 20 min at room temperature. Cells were then incubated for two hours at room temperature with the primary antibody diluted in PBS (hNIS 1/20 and flag 1/1000). Fluorescently labeled secondary antibody (1:500, Alexa Fluor donkey anti-rabbit 488 or donkey anti-mouse 555, Invitrogen) was incubated for 30 min at room temperature. Nuclei were counterstained using DAPI and sections were mounted using anti-fade mounting medium (Dako). Images were acquired using a Nikon Eclipse 80 i Fluorescence microscope equipped with a Nikon DS-2 MB Wc digital camera (Nikon Tokyo, Japan). Stainings in which primary antibodies were omitted, were used as a negative control.

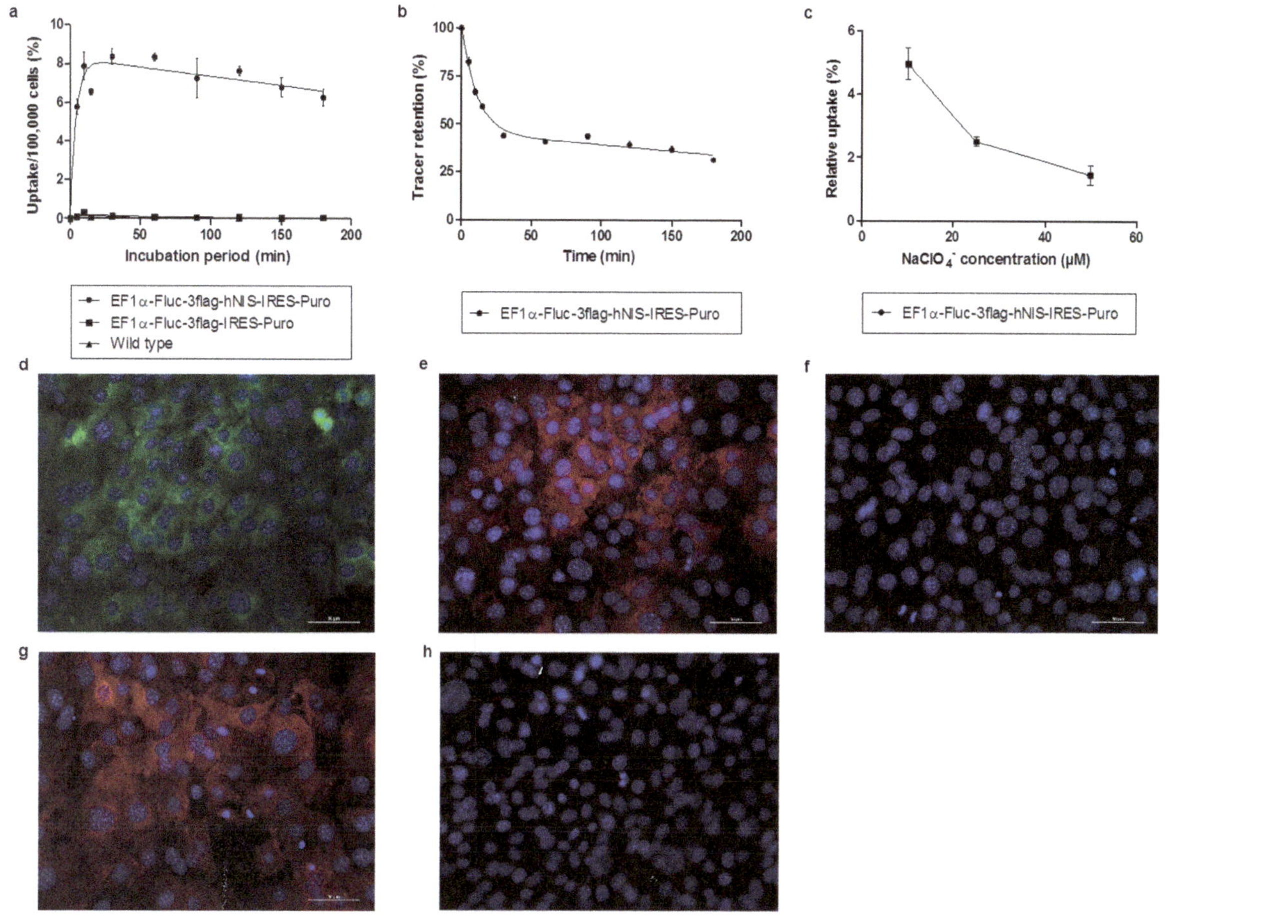

Figure 4. Validation of the multicistronic LV contruct. $^{99m}TcO_4^-$ uptake kinetics in MSCs transduced with the two different multicistronic LV constructs or wild type MSCs (a). Tracer elution from Fluc-hNIS expressing MSCs (b). The effect of different concentrations of $NaClO_4$ on the uptake of $^{99m}TcO_4^-$ in Fluc-hNIS expressing MSCs (c). Immunocytochemistry against hNIS on Fluc-hNIS expressing MSCs and DAPI (d), against 3flag in 3flagFluc-hNIS expressing MSCs and DAPI (e) and negative control (f). Immunocytochemistry against 3flag in 3flagFluc expressing MSCs and DAPI (g) and negative control (h).

MSC differentiations

For adipogenic differentiations, cells were plated at a density of 10400 cells/cm^2 in differentiation medium, containing αMEM (Gibco), 10% FBS, 100 units of penicillin and 1000 units of streptomycin, 1 μM dexamethasone, 10 μg/mL human insulin, 100 μM indomethacin and 25 μM methyl-isobutylxanthine (all from Sigma-Aldrich, St Louis, MO, USA). Medium was changed twice a week for 14 days (n = 3).

After 14 days, cells were rinsed with PBS, fixed using unifix (Klinipath, Duiven, The Netherlands) for 20 minutes and stained with fresh Oil red O solution (Sigma-Aldrich) for 10 minutes. Images were taken using an inverted Zeiss axiovert microscope (Hertfordshire, UK). For quantification of the lipid droplet staining, the dye was extracted from the cells using 100% ethanol and aliquots of 200 μL were transferred to a 96-well plate in triplicate. Absorbance was measured at 450 nm using a Victor 1420 plate reader (Perkin Elmer).

To assess the effect on osteogenic differentiation capacity, MSCs were seeded at 8400 cells/cm^2 in normal growth medium and allowed to become confluent for 48 h before adding osteogenic differentiation medium. Differentiation medium contained DMEM with sodium pyruvate, 10% FBS, 1% penicillin/streptomycin, 50 μg/mL L-Ascorbic acid 2-phosphate sequimagnesium salt hydrate (AA-P; Sigma-Aldrich), 100 nM dexamethasone and 10 mM glycerol-2-phosphate disodium salt hydrate (bGP). Medium was changed twice a week for 3 weeks (n = 3).

At day 21, cells were rinsed with PBS, fixed with ice-cold 70% ethanol for 1 h and stained with Alizarin red S solution (Sigma) for 30 minutes. Images were taken using an inverted Zeiss axiovert microscope. Afterwards, the dye was extracted from the cells using a 10% cetylpiridinium chloride solution for 1 h. Aliquots of 200 μL were transferred to a 96-well plate in triplicate, and absorbance was measured at 560 nm.

For chondrogenic differentiation, micromasses were generated by seeding 20 μL droplets of cell suspension, each containing 200,000 cells, into separate wells of a 24-well plate. After attachment for 3 hours expansion medium was added. The next day, chondrogenic differentiation medium was added containing DMEM/F12, 2.5% FBS, 100 nM dexamethasone, 1× ITS+, 50 μg/mL AA-P, 10 ng/mL Transforming Growth Factor-β (TGF-β), 10 μM ROCK inhibitor and 40 μg/mL L-proline. Every 2 days, medium was changed for 21 days (n = 3).

At day 21, cells were fixed using unifix and stained overnight with Alcian Blue. The dye was extracted using 6M guanidine HCL. The optical density of the extracted dye was measured at 595 nm in triplicates.

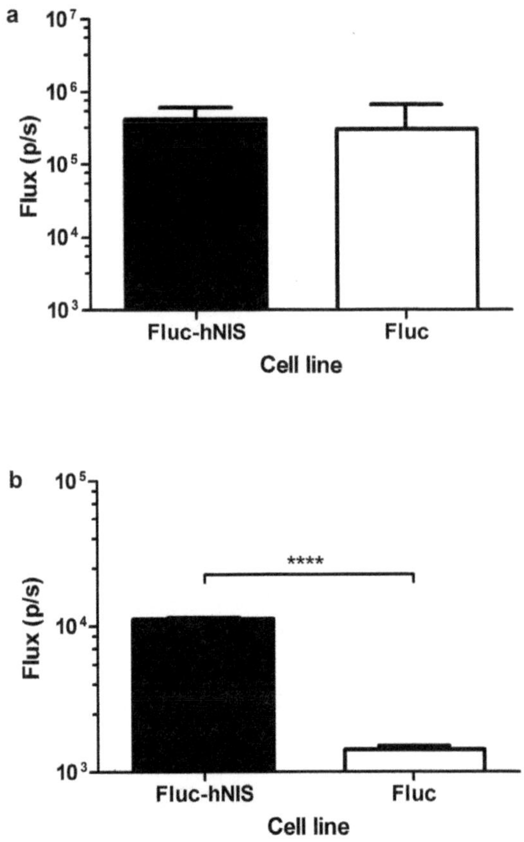

Figure 5. In vitro optical validation of the imaging reporter genes. In vitro BLI (a) and CLI (b) of Fluc-hNIS expressing MSCs and Fluc expressing MSCs. ****: $p<0.0001$.

Animal preparations

The protocols used in this study were approved by the Ethical Committee of the KU Leuven (Permit Number: P145-2010).

For the intravenous injections, animals were anesthetized with 2% isoflurane (Isoflurane ISP®, Rothacher, Basel, Switzerland) in 100% oxygen, at a flow rate of 2 L/minute. Different cell numbers of MSCs expressing Fluc-hNIS ranging from 10,000 to 1,000,000 were injected in the tail vein of healthy C57BL/6 mice, and 30 minutes after cell injections BLI was performed ($n=3$).

For xenografts, both Fluc expressing MSCs (as a control), and Fluc-hNIS expressing MSCs were injected subcutaneously in anesthetized nude athymic mice (nu/nu) ($n=3$). Fluc expressing MSCs were injected on the left side of the body, whereas Fluc-hNIS expressing cells were injected on the right side of the body. Each mouse was injected with two different cell numbers per cell line: 10,000 and 1,000,000 mixed with matrigel in a 1:1 ratio (BD biosciences, New Jersey, USA).

Small-animal PET

The imaging of xenografts was performed on day 2 after xenograft generation with 11 MBq of ^{124}I (Perkin Elmer) per mouse injected intravenously, with dynamic acquisitions of 90 minutes ($n=3$) using a Focus 220 small-animal PET system (Siemens Medical Solutions USA, Knoxville,TN). Images were reconstructed with a maximum a posteriori (MAP) image reconstruction algorithm, and analyzed in PMOD 3.0 (PMOD technologies, Zürich, Switzerland). Images were converted to standardized uptake value (SUV) according to the standard formula: SUV = Activity concentration in organ/[injected activity/animal weight].

A manually delineated volume of interest (VOI) was positioned on the dynamic images over the Fluc-hNIS-expressing xenografts, the control Fluc-expressing xenografts, and muscle tissue and the brain as background tissues to generate time activity curves. Ratios were calculated, comparing the signal in the Fluc-hNIS expressing xenograft with background signals from control Fluc-expressing xenografts, muscle tissue and brain.

Cerenkov luminescence imaging

For the *in vitro* CLI, cells were plated in triplicate in 24-well plates at a density of 1×10^5 cells per well in normal growth medium, and kept under standard incubation conditions. After 24 hours, cells were washed with PBS and incubated with 250 μL of tracer solution (0.74 MBq/mL ^{124}I in DMEM; Gibco) for one hour. Cells were washed 3 times with PBS, and placed in the BLI chamber for the acquisition of 1 minute scans.

The *in vivo* CLI scans on xenografts were performed daily after performing the ^{124}I small-animal PET scans. CLI protocols were executed as follows: animals were anesthetized using isoflurane in 100% oxygen, at a flow rate of 2 L/minute and positioned in the BLI chamber without prior injection of D-luciferin. Images were acquired using an IVIS 100 system (Perkin Elmer) and CLI acquisition was done by acquiring one minute frames. The data are reported as total photon flux (p/s) from a circular region of interest (ROI). For all animals, first a daily CLI scan was performed to measure the ^{124}I distribution, and thereafter D-luciferin was injected to measure the BLI signal intensity.

Bioluminescence imaging

For the *in vitro* BLI, cells were plated in triplicate in 24-well plates at a density of 1×10^5 cells per well in normal growth medium, and kept under standard incubation conditions. After 24 hours, cells were washed with PBS and incubated with 250 μL of D-luciferin (0.3 μg/mL; Promega, Benelux, Leiden, The Netherlands). Cells were placed in the BLI chamber immediately for the acquisition of 1 minute scans.

Animals were anesthetized with 2% isoflurane in 100% oxygen, at a flow rate of 2 L/minute, after which D-luciferin, dissolved in PBS (15 mg/mL), was injected intravenously (126 mg/kg body weight). Images were acquired using an IVIS 100 system (Perkin Elmer). Consecutive 1 minute frames were acquired until the maximum signal intensity was reached. Each frame depicts the bioluminescence signal intensity as a pseudocolor image superimposed on the gray-scale photographic image. The data are reported as total photon flux (p/s) from a circular region of interest (ROI). BLI signal intensity was monitored over 8 days after xenograft generation.

For the quantification of BLI data in the mouse xenograft model, values from according ROIs measured with CLI were subtracted from the raw BLI ROI values to obtain specific BLI signal intensities.

Statistical Analysis

Data are presented as mean ± standard error of the mean (SEM). For the uptake experiments using the unicistronic, bicistronic and multicistronic vector after puromycin selection, two-way analysis of variance (ANOVA) statistical tests were performed with Bonferroni post-hoc tests. P-values<0.05 were considered statistically significant. BLI and CLI data were tested using an unpaired two-sided t-test after log transformation. Differentiation analysis was performed using a one-way ANOVA with Tukey post-hoc tests. The *in vivo* BLI data after intravenous

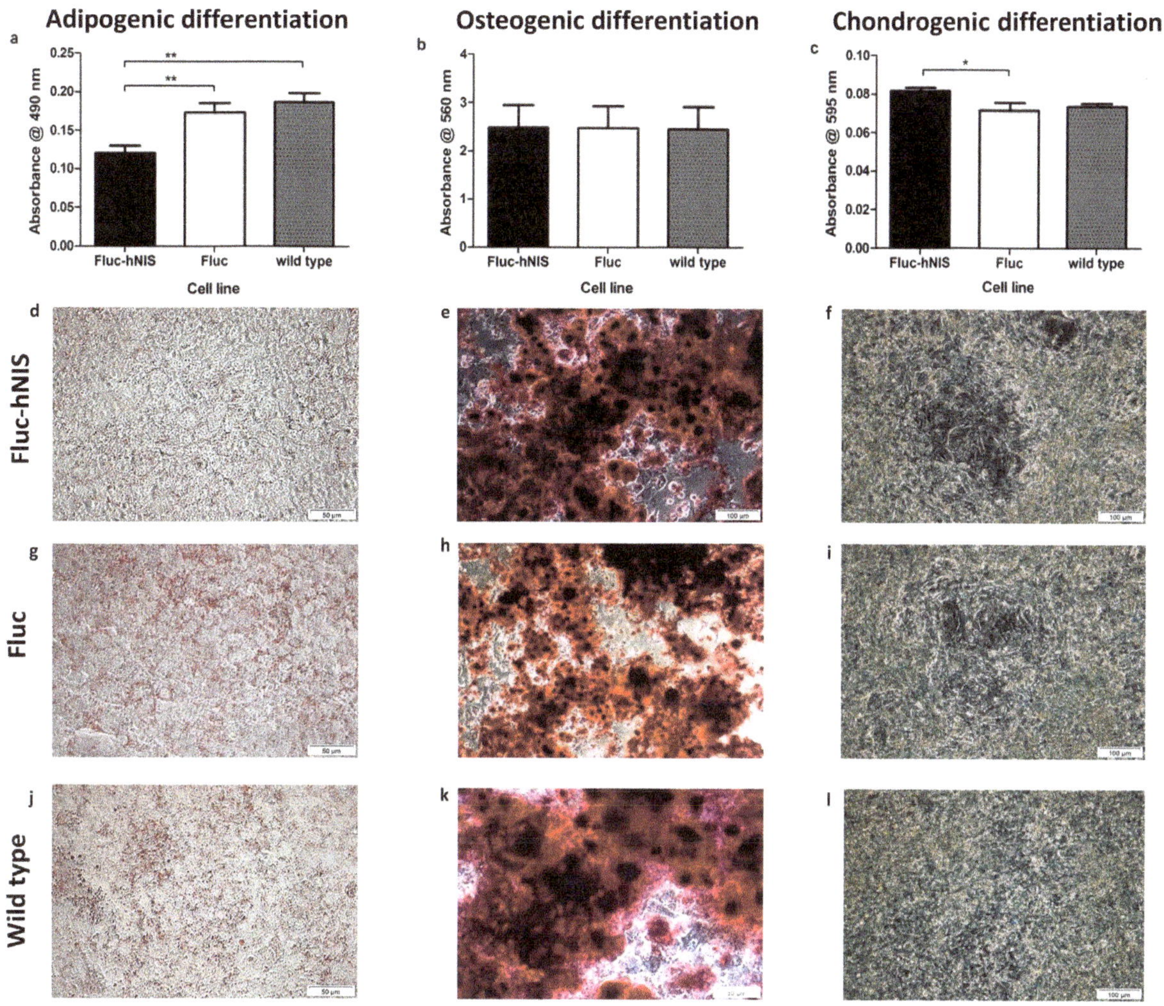

Figure 6. Adipogenic, osteogenic and chondrogenic differentiation capacity of Fluc-hNIS expressing MSCs, Fluc expressing MSCs and wild types. Extraction of the Oil red O dye following adipogenic differentiation measured at 490 nm (a). Extraction of the Alizarin red S dye from differentiated osteoblast matrix, measured at 560 nm (b). Extraction of Alcian Blue dye from differentiated chondrocyte matrix, measured at 595 nm (c). *: $p<0.05$, **: $p<0.01$. Images of the Oil red O stainings (d,g,j), the Alizarin red S stainings (e,h,k) and the Alcian Blue stainings (f,i,l) for all conditions: Fluc-hNIS, Fluc expressing cells and wild types.

injections were tested for gaussion distribution using the Kolmogorov-Smirnov test and the Pearson's R coefficient was calculated as a measure of linear correlation. The *in vivo* xenograft BLI data were evaluated using an unpaired t-test.

Data were processed using GraphPad Prism version 5.00 for Windows (GraphPad Software, San Diego, California, USA).

Results

hNIS is an efficient PET reporter in cell culture

Different LV vector types (Table 1, Fig. 1) were tested in HEK293T cells for the optimization of the imaging reporter gene expression. Stably hNIS expressing cells or control cells were incubated with $^{99m}TcO_4^-$. All transfections and transductions (unicistronic LV and bicistronic LV, Fig. 2a,b) resulted in a 100-fold to 300-fold higher uptake of tracer in cells expressing the hNIS gene, compared to the controls (unicistronic LV: $p<0.001$; bicistronic LV: $p<0.0001$).

EF1α as most potent promoter in MSC

The optimal promoter allowing the highest expression in MSCs was assessed by comparing transduction of MSCs with LVs encoding eGFP driven by human promoters (EF1α and CypA), and viral promoters (CMVie, SFFV) (Fig. 3a,b). eGFP fluorescence was monitored with FACS, and there was a significantly higher expression level in cells transduced with the EF1α promoter, both without (Fig. 3a) and with sorting (Fig. 3b) of the 5 percent highest expressing cells ($p<0.0001$).

Puromycin selection increases reporter gene expression

Cells transduced with the multicistronic LV constructs using the EF1α promoter were either kept on normal growth medium or on growth medium with puromycin. To determine the effect of puromycin selection of the expression of the imaging reporter genes, the uptake of $^{99m}TcO_4^-$ was assessed in both selected and non-selected cells (Fig. 3c). A significantly higher uptake of the radioligand was observed in Fluc-hNIS expressing MSCs upon selection with puromycin ($p<0.0001$), compared to nonselected and Fluc expressing MSCs.

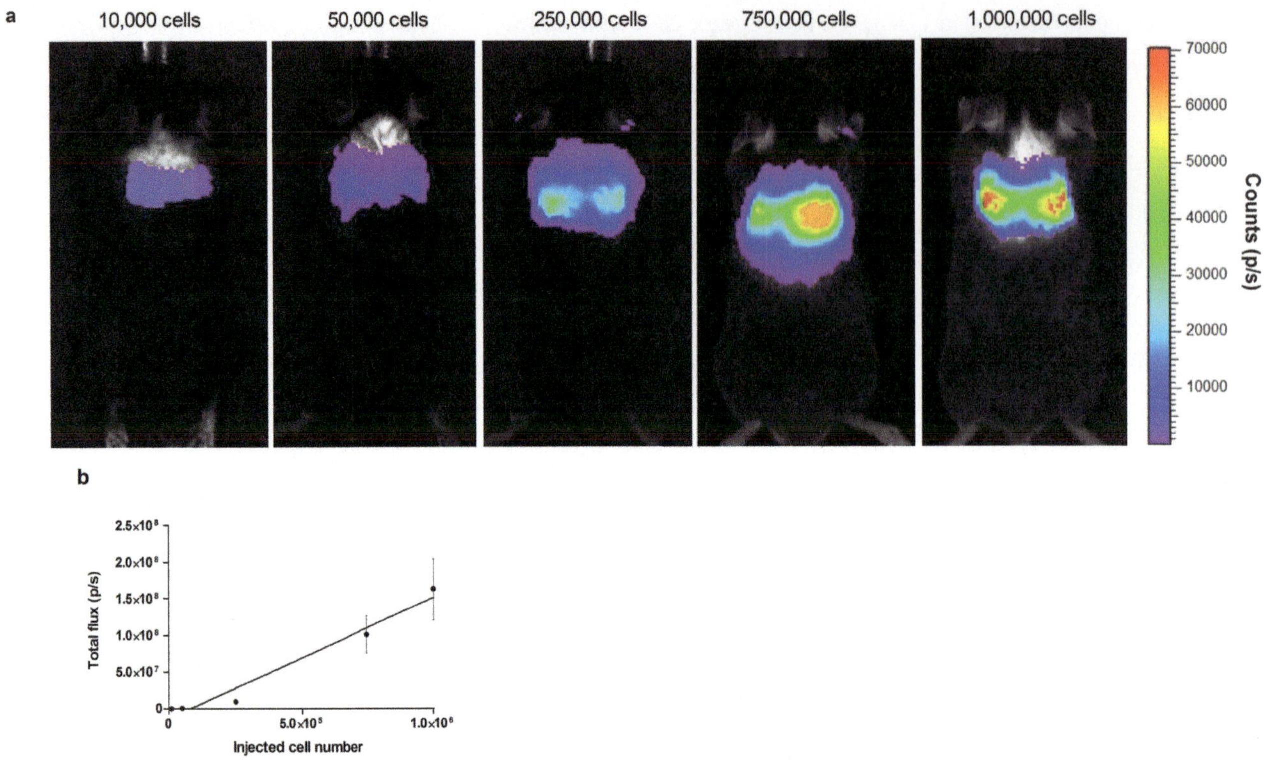

Figure 7. BLI data of intravenously injected Fluc-hNIS expressing MSCs. An accumulation of cells could be visualized in the lungs. Different cell numbers were injected, and total flux was measured. Both parameters were strongly correlated with an R^2 of 0.97.

Validation of hNIS expression in MSCs

MSCs were transduced with the multicistronic LV, and the uptake kinetics of $^{99m}TcO_4^-$ were assessed. Untransduced MSCs and Fluc expressing MSCs were included as a negative control. Decay-corrected uptake data (Fig. 4a) showed a strong increase in uptake within the first 30 minutes of incubation in Fluc-hNIS expressing MSCs, which was followed by a state of equilibrium between the concentration inside and outside of the cells. A significantly higher uptake (75 and 120 fold increase) was observed in Fluc-hNIS expressing MSCs compared to either Fluc expressing MSCs or wild type MSCs ($p<0.001$).

The elution of the tracer from the cells after initial uptake was assessed in Fluc-hNIS expressing MSCs (Fig. 4b). After labeling, cells were incubated in tracer-free medium and the relative elution was calculated. A two-phased elution was observed, with a substantial tracer elution occurring within 30 minutes, followed by a slow elution of the tracer from the cells ($p<0.001$). After 3 hours, 31%±1.3% of the initially bound tracer was still present within the cells.

To determine the specificity of the uptake, the hNIS inhibitor $NaClO_4$ was used (Fig. 4c). Cells were incubated with the tracer solved in the blocking solution and the uptake of $^{99m}TcO_4^-$ was measured. A significant decrease in the uptake of the tracer could be observed ($p<0.01$) after the administration of the blocker. Furthermore, the uptake of $^{99m}TcO_4^-$ decreased with increasing concentrations of $NaClO_4$ ($p<0.01$). A decrease of 95%, 97.5% and 98.6% was observed with concentrations of 10, 20 and 50 μM of $NaClO_4$, respectively. Data are represented as relative values compared to incubation without perchlorate.

The expression of the imaging reporter genes was confirmed by fluorescence immunocytochemistry. Fluc-hNIS expressing MSCs and Fluc expressing MSCs were stained with primary antibodies against hNIS and 3flag, and the expression of both reporter genes was observed (Fig. 4d,e,g). Negative controls did not show an aspecific staining or background signal (Fig. 4f,h).

In vitro BLI and CLI

In vitro BLI was performed to determine the expression of Fluc in the cells after transduction with both the Fluc-hNIS LV and the Fluc LV (Fig. 5a). Fluxes (in p/s) of log 5.63±0.42 and 5.49±0.9 were measured for the Fluc-hNIS and Fluc expressing cells, respectively. Hence, there was no significant difference in Fluc expression between both cell lines.

Both cell lines were also incubated with I^{124} and *in vitro* CLI was performed (Fig. 5b). A significant difference was observed between the fluxes of both cell lines ($p<0.0001$), with noted log values of 4.05±0.02 for the Fluc-hNIS expressing cells, and 3.15±0.06 for the Fluc expressing cells.

MSC differentiations

Fluc-hNIS expressing MSCs, Fluc expressing MSCs and wild type MSCs were differentiated towards the adipogenic, chondrogenic and osteogenic lineage (Fig. 6).

Quantification of the extracted oil red O dye resulted in following absorbance values (measured at 490 nm): 0.12, 0.17 and 0.19 for the Fluc-hNIS expressing MSCs, Fluc expressing MSCs or wild type MSCs, respectively, resulting in a respective difference of 35.3% and 7.4% compared to wild type MSCs (Fig. 6a,d,g,j). A significant difference between cells transduced with the Fluc-hNIS expressing MSCs and the two other cell lines was observed ($p<0.01$).

Following osteogenic differentiation, quantification of matrix mineralization resulted in following values after measuring the

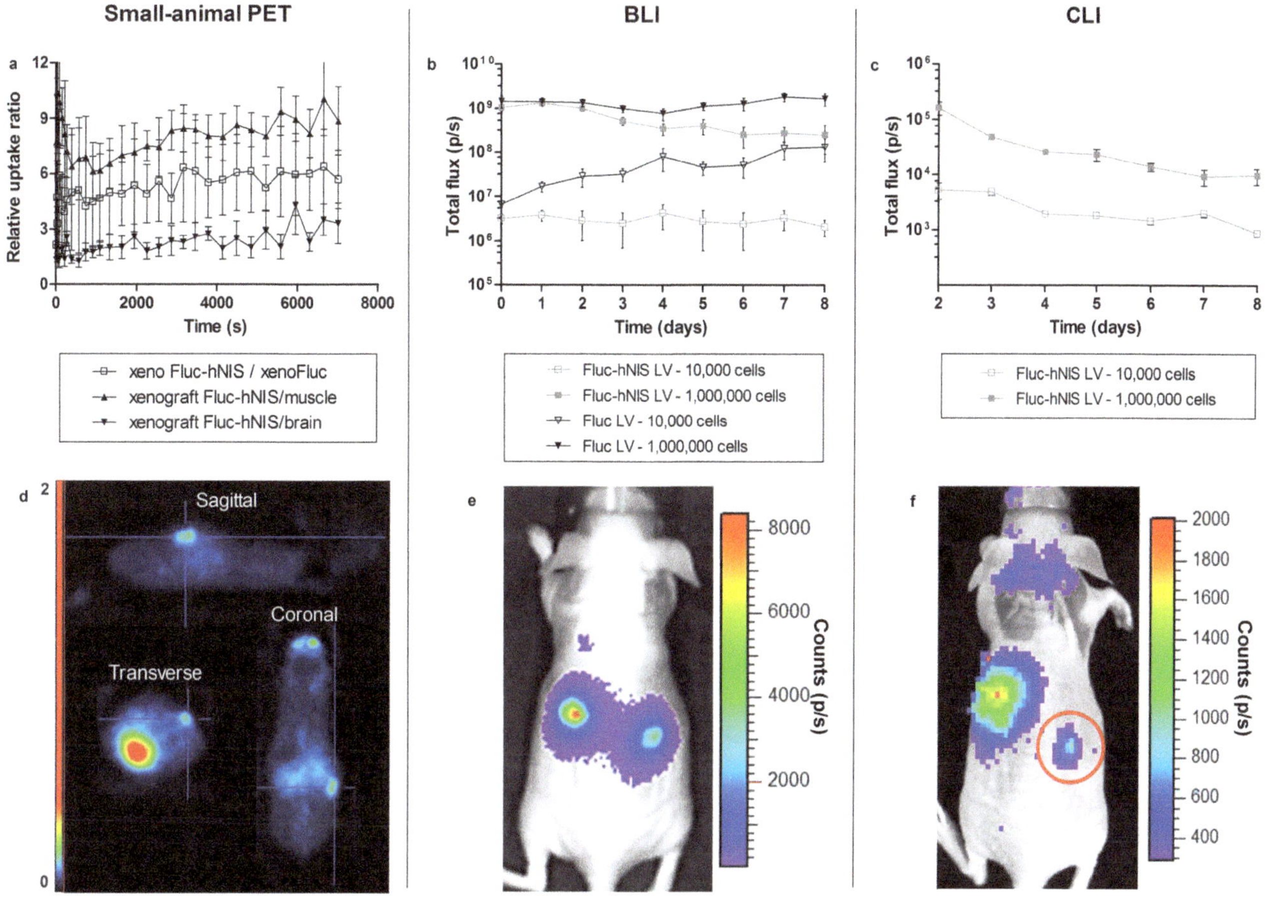

Figure 8. Visualization and follow-up of MSC xenografts. Fluc expressing MSCs were injected on the left flank of the body, and Fluc-hNIS expressing MSCs were engrafted on the right side of the body. Xenografts of 10,000 injected cells were injected near the front, and 1,000,000 cells were injected at the back for both conditions. Time activity curves of the ratios comparing the Fluc-hNIS expressing xenograft to background up tissues, showing higher expressions in the Fluc-hNIS expressing xenograft compared to background signals (a,d). BLI was also performed to follow the xenografts over time, with robust expression within all xenografts (b,e). CLI was performed to follow up the mice after ^{124}I injection. The Fluc-hNIS expressing MSCs could be visualized and monitored over time (c,f).

absorption of the samples at 560 nm: 2.49, 2.48 and 2.45 for the Fluc-hNIS expressing MSCs, Fluc expressing MSCs or wild type MSCs, respectively (Fig. 6b,e,h,k). Hence, no statistical difference was obtained ($p>0.05$).

Alcian blue staining following chondrogenic differentiation was quantified by dye extraction and measuring the absorption at 595 nm. The following values were obtained: 0.082, 0.072 and 0.074 for Fluc-hNIS expressing MSCs, Fluc expressing MSCs or wild type MSCs, respectively (Fig. 6c,f,i,l). A statistically significant increase in alcian blue dye incorporation within the chondrogenic micromasses was observed in the Fluc-hNIS expressing MSCs ($p<0.05$).

Imaging of intravenously injected MSCs using BLI

Different numbers of cells, ranging from 10,000 to 1,000,000, were injected in the tail vein of healthy C57BL/6 mice, and 30 minutes after cell injection, either BLI or small-animal PET was performed as a proof-of-principle for cell visualization. Using BLI, a robust signal could be obtained in the lungs. The signal intensity in the lungs increased when increasing cell amounts were injected intravenously. A significant correlation between injected amount of cells and total photon flux was obtained ($p<0.01$) with an R^2 value of 0.97 (Fig. 7 a,b).

Imaging of MSC xenografts using BLI, ^{124}I small-animal PET and CLI

MSCs expressing Fluc-hNIS or Fluc were engrafted subcutaneously in nude mice ($n=3$), and imaged non-invasively using BLI, Cherenkov imaging and ^{124}I small-animal PET (Fig. 8). On day 2 after cell injection, a ^{124}I small-animal PET scan was performed to visualize the engrafted cells expressing Fluc-hNIS. A clear focus of increased tracer concentration was observed at the site where the 1,000,000 MSCs expressing Fluc-hNIS were injected. The xenograft resulting from the 10,000 cells could not be detected on the small-animal PET images (Fig. 8d).

The xenograft to background ratio was 8 when comparing the Fluc-hNIS expressing xenograft with muscle tissue. Five times more radiotracer uptake was seen compared to the control xenograft of 1,000,000 cells expressing Fluc. The brain showed 3 times less accumulation of the tracer compared to the Fluc-hNIS expressing xenograft (Fig. 8a).

The BLI signal in xenografts derived from 1,000,000 cells was significantly higher than in xenografts derived from 10,000 cells in both MSC conditions ($p<0.001$). However, robust and specific BLI signals could be obtained in all conditions over time, with no decrease in signal being perceived in the monitored time span of 8 days (Fig. 8b,e). The BLI signals coming from the Fluc-hNIS expressing MSCs and the Fluc expressing MSCs were within the same range.

Two days after xenograft injection, ^{124}I small-animal PET was performed. In addition, ^{124}I accumulation was monitored over time using CLI for 8 days. (Fig. 8c,f). Accumulation of ^{124}I was detected in the thyroid, in the stomach and in the Fluc-hNIS expressing xenograft situated on the right side of the body (red circle). No signal was observed emanating from the bladder.

The Cherenkov signal detected in the xenograft expressing Fluc-hNIS with 1,000,000 cells persisted over the 8 days of measurements, decreasing according to the radioactive decay time of ^{124}I (T1/2: 4.2 days) as was shown after an exponential fit with an R^2 value of 0.63, yielding a half-life of 4.4 days.

Discussion

The emergence of stem cell therapies in regenerative medicine urges the development of methodology for non-invasive tracking of engrafted cells. The ability to image MSCs would be beneficial as they hold clinical promise both due to their differentiation capacity as well as their possibility to exert immune modulation [5–7].

Stem cells expressing PET and SPECT imaging reporter genes can be visualized longitudinally *in vivo* using nuclear medicine imaging techniques. In this context, hNIS has been used already in several studies both in the context of diagnosis and therapy (for review [24]).

Tracking stem cells using imaging reporter genes requires the insertion of a functionally active transgene into the stem cells to obtain a functionally active reporter protein. LV are good candidates because the genes of interest are introduced into the genomic DNA of the host cells, and both dividing and non-dividing cells can be targeted [25].

For optimal expression of our reporter genes, we first optimized the transduction process of the MSCs as well as the promoter to drive the final construct. Here, the EF1α promoter resulted in the highest expression of eGFP, and was therefore chosen to drive our reporter construct. The optimization of a promoter for transductions has been shown to be of great importance. Other groups have also determined the optimal promoter for several purposes, and in different cell types the EF1α promoter proved to be a potent promoter with long-term expression on a high level outperforming other commonly used promoters [26–31].

The introduction of genes into the host cells' genetic material can result in adverse effects resulting in the disruption of biological processes. We evaluated the effect of the lentiviral transduction process on MSCs. Although some statistically significant differences were observed, the biological relevance is likely nihil. Therefore, we can conclude that lentiviral transduction with the presently used imaging reporter gene constructs does not hamper the differentiation capacity of the MSCs. Terrovitis et al have also shown that ectopic expression of hNIS using lentiviral vectors does not interfere with cell viability and function in cardiac-derived stem cells [32]. This was also shown by Hu et al, who showed that adenoviral vector transduction of the hNIS and rNIS gene had no adverse effects on MSCs [33].

We demonstrated that hNIS and Fluc expression is functional both *in vitro* and *in vivo*. *In vitro*, a robust radiotracer uptake of $^{99m}TcO_4^-$ was obtained with a steady state already at 30 minutes of incubation. The uptake was 75 or 120 times higher compared to control cells. Furthermore, $NaClO_4$, an inhibitor of hNIS, was able to block the uptake of $^{99m}TcO_4^-$ in a dose-dependent manner.

However, a substantial elution from the cells was observed after labeling, with 31%±1.29% of the tracer being retained within the cells after 3 hours. This might be due to the initial strong concentration gradient that is generated while putting the cells containing the tracer on tracer-free medium, leading to a rapid washout of tracer molecules from the cells back into the medium, until the gradient between the cells and the new supernatant reaches a state of equilibrium. Furthermore, no organification of the $^{99m}TcO_4^-$ occurs after entry into the cells, as is the case in thyroidal tissue, and therefore no trapping of the molecules takes place. Nevertheless, 31% of the initially bound tracer molecules remained trapped within the cells after 3 hours.

As a proof-of-principle, Fluc-hNIS expressing MSCs were injected into the tail vein of healthy mice, and BLI signals in the lungs were measured. Due to the first pass mechanism, cells injected intravenously will first encounter the pulmonary capillary bed and will transiently reside in the lung (pre)capillary bed. The BLI signal correlated with the amount of cells injected, allowing non-invasive detection of the cells.

Finally, long-term noninvasive multimodality imaging was performed. For this purpose, 10,000 or 1,000,000 MSCs expressing Fluc or Fluc-hNIS were injected subcutaneously on the left flank and the right flank, respectively. All xenografts could be visualized using BLI, and monitored over time with a robust signal in all conditions. On day 2 after engraftment of the MSCs, ^{124}I small-animal PET was performed, and the xenograft of 1,000,000 Fluc-hNIS expressing MSCs was clearly visible. The injection site of 10,000 cells was not visible because of the partial volume effect and the relatively higher background activity resulting from hNIS expressing organs and circulating tracer.

The accumulation of ^{124}I was also monitored using CLI. These scans confirmed the information that was obtained with small-animal PET, because the cerenkov signal detected in the xenograft expressing Fluc-hNIS after injection of 1,000,000 cells persisted over the 8 days of measurement, decreasing according to the radioactive decay time of ^{124}I (T1/2: 4.2 days) as shown by the exponential fit with an R^2 value of 0.63, yielding a half-life of 4.4 days. From these data we can conclude that the ^{124}I is trapped within the cells, and can be visualized using CLI. Besides the xenograft, also the stomach and the thyroid accumulated ^{124}I, due to the endogenous hNIS expression in these tissues [1].

From these data we can conclude that imaging reporter genes are a suitable tool for the non-invasive visualization of stem cells after *in vivo* administration. Combining multiple imaging reporter genes in the same construct enables us to perform multimodality imaging to confirm the obtained data from one single imaging modality. hNIS was used as an imaging reporter gene together with Fluc. Using Fluc, longitudinal BLI can be performed as a traditional optical imaging modality. BLI features a very high sensitivity, almost no background signal, short scanning time, and thus high throughput imaging.

hNIS is a human reporter gene reducing the chances of immune responses against the reporter gene product. Furthermore, the expression of hNIS is restricted to a limited number of tissues, implying a relatively low background signal. hNIS can be used for both therapeutic and diagnostic purposes. Imaging can be performed with tracers for gamma cameras ($^{99m}TcO_4^-$), which is widely available in every nuclear medicine department worldwide. The physical advantages of a PET camera can also be used as we have shown by using ^{124}I. Furthermore, the gene can be used for therapeutic applications through the use of ^{131}I.

When using multimodality imaging, it is also necessary to consider both the advantages and the disadvantages of each imaging modality. Optical imaging devices can be used for several purposes such as fluorescence, bioluminescence and more recently Cerenkov radiation. Here, CLI and BLI were used for cell tracking, and higher signal intensity could clearly be seen in BLI compared to CLI. Furthermore, BLI results in a very high signal to noise ratio due to the low background signals. In contrast, CLI images are noisy but CLI is a more translational optical imaging system than BLI, because it is using tracer molecules that are often already used in a clinical setting. This is particularly the case in preclinical imaging, as not every institute is equipped with costly dedicated small-animal nuclear imaging instruments.

For the imaging of radioactive tracer molecules, both CLI and small-animal PET can be used. The optical imaging equipment used for CLI allows a quick acquisition of data, and devices are more available compared to small-animal PET devices due to the lower costs of the hardware. However, PET imaging can also be performed in a relatively quick way, and results in the generation of tomographic images that are quantitatively more reliable. Furthermore, despite the efforts that have been taken for 3D optical imaging, there is a major lack of anatomical and tomographic data. Also, PET does not generate anatomical information, but this can be overcome relatively easily by co-registration of PET images to anatomical computed tomography (CT) images or magnetic resonance imaging (MRI) data. Indeed, the development of hybrid systems combining PET and either CT or MRI is an effective answer to this matter. Above all, PET is the only modality that can be used in patients, and therefore has the highest translational capacity of all the modalities.

For cell tracking studies, the combination of CLI and nuclear imaging might form a translational bridge between optical imaging and nuclear imaging modalities. Hence, the advantages of both techniques can be combined: quantitative information in a tomographic manner, and the high throughput and sensitivity of the optical imaging. We here show that hNIS is a suitable reporter gene for molecular imaging with PET and CLI. This reporter gene can be used in humans and is therefore a good candidate in translational studies.

Hence, future studies will include further reproducibility tests and the application of this system in disease models.

Acknowledgments

CMD is a post-doctoral fellow of the Clinical Research Fund of the University hospitals Leuven. The authors thank Manja Muijtjens, Pieter Berckmans, Jeanine Santermans and Ann Van Santvoort for their help in data acquisition and processing. The radiopharmacy team from the division for Nuclear Medicine of the University Hospitals/KU Leuven is acknowledged for $^{99m}TcO_4^-$ tracer preparations. Additionally, the Leuven Viral Vector Core is acknowledged for the production of the lentiviral vector constructs.

Author Contributions

Conceived and designed the experiments: EW CMV CMD. Performed the experiments: EW BH TS. Analyzed the data: EW BH CC CMD. Contributed reagents/materials/analysis tools: RG SJR TS MM AI ZD. Wrote the paper: EW CMD CMV KVL. Revised the manuscript: EW BH RG CC SJR TS MM AI ZD KVL CMV CMD.

References

1. Dohan O, De la Vieja A, Paroder V, Riedel C, Artani M, et al. (2003) The sodium/iodide Symporter (NIS): characterization, regulation, and medical significance. Endocr Rev 24: 48–77.
2. Van Sande J, Massart C, Beauwens R, Schoutens A, Costagliola S, et al. (2003) Anion selectivity by the sodium iodide symporter. Endocrinology 144: 247–252.
3. Massoud TF, Gambhir SS (2003) Molecular imaging in living subjects: seeing fundamental biological processes in a new light. Genes Dev 17: 545–580.
4. Ahn BC (2012) Sodium iodide symporter for nuclear molecular imaging and gene therapy: from bedside to bench and back. Theranostics 2: 392–402.
5. Caplan AI (1991) Mesenchymal stem cells. J Orthop Res 9: 641–650.
6. Pittenger MF, Mackay AM, Beck SC, Jaiswal RK, Douglas R, et al. (1999) Multilineage potential of adult human mesenchymal stem cells. Science 284: 143–147.
7. Barry FP, Murphy JM (2004) Mesenchymal stem cells: clinical applications and biological characterization. Int J Biochem Cell Biol 36: 568–584.
8. Uccelli A, Pistoia V, Moretta L (2007) Mesenchymal stem cells: a new strategy for immunosuppression? Trends Immunol 28: 219–226.
9. Caplan AI (2007) Adult mesenchymal stem cells for tissue engineering versus regenerative medicine. J Cell Physiol 213: 341–347.
10. Dwyer RM, Khan S, Barry FP, O'Brien T, Kerin MJ (2010) Advances in mesenchymal stem cell-mediated gene therapy for cancer. Stem Cell Res Ther 1: 25.
11. Kumar S, Chanda D, Ponnazhagan S (2008) Therapeutic potential of genetically modified mesenchymal stem cells. Gene Ther 15: 711–715.
12. Nakamizo A, Marini F, Amano T, Khan A, Studeny M, et al. (2005) Human bone marrow-derived mesenchymal stem cells in the treatment of gliomas. Cancer Res 65: 3307–3318.
13. Hare JM, Traverse JH, Henry TD, Dib N, Strumpf RK, et al. (2009) A randomized, double-blind, placebo-controlled, dose-escalation study of intravenous adult human mesenchymal stem cells (prochymal) after acute myocardial infarction. J Am Coll Cardiol 54: 2277–2286.
14. Bernardo ME, Ball LM, Cometa AM, Roelofs H, Zecca M, et al. (2011) Co-infusion of ex vivo-expanded, parental MSCs prevents life-threatening acute GVHD, but does not reduce the risk of graft failure in pediatric patients undergoing allogeneic umbilical cord blood transplantation. Bone Marrow Transplant 46: 200–207.
15. Kebriaei P, Isola L, Bahceci E, Holland K, Rowley S, et al. (2009) Adult human mesenchymal stem cells added to corticosteroid therapy for the treatment of acute graft-versus-host disease. Biol Blood Marrow Transplant 15: 804–811.
16. Le Blanc K, Frassoni F, Ball L, Locatelli F, Roelofs H, et al. (2008) Mesenchymal stem cells for treatment of steroid-resistant, severe, acute graft-versus-host disease: a phase II study. Lancet 371: 1579–1586.
17. Reinders ME, de Fijter JW, Roelofs H, Bajema IM, de Vries DK, et al. (2013) Autologous bone marrow-derived mesenchymal stromal cells for the treatment of allograft rejection after renal transplantation: results of a phase I study. Stem Cells Transl Med 2: 107–111.
18. Ringden O, Uzunel M, Rasmusson I, Remberger M, Sundberg B, et al. (2006) Mesenchymal stem cells for treatment of therapy-resistant graft-versus-host disease. Transplantation 81: 1390–1397.
19. Thorek D, Robertson R, Bacchus WA, Hahn J, Rothberg J, et al. (2012) Cerenkov imaging - a new modality for molecular imaging. Am J Nucl Med Mol Imaging 2: 163–173.
20. Xu Y, Liu H, Cheng Z (2009) Harnessing the power of radionuclides for optical imaging: Cerenkov luminescence imaging. J Nucl Med 52: 2009–2018.
21. Peister A, Mellad JA, Larson BL, Hall BM, Gibson LF, et al. (2004) Adult stem cells from bone marrow (MSCs) isolated from different strains of inbred mice vary in surface epitopes, rates of proliferation, and differentiation potential. Blood 103: 1662–1668.
22. Ibrahimi A, Vande Velde G, Reumers V, Toelen J, Thiry I, et al. (2009) Highly efficient multicistronic lentiviral vectors with peptide 2A sequences. Hum Gene Ther 20: 845–860.
23. Geraerts M, Willems S, Baekelandt V, Debyser Z, Gijsbers R (2006) Comparison of lentiviral vector titration methods. BMC Biotechnol 6: 34.
24. Cho JY (2002) A transporter gene (sodium iodide symporter) for dual purposes in gene therapy: imaging and therapy. Curr Gene Ther 2: 393–402.
25. Sakuma T, Barry MA, Ikeda Y (2012) Lentiviral vectors: basic to translational. Biochem J 443: 603–618.
26. Dupuy FP, Mouly E, Mesel-Lemoine M, Morel C, Abriol J, et al. (2005) Lentiviral transduction of human hematopoietic cells by HIV-1- and SIV-based vectors containing a bicistronic cassette driven by various internal promoters. J Gene Med 7: 1158–1171.
27. Kim S, Kim GJ, Miyoshi H, Moon SH, Ahn SE, et al. (2007) Efficiency of the elongation factor-1alpha promoter in mammalian embryonic stem cells using lentiviral gene delivery systems. Stem Cells Dev 16: 537–545.
28. Petkov S, Hyttel P, Niemann H (2013) The choice of expression vector promoter is an important factor in the reprogramming of porcine fibroblasts into induced pluripotent cells. Cell Reprogram 15: 1–8.
29. Ramezani A, Hawley TS, Hawley RG (2000) Lentiviral vectors for enhanced gene expression in human hematopoietic cells. Mol Ther 2: 458–469.

30. Varma NR, Janic B, Ali MM, Iskander A, Arbab AS (2011) Lentiviral Based Gene Transduction and Promoter Studies in Human Hematopoietic Stem Cells (hHSCs). J Stem Cells Regen Med 7: 41–53.
31. Wang R, Liang J, Jiang H, Qin LJ, Yang HT (2008) Promoter-dependent EGFP expression during embryonic stem cell propagation and differentiation. Stem Cells Dev 17: 279–289.
32. Terrovitis J, Kwok KF, Lautamaki R, Engles JM, Barth AS, et al. (2008) Ectopic expression of the sodium-iodide symporter enables imaging of transplanted cardiac stem cells in vivo by single-photon emission computed tomography or positron emission tomography. J Am Coll Cardiol 52: 1652–1660.
33. Hu S, Cao W, Lan X, He Y, Lang J, et al. (2011) Comparison of rNIS and hNIS as reporter genes for noninvasive imaging of bone mesenchymal stem cells transplanted into infarcted rat myocardium. Mol Imaging 10: 227–237.

13

Extracts of Adipose Derived Stem Cells Slows Progression in the R6/2 Model of Huntington's Disease

Wooseok Im[a], Jaejun Ban[a], Jiyeon Lim, Mijung Lee, Soon-Tae Lee, Kon Chu, Manho Kim*

Department of Neurology, Biomedical Research Institute, Seoul National University Hospital, Seoul, Korea

Abstract

Stem cell therapy is a promising treatment for incurable disorders including Huntington's disease (HD). Adipose-derived stem cell (ASC) is an easily available source of stem cells. Since ASCs can be differentiated into nervous stem cells, it has clinically feasible potential for neurodegenerative disease. In addition, ASCs secrete various anti-apoptotic growth factors, which improve the symptoms of disease from transplanted ASCs. Thus, cell-free extracts of ASCs (ASCs-E) could be a potential candidate for treatment of HD. Here, we investigated effects of ASCs-E on R6/2 HD mouse model and neuronal cells. In R6/2 HD model, injection of ASCs-E improved the performance in Rotarod test. ASCs-E also ameliorated striatal atrophy and mutant huntingtin aggregation in the striatum. In Western blot increased expressions of p-Akt, p-CREB and PGC1α were noted by injection of ASCs-E, when comparing to the R6/2 HD model. Neuro2A neuroblastoma cells treated with ASCs-E showed increased expression of p-CREB and PGC1α. In conclusion, ASCs-E delayed disease progression in animal model of HD by restoring of CREB-PGC1α pathway and could be a potential resource for treatment of HD.

Editor: Hemachandra Reddy, Oregon Health & Science University, United States of America

Funding: This work was supported by grants from the Korea Health 21 R& D Project, Ministry of Health & Welfare (A092058), WCU-Neurocytomics program and National Research Foundation of Korea (NRF) (2011-0012728). The funders had no role in study design, data collection and analysis, decision to publish, or preparation of the manuscript.

Competing Interests: The authors have declared that no competing interests exist.

* E-mail: kimmanho@snu.ac.kr

[a] These authors contributed equally to this work.

Introduction

Huntington's disease (HD) is an autosomal dominant neurodegenerative disorder characterized by loss of striatal projection neurons. HD patients undergo involuntary motor movement, psychiatric manifestation and cognitive decline [1–4]. The huntingtin gene encodes a protein called huntingtin and HD patient has an expanded CAG repeats in exon 1 of the huntingtin gene. These CAG repeats express polyglutamine and this protein repeat is related with the aggregation of mutant huntingtin in neuron, which causes the cell death of striatum in HD [1,2,4,5]. There is ongoing research for a cure for HD, however only a few medications can relieve symptoms in HD [3,6].

Autologous transplantation of stem cells has been suggested as an ideal source for cell therapy, since problem of immune rejection is less likely. Among the tissue derived stem cells, adipose stem cell (ASC) is a feasible source for stem cell-based therapy, considering their abundance, multipotency, and ethical consideration [7–10]. ASCs transplantation in vivo models have been reported to be beneficial. However, whether transplanted ASCs replaced damaged cells is unclear [11,12]. Stem cells secrete various factors which can modulate a hostile microenvironment of diseases [13–16]. ASCs also express multiple growth factors and could be used to treat diseases via secretion factors, called paracrine mechanism [10,17–19].

One of the pathomechanisms of HD progression is transcriptional and mitochondrial dysregulation induced by mutant huntingtin (mHtt) [4,20]. In HD, peroxisome proliferator-activated receptor γ coactivator-1α (PGC-1α), which affects mitochondrial energy metabolism, is known to be imbalanced by mHtt and, thus accelerating cell death [21–23]. Expression of PGC-1α is known to be controlled by cAMP response element binding protein (CREB) [24] which is also altered in HD [22,25].

In this study, we attempted to test whether or not the extracts of ASCs (ASCs-E) can be an alternative to direct ASCs transplantation. To investigate effects of ASCs-E in HD transgenic mice, ASCs-E was intraperitoneally injected and changes of disease progression were monitored. We also examined whether ASCs-E could modulate Akt, CREB and PGC-1α expression in brain, which is imbalanced in HD. To confirm the effects of cell-free ASCs-E in vitro, neuro2A (N2A), mouse neuroblast cell, was treated with ASCs-E and examined p-Akt, p-CREB and PGC-1α expression.

Materials and Methods

Ethics Statement

Subcutaneous adipose samples were obtained from a normal human who provides the written informed consent to participate in this study and Institutional Review Board in Seoul National University Hospital approved this study. All animal studies were carried out with the approval of the Institutional Animal Care and Use Committee (IACUC) of Seoul National University Hospital, which was accredited by the Association for the Assessment and Accreditation of Laboratory Animal Care International.

Preparation of human ASCs-E

The adipose tissues were minced and digested in 0.075% collagenase type I solution (Invitrogen, USA) in phosphate-buffered saline (PBS) with gentle agitation for 1 hour at 37°C. Digested tissues were centrifuged at 200×g for 10 minutes and upper mature adipocyte fractions were removed from stromal fractions. The remaining stromal fractions (pellet) were treated with red blood cell lysis buffer (Sigma, USA) for 10 minutes at room temperature, filtered through a 100-μm nylon mesh, and centrifuged at 200×g for 10 minutes. The combined stromal fractions of the samples were re-suspended and cultured in endothelial growth medium -2 MV (EGM-2 MV; Lonza, USA). For the preparation of human ASCs - E, the cultured ASCs were harvested in cold PBS and centrifuged at 200×g for 5 minutes. The ASCs were suspended with an extraction buffer (1 mM DTT, 1 mM EDTA, protease inhibitor cocktail P8340, 0.1% DEPC in PBS), and lysed by several mild mechanical air pressure. The lysate was centrifuged at 14,000×g for 15 minutes, and the supernatant was passed through a syringe filter unit (0.45 μm). Dulbecco's modified Eagle's medium (DMEM) (Welgene, Korea) was used as extraction buffer for in vitro treatment.

Neuro2A culture and ASCs-E treatment

N2A cells were seeded onto 6-well culture plates (SPL, Korea) with DMEM containing 10% fetal bovine serum. Two days later, N2A cells were exposed to 10 μg/ml of ASCs-E for 1day. Following incubation with ASCs-E, cells were harvested and lysed in Radio Immuno Precipitation Assay (RIPA) buffer (Thermo, USA) with freshly added protease inhibitor and phosphatase inhibitor (Roche, USA). After cellular debris was cleared by centrifugation, protein concentration was determined by the BCA protein assay (Pierce, USA) according to the manufacturer's instructions.

Huntington's disease animal model and ASCs-E injection schedule

We used transgenic HD mice of the R6/2 line (B6CBA-Tg(HDexon1)62Gpb/3J, 111 CAGs) and their WT littermates (Jackson Laboratories, USA). The R6/2 transgenic mouse model expresses exon 1 of a human mHtt and is the most widely used animal model for studying HD [5]. These mice were obtained by crossing ovarian transplant hemizygote females with B6CBAF1/J males. The mice were housed in groups with ad libitum access to food and water and a 12 hours light/12 hours dark cycle. The genotype was assessed using a PCR assay.

In R6/2 mice, disease phenotype appears at 8 weeks of age. Considering protective effect of ASCs-E, injection was started at 6 weeks old. ASCs-E was injected intraperitoneally two times a week (40 mg/kg) and control mice were injected with vehicle (equal volume of extraction buffer). The ASCs-E was freshly prepared just before administration and the injection dosage was selected according to our previous published reports [26,27].

Rotarod performance and weight measurement

The Rotarod test was evaluated using a rotarod apparatus (Jungdo Instruments, Korea). Mice were placed on the rod with an accelerating rotating speed from 4 to 40 rpm over a period of 3 minutes with a 15 minutes rest between trials. Mice were trained on three consecutive days for three trials per day at 4 weeks of age. Three trials were performed and the mean latencies to fall were used to analyze data. Rotarod evaluation and weight measurement were performed every week from 5 to 12 weeks (Figure 1A).

Immunohistochemistry and striatal volume analysis

For immunohistochemistry, mice were anesthetized and perfused through the heart with 10 ml of cold saline and 4% paraformaldehyde in 0.1 M PBS at 12 weeks of age. Brains were removed from the skull, cryoprotected in 30% sucrose solution at 4°C, and sectioned 20 μm. Free-floating sections were washed and blocked with normal goat serum, then stained with the EM48 mHtt antibody (1:300; Millipore). On the following day, the sections were washed in PBS with three times and incubated with FITC conjugated anti-rabbit IgG (1:50; Jackson ImmunoResearch Laboratories) for 2 hours.

To examine the volume of striatum in R6/2 mice, serial-cut 200-μm coronal tissue sections from the rostral segment of the neostriatum to the level of the anterior commissure (bregma 1.54 to 0.10 mm) were used for Nissl staining. The areas of the striatum and the peristriatum were defined from each serial section, and gross volumes were measured with integrating each sectional area using Image-Pro Plus (Media Cybernetics, USA).

Protein extracts and western blot analysis

Brains of R6/2 mice were isolated, immediately frozen on liquid nitrogen, and stored at −70°C until protein extraction. Cultured N2A cells were washed and harvested in PBS using a cell scraper. Protein extracts were prepared using RIPA buffer (Thermo, USA) with freshly added protease inhibitor and phosphatase inhibitor (Roche, USA). 30 μg protein samples were separated on 10% SDS-PAGE and transferred onto PVDF membrane. The blots were probed with primary antibodies: PGC-1α (1:1000; Santa Cruz, CA), p-Akt (1:1000; Cell Signaling), CREB (1:1000; Cell signaling), EM48 (1:200; Millipore), followed by horseradish peroxidase conjugated secondary anti-mouse or rabbit antibody (1:5000; GE Healthcare), and blots were developed using ECL. Western blots were scanned and intensity values were obtained using ImageJ software.

Statistical analysis

All values shown in the figures are presented as mean ± standard error. Rotarod performance and weights were analyzed by ANOVA with Fisher's post hoc test at each week of age. Western blot and histological results were analyzed by Student's t-test. A 2-tailed probability value below 0.05 was considered statistically significant. Data were analyzed using SPSS version 17.0 (SPSS Inc., USA).

Results

ASCs-E slowed behavioral phenotypes of R6/2 mice model

To investigate effects of ASCs-E on behavioral deficits of R6/2 mice, ASCs-E was injected from 6 weeks old and phenotypes were examined. Mice showed gradual weight loss from 10 to 12 weeks of age. In contrast with vehicle treated R6/2 mice, ASCs-E injected R6/2 showed delayed progression of weight loss at 12 weeks old (22.1±0.2 vs. 24±0.9, $p<0.01$) (Figure 1B). In rotarod test, ASCs-E treated group showed lowered latency to fall compared with vehicle treated group 10, 11 ($p<0.05$) and 12 weeks ($p<0.01$) of age (Figure 1C).

ASCs-E reduced striatal atrophy and mutant Htt aggregation of R6/2 mice

R6/2 mouse shows striatal atrophy and mHtt aggregation in striatum and cortex during disease progression [28-30]. To examine histological changes of the brain, volume of striatum and

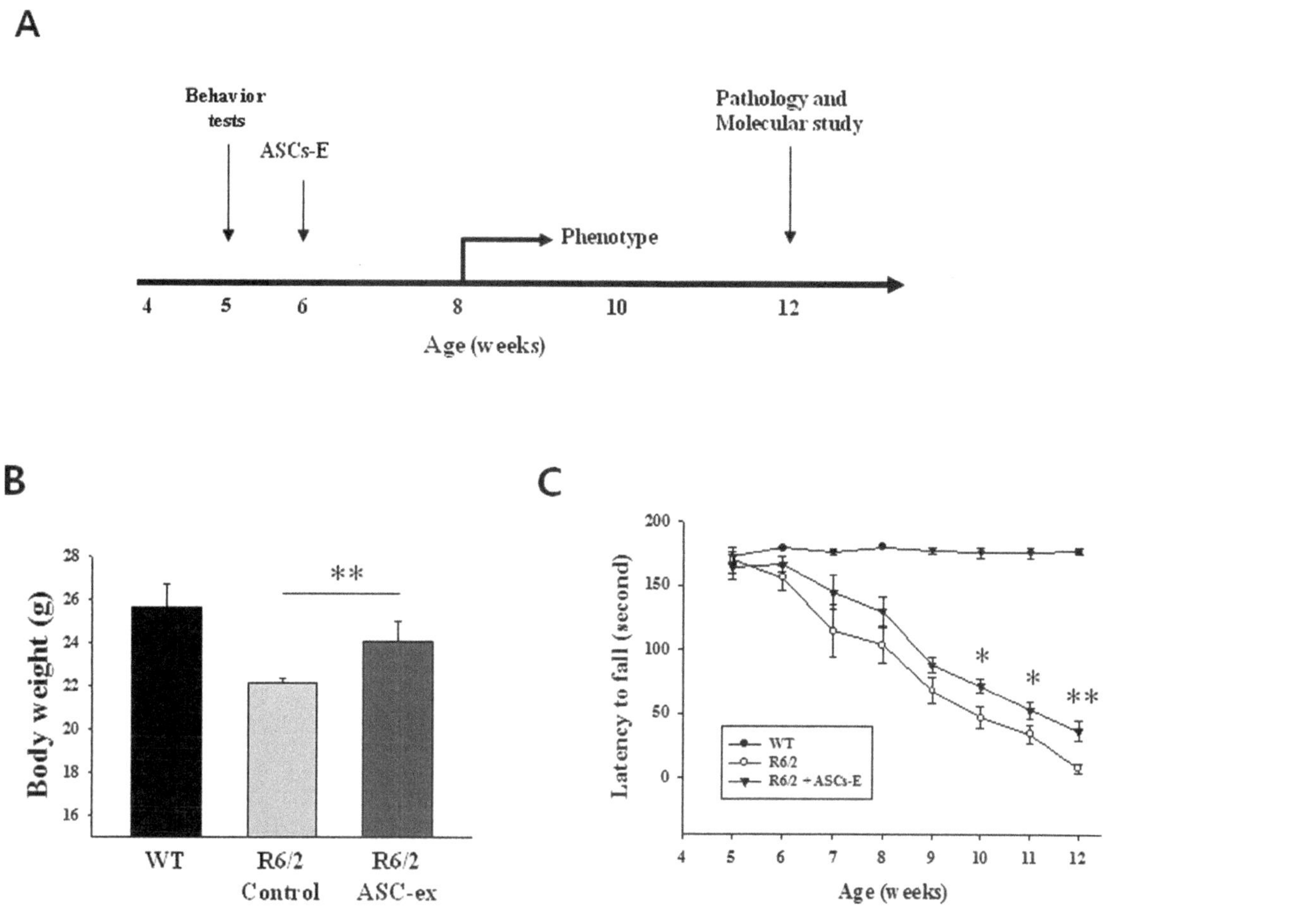

Figure 1. Experimental schedule and disease phenotypes of R6/2 mice. (A) Schedule for ASCs-E injection, behavior test, weight measure and brain sampling. (B) ASCs-E injection mitigated weight loss in R6/2 mouse at 12 weeks age. (C) Rotarod test showed better motor performance at 10, 11 and 12 weeks of age in ASCs-E treated R6/2 compared with control. * $p<0.05$, ** $p<0.01$.

mHtt aggregation were evaluated at 12 weeks of age. To evaluate effect on mHtt aggregation, brains were sectioned and sliced tissues were stained with an EM48 antibody which detects aggregation of mutant huntingtin. As shown in figure 2A, injection of ASCs-E reduced mHtt aggregation in the striatum but not in the cortex. We extracted proteins from R6/2 mouse brain and mHtt aggregation was measured by western blot. ASCs-E treated group showed decreased mHtt aggregation in the brain compared with vehicle treated group (Figure 2B, C). Volumes of striatum and peristriatum were measured respectively and the ratio of striatum to peristriatum was calculated. We found that ASCs-E injected R6/2 brain showed an increased ratio of striatum to peristriatum compared with vehicle injected R6/2 brain (0.22 ± 0.01 vs. 0.25 ± 0.01, $p<0.01$, $n=7$) (Figure 2D, E).

ASCs-E activates CREB and PGC-1α in R6/2 mice and neuronal cells

Dysfunction of CREB-PGC-1α pathway has been regarded as the key molecules for HD progression [22,25]. To examine effects of ASCs-E injection on this pathway, 12 weeks old R6/2 mice treated with control vehicle or ASCs-E for 6 weeks were sacrificed and western blot analysis was performed. Transplantation of ASCs-E promoted expression of p-Akt, p-CREB and PGC-1α ($p<0.05$ vs. R6/2 control), whereas they were decreased in R6/2 control mice (Figure 3A). To examine in vitro effects of ASCs-E, neuro2A cells were cultured and incubated with ASCs-E for 1 day. Then cells were harvested and levels of p-Akt, p-CREB and PGC1α were measured by western blot. In ASCs-E treated cells, levels of p-CREB and PGC1α were significantly increased (Figure 3B). Overall, neuroprotective p-CREB-PGC1α pathway was activated by treatment of ASCs-E.

Discussion

In this study, we investigated whether treatment of ASCs-E affects expression of HD related molecules in neuronal cell line and disease progression in R6/2 transgenic mouse model. In neuronal cells, we observed that ASCs-E upregulates expression of p-CREB and PGC-1α, which could modify HD progression. We found that ASCs-E reduced mHtt aggregate and striatal atrophy in the brain of R6/2 mice, and slowed progression of HD phenotype, such as weight loss or decline of rotarod performance. We observed that lowered activation of Akt-CREB-PGC-1α pathway was recovered by injection of ASCs-E. Our study suggests that systemic injection of ASCs-E could slow ongoing HD progression.

Recent studies reported that mitochondrial dysfunction caused by imbalance of CREB-PGC-1α pathway is critically involved in HD progression. Inhibition of CREB activity causes PGC-1α impairment, inducing cell death by glutamate toxicity in striatum in HD [21–23,25]. Considering that activation of PGC-1α

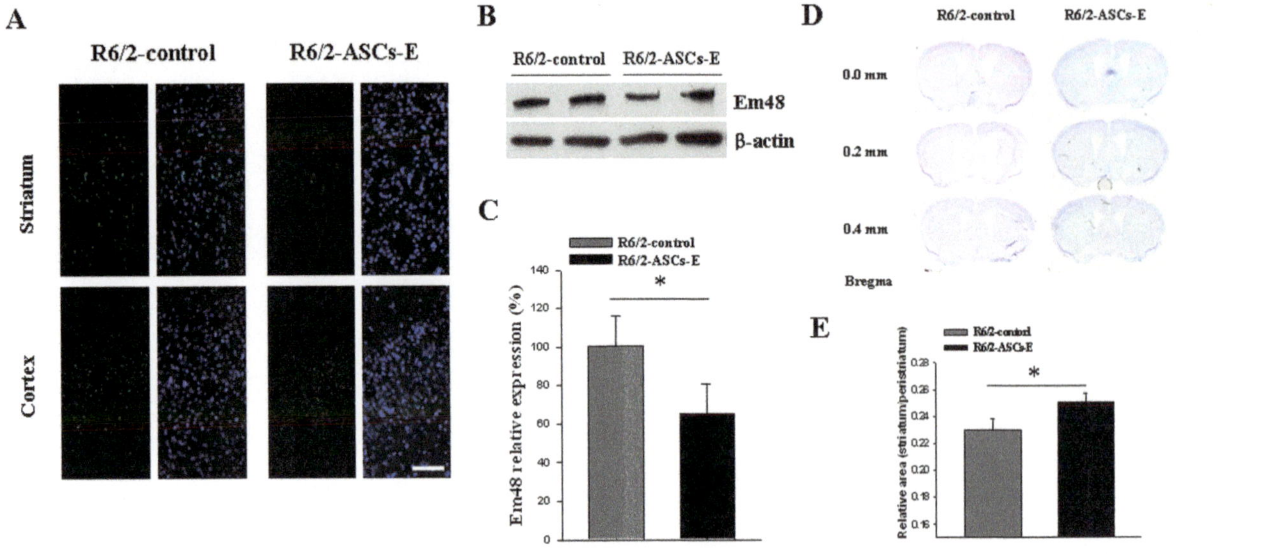

Figure 2. Striatal atrophy and mHtt aggregation. (A) Striatal mHtt aggregation was mitigated in ASCs-E treated group. (B) mHtt aggregation was visualized by western blot. (C) mHtt aggregation level was higher in ASCs-E treated R6/2 mouse brain compared with R6/2 control. (D) Striatum was sectioned and stained with Nissl in R6/2 mouse treated with ASCs-E or vehicle. (E) Striatum/peristriatum ratio is higher in ASCs-E injected group compared with vehicle treated group. Bar =100 μm, * $p<0.01$.

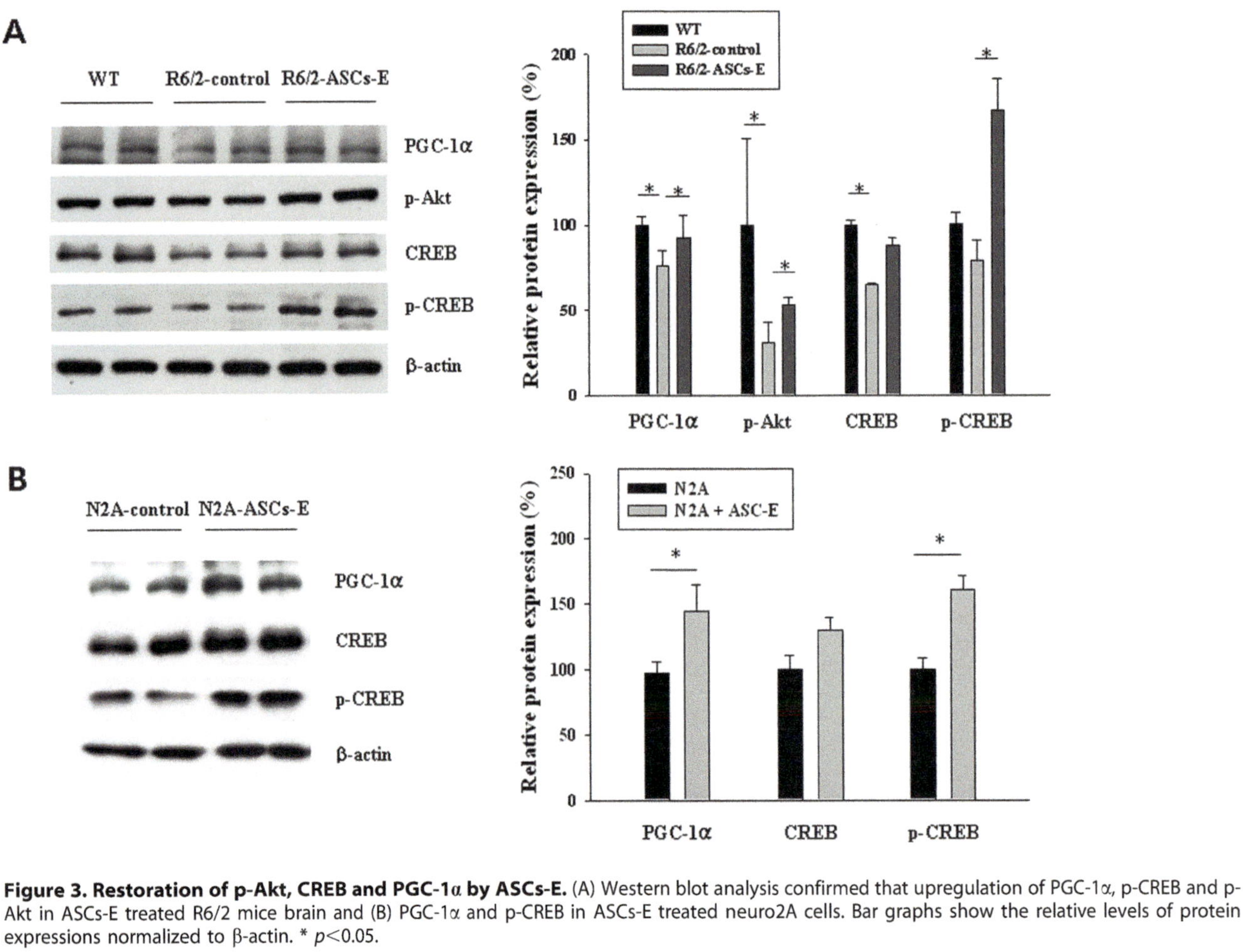

Figure 3. Restoration of p-Akt, CREB and PGC-1α by ASCs-E. (A) Western blot analysis confirmed that upregulation of PGC-1α, p-CREB and p-Akt in ASCs-E treated R6/2 mice brain and (B) PGC-1α and p-CREB in ASCs-E treated neuro2A cells. Bar graphs show the relative levels of protein expressions normalized to β-actin. * $p<0.05$.

protects against mitochondrial dysfunction and cell toxicity induced by mHtt, activation of CREB-PGC-1α pathway can have a key role to inhibit the progression of HD. P-Akt is believed to implicate CREB-PGC-1α expression and HD progression [31,32]. In this study, we found ASCs-E activates Akt, CREB and PGC-1α, indicating that therapeutic effects of ASCs-E may be from upregulation of neuroprotective CREB-PGC1α pathway.

There are concerns about adverse effects in cell transplantation such as a risk of microvascular embolism or teratoma formation [33]. However, stem cell can have paracrine effects on a disease related hostile microenvironment. For example, stem cells could alleviate various diseases by modulating host environment such as immune regulation and neuroprotection via paracrine mechanism [10,13–19]. Growth factors secreted by ASCs include EGF, FGF, VEGF, HGF and IGF-1, implying that these cells are sufficient for paracrine effect [10,17–19]. These growth factors are believed to exist in the brain and exert important functions in neuronal cell development, survival and maintenance [34]. Injection of ASCs improved symptoms of R6/2 mice, suggesting the paracrine effects of ASCs as one of the underlying mechanisms in cell transplantation [18]. In this study, ASCs-E was beneficial on HD, similarly to ASCs injection, confirming the paracrine effects of ASCs.

To use ASCs-E clinically, however, further researches are required to elucidate some issues. For instance, we do not know what molecule of ASCs-E exert therapeutic effects. Previous reports showed that therapeutic effect of ASCs-E disappeared by heat inactivation [26,27], which suggests that protein components could be the key molecules of ASCs-E. Studies on pharmacokinetics and distribution of ASCs-E in vivo are also required for clinical application. To use ASCs-E on neurodegenerative diseases, measurement of blood-brain barrier permeability is needed. Future studies will help resolving these questions.

In summary, stem cell transplantation has been investigated as a therapeutic strategy to treat the HD. Considering clinical feasibility and safety, injection of ASCs-E could be effective and safe strategy for HD patients, comparable to the stem cell therapy. Findings from our study provide possible strategy to inhibit HD progression and reasonable resources for treatment.

Author Contributions

Conceived and designed the experiments: KC MK. Performed the experiments: WI JB JL ML. Analyzed the data: WI JB STL. Contributed reagents/materials/analysis tools: WI JL ML. Wrote the paper: WI JB MK.

References

1. Landles C, Bates GP (2004) Huntingtin and the molecular pathogenesis of Huntington's disease. Fourth in molecular medicine review series. EMBO Rep 5: 958–963.
2. Walker FO (2007) Huntington's disease. Lancet 369: 218–228.
3. Bonelli RM, Wenning GK, Kapfhammer HP (2004) Huntington's disease: present treatments and future therapeutic modalities. Int Clin Psychopharmacol 19: 51–62.
4. Munoz-Sanjuan I, Bates GP (2011) The importance of integrating basic and clinical research toward the development of new therapies for Huntington disease. J Clin Invest 121: 476–483.
5. Mangiarini L, Sathasivam K, Seller M, Cozens B, Harper A, et al. (1996) Exon 1 of the HD gene with an expanded CAG repeat is sufficient to cause a progressive neurological phenotype in transgenic mice. Cell 87: 493–506.
6. Frank S, Jankovic J (2010) Advances in the pharmacological management of Huntington's disease. Drugs 70: 561–571.
7. Mizuno H (2010) Adipose-derived stem and stromal cells for cell-based therapy: current status of preclinical studies and clinical trials. Curr Opin Mol Ther 12: 442–449.
8. Schaffler A, Buchler C (2007) Concise review: adipose tissue-derived stromal cells – basic and clinical implications for novel cell-based therapies. Stem Cells 25: 818–827.
9. Zuk PA, Zhu M, Ashjian P, De Ugarte DA, Huang JI, et al. (2002) Human adipose tissue is a source of multipotent stem cells. Mol Biol Cell 13: 4279–4295.
10. Nakagami H, Maeda K, Morishita R, Iguchi S, Nishikawa T, et al. (2005) Novel autologous cell therapy in ischemic limb disease through growth factor secretion by cultured adipose tissue-derived stromal cells. Arterioscler Thromb Vasc Biol 25: 2542–2547.
11. Atsma DE, Fibbe WE, Rabelink TJ (2007) Opportunities and challenges for mesenchymal stem cell-mediated heart repair. Curr Opin Lipidol 18: 645–649.
12. Prockop DJ (2009) Repair of tissues by adult stem/progenitor cells (MSCs): controversies, myths, and changing paradigms. Mol Ther 17: 939–946.
13. Lee JW, Fang X, Krasnodembskaya A, Howard JP, Matthay MA (2011) Concise review: Mesenchymal stem cells for acute lung injury: role of paracrine soluble factors. Stem Cells 29: 913–919.
14. Li F, Niyibizi C (2012) Differentiating multipotent mesenchymal stromal cells generate factors that exert paracrine activities on exogenous MSCs: Implications for paracrine activities in bone regeneration. Biochem Biophys Res Commun.
15. Gnecchi M, Zhang Z, Ni A, Dzau VJ (2008) Paracrine mechanisms in adult stem cell signaling and therapy. Circ Res 103: 1204–1219.
16. Martino G, Pluchino S (2006) The therapeutic potential of neural stem cells. Nat Rev Neurosci 7: 395–406.
17. Rehman J, Traktuev D, Li J, Merfeld-Clauss S, Temm-Grove CJ, et al. (2004) Secretion of angiogenic and antiapoptotic factors by human adipose stromal cells. Circulation 109: 1292–1298.
18. Lee ST, Chu K, Jung KH, Im WS, Park JE, et al. (2009) Slowed progression in models of Huntington disease by adipose stem cell transplantation. Ann Neurol 66: 671–681.
19. Heo SC, Jeon ES, Lee IH, Kim HS, Kim MB, et al. (2011) Tumor necrosis factor-alpha-activated human adipose tissue-derived mesenchymal stem cells accelerate cutaneous wound healing through paracrine mechanisms. J Invest Dermatol 131: 1559–1567.
20. Tabrizi SJ, Workman J, Hart PE, Mangiarini L, Mahal A, et al. (2000) Mitochondrial dysfunction and free radical damage in the Huntington R6/2 transgenic mouse. Ann Neurol 47: 80–86.
21. Chaturvedi RK, Calingasan NY, Yang L, Hennessey T, Johri A, et al. (2010) Impairment of PGC-1alpha expression, neuropathology and hepatic steatosis in a transgenic mouse model of Huntington's disease following chronic energy deprivation. Hum Mol Genet 19: 3190–3205.
22. McGill JK, Beal MF (2006) PGC-1alpha, a new therapeutic target in Huntington's disease? Cell 127: 465–468.
23. Cui L, Jeong H, Borovecki F, Parkhurst CN, Tanese N, et al. (2006) Transcriptional repression of PGC-1alpha by mutant huntingtin leads to mitochondrial dysfunction and neurodegeneration. Cell 127: 59–69.
24. St-Pierre J, Drori S, Uldry M, Silvaggi JM, Rhee J, et al. (2006) Suppression of reactive oxygen species and neurodegeneration by the PGC-1 transcriptional coactivators. Cell 127: 397–408.
25. Okamoto S, Pouladi MA, Talantova M, Yao D, Xia P, et al. (2009) Balance between synaptic versus extrasynaptic NMDA receptor activity influences inclusions and neurotoxicity of mutant huntingtin. Nat Med 15: 1407–1413.
26. Jeon D, Chu K, Lee ST, Jung KH, Ban JJ, et al. (2013) Neuroprotective effect of a cell-free extract derived from human adipose stem cells in experimental stroke models. Neurobiol Dis.
27. Jeon D, Chu K, Lee ST, Jung KH, Kang KM, et al. (2011) A cell-free extract from human adipose stem cells protects mice against epilepsy. Epilepsia 52: 1617–1626.
28. Meade CA, Deng YP, Fusco FR, Del Mar N, Hersch S, et al. (2002) Cellular localization and development of neuronal intranuclear inclusions in striatal and cortical neurons in R6/2 transgenic mice. J Comp Neurol 449: 241–269.
29. Dai Y, Dudek NL, Li Q, Fowler SC, Muma NA (2009) Striatal expression of a calmodulin fragment improved motor function, weight loss, and neuropathology in the R6/2 mouse model of Huntington's disease. J Neurosci 29: 11550–11559.
30. Hockly E, Richon VM, Woodman B, Smith DL, Zhou X, et al. (2003) Suberoylanilide hydroxamic acid, a histone deacetylase inhibitor, ameliorates motor deficits in a mouse model of Huntington's disease. Proc Natl Acad Sci U S A 100: 2041–2046.
31. Colin E, Regulier E, Perrin V, Durr A, Brice A, et al. (2005) Akt is altered in an animal model of Huntington's disease and in patients. Eur J Neurosci 21: 1478–1488.
32. Humbert S, Bryson EA, Cordelieres FP, Connors NC, Datta SR, et al. (2002) The IGF-1/Akt pathway is neuroprotective in Huntington's disease and involves Huntingtin phosphorylation by Akt. Dev Cell 2: 831–837.
33. Leeper NJ, Hunter AL, Cooke JP (2010) Stem cell therapy for vascular regeneration: adult, embryonic, and induced pluripotent stem cells. Circulation 122: 517–526.
34. Abe K (2000) Therapeutic potential of neurotrophic factors and neural stem cells against ischemic brain injury. J Cereb Blood Flow Metab 20: 1393–1408.

Evaluation of a Novel Non-Destructive Catch and Release Technology for Harvesting Autologous Adult Stem Cells

Nicholas Bryan[1]*, Fiona C. Lewis[1], Damian Bond[2], Christopher Stanley[2], John A. Hunt[1]

1 Clinical Engineering, United Kingdom Centre for Tissue Engineering, Institute of Ageing and Chronic Disease, University of Liverpool, Liverpool, United Kingdom,
2 CellCap Technologies Ltd., The Greens Farm, Todmorden, Bacup, United Kingdom

Abstract

Background: Cell based therapies are required now to meet the critical care needs of paediatrics and healthy ageing in an increasingly long-lived human population. Repair of compromised tissue by supporting autologous regeneration is a life changing objective uniting the fields of medical science and engineering. Adipose stem cells (adSCs) are a compelling candidate for use in cell based medicine due to their plasticity and residence in numerous tissues. Adipose found in all animals contains a relatively high concentration of stem cells and is easily isolated by a minimally invasive clinical intervention; such as liposuction.

Methods: This study utilised primary rat adipose to validate a novel strategy for selecting adult stem cells. Experiments explored the use of large, very dense cell-specific antibody loaded isolation beads (diameter 5x–10x greater than target cells) which overcome the problem of endocytosis and have proved to be very effective in cell isolation from minimally processed primary tissue. The technique also benefited from pH mediated release, which enabled elution of captured cells using a simple pH shift.

Results: Large beads successfully captured and released adSCs from rat adipose, which were characterised using a combination of microscopy, flow cytometry and PCR. The resultant purified cell population retains minimal capture artefact facilitating autologous reperfusion or application in *in vitro* models.

Conclusion: Although evidenced here for adSCs, this approach provides a technological advance at a platform level; whereby it can be applied to isolate any cell population for which there is a characterised surface antigen.

Editor: Irina Kerkis, Instituto Butantan, Brazil

Funding: This research was funded by Strategic Projects to Access Research and Knowledge (SPARK), United Kingdom (grant number not available)(SPARK is a specific funding call offered by Knowledge Transfer Network) United Kingdom). The funders had no role in study design, data collection and analysis, decision to publish, or preparation of the manuscript.

Competing Interests: DB and CS are employees of CellCap Technologies Ltd. NB and JH are founding members of CellCap although currently have no financial benefit from this company and are presently employed by the University of Liverpool. FL has no affiliation to CellCap. There are no patents, products in development or marketed products to declare.

* E-mail: n.bryan@liv.ac.uk

Introduction

Stem cell niches exist within almost all tissues of an adult organism; their function to specifically localise and differentiate into a specific type of cell to renew and repair the tissue in which they reside has been realised scientifically [1,2]. However, a fundamental cellular and biochemical understanding of the precise mechanisms behind their physiological functions are yet to be defined, and therefore hampers our ability to harness their potential in efficacious and cost effective medicine [3]. Stem cells have been successfully isolated from a diverse range of tissues, including bone marrow [4–6], pancreas [7], adipose [8,6], dental pulp [9–11] and umbilical tissues [12–13] and their multilineage potential demonstrated through directed differentiation and functionalisation into representatives from all three developmental germ layers; a characteristic historically reserved solely for stem cells of embryonic origin [14–16].

Extracting stem cells from their associated tissue in a manner which renders them viable, phenotypically stable and suitable for therapeutic application has presented a major challenge to the field of cell biology but offers a tantalising omnipotent cell source for regenerative medicine [17]. When considering sources of stem cells, lipoaspirate presents itself as a favourable, readily accessible supply, which can be obtained through minimally invasive procedures, without donor site morbidity [18–19]. Additionally, the concentration of stem cells within adipose has been reported to be significantly higher than bone marrow [20]. Coupled with the large quantities of lipoaspirate that can be harvested at any one time, adipose may be considered as a future gold standard stem cell source. Immunophenotyping of cultured adSCs has also revealed >90% similarity with bone marrow-derived stem cells including CD90, CD29, CD44, CD73 and CD105 cell surface antigens [20–21].

Isolation of stromal vascular fraction (SVF) from rat adipose was first achieved by Rodbell *et al.* in the 1960 s. Despite this, the notion of adSCs was not widely recognised until 2001 when Zuk *et al.* demonstrated SVF contained large numbers of adSCs, which could differentiate into osteogenic, chondrogenic, adipogenic and myogenic lineages [22]. To date, the majority of approaches for isolating adSCs from SVF are based on the same basic principle; exploiting the ability of adSCs to adhere to plasma treated tissue culture polystyrene substrates. However, this simple but crude approach leads to a heterogeneous culture that contains a variety of adherent cells which includes fibroblasts and endothelial cells, ultimately resulting in a population of which adSCs are a minority [23].

The phenotypic, functional and particularly immunomodulatory effects of prolonged adSC *in vitro* culture are not fully understood, therefore robust and reproducible characterisation of freshly isolated adSCs would present a breakthrough in interpreting complex adSC cell biology. However this has largely been hindered by their rarity and the ability to isolate substantial numbers from fresh tissue to perform immediate and reproducible molecular biology. Several methods are available to isolate adSCs and other primary cells. Currently the two most commonly applied techniques are cell sorting by flow cytometry and paramagnetic particle isolation, both of which allow selection of cells based on antibody/antigen immunolabelling. Flow cytometry utilises fluidic processing to localise target cells into drops or diverted pathways. There are however significant hydrodynamic forces associated with this, which stem cells in particular are affected by. Magnetic particles currently in use are 50 nm−4.5 µm diameter, to which cell-specific antibodies are attached. These bind cells, which then become decorated with the particles; the complexes are subsequently exposed to a magnetic field resulting in separation of specific tagged cells from a heterogeneous cell population. This provides a convenient method of selecting cells; however the very small size of the paramagnetic particles means they are typically internalised into the cell, resulting in potential phenotypic changes [23]. Additionally, these small particles are not compatible with the dense proteinaceous matrix of primary tissue where they are observed to bind strongly to tissue materials and even air bubbles (unpublished observations): therefore extensive tissue pre-processing to create a simpler matrix for cell capture is required.

Commonly flow cytometry or immunomagnetic selection relies on negative depletion to remove non-stem cells from the culture milieu. This technique leaves stem cells without phenotypic compromise by selection chemistry; however the heterogeneity of primary tissue renders it impossible to attain homogeneous stem cell cultures even after several rounds of depletion [24–25]. Positive selection of stem cells is possible, however due to the nature of flow cytometry both positive and negative populations of cells will subject to the physical forces of flow cytometric sorting. In the case of paramagnetic bead isolation, the positive cell fraction runs the risk of endocytosis of small magnetic particles, which may render them transcriptionally modified. The derivation of pure stem cell populations for *in vitro* characterisation, pre-clinical research and ultimately translation into regenerative medical therapeutics remains a rate limiting challenge to the progression of stem cell medicine and biology.

Here we provide experimental evidence for a novel cell isolation system for high affinity catch and release of adSCs from minimally processed adult tissue. This system utilises large, dense separation beads populated with an antibody binding ligand. The ligand binds cell-specific antibody in a pH dependent manner permitting simple cell release with a small shift in reaction pH. Herein this system was utilised to isolate and release adSCs from rat adipose SVF.

Materials and Methods

Ethics statement

All studies adhered to UK home office use of animals in scientific procedures guidelines and were approved by the Institutional Review Board of the University of Liverpool.

Isolation of stromal vascular fraction (SVF) from rat adipose tissue

Subcutaneous and visceral adipose were dissected from adult Wister rats. Primary tissue was washed 3x using PBS, coarsely macerated using sterile dissection scissors and liquidised by forcing through a 10 ml syringe. Digestion was achieved by incubation in 0.2% collagenase/PBS (Sigma-Aldrich, UK) (37°C, 90 mins, 50% v/v collagenase solution/tissue homogenate). After this time had elapsed the reaction was neutralised by addition of 10% fetal calf serum. The digest was passed through a 100 µm cell strainer then centrifuged (400 g, 10 mins). To remove residual erythrocytes, cells were suspended in 200 µl PBS with 1 ml Optilyse C (Beckman Coulter, RT, 10 mins). 10 ml PBS was then added to the erythrolysed cell suspension before a final centrifugation to recover SVF cells (400 g 10 mins). Resulting cells were suspended in an appropriate volume of PBS and numerated using a hemocytometer.

Immunofluorescent staining and FACS analysis

SVF was labelled with FITC conjugated mouse anti rat CD90, CD29, CD44, CD45, and CD31 (15 mins, 4°C, 1 µg antibody/10^5 cells). A FITC conjugated isotype control (IgG_1) was used at the same concentration to allow post-hoc subtraction of non-antigen-specific fluorescence. The percentage cells in the SVF fraction expressing these antigens was quantified using flow cytometry to numerate cells with associated antibody mediated fluorescence.

$CD90^+$ isolation: protein A-coated beads (non-reversible antibody binding)

$CD90^+$ cell capture was achieved by labelling cells and loading Protein A-coated capture beads (50–200 µm diameter, CellCap Technologies Ltd) with CD90 antibody at the following concentrations: 1 µg antibody/10^5 cells and 1 µg antibody/10 µL beads. Equal volumes of cell suspensions and beads were incubated in a final volume of 1 ml PBS with gentle rolling in 1.5 ml polypropylene tubes on a roller table (30 mins, 4°C). Reactions in which neither cells nor beads received antibody were performed as a negative control.

Post bead/cell interaction, the percentage of cells specifically depleted by specific capture was quantified using flow cytometry, again based on cellular events associated with antibody mediated FITC fluorescence.

RNA isolation

The following solutions were prepared prior to RNA isolation (all reagents Qiagen, UK unless stated otherwise). 44 ml of ACS grade 100% ethanol was added to 6 ml wash buffer (RPE), while 10 µl of 1 M β-mercaptoethanol (Sigma-Aldrich UK) was added to 1 ml lysis buffer (RLT).

Prior to RNA extraction cells were washed with PBS (3×5 minutes, room temperature). Following this, 350 µl of buffer RLT was added to each sample and incubated for 5 minutes at room temperature. Resulting lysates were transferred to QIAshredder columns and spun at 13400 g for 2 minutes. 250 µl of 100% ethanol was added to the elutant, transferred to an RNeasy mini column and spun at 13400 g for 15 seconds. 500 μl wash buffer (RW1) was then added to the column and incubated for 5 minutes at room temperature before being spun at 13400 g for 15 seconds. Following this, 500 μl pre-warmed buffer RPE was added to the column and spun for 13400 g for 15 seconds. This step was repeated a second time and spun at 13400 g for 2 minutes. Finally, to elute RNA columns were transferred to RNase free tubes and 30 μl RNase free ddH_2O added, incubated at room temperature for 2 minutes then spun at 13400 g for 1 minute. Quantity and purity of RNA was determined by spectrophotometry (260/280 nm absorbance). Only samples that had a 260/280 nm absorbance between 1.9 and 2.1 were used in subsequent experiments.

Real-time PCR (qRT-PCR)

Prior to reverse transcription the following stock solutions were created. Stock 1:1 μl 50 μM $Oligo(dt)_{20}$, 1 μl 10 mM dNTP cocktail, 9 μl RNAse free ddH_2O. Stock 2:4 μl 5x first strand buffers, 1 μl 0.1 M DTT, 1 μl RNaseOUT recombinant RNAse inhibitor (40 U/μL) and 1 μl superscript III RT (200 U/μl). The above stock solutions were suitable quantities for the reverse transcription of 2 μl of RNA (15–200 μg/ml RNA).

2 μl of RNA was added to stock 1 denatured at 65°C for 5 minutes followed immediately by a 1 minute chill at −20°C. Stock 2 was then added, heated for 40 minutes at 50°C, followed by a further 15 minutes at 70°C. All reagents were purchased from Invitrogen, UK.

qRT-PCR reactions were assembled containing 2 μl cDNA diluted 100 fold using molecular biology grade ddH_2O, 0.5 μl sense primer (100 μM), 0.5 μl antisense primer (100 μM), 7.5 μl Sybr green single tube PCR master mix (Bio-Rad, UK) and 4.5 μl molecular biology grade ddH_2O.

Primers for gene of interest (CD90) and reference gene (β-actin) were designed in house. (CD90 sense: CTGCTGAGCCTTTGTGGAC, anti-sense; GCATCTT-TATTGAGTGTG, $T_{A:}$ 50.1°C. β-actin: sense; GGGACCT-GACTGACTACCTC, anti-sense; GCCATCTCTTGCTC-GAAG, $T_{A:}$ 53.9°C).

Reactions were denatured for 3 minutes at 95°C then cycled 40 times at 95°C for 30 seconds, followed by 40 cycles of annealing; 55°C for 30 seconds, 95°C for 30 seconds and finally 40 cycles at 55°C for 10 seconds. For all qRT-PCR reactions n = 3/sample. Using inter-experimental variations CD90 transcript could be calculated by normalisation to the ubiquitously expressed β-actin reference gene based on standard threshold cycle (C_T) analysis: $2^{-(\Delta CTsample\ -\ \Delta CTcontrol)}$ where $\Delta C_T = C_T$ gene of interest-C_T reference gene.

Primer design

Coding strand cDNA sequences (CDS) of genes of interest were identified using the genome search platform www.ncbi.nlm.nih.gov. The CDS sequence was copied into the primer design platform, Beacon Designer V.7.21 (Premier Biosoft International, USA). Amplicons of 75–200 bp were selected for optimal compliance with SYBR green chemistry, along with low guanine-cytosine (GC) content and an annealing temperature of 55.0+/−5.0°C. In addition, all primers were designed between 18–24 bp in length. The proposed primers were verified for tertiary structures using the DNA mfold server provided by M. Zuker at http://frontend.bioinfo.rpi.edu/applications/mfold/cgi-bin/dna-form1.cgi. Primers that formed complex hairpin loops at the annealing temperature identified by the primer design platform were discarded as it was unlikely that they would anneal correctly.

Cross sequence homology was investigated by perfoming a basic local alignment search (BLAST) was then used to interrogate the rat genome to identify regions complementary to the designed primers outside of the target gene. Primers displaying any cross sequence homology were rejected.

Primers were synthesised at a production scale of 25 nM by Invitrogen.

$CD90^+$ isolation: mixed-mode ligand-coated beads (reversible antibody binding)

In order to optimise loading of CD90 FITC-conjugated antibodies on mixed-mode (i.e. containing both aromatic and acidic groups) ligand beads (50–200 μm diameter, supplied by CellCap Technologies Ltd) several buffer configurations were explored; 200 mM TRIS or 0.1 M phosphate buffer adjusted to pH 5, 6 or 7.4. In each case beads were washed 3x in buffer before addition of 1 μg antibody/1 ml beads (15 mins, 4°C). Beads were then washed three times in the corresponding buffer to remove unbound antibody and antibody loading confirmed using fluorescent microscopy. Release was achieved by incubating beads for 15 mins at 4°C, at either pH 7.4 or 8.4, an additional blocking variable was also added which included incubation with 10% rabbit serum for 15 min prior to transfer to release buffer.

Results

Rat SVF characterisation and capture antigen selection

Prior to cell isolation SVF was characterised using a panel of surface antigens (CD90, CD29, CD44, CD45 and CD31) which validated $CD90^+$ as the most appropriate target for adSC isolation from this tissue. Flow cytometry allowed conclusion that of the antigens analysed, CD90 was the most abundant ranging between 5–10% of total SVF cells, n = 7 (inter-animal repeats) (Fig. 1). It was also found that the percentage of $CD90^+$ adSCs in SVF did not vary as a function of anatomical origin of the source material, with inter-abdominal and subcutaneous adipose yielding identical adSC ($CD90^+$) concentrations.

$CD90^+$ isolation: protein A-coated beads (non-reversible antibody binding)

Initial experiments utilised protein A-coated beads (50–200 μm) to deplete $CD90^+$ cells from SVF. Before cell isolation, loading of CD90 antibody onto beads was confirmed by exploiting the antibodies FITC conjugation to visualise antibodies co-localised with beads using fluorescent microscopy (Fig. 2).

Flow cytometric analysis after cell/bead interaction concluded that labelling/loading of both cells and beads with antibody provided the greatest $CD90^+$ depletion and that the antibody concentration used to load the beads could be considerably reduced without compromising $CD90^+$ cell depletion. A mean depletion of 80% was recorded when cells were pre-labelled with 1 μg antibody/10^5 cells and beads loaded with a very low antibody concentration; 0.001 μg (Fig. 3). To further confirm $CD90^+$ capture, RNA was isolated from both components of the capture reaction; beads and surrounding capture supernatents, and qRT-PCR performed to compare CD90 presence relative to a negative control capture in which no antibody was added to the system. This demonstrated that cells bound to the bead surface expressed CD90 while the contrary was true of the cells which remained unassociated with beads in the surrounding reaction supernatant. This showed that CD90+ cells in the reaction mixture were associated with the bead surface post CD90 targeted cell/bead interaction (Fig. 4).

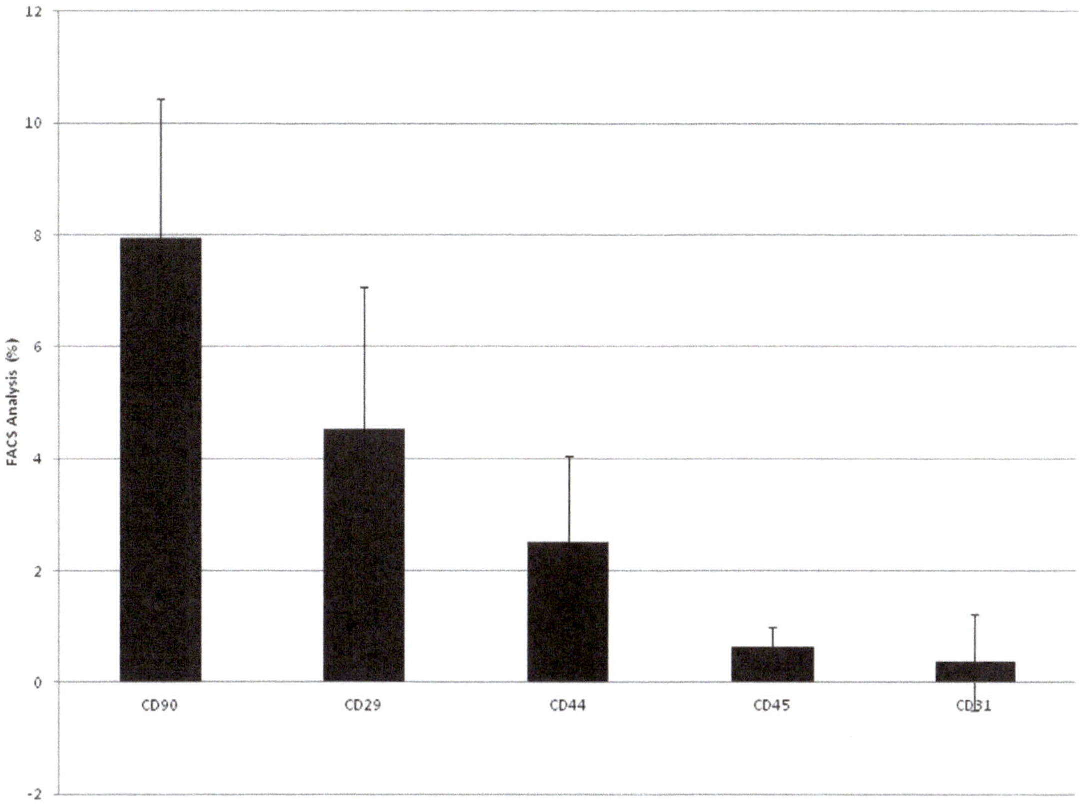

Figure 1. Flow cytometric characterisation of rat primary adipose. A: SVF Characterisation. Error bars represent 1 standard deviation from the mean, n = 7 (inter-animal repeats). B: Percentage CD90+ cells present in subcutaneous and intra-abdominal adipose tissue. Error bars represent 1 standard deviation from the mean, n = 3 (inter-animal repeats).

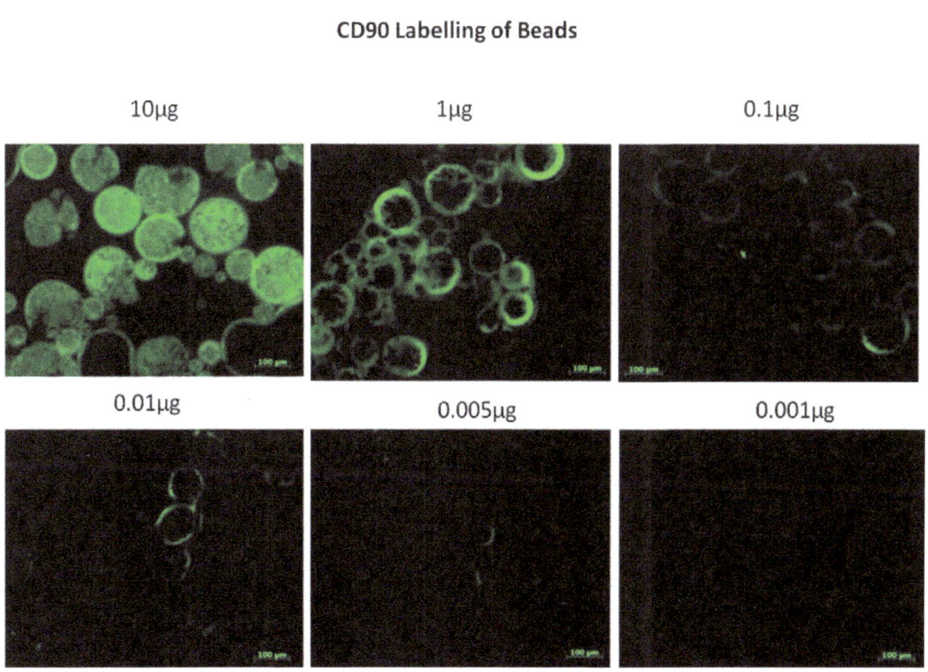

Figure 2. Fluorescent microscopic observation of successful CD90 FITC antibody loading of protein A beads.

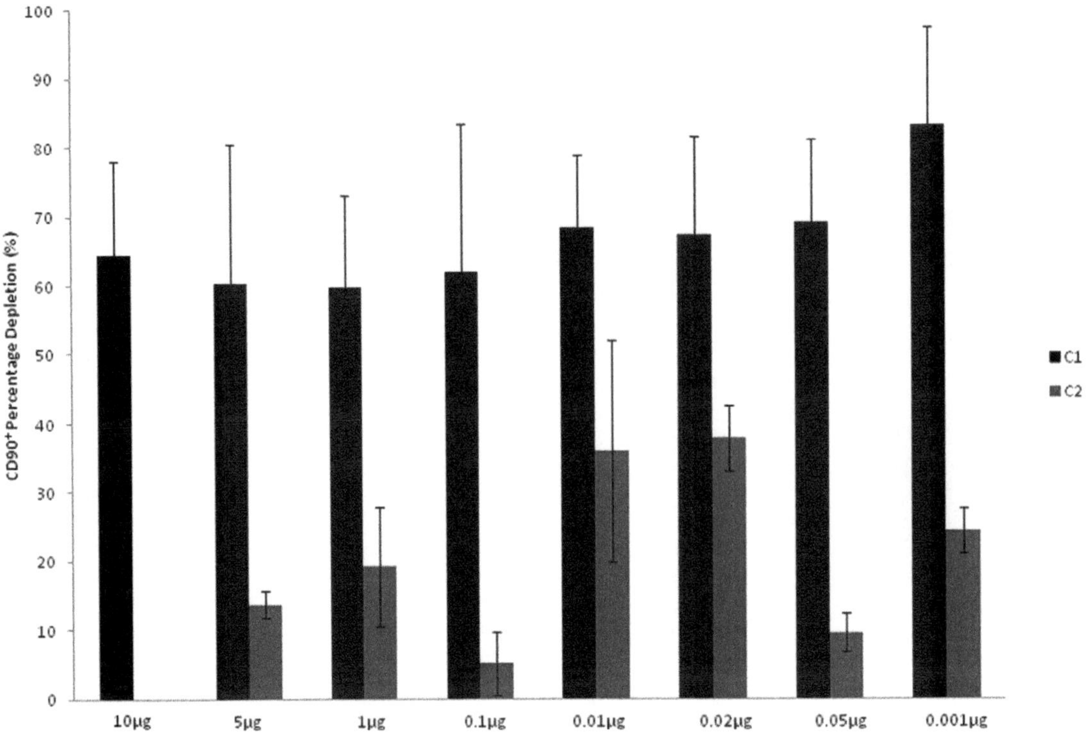

Figure 3. Percentage CD90+ cell depletion as a function of decreasing antibody concentration on capture beads. Antibody concentration on the cells remained constant; 1 µg antibody/10^5 cells. X; bead antibody concentration, Y: percentage CD90 cell depletion. Error bars represent 1 standard deviation, n = 5 (technical replicates). C1 – both cells and beads labelled, C2 – beads alone labeled.

$CD90^+$ isolation: mixed-mode ligand-coated beads (reversible antibody binding)

To evaluate this bead coating it was necessary to replicate loading of FITC-conjugated CD90 antibody to the surface of mixed-mode ligand coated beads. These beads are decorated with an aromatic acid ligand which has the ability to bind and release Ig in a pH-mediated manner. This suggested them as ideal candidates for isolation and recovery of cells using antibody reliant mechanisms [26]. Briefly, antibody loading was investigated using both TRIS and phosphate buffers at pHs 5, 6 and 7.4 with Fluorescent microscopy to observe FITC Ig/bead co-localisation. pHs 5 and 6 facilitated successful antibody loading whilst pH 7.4 did not (Fig. 5). Release of bound antibody was interrogated using both TRIS and phosphate buffers at pHs 7.4 and 8.4+/– a pre-release incubation with 10% rabbit serum. optimal release (leaving no visualisable FITC fluorescence) was observed after pre-incubating antibody-coated beads with rabbit serum before transfer to TRIS pH 8.4. This mechanism was found to be instantaneous.

Ligand beads loaded with either 10 µg or 0.001 µg/ml CD90 antibody were combined with CD90 labelled SVF (1 µg antibody/10^5 cells) in 200 mM TRIS, pH 5. Following incubation, supernatants were subject to flow cytometric analysis, which revealed depletion of $CD90^+$ cells to an extent comparable to their Protein A-coated equivalent. Cell laden beads were incubated with rabbit serum for 15 minutes at 4°C before washing in 200 mM TRIS pH 8.4 to elute bound cells. The beads were allowed to settle, supernatant removed and centrifuged for 10 minutes at 1500 rpm. The mean number of recovered cells was 5.6×10^3, which correlated with the initial population of 5.0×10^3 $CD90^+$ cells/10^5 cells, n = 4 (Fig. 6). Post-elution the identity of released cells was confirmed by fluorescent microscopic visualisation of CD90 FITC conjugated antibody associated with the cell surface (Fig. 6).

Discussion

adSCs present an untapped source of pluripotent cells for future medicine. However their isolation and characterisation requires significant advancement to enable their potential to be realised. An ideal isolation procedure would facilitate straight forward clinical translation from a regulatory perspective whilst also fundamentally increasing first principles biological understanding of the role of adSCs in normal tissue homeostasis, ageing, wound healing and chronic disease.

Development of a cell isolation system that provides high purity and yield whilst avoiding receptor mediated endocytosis of small immunomagnetic particles and fluidic shear and extreme hydrodynamic forces of flow cytometry will provide a superior and urgently required alternative to today's widely applied selection methods. Ideally, an advanced new approach should provide a homogeneous population without compromising the immunoprivilege of autologous cells. The system should perform in complex, matrix rich primary tissues with minimal preparatory pre-processing, which can induce cellular transcriptomic and secretomic modifications.

Immunophenotyping has identified a number of adSC surface molecules, which provide targets to exhume these cells from the heterogeneous milieu of stromal tissue. We have evidenced using a relevant animal model that CD90 is the most appropriate molecule for isolating adSCs. In this rat model CD90 expression was relatively low ranging between 5–10% compared to previous reports investigating human adipose adSCs, where approximately

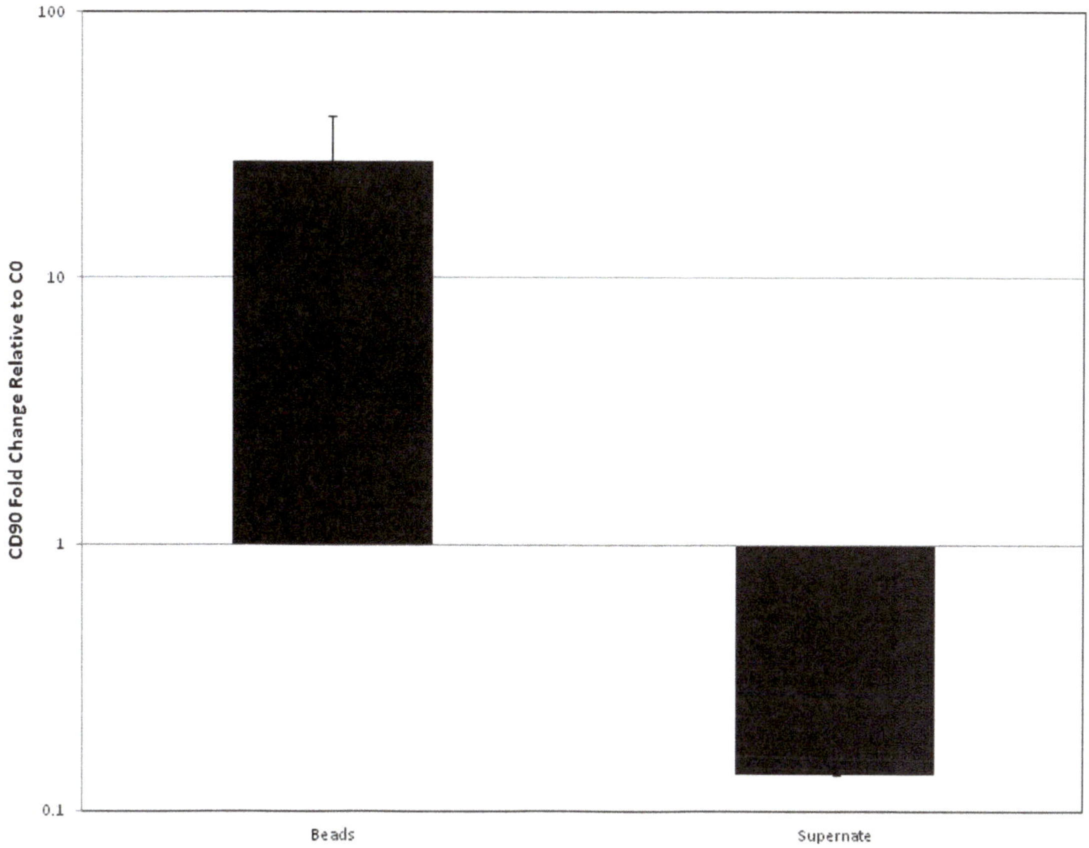

Figure 4. qRT-PCR analysis of CD90 fold change relative to a control capture containing no antibody after normalisation to a beta-actin control. Error bars represent 1 standard deviation from the mean, n = 3 (technical replicates).

50% of cells are $CD90^+$ [27–28]. Nevertheless $CD90^+$ cells were repeatedly identified in the SVF of rat adipose tissue from both the intra and extra-abdominal regions, which is highly clinically relevant, demonstrating minimally invasive subcutaneous adipose (lipoaspirate) is no less potent a source of stem cells than visceral adipose. This reduces the potential for donor site morbidity during tissue harvesting associated with considerably more invasive obtaining of ASCs by bone marrow aspiration.

In this study we used a preparation of large, dense antibody-loaded beads with diameters ranging from 50–200 μm (manufacturer's specification) as the core of the cell-isolation system. Previously the use of beads in this size range has not been considered favourable for isolation of mammalian cells due to an expectation that binding kinetics would be too slow [29], the cell loading capacity of the beads would be too low due to a lack of surface area [29] or there would simply be a lack of interaction between cells and beads of this size [30]. A previous report on an end-over-end mixed batch adsorption process using large beads (mean diameter 61 μm) did not perform well and this was attributed to poor suspension of the beads, a lack of contact of cells with the beads and the possibility of mechanical disruption of the cells on the bead surface [31–32]. In contrast to these earlier reports we have shown that the use of large beads in a 'roller bottle' format has proved to be a very effective method of cell capture. We attribute this to the ease with which the large number of high density beads move through SVF without mechanical restriction, the repeated sedimentation and resupension cycles that they go through and the very high affinity of the interaction when both beads and cell surfaces are populated with antibody. This implies 2 levels of interaction during capture; the antibody on the beads interacting with cell surface antigen in addition to the antibody on the cells binding the Ig specific ligand on the beads.

Initial experiments were with Protein A-coated large beads: this is a recombinant microbial protein that binds immunoglobulin at neutral pH; elution can only be achieved at pH <3 [33]. Protein A binding affinity can vary between antibody species and sub-class, however previous studies have demonstrated that Protein A binds murine IgG with high affinity [34] therefore it was not surprising

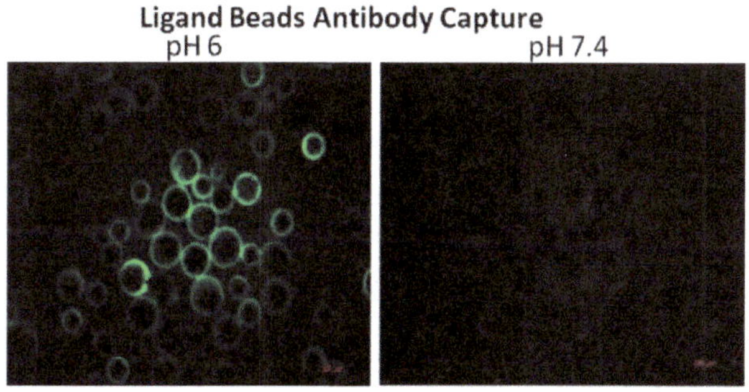

Figure 5. Optimisation of appropriate reaction conditions for cell capture using ligand beads using fluorescent microscopy to observe ligand particles co-localised with FITC conjugated antibody across a pH gradient.

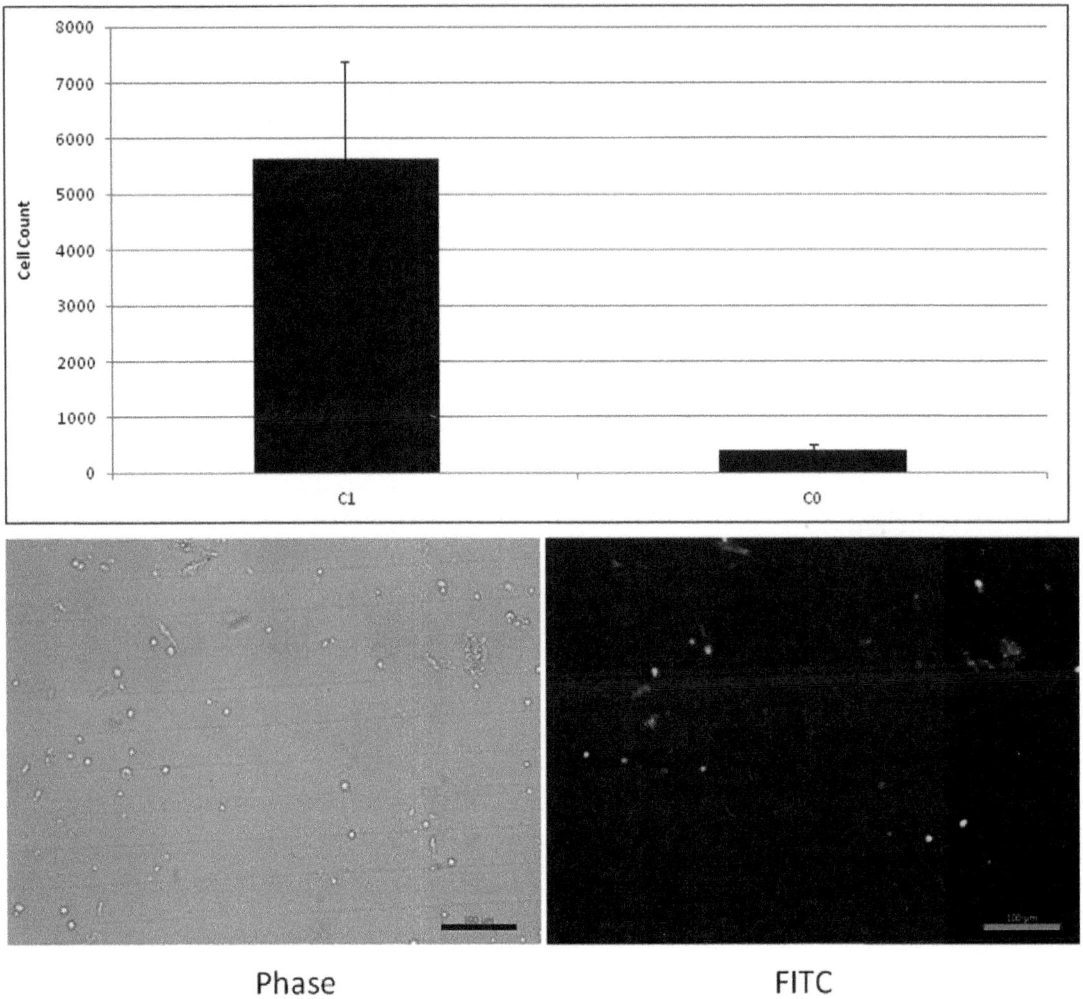

Figure 6. A: Quantification of released cells using a hemocytometer. Error bars represent 1 standard deviation from the mean, n=4 (technical replicates) and fluorescent microscopic characterisation of the released cell fraction by observation of CD90 FITC labeling. scale bar represents 100 μm.

that the murine IgG anti-rat CD90 bound to these Protein A coated beads. Efficient cell capture was still obtained when beads were loaded with an incredibly small quantity of antibody. $CD90^+$ cell depletion was reported using flow cytometry, as the essential technique at the analytical stage to quantify cell capture on a population scale supported by fluorescent microscopy and qRT-PCR which identified CD90 transcript specifically associated with the beads. Collectively this cross referenced and validated this technique for isolation of $CD90^+$ cells from heterogeneous SVF.

Although isolation of cells with Protein A beads was demonstrated to be of high efficiency an effective cellular release method was still required and so an alternative cell capture bead was explored for this purpose. This bead was modified with a covalently bound mixed-mode ligand coating based on an aromatic acid moiety; these ligands bind and release immunoglobins in a pH dependant manner over a narrow range and have been used in preparative chromatographic processes [26]. Otherwise, the overall structure and density of this mixed mode ligand bead was identical to the Protein A counterpart. This investigation successfully demonstrated loading of FITC-conjugated antibody onto mixed mode ligand beads at pH5-6, whilst raising the pH to 8.4 instantly released the antibody and subsequent bound cells. Pre-incubating ligand beads with an excess of polyclonal IgG (rabbit serum) prior to release significantly increased release efficiency. This suggested that saturating the ligand binding sites on the beads with non-specific IgG reduced the possibility of multiple interactions with the cell-specific antibody leading to optimal release kinetics.

This study therefore presents the initial steps in the validation of a new, minimally invasive stem cell harvesting system. Future research will focus on confirming and quantifying cell viability and phenotype maintenance in response to subjection to this novel pH mediated sorting strategy.

Conclusion

A new approach to isolating highly purified populations of cells from primary complex mammalian tissues has been experimentally evidenced and validated. The technique marks a technical breakthrough in derivation of specific cell types based on high affinity selection. The cell isolation target molecule can be freely chosen on the basis of what is known about the cell type of choice. This approach provides a platform technology breakthrough that has been demonstrated here for adSCs but can by design and target molecule be tailored to any cell and tissue as required. It was not the authoring research teams intention to "invent" a new technique for cell isolation however limitations in the currently available techniques forced new approaches to be considered and developed to progress delivery of cell based therapies into man and animals and provide the most relevant *in vitro* 3D tissue analogues. The value and novelty in this approach lies in its non-invasive nature at a cellular level in addition to its ideal physical properties to be used with minimally processed primary tissue. The ability to tailor this approach by simply varying the antibody used makes its applications across cell biology and biomedicine vast. As demonstrated here this strategy may be used to exploit adipose as a clinical cell source with straight forward translation based on a cell population with an absolute minimum of exogenous manipulation prior to delivery.

Author Contributions

Conceived and designed the experiments: FL NB DB CS JH. Performed the experiments: FL NB. Analyzed the data: FL NB. Contributed reagents/materials/analysis tools: CS DB JH. Wrote the paper: FL NB CS DB JH.

References

1. Ohlstein B, Kai T, Decotto E, Spradling A (2004) The stem cell niche: theme and variations. Curr Op Cell Biol 6: 693–699.
2. Mitsiadis TA, Barrandon O, Rochat A, Barrandon Y, De Bari C (2007) Stem cell niches in mammals. Exp Cell Res 313: 3377–3385.
3. Flake A (2004) The conceptual application of systems theory to stem cell biology: a matter of context. Blood Cell Mol Dis 32: 58–64.
4. Friedenstein AJ, Gorskaja JF, Kulagina NN (1976) Fibroblast precursors in normal and irradiated mouse hematopoietic organs. Exp Hematol 4: 267–274.
5. Dvorakova J, Hruba A, Velebny V, Kubala L (2008) Isolation and characterisation of mesenchymal stem cell populations entrapped in bone marrow collection sets. Cell Biol Int 32: 1116–1125.
6. Romanov YA, Darevskaya AN, Merlikina NV, Buravkova LB (2005) Mesenchymal stem cells from human bone marrow and adipose tissue: isolation, characterisation and differentiation potentialities. Bull Exp Biol Med 140: 138–143.
7. Kruse C, Kajahn J, Petschnik AE, Maab A, Klink E, et al. (2006) Adult pancreatic stem/progenitor cells spontaneously differentiate in vitro into multiple cell lineages and form teratoma-like structures. Annals of Anatomy 188: 503–506.
8. Zuk PA, Zhu M, Mizuno H, Huang J, Futrell JW, et al. (2001) Multilineage cells from human adipose tissue: implications for cell-based therapies. Tissue Eng 7: 211–228.
9. Gronthos S, Mankani M, Brahim J, Robey PG, Shi S (2000) Postnatal human dental pulp stem cells (DPSCs) in vitro and in vivo. Proc Natl Acad Sci 97: 13625–13630.
10. Gronthos S, Brahim J, Li W, Fisher LW, Cherman N, et al. (2002) Stem cell properties of human dental pulp stem cells. J Dent Res 81: 531–535.
11. Huang AHC, Chen YK, Lin LM, Shieh TY, Chan AW (2008) Isolation and characterisation of dental pulp stem cells from a supernumerary tooth. J Oral Pathol Med 37: 571–574.
12. Qiao C, Xu W, Zhu W, Hu J, Qian H, et al. (2008) Human mesenchymal stem cells isolated from the umbilical cord. Cell Biol Int 32: 8–15.
13. Kestendjieva S, Kyurkchiev D, Tsvetkova G, Mehandjiev T, Dimitrov A, et al. (2008) Characterisation of mesenchymal stem cells isolated from the human umbilical cord. Cell Biol Int 32: 724–732.
14. Jiang Y, Jahagirdar BN, Reinhardt RL, Schwartz RE, Keene CD, et al. (2002) Pluripotency of mesenchymal stem cells derived from adult marrow. Nature 418: 41–49.
15. Banas A, Yamamoto Y, Teratani T, Ochiya T (2007) Stem cell plasticity: learning from hepatogeneic differentiation strategies. Dev Dynam 236: 3228–3241.
16. Arthur A, Rychkov G, Shi S, Koblar SA, Gronthos S (2008) Adult human dental pulp stem cells differentiate toward functionally active neurons under appropriate environmental cues. Stem Cells 26: 1787–1795.
17. Pountos I, Corscadden D, Emery P, Giannoudis PV (2007) Mesenchymal stem cell tissue engineering: techniques for isolation, expansion and application. Injury 38: 23–33.
18. Zuk PA, Zhu M, Mizuno H, Huang J, Futrell JW, et al. (2001) Multilineage cells from human adipose tissue: implications for cell-based therapies. Tissue Eng 7: 211–228.
19. Zhu Y, Liu T, Song K, Fan X, Ma X, et al. (2008) Adipose-derived stem cells: a better stem cell than BMSC. Cell Bioch Func 26: 664–675.
20. Kern S, Eichler H, Stoeve J, Kluter H, Bieback K (2006) Comparative analysis of mesenchymal stem cells from bone marrow, umbilical cord, or adipose tissue. Stem Cells 24: 294–1301.
21. Musina RA, Bekchanova ES, Sukhikh GT (2005) Comparison of mesenchymal stem cells isolated from different human tissues. Bull Exp Biol Med 139: 504–509.
22. Rodbell M, Jones AB (1966) The metabolism of isolated fat cells. The similar inhibitory actions of phospholipase C (clostridium perfringens alpha toxin) and insulin on lipolysis stimulated by lipolytic hormones and theophylline. J Biol Chem 241: 140–142.
23. Handley ME, Pollara G, Chain BM, Katz DR (2005) The use of targeted microbeads for quantitative analysis of the phagocytic properties of human monocyte-derived dendritic cells. J Immunol Meth 297: 27–38.
24. Mödder UI, Roforth MM, Nicks KM, Peterson JM, McCready LK, et al. (2012) Characterization of mesenchymal progenitor cells isolated from human bone marrow by negative selection. Bone 3:804–10.
25. McKenzie KP, Mayer DC, Aubin JE (2012) Osteogenesis and expression of the bone marrow niche in endothelial cell-depleted HipOPs. J Cell Biochem. Epub ahead of print.
26. Knudsen KL, Hansen MB, Henriksen LR, Andersen BK, Lihme A (1992) Sulphone aromatic ligands for theophilic adsorption chromatography: Purification of human and mouse immunoglobulins. Analytical Biochemistry 201: 170–177.
27. Zhao G, Dong XY, Sun Y (2009) Ligands for mixed-mode protein chromatography: principles, characteristics and design. J Biotech 144: 3–11.
28. Francesco FD, Tirino V, Desiderio V, Ferraro G, D'Andrea F, et al. (2009) Human $CD34^+/CD90^+$ ASCs are capable of growing as sphere clusters, producing high levels of VEGF and forming capillaries. PLoS ONE 4: e6537.
29. Ishimura D, Yamamoto N, Tajima K, Ohno A, Yamamoto Y, et al. (2008) Differentiation of adipose-derived stromal vascular fraction culture cells into chondrocytes using the method of cell sorting with a mesenchymal stem cell marker. Tohoku J Exp Med 216: 149–156.
30. Willoughby N (2009) To big to bind? Will the purification of large and complex therapeutic targets spell the beginning of the end for column chromatography. J Chem Technol Biot 84: 145–150.
31. Balmayor ER, Pashkuleva L, Frias AM, Azevedo HS, Reis RL (2011) Synthesis and functionalization of superparamagnetic poly-{varepsilon}-caprolactone microparticles for the selective isolation of subpopulations of human adipose-derived stem cells. J R Soc Interface 8: 896–908.
32. Ujam LB, Clemmitt RH, Clarke SA, Brooks RA, Rushton N, et al. (2003) Isolation of monocytes from human from human peripheral blood using immune-affinity expanded bed adsorption. Biotechnol Bioeng 83: 554–566.
33. Lindmark R, Thorén-Tolling K, Sjöquist J (1983) Binding of immunoglobulins to protein A and immunoglobulin levels in mammalian sera. J Immunol Meth 62: 1–13.
34. Ey PL, Prowse SJ, Jenkin CR (1978) Isolation of pure IgG1, IgG2a, and IgG2b immunoglobulins from mouse serum using protein A-sepharose. Immunochemistry 15: 429–436.

15

In Vitro Hepatic Trans-Differentiation of Human Mesenchymal Stem Cells Using Sera from Congestive/Ischemic Liver during Cardiac Failure

Dillip Kumar Bishi[1,2], Santosh Mathapati[1,2], Kotturathu Mammen Cherian[2], Soma Guhathakurta[3]*, Rama Shanker Verma[1]*

1 Stem Cells and Molecular Biology Laboratory, Department of Biotechnology, Indian Institute of Technology Madras, Chennai, India, **2** Stem Cells and Tissue Engineering Laboratory, International Centre for Cardiothoracic and Vascular Diseases, Frontier Lifeline, Chennai, India, **3** Department of Engineering Design, Indian Institute of Technology Madras, Chennai, India

Abstract

Cellular therapy for end-stage liver failures using human mesenchymal stem cells (hMSCs)-derived hepatocytes is a potential alternative to liver transplantation. Hepatic trans-differentiation of hMSCs is routinely accomplished by induction with commercially available recombinant growth factors, which is of limited clinical applications. In the present study, we have evaluated the potential of sera from cardiac-failure-associated congestive/ischemic liver patients for hepatic trans-differentiation of hMSCs. Results from such experiments were confirmed through morphological changes and expression of hepatocyte-specific markers at molecular and cellular level. Furthermore, the process of mesenchymal-to-epithelial transition during hepatic trans-differentiation of hMSCs was confirmed by elevated expression of E-Cadherin and down-regulation of Snail. The functionality of hMSCs-derived hepatocytes was validated by various liver function tests such as albumin synthesis, urea release, glycogen accumulation and presence of a drug inducible cytochrome P450 system. Based on these findings, we conclude that sera from congestive/ischemic liver during cardiac failure support a liver specific microenvironment for effective hepatic trans-differentiation of hMSCs *in vitro*.

Editor: Antonio Paolo Beltrami, University of Udine, Italy

Funding: No current external funding sources for this study. The funders had no role in study design, data collection and analysis, decision to publish, or preparation of the manuscript.

Competing Interests: The authors have declared that no competing interests exist.

* E-mail: husun1971@gmail.com (SG); vermars@iitm.ac.in (RSV)

Introduction

Liver disease is one of the leading causes of mortality worldwide, accounting for two million deaths every year [1]. Liver, an organ with highest inbuilt regenerative capacity, cannot restore its damaged tissues in end-stage liver diseases. Although most common causes of hepatic dysfunctions leading to liver failure are excessive alcohol usage, toxic injury and viral infection, there are incidences of hepatic abnormality observed in cardiac failure [2]. Major causes of such complications are chronic congestion due to heart failure and ischemic hepatitis from acute circulatory impairment. The prevalence of liver dysfunction in cardiac failure is approximately 15–65% [3]. There are cases of cardiac-failure-associated liver congestion/ischemia leading to hyperbilirubinemia, sometime referred as "cardiac jaundice" [4,5].

Present status of liver salvation in end-stage liver diseases is orthotopic liver transplantation, but it has serious limitations like donor availability and organ rejection [6]. Cellular therapy using hepatocytes are promising, but inadequate cell source for hepatocyte transplantation is a major hurdle [7]. Therefore, regenerative therapy using stem cells looks promising as it can provide a clinically feasible alternative to liver transplantation due to the unique plasticity, clonality and differentiation potential of stem cells towards functional hepatocytes *in vitro* and *in vivo* [8,9]. The major categories of stem cells that have been implicated for hepatic regeneration are resident hepatic stem cells (oval cells, hepatocytes, foetal liver cells, etc.) and extra-hepatic stem cells [such as embryonic stem (ES) cells, induced pluripotent stem cells (iPSCs), adult stem cells from bone marrow, cord blood and adipose tissue)] [10]. Resident hepatic stem cells (oval cells) are the best candidates to repopulate damaged livers [11], but they are scarce and are unfavourable for large-scale expansion during clinical application [12]. ES cells possess tremendous hepatogenic potential, but ethical issues and risk of tumour formation restricts their therapeutic application [8]. Similarly, iPSCs though hold great promises for hepatocyte generation, safe and effective iPSCs for use in cell therapy are yet to be developed [13]. Adult bone marrow stem cells are reported as a potential cell source for liver regeneration [14] and can generate hepatocytes *in vitro* and *in vivo* [15,16]. Clinical trials involving autologous bone marrow mesenchymal stem cell (MSC) therapy for liver diseases showed substantial improvements in liver functionality [17–19]. Since hMSCs could be harvested from the patient's own marrow and can effectively give rise to functional hepatocytes, it has tremendous potential for end-stage liver failure treatment.

MSCs have been trans-differentiated effectively towards hepatocyte-like cells *in vitro* using an array of commercially available recombinant growth factors [hepatocyte growth factor (HGF), epidermal growth factor (EGF), fibroblast growth factor (FGF)], cytokines [Oncostatin M (OSM)] and chemical compounds (nicotinamide, dexamethasone, insulin etc.) by inducing either as a cocktail [20] or in a sequential manner [21,22]. In fact, HGF alone is found to be sufficient for hepatic differentiation of MSCs [23]. However, hepatic inductions with such recombinant growth factors are not optimal for clinical applications due to their bacterial origin and in most cases they are not free of endotoxins. Thus a natural source of hepatogenic factors, readily available from patients, would be ideal as a conditioned culture system to augment hepatic differentiation of stem cells with suitable clinical relevance. There have been well known reports of usage of liver failure sera and cholestatic sera upon hepatogenic induction of bone marrow stem cells [24–28], which describe the potential role of hepatogenic factors (including HGF) released from hepatocytes during liver damage or cholestasis. Serum levels of HGF increase in patients with a variety of liver diseases [29,30] as well as in cardiovascular diseases such as acute myocardial infarction, hypertension and congestive heart failure [31–34]. In the present study, our primary goal was to evaluate the effectiveness of a novel hepatogenic conditioned sera collected from patients with cardiac-failure-associated secondary hyperbilirubinemia (jaundice) on hepatic trans-differentiation potential of human bone marrow MSCs. The patient sera used for hepatic induction were assessed for the presence of hepatogenic factors (such as HGF) and we could achieve functional hepatocytes with such novel hepatogenic conditioned culture system.

Materials and Methods

Assessment of Clinical and Biochemical Profiles of Patients with Cardiac-failure -associated Congestive/ ischemic Liver

Study Approval. This study was reviewed and approved by the Institutional Ethics Committee of International Centre for Cardiothoracic and Vascular Diseases, Frontier Lifeline hospital, Chennai, India.

Patient and Control Cohorts. Twenty seven patients with cardiac-failure-associated congestive/ischemic liver with symptoms of secondary jaundice (hyperbilirubinemia) were recruited for this study from the critical care unit of International centre for cardiothoracic and vascular disease, Frontier Lifeline hospital, Chennai. In addition a control group, consisting of 27 volunteers who were age, gender and ethnically matched to the patient group was recruited. The investigation conforms to the principles outlined in the Declaration of Helsinki [35]. Written informed consents were obtained from all participants before inclusion in the study and the initiation of any study related procedures. The inclusion criteria of the patient group were: presence of chronic cardiac complications leading to heart failure and have developed jaundice (total bilirubin $\geq$ 3.0). The inclusion criteria for the control group were: absence of a known coronary, valvular, or myocardial disease. Co-morbidities for coronary artery disease such as diabetes mellitus, hypertension, hyperlipidaemia, and smoking did not preclude recruitment. Exclusion criteria for all participants were: pregnancy, dialysis, and known or treated malignancies, viral infection, or drug induced liver dysfunction, hepatobiliary diseases, cirrhosis or alcoholic hepatitis. Patients were excluded if they had pre-existing liver disease or injury to the liver during trauma, any preexisting chronic medical condition (including hepatitis, organ system failure). Sera from both patients as well as control groups were collected and screened for microbial infections, hepatitis and endotoxin and stored in −80°C for further experiments.

Clinical and Biochemical Profile of Patients with Cardiac-failure-associated Liver Dysfunction

All the clinically relevant data such as patient demographics, history of cardiac disease, etiology and the main precipitating cause of cardiac-failure-associated hyperbilirubinemia, cardiovascular risk factors as well as results of X-ray, echocardiographic and laboratory tests were collected from medical records. The patients were categorized into various cardiac disease groups such as ischemic heart disease (IHD), valvular heart disease (VHD), dilated cardiomyopathy (DCM) and congenital heart disease (CHD) based upon their signs and symptoms. Baseline laboratory data, particularly serum bilirubin (total and direct), albumin, alkaline phosphatase, alanine aminotransferase (ALT), aspartate aminotransferase (AST), lactate dehydrogenase (LDH), γ-glutamyl transpeptidase (GGT), alkaline phosphatase (ALP), urea, creatinine and haemoglobin tests were collected from medical record for each individuals on admission to hospital or during follow up. Based on liver function tests for serum enzymes (ALT, AST, GGT and ALP) and bilirubin level, the patients were categorized to have either hepatocellular or cholestatic profile as per earlier reports [36]. The concentration of HGF present in patient sera and normal sera was measured by ELISA (Human HGF Quantikine ELISA kit, R&D systems MN, USA) as per the manufacturer's protocol.

Hepatic Trans-differentiation of Human MSCs

Growth Characteristics of hMSCs in Presence of Cardiac-failure-associated Jaundiced Sera (in vitro Cytotoxicity Assay). Three different sets of human MSC cell line were evaluated in this study, one acquired from Lonza (Allendale, NJ, USA) and two others isolated and expanded from patient bone marrow in our laboratory. Cell proliferation assay of hMSCs grown in various hepatogenic culture conditions reveals the overall impact of sera components (particularly bilirubin) on cell growth. Therefore, we evaluated the *in vitro* cytotoxic effect of different concentrations of patient sera (10% vs. 5% in DMEM) upon proliferation potential of MSCs. As a control, the impact of normal volunteer donor's sera (10% and 5% in DMEM) on hMSC proliferation was evaluated. Cells grown in 10% FBS was used as positive control. Cell death assay was performed by annexin V-FITC/PI apoptosis detection kit (Cell Signaling Technology) as per manufacturer's protocol. Cell proliferation assay was done by direct cell counting using trypan blue on every 3rd day until 9 days and also by MTT based *in vitro* toxicology assay kit (Sigma) as per manufacturer's protocol. An increase or decrease in cell number results in a concomitant change in the amount of formazan formed as measured spectrophotometrically, indicating the degree of cytotoxicity caused by the serum components.

Hepatic Trans-differentiation - Protocol Optimization. MSCs of passage 3 were expanded on collagen type-I coated 35-mm diameter 6-well plates (Nunc, Wiesbaden-Biebrich, Germany) till 80% confluence in Isocove's modified Dulbecco's medium (IMDM) containing 10% FBS and antibiotics. Hepatic transdifferentiation was carried out as per the standard protocol [21] with minor modifications. Briefly, the expanded hMSCs (1.5×10^4 cells/cm^2) were subjected to endoderm specification with 20 ng/ml EGF and 10 ng/ml FGF-4 for 48 hrs in serum free IMDM. Hepatic induction was carried out by treating hMSCs with either of the following group of hepatogenic culture systems/control group:

Group I: Pooled sera from patients with cardiac-failure-associated congestive/ischemic liver leading to jaundice (5% in IMDM)

Group II: Pooled sera from normal volunteers (5% in IMDM)

Table 1. Baseline characteristics of patients and control group[a].

Parameters	Cardiac-failure-associated hyperbilirubinemia patient group		Control group (without cardiovascular disease)		P-value
	Mean ± S.D. (or %)	IQR	Mean ± S.D. (or %)	IQR	(Mann-Whitney U test)
Gender (male)	19 (70.3%)		22 (81%)		0.349
Age (years)	42.2±13.2	14–63	44.4±12.2	19–68	0.531
Underlying heart disease			NA		NA
Ischemic heart disease (%)	8 (29.6%)				
Valvular heart disease (%)	11 (40.7%)				
Dilated cardiomyopathy (%)	3 (11.1%)				
Congenital heart disease (%)	5 (18.5%)				
NYHA class I (%)	3 (11.1%)				
NYHA class II (%)	10 (37%)				
NYHA class III/IV (%)	14 (51.85%)				
Symptoms and signs					
Heart rate (bpm)	79±12	66–104	72±6.3	65–82	0.01
Systolic blood pressure (mm Hg)	129.3±18.3	100–170	119.1±9.8	100–140	0.011
Diastolic blood pressure (mm Hg)	82.1±16.8	70–120	73±8.3	56–80	0.015
ECG (sinus rhythm)	15 (55.5%)		NA		
Atrial Fibrillation	8 (29.6%)		NA		
LV-EF (%)	38.5±18		NA		
LV-FS (%)	22.3±11.7		NA		
Congestive hepatomegaly with ascites	7 (25.9%)		NA		
Laboratory parameters					
Serum bilirubin total (mg/dL)	8.5±6.7	2.3 - 30.5	0.73±0.24	0.4–1.2	0
Serum bilirubin direct (mg/dL)	1.56±2.24	0.4 - 11.5	0.18±0.08	0.1 -0.4	0.002
Serum albumin(mg/dL)	2.9±0.39	2 - 3.4	4±0.5	3.1–4.8	0
Serum ALP (IU/L)	176±50.1	113–340	148±48.4	90–260	0.031
Serum ALT (IU/L)	222.7 ±182.38	23–740	20.7±8.7	8 to 38	< 0.001
Serum AST (IU/L)	176±154.4	24–598	19.4±8.3	9–42	< 0.001
Serum LDH (IU/L)	524.8±215.5	210–1071	313±61.2	218–456	< 0.001
Serum Urea (g/L)	86.7±47.8	29–194	31.2±12.9	12–54	< 0.001
Serum creatinin (mg/dL)	1.58±0.74	0.7–3.2	0.65±0.22	0.3– 1	< 0.001
Haemoglobin (g/dL)	10.2±2.3	7–15.1	14.8±1.6	12–17.4	< 0.001
HGF (ng/mL)	**10.09**±**4.84**	2.68 - 21.21	**0.52**±**0.34**	0.033 -1.015	< 0.001

[a] (IQR, Inter-quartile range; NYHA, New Y°rk Heart Association; JVP, jugular venous pressure; ECG, electrocardiogram; LV-EF, left ventricular ejection fraction; LV-FS, left ventricular fractional shortening).

Group III: Commercially available well defined growth factor cocktail of HGF (20 ng/ml) + FGF-4 (10 ng/ml) + Nicotinamide (0.61 gm/L) (in serum free IMDM, positive control group)

Group IV: FBS (10% in IMDM, negative control group of uninduced hMSCs)

This step was continued for a period of 7 days with fresh induction media replenished twice weekly. The differentiating cells were then treated with maturation medium, consisting of IMDM supplemented with 20 ng/mL oncostatin M, 1 μmol/L dexamethasone, and 50 mg/mL ITS^{+} premix (Insulin, Transferrin, selenium, linoleic acid) till day 28.

Characterization of hMSC-derived Hepatocytes. *Morphological changes of trans-differentiated cells:* During hepatic trans-differentiation of hMSCs changes in morphology of differentiating cells in various induction groups were observed at various time points by phase contrast microscopy (Olympus, CX-41).

Immunofluorescence Analysis of Hepatic Markers Expression. The expression of hepatocyte specific markers (ALB, CY-18 and AFP) was analyzed for both induced and uninduced cells after 28 days of induction. HepG2 cells were used as positive control and uninduced hMSCs as negative control. The hMSCs–derived hepatocytes fixed in 100% methanol for 10 minutes at −20°C, washed and incubated in 3% BSA for 1hr at RT. The cells were then incubated with primary antibodies raised in mice against human albumin (1:100) or cytokeratin-18 (1:50) or α–fetoprotein (1:200) overnight at 4°C. To confirm mesenchymal to epithelial transition happening during hepatic trans-differentiation of hMSCs, they were stained with primary antibody raised against human E-Cadherin raised in mice (a gift from Prof.

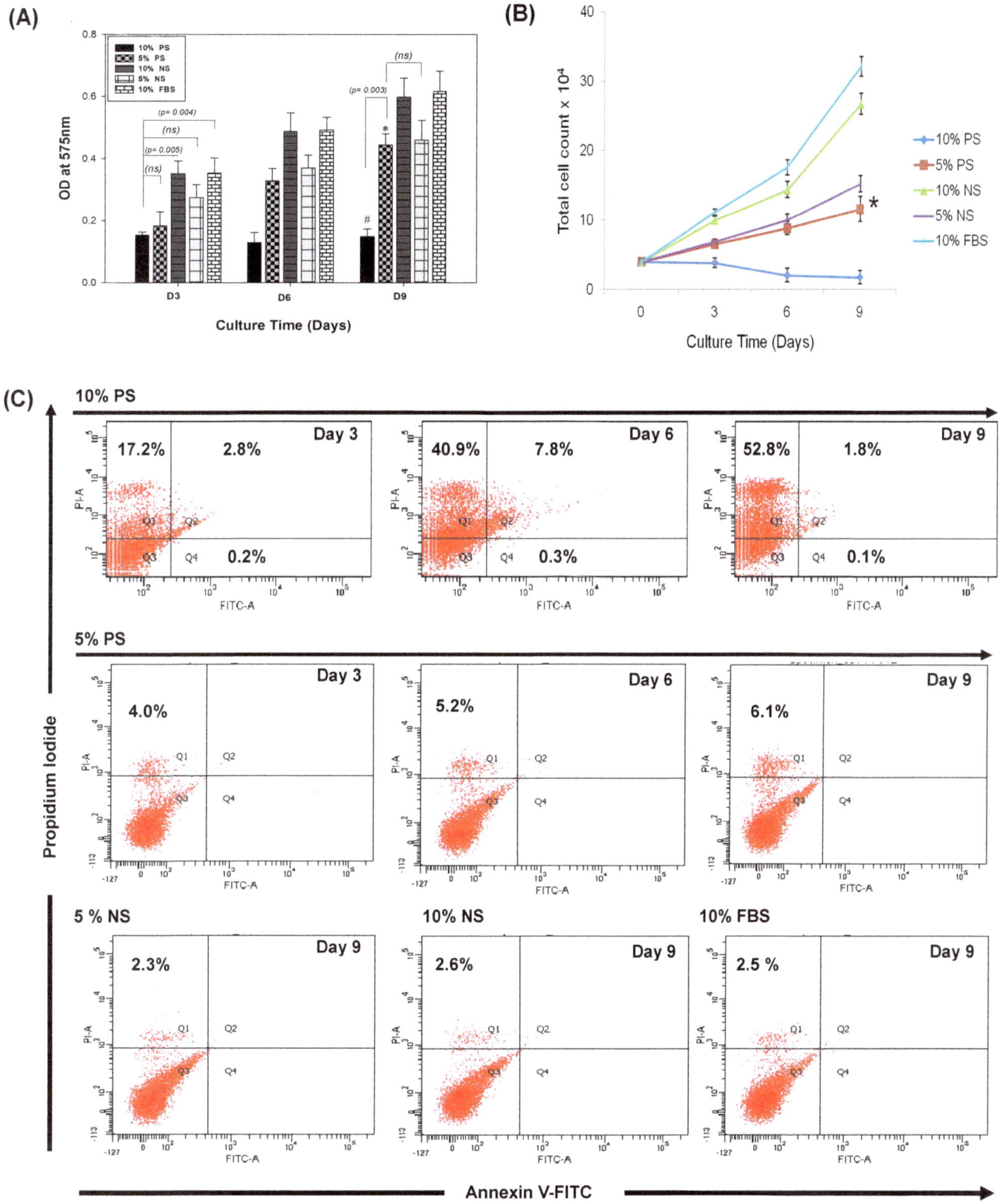

Figure 1. Cell proliferation assay for hMSCs in various hepatogenic culture conditions. (A) *In vitro* cytotoxicity assay showing MTT activity across 9 days of induction in different culture conditions. Error bars represents mean ± S.D. The differences in cell proliferation between the groups were considered statistically significant at $p<0.05$ and p-values are indicated on the graph. (*) indicates significant increase in cell proliferation at day 9 compared to day 3 and day 6 in 5% patient sera group. (#) indicates no significant difference in hMSC proliferation at different time points across 9 days of culture. (B) Cell proliferation assay by direct cell counting reveals higher proliferation rate in 10% FBS, which was comparable to that of 10% normal sera (NS) induction group. 10% patient sera (PS) induction group did not show any increase in cell proliferation. However, 5% PS group showed increase in proliferation, which was significantly more compared to 10% PS group (*$p<0.05$). (C) Cell death assay by annexinV-FITC/PI flow cytometric quantification revealed that 10% patient sera caused mostly necrotic cell death with increase in cell death across 9 days, whereas in 5% PS group cell death was comparably minimal. However, there was negligible cell death in control groups.

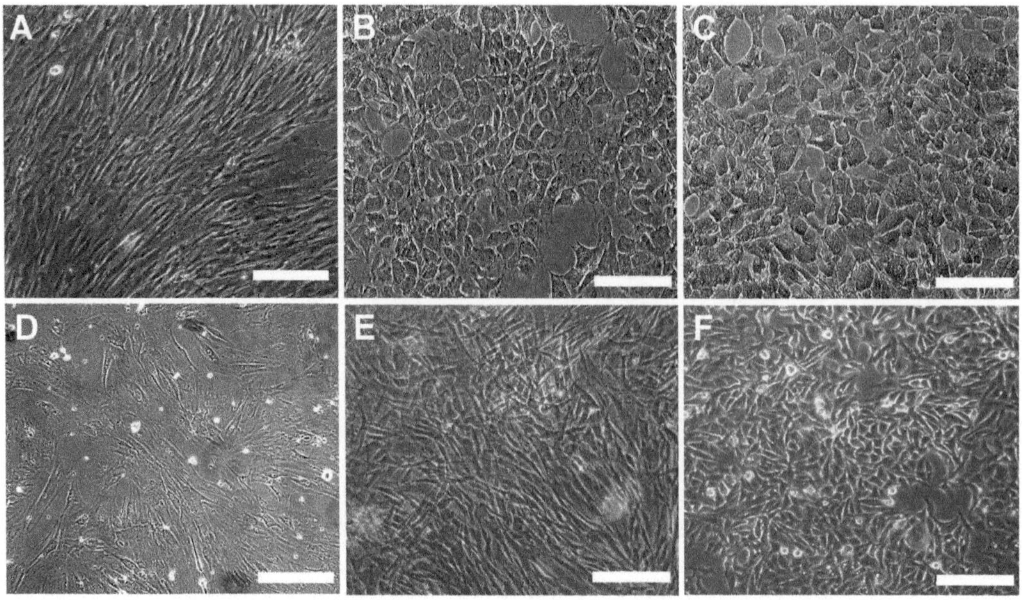

Figure 2. Phase contrast micrographs showing morphological changes observed during hepatic trans-differentiation of hMSCs. (A) hMSCs before hepatic induction, (B) a colony of hMSC derived hepatocytes generated by hepatogenic cocktail induction showing epitheloid morphology after 28 days, (C) polygonal hepatocyte-like colony obtained after induction with patient sera, (D) hMSCs grown in normal sera showing undifferentiated, broad, fibroblastic cells, (E) Untreated hMSCs grown in 10% FBS, showing fibroblastic morphology and (F) an epitheloid colony of HepG2 cells. (Original magnifications × 100) (Scale bar = 100 μM).

Jonathan M. G. Higgins, Harvard Medical School, USA). Next day, cells were washed and incubated with secondary anti-mouse Alexafluor 488 (for albumin and cytokeratin-18) (1:200) or Alexafluor 594 (for α-fetoprotein and E-Cadherin) (1:200) for 1 hr in dark and counterstained with nuclear stain (Hoechst 33342) (1:250). The stained cells were observed under a fluorescent microscope (Nikon TiE) and images were captured.

RNA Isolation and Real time-qPCR (RT-qPCR) Analysis for Hepatic Genes Expression.. Total RNA was extracted from hMSCs-derived hepatocytes under various culture conditions, HepG2 cells and untreated hMSCs using TRI Reagent (Sigma) as per manufacturer's protocol and quality and quantity of RNA was measured spectrophotometrically. For cDNA synthesis, 5 μg of total RNA was transcribed with MMLV-reverse transcriptase (Sigma). Quantitative real time-PCR for early hepatic genes such as AFP, HNF-4α and late hepatic genes such as CY-18 and ALB were carried out with Mastercycler ep *realplex* (Eppendorf, Germany) using SYBR Green JumpStart TaqReady-Mix (Sigma). Assessment of mesenchymal-to-epithelial transition during hepatic differentiation of hMSCs was done by evaluating changes in expression of mesenchymal marker (SNAIL) and epithelial marker (CDH1, E-Cadherin). The primer sequence details are specified in Table S2. The experiment was performed in a 20 μl reaction mixture of 2 × SYBR Green PCR Master Mix, 30 ng template cDNA and 5 pmol forward and reverse primers, for a total of 40 cycles with annealing temperature of 60°C. Relative quantification of gene expression in unknown samples was performed using the comparative C_T ($2^{-\Delta\Delta CT}$) method by normalizing with a housekeeping gene (glyceraldehyde-3-phosphate dehydrogenase) and untreated hMSCs were used as the calibrator. All reactions were performed in triplicate for statistical validation.

Flow Cytometry Analysis of hMSCs-derived Hepatocytes. To evaluate the efficiency of hepatic trans-differentiation, the total number of ALB^+ cells in the hMSCs-derived hepatocytes and non-differentiating cell populations were quantified by flow cytometry. HepG2 cells were used as positive control and untreated hMSCs as negative control. Briefly, the cells after 28 days of induction were trypsinized and fixed with 4% para-formaldehyde (PFA) solution. Approximately, 10^6 cells were aliquoted into FACS tube and permeabilized with 0.3% Triton-X, blocked with 3% BSA for 1 hr and incubated with primary monoclonal antibody against human albumin raised in mouse (Sigma) (1:100) for 1hr at RT. The cells were washed and incubated further with FITC-conjugated anti-mouse secondary antibody (Sigma) (1:400) for 1hr in dark. Similar procedure was adopted for quantifying endoglin$^+$ (CD105) cells in the hMSCs-derived hepatocytes to crosscheck the status of undifferentiated cells. The stained cells were quantified by a flow cytometer (BDFACS Calibur) with 10000 events recorded for each sample and compared with the isotype control (human IgG/IgM).

Functional Characterization of Trans-differentiated Hepatocytes. Liver function tests were carried out to validate the functionality of hMSCs-derived hepatocytes. The amount of albumin secreted in the culture media collected at various time points (7th day, 14th day, 21st day and 28th day) was determined using human albumin ELISA quantitation set (Bethyl Laboratory, Montgomery, TX, USA) according to the manufacturer's instructions. Since patient sera and control sera used in the conditioned media already contains albumin, its concentration was normalized and only the secreted albumin concentration was taken into account. The presence of insoluble stored glycogen content in the trans-differentiated cells was assessed by per-iodic acid Schiff's (PAS) staining, for which the cells at day 28 were fixed with 4% PFA for 10 minutes. It was then oxidized with 1% per-iodic acid for 5 min, washed and incubated with Schiff's reagent for 15 min. The cells were washed and counter stained with Mayer's hematoxylin and observed under light microscope. Uninduced hMSCs were used as negative control. Urea concentrations in the hMSCs derived hepatocytes in different culture

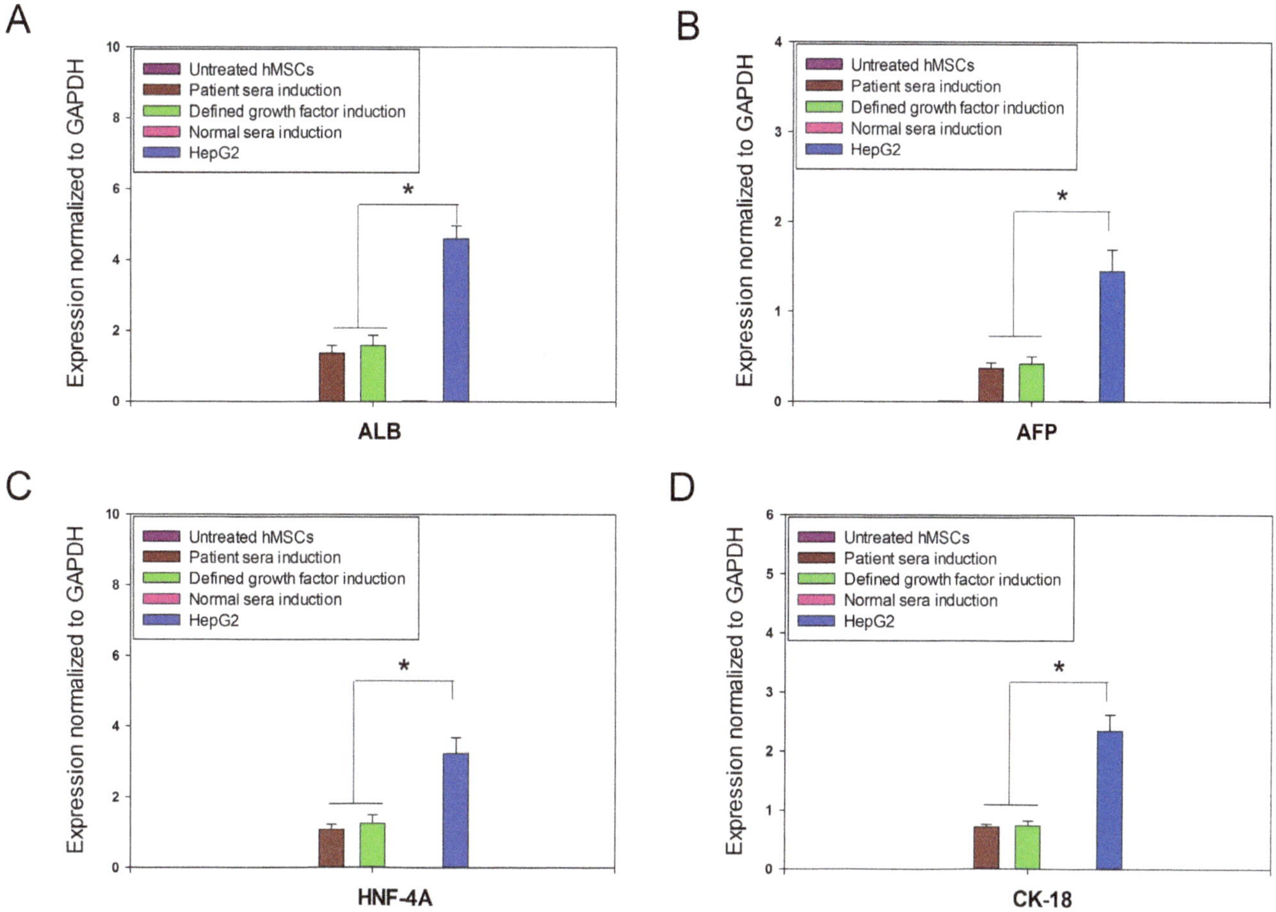

Figure 3. Quantitative real time PCR expression of hepatocyte specific genes. Positive expression of ALB, AFP, HNF-4α and CK18 were observed in patient sera induction as well as hepatogenic cocktail induction group, compared to untreated hMSCs and normal sera induction group. No differences in expressions were observed between the two positive induction groups, but it was significantly lesser compared to HepG2 (**p*<0.05). The expressions of genes were normalized to GAPDH and mRNA of untreated hMSCs was used as calibrator.

conditions and in controls were measured spectrophotometrically (Urea assay kit, Sigma) as per manufacturer's instruction. To assess the presence of an inducible cytochrome P450 (CYP) system in the trans-differentiated hepatocytes, cells were induced with phenobarbital (1 mM) for 3 days and cytochrome P450-dependent pentoxyresorufin o-dealkylase activity (PROD) was performed. PROD assay without phenobarbital induction was used as experimental control. Fluorescence detection of PROD metabolism in the trans-differentiated hepatocytes and control cells were observed under microscope, which determined the CYP activity. Phenobarbital induced PROD activity (resorufin fluorescence) was quantified spectrophotometrically at excitation and emission wavelengths of 530 and 590 nm respectively [37].

Statistical Analysis

The clinical data were presented as mean ± SD along with inter quartile range (IQR) or percentage (%), whichever is applicable. Continuous clinical variables were compared between the two groups by a two-sided unpaired t-test, whereas laboratory findings as well as HGF level was compared using Mann-Whitney U test, as the data were not normally distributed. The correlation between HGF level and the total bilirubin level was measured using the Spearman rank coefficient as the relationship was not linear. All the experiments were done in triplicate for statistical validation and reproducibility was assessed by repeating each experiment at least three times. Statistical differences among various groups were analysed using one- or two-way analysis of variance (ANOVA) followed by Tukey's post hoc test with statistical package for social sciences (IBM SPSS statistics, version 19, NY, USA). Differences were considered statistically significant at $p \leq 0.05$.

Results

Clinical and Biochemical Profiles of Patients

Clinical Parameters. The baseline clinical characteristics and demographics of the patient and control groups are shown in Table 1. Among the patient group, 40.7% of subjects were presented with a history of ischemic heart disease (IHD) that includes most cases of coronary artery disease (double vessel disease and triple vessel disease) and ischemic cardiomyopathy. Around 66.6% of patients were having valvular heart diseases (VHD), represented mostly by rheumatic heart disease, severe mitral stenosis or mitral regurgitations. Remaining patients were either affected by dilated cardiomyopathy (DCM) (18.5%) or congenital heart diseases (CHD) (29.6%). A progressive deterioration of the cardiac function (chest pain, dyspnoea, ankle edema) was noted shortly before admission in 17 patients. Chest X-ray

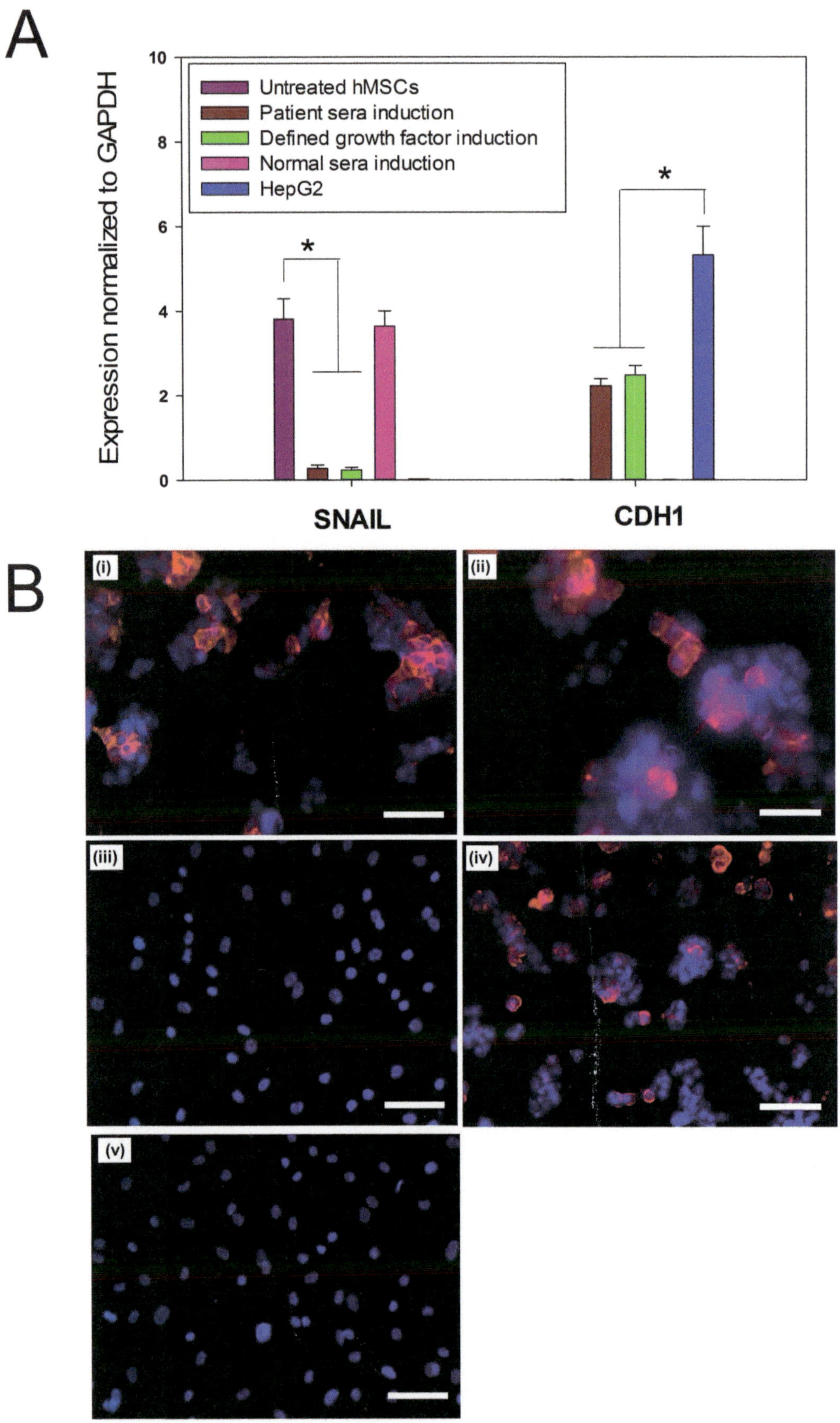
A
Expression normalized to GAPDH
10
8
6
4
2
0
Untreated hMSCs
Patient sera induction
Defined growth factor induction
Normal sera induction
HepG2
*
*
SNAIL
CDH1
B
(i)
(ii)
(iii)
(iv)
(v)

Figure 4. Evaluation of mesenchymal-to-epithelial transition during hepatic trans-differentiation of hMSCs. (A) Real time qPCR analysis for relative quantification of Snail (a mesenchymal marker) and CDH 1 (an epithelial marker) showed that hMSCs-derived hepatocytes in patient sera induction group and growth factor cocktail induction group with a higher expression of CDH1 had low Snail expression. In contrast, cells in normal sera induction group and untreated hMSCs had high expression levels of Snail with very low or no CDH 1 expression. HepG2 cells with high CHD 1 expression had no Snail expression. Statistically significant difference in CDH1 expression was observed between hMSC-derived hepatocytes and HepG2 cells (*$p<0.05$). Similarly, the difference in Snail expression was statistically significant between hMSCs-derived hepatocytes and untreated hMSCs or normal sera treated hMSCs. (B) Immunofluorescence studies showed positive expression of E-Cadherin in hMSCs derived hepatocytes in growth factor cocktail induction group (i) and patient sera induction group (ii). Normal sera treated hMSCs (iii) and untreated hMSCs (v) were negative for E-Cadherin expression. HepG2 cells were used as positive control (iv). (Original magnifications × 100) (Scale bar = 100 μM).

revealed that 5 patients had symptoms of cardiomegaly and three patients had levocardia and boot-shaped heart. Half of the patients were presented with severe CHF (chronic heart failure) (NYHA class III–IV, 51.85%) out of which five patients (18.5%) died of cardiac-failure-driven multi-organ failure with septic shock during hospitalization. The potential risk factors involved Type II diabetes mellitus, hypertension and hyperlipidaemia. Congestive hepatomegaly with ascites was observed in 7 patients, who were having no history of liver disease. Both the patient sera and control sera were complete microbe free and endotoxin free. Significant differences were observed in the laboratory findings (both haematological and biochemical parameters for liver function test) as well as clinical parameters (BP, heart rate and co-morbidities) between the patients and the control group.

Liver Function Tests. Significant differences were observed in the serum concentrations of liver enzymes, bilirubin (total and direct), urea, creatinine, haemoglobin and HGF between the patient and the control group ($p<0.05$). Hyperbilirubinemia, a feature hepatic dysfunction, was hallmark of almost all patients in this study. Nine patients were presented with total bilirubin level of ≥10 mg/dL with the concentration as high as 30.5 mg/dL and remaining patients had 3–10 mg/dL. Almost all the patients were having hypoalbuminemia, which is a feature of hepatic dysfunction. Based upon liver enzyme patterns in patient sera as well as bilirubin level, the patient group with hepatic abnormality was grouped as either having a hepatocellular profile (n = 19) or a cholestatic profile (n = 8). The amount of HGF in patient sera was significantly higher compared to those of control group ($p<0.001$). Within the patient group, higher concentration of HGF was observed in patients with IHD, which was not significantly different from other patients having VHD, DCM or CHD ($p = 0.649$) (Table S1). However, a strong and significant correlation was observed between HGF level and the total bilirubin level ($r = 0.854$, $p<0.001$) as assessed by the non-parametric Spearman rank correlation test (Figure S1).

Hepatic Trans-differentiation of hMSCs in Hepatogenic Conditioned Culture System

Cell Proliferation Assay for hMSCs in various Culture Conditions. MTT assay revealed no significant difference in hMSCs grown in 10% patient sera and 5% patient sera ($p = 0.505$) as well as between 10% patient sera and 5% normal sera ($p = 0.175$) on day 3 (Figure 1A). However, differences were observed in hMSCs proliferation between 10% patient sera and 10% normal sera as well as with 10% FBS on day 3. As the culture time progressed, culture condition using 10% patient sera was

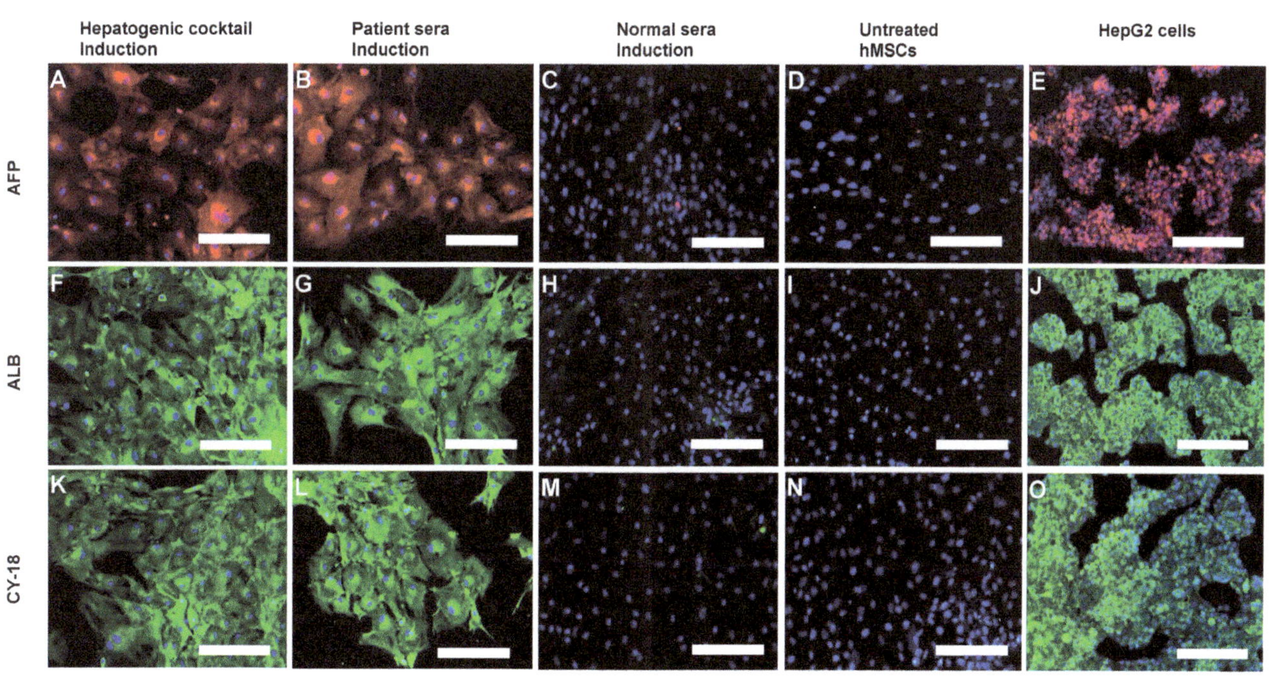

Figure 5. Immunofluorescence analysis of hepatic markers. Immunofluorescence micrographs (merged) showing positive expression of hepatocyte specific markers, AFP (Alexa Fluor 594: red colour), ALB (Alexafluor 488: green colour) and CY-18 (Alexafluor 488: green colour) in hMSCs-derived hepatocyte-like colonies after 28 days of differentiation and negative expression in undifferentiated hMSCs. HepG2 cells were used as positive control. Nuclei were stained with DAPI (blue colour). (Original magnifications × 100) (Scale bars = 100 μM).

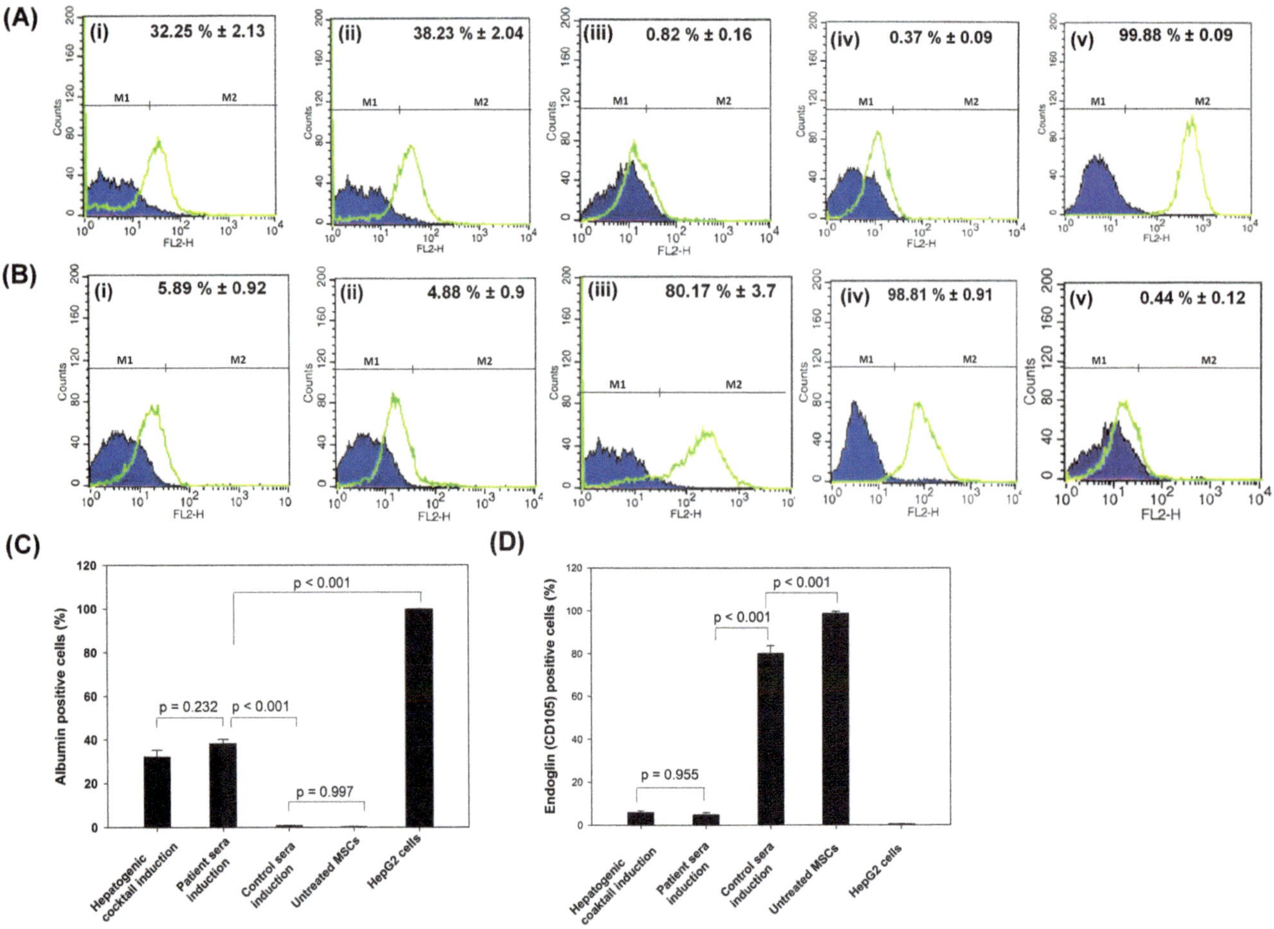

Figure 6. Assessment of hepatic trans-differentiation efficiency of hMSCs. Representative flow cytometry data showing average number of albumin positive cells (A) and endoglin (CD105) positive cells (B) after hepatic trans-differentiation of hMSCs in hepatogenic cocktail induction group (i), patient sera induction group (ii) and normal sera induction group (iii). Untreated hMSCs grown in FBS (iv) and HepG2 cells (v) were used as negative and positive control respectively. Histograms showing isotype control (blue area, quantified by M1) and the positive expression of either albumin or endoglin (green area, quantified by M2). Results are representative of three different experiments. Comparison of the total number of albumin positive cells (C) and endoglin positive cells (D) in different hepatogenic culture conditions as well as controls represents the trans-differentiation efficiency. Differences in cell number between the groups were considered statistically significant at $p < 0.05$ and p-values are indicated on the graph.

found to be cytotoxic, as it did not support further proliferation of hMSCs across 9 days. It was further confirmed by performing direct cell counting under different culture conditions (Figure 1B). Except, 10% patient sera induction group, all other culture conditions, including 5% patient sera induction group, supported significant proliferation ($p<0.05$) over 9 days of culture. No significant difference was observed between cells grown in 5% patient sera induction group and 5% normal sera group. Further, cell death assay was performed by annexin V-FITC/PI flow cytometric assay (Figure 1C), which showed that 10% patient sera caused mostly necrotic cell death revealed by more of propidium iodide (PI)-positive cells and as the day progressed the amount of cell death increased significantly ($p<0.05$). Although there was meagre cell death in 5% patient sera, it was negligible compared to 10% patient sera. From the above findings, we inferred that culture condition containing 10% patient sera was cytotoxic and 5% sera suitably allowed hMSCs proliferation, comparable to that of control sera.

Morphological Features of hMSCs-derived Hepatocytes. In patient sera induction as well as hepatogenic growth factor cocktail induction group, the process of trans-differentiation was evident by increased nucleo-cytoplasmic ratio, accumulations of cytoplasmic granules and most prominently change in fibroblastic morphology of hMSC to cuboidal, epitheloid morphology of hepatocytes (Figure 2). Islands of compact, round or polygonal-shaped cells (more than 85%) surrounded by spindle-shaped cells with retracting ends and with dense granular cytoplasm were observed during maturation step (Figure 2B and 2C). Approximately, 16±4 hepatocyte-like colonies were formed per 1.5×10^4 hMSCs in hepatogenic growth factor induction group, whereas 13±5 hepatocyte-like colonies were formed in patient sera induction group after day 28. During the process of trans-differentiation in these two groups, the cell proliferation rate declined as the hMSCs underwent mesenchymal–epithelial transition confirming hepatic specification. However, cells grown in normal control serum as well as in 10% FBS did not show any morphological changes and kept on proliferating (Figure 2D and 2E). HepG2 cells showing epitheloid colonies were used as control (Figure 2F).

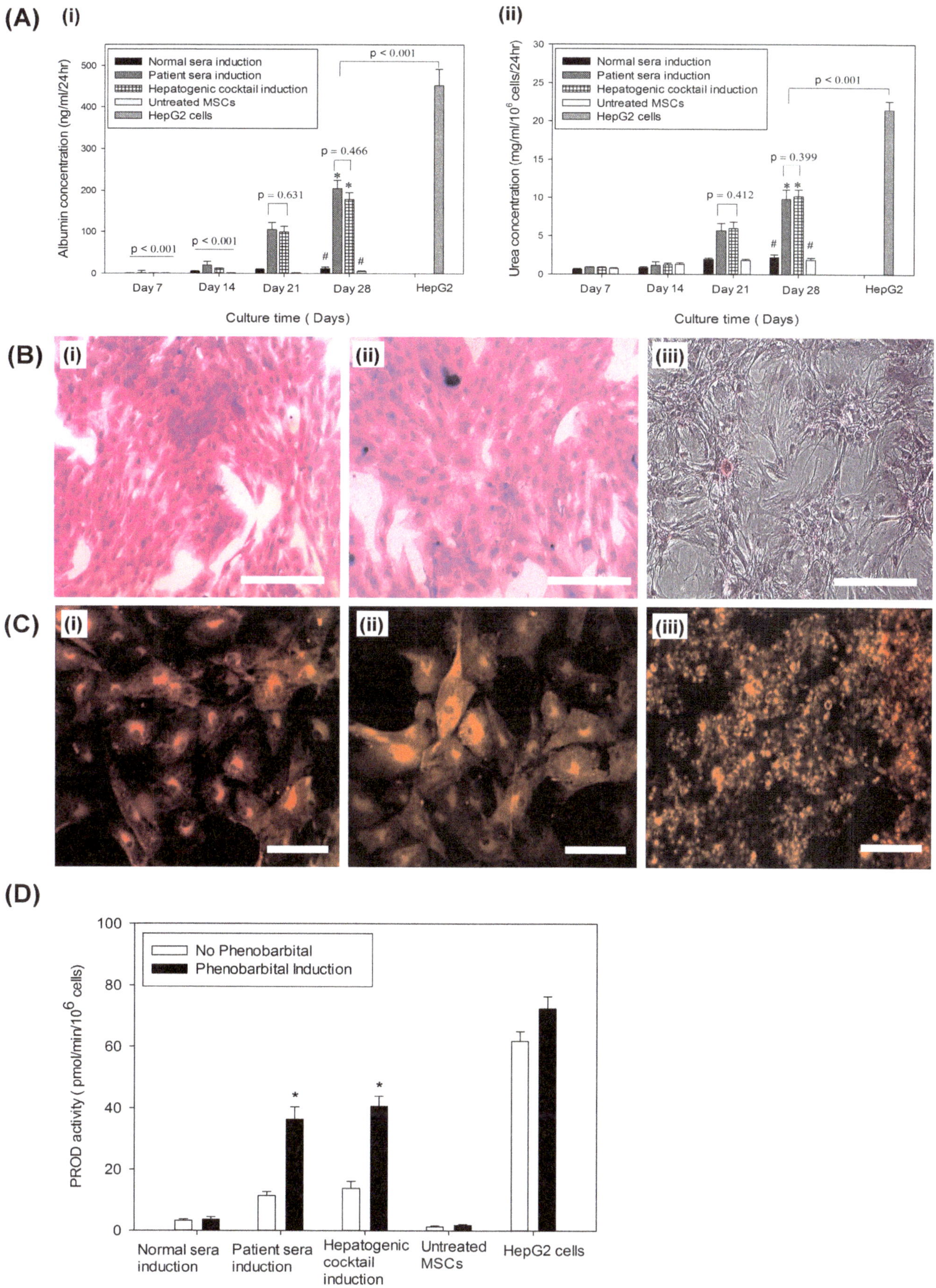
(A) (i)
Normal sera induction
Patient sera induction
Hepatogenic cocktail induction
Untreated MSCs
HepG2 cells
Albumin concentration (ng/ml/24hr)
p < 0.001
p < 0.001
p = 0.631
p = 0.466
p < 0.001
Day 7
Day 14
Day 21
Day 28
HepG2
Culture time (Days)
(ii)
Urea concentration (mg/ml/10^6 cells/24hr)
p = 0.412
p = 0.399
p < 0.001
(B) (i) (ii) (iii)
(C) (i) (ii) (iii)
(D)
No Phenobarbital
Phenobarbital Induction
PROD activity (pmol/min/10^6 cells)
Normal sera induction
Patient sera induction
Hepatogenic cocktail induction
Untreated MSCs
HepG2 cells

Figure 7. Functional characterization of hMSCs-derived hepatocytes. (Ai) Albumin ELISA of the culture media collected from various hepatic induction groups and controls. Differences in albumin concentration between the groups were considered statistically significant at $p<0.05$ and p-values are indicated on the graph. (*) indicates significant increase in albumin release by hepatocytes in patient sera induction group as well as cocktail induction group at day 28 compared to their day 21 counterparts. (#) represents no significant difference in albumin concentration in undifferentiated hMSCs of normal sera induction group and untreated hMSCs at 28 compared to earlier days. (Aii) Urea assay showed higher concentration in hMSCs derived hepatocytes compared to controls on day 28, but was less than that of HepG2 cells. (B) Glycogen accumulation was represented by PAS staining, which showed positive expression (magenta color) in hMSCs-derived hepatocyte colonies of patient sera induction group (i) and defined growth factor induction group (ii). Normal sera treated hMSCs were PAS negative (iii). (Original magnifications, ×100) (Scale bars = 100 μM). (C) PROD Assay for phenobarbital inducible cytochrome P450 was confirmed by positive expression (red fluorescence) in hMSC-derived hepatocytes in patient sera induction group (i) and growth factor cocktail induction group (ii). HepG2 cells were used as positive control (iii). (Original magnifications × 100) (Scale bars = 100 μM). (D) Spectrofluorimetric quantification of PROD activity revealed a significant increase in fluorescence intensity after phenobarbital induction in hMSC-derived hepatocytes in patient sera induction group and defined growth factor induction group (* $p<0.05$). However, negligible PROD activity was observed in untreated hMSCs and normal sera treated hMSCs even after phenobarbital induction. HepG2 cells with higher level of fluorescence activity compared to hMSC-derived hepatocytes did not show significant elevation in PROD activity after phenobarbital induction.

Real time qPCR Analysis of Hepatocyte Specific Markers in hMSCs- derived Hepatocytes. Relative quantification of hepatocyte specific genes (AFP, ALB, HNF-4α, and CY-18) by qRT-PCR analysis showed higher expression levels in hMSC-derived hepatocytes in patient sera treatment group as well as defined growth factor treatment group (Figure 3), with no expression observed in untreated hMSCs and normal sera treatment group. Although the difference in expression levels of these hepatic markers was not significant between hMSC-derived hepatocytes in both positive induction groups, it was significantly lower than that of HepG2 cells ($p<0.05$). To confirm the mesenchymal-epithelial transition during hepatic trans-differentiation of hMSCs, we quantified relative expression levels of Snail and CDH 1 mRNA by qRT-PCR. The results revealed significantly higher expression level of Snail in untreated hMSCs and normal sera treated hMSCs compared to hMSCs-derived hepatocytes in patient sera induction group and defined growth factor induction group ($p<0.05$) (Figure 4A). Snail was not detectable in HepG2 cells. In contrast, CDH 1 expression was observed in hMSCs-derived hepatocytes in patient sera induction group and defined growth factor induction group, which was significantly lesser than that of HepG2 cells ($p<0.05$). Untreated hMSCs and normal sera treated hMSCs showed no expression of CDH1.

Immunofluorescence and Flow Cytometry Analysis of Hepatic markers. Strongly positive expression of all three hepatocyte markers (AFP, ALB and CY-18) was observed in hMSCs-derived hepatocytes in patient sera induction group and hepatogenic cocktail induction group, which was comparable to that of HepG2 cells (Figure 5). However, expression of these hepatocytes-specific markers in untreated hMSCs as well as those grown in normal sera was negligible. Expression of E-Cadherin was observed in hMSCs derived-hepatocytes in both patient sera induction group and growth factor cocktail induction group (Figure 4B), but not in normal sera induction group and untreated hMSCs. Flow cytometric analysis showed that 32±2.13% cells were albumin positive in hepatogenic cocktail group (Figure 6Ai) and 38.23±2.05% (Figure 6Aii) cells in patient sera induction group and the difference between them was insignificant (Figure 6C). The hMSCs grown in normal sera and untreated hMSCs were negligibly albumin-positive, accounting for 0.82% and 0.37% of total cells respectively (Figure 6Aiii and 6Aiv). Since, hepatic transdifferentiation efficiency was only 30–35%, expression of endoglin (CD105), a cell surface marker of undifferentiated mesenchymal stem cell, was further assessed. In hepatocytes observed, Approximately, 5.89% and 4.88% endoglin-positive cells were observed in hepatogenic cocktail induction group and patient sera induction group respectively (Figure 6Bi and Bii), and difference between them was statistically insignificant (Figure 6D). HepG2 cells, an endoglin-negative human hepatoma cell line [38], showed only 0.44% endoglin expression (Figure 6Bv), which was significantly different from that of hepatocytes in patient sera induction group ($p = 0.077$), but not from hepatogenic cocktail induction group ($p = 0.026$). However, hMSCs grown in normal sera as well as untreated hMSCs showed 80.17% and 98.81% endoglin-positive cells, after 28 days of culture (Figure 6Biii and 6Biv).

Liver Function Tests for hMSC-derived Hepatocytes. Albumin release was negligible during first two weeks, which increased significantly on day 21 in patient sera induction group and growth factor induction group compared to normal sera induction group and untreated MSCs ($p<0.05$) (Figure 7Ai). Further increase in albumin release was observed on day 28 compared to day 21 in both patient sera induction group and growth factor induction group, although difference between them was insignificant ($p = 0.466$). Urea assay also reveals that hMSCs derived hepatocytes in both patient sera induction group and growth factor cocktail induction group had higher concentration on day 28 compared to controls, but no significant difference was observed between them ($p<$ 0.05) (Figure 7Aii). Similarly, accumulation of intracellular glycogen was evident by positive PAS staining in hepatocyte colonies obtained in patient sera induction group and growth factor treatment group after day 28, whereas normal sera treated hMSCs were PAS negative (Figure 7B). The number of PAS-positive cells in patient sera induction group was 36±7% compared to 31±3% in hepatogenic growth factor induction group, exhibiting almost similar pattern of maturation. The liver function tests of hMSCs-derived hepatocytes were consistent with the morphological changes and gene expression study. To assess whether the hMSCs-derived hepatocytes in patient sera induction group and growth factor induction group possess inducible cytochrome P450 variant for drug metabolism, PROD assay was carried out before and after phenobarbital induction. Higher number of PROD positive hepatocytes were observed in patient sera induction group and growth factor induction group under fluorescent microscope (Figure 7C), whereas very low or no PROD activity was observed in untreated hMSCs and normal sera induction group (data not shown). Spectrofluorimetric quantification of PROD activity showed a significant increase in fluorescence intensity after phenobarbital induction in hMSCs-derived hepatocytes in patient sera induction group and growth factor cocktail induction group ($p<0.05$) (Figure 7D), revealing presence of inducible cytochrome P450 system. Although HepG2 cells had higher level of resorufin fluorescence compared to hMSCs-derived hepatocytes, it was not elevated significantly after phenobarbital induction, may be due to drug insensitivity. No increase in PROD activity was observed in untreated hMSCs and normal sera treated hMSCs even after phenobarbital induction.

Discussion

In this study, severe liver function abnormalities were observed in a well-defined cohort of patients with various cardiac diseases leading to chronic heart failure. The LFTs are mostly of cholestatic pattern (70.3% of patients) with severe increase in bilirubin and ALP levels (Table 1), which corroborates with earlier findings [4,39]. Remaining patients showed hepatocellular LFTs represented by elevated serum ALT and AST levels and hyperbilirubinemia. Such cardiac hyperbilirubinemia has been associated with poor outcome in chronic cardiac failures [40,41], which may arise due to hepatocellular damage from decreased perfusion, cholestatic liver damage from increased central venous pressure (CVP), sluggish flow of bile, inspissated bile and bile duct calculi and low left ventricular ejection fraction (LV-EF) leading to venous congestion [4,5,42]. Interestingly, high level of bilirubin is associated with a decreased risk of coronary heart disease (CHD) and cardiovascular disease (CVD) [43]. While the full spectrum by which bilirubin acts to protect against CVD is not fully understood, earlier reports suggests that bilirubin protects against oxidative stress by acting as an anti-oxidant and free radical scavenger with anti-atherogenic properties [44]. In this study, the increased level of this component serves as a potent prognostic biomarker for complex cardiovascular diseases like IHD, VHD, CHD and DCM. Sera obtained from these patients also exhibited increased level of HGF, as reported in many other liver diseases such as hepatitis, cirrhosis and hepatic failure [29,30,45]. HGF promotes liver regeneration and prevent hepatic fibrosis caused due to tissue damage [46]. Moreover, a strong correlation was observed between the total bilirubin and the HGF levels in the present study, which signifies that elevation in serum HGF level was directly related to the severity of cardiovascular disease. Besides the liver, serum HGF elevation occurs during the early phases of both acute myocardial infarction (AMI) and heart failure (HF) [31,47,48]. Therefore, it is important to determine the sources of HGF in this study, as both the heart and liver showed abnormal function and would have released HGF upon damage. It is known that enhanced level of HGF in serum may be either due to low hepatic clearance or elevated synthesis of HGF [49]. The elevated serum HGF level in this study may be caused by either the attempt to regenerate new hepatocytes in damaged liver, or the decrease rate of clearance by liver, which require further studies for confirmation. Since there is marked increase in serum HGF level in this study, we hypothesize that HGF may play a key role as a prognostic biomarker of cardiac failure-associated jaundice (hyperbilirubinemia).

Cardiac-failure-associated jaundiced sera containing higher level of HGF was optimized as a hepatogenic conditioned culture system for hMSCs trans-differentiation. Our initial findings demonstrate that 10% sera were cytotoxic, while 5% sera suitably supported proliferation as well as hepatic trans-differentiation of hMSCs, which corroborates earlier reports [24,26–28,50]. It has also been reported that conditional selective medium containing cholestatic serum provides selective proliferative signals for hepatocytes, simultaneously eliminating non-hepatocyte-like cells [26,50]. Interestingly, HGF is known to have potential anti-necrotic and anti-apoptotic effect on hepatocytes in cholestatic jaundice and rescues them from damage [51]. Since the cardiac jaundiced sera possess higher concentration of HGF, we assume that 5% sera induced hepatic differentiation of hMSCs with no significant effect on cell viability. Although, 10% sera also contain higher concentration of HGF, its anti-apoptotic and anti-necrotic effects would have been suppressed by the higher concentration of bilirubin and bile acids [52,53]. Nevertheless, high serum level of HGF released from damaged heart/liver in such cases could effectively potentiate hepatic induction, comparable to that of known recombinant hepatogenic factors [20–22].

Quantitative gene expression studies as well as immunofluorescence analysis revealed that hMSCs-derived hepatocytes expressed mature markers like albumin and cytokeratin-18 by 28 days, which were not detected by day 14. The hMSCs treated with jaundiced sera generated approximately 38% albumin-positive cells, which was comparable to 32% in case of sequential induction with known hepatogenic growth factors and cytokines. Our data were almost consistent with earlier reports of hepatic trans-differentiation of commercially available hMSCs [54]. However, we could not achieve albumin positive cells as high as $\geq 81\%$ in case of ESCs/iPSCs-derived hepatocytes [55]. This decrease in hepatic trans-differentiation efficiency of hMSCs may be due to mesenchymal-to-epithelial transition followed by hMSCs compared to highly efficient hepatic differentiation of ESCs/iPSCs mimicking normal liver development process [56]. This phenomenon of mesenchymal-to-epithelial transition was confirmed by decreased expression of Snail, a mesenchymal marker and increased expression of CDH1, an epithelial marker in the hMSCs-derived hepatocytes in patient sera and cocktail induction groups. Since the trans-differentiation efficiency was below 40%, the obvious question arises regarding the fate of albumin-negative cells as to whether they still possess the hMSC phenotype. In this regard, endoglin (CD105)-positive fractions were quantified in hepatocyte-like colonies, which represent as low as around 5% populations in both patient sera induction as well as cytokine induction group. This implies that hMSC phenotype was almost lost during hepatic differentiation and these alb^{-}/$endoglin^{-}$ cells might be the precursors of hepatocytes, which requires further characterization. The hMSCs-derived hepatocytes using patient sera in this study showed *in vitro* functional features characteristics of liver cells such as albumin synthesis, urea release, glycogen storage with phenobarbital-inducible cytochrome P450 activity. One major advantage of using patient sera over commercially available growth factors might be clinical transplantation of hMSCs-derived hepatocytes. It is because the commercially available growth factors are of bacterial origin and presence of endotoxin (as lipopolysaccharide) might make their clinical usage unsuitable. It is reported that endotoxins present in commercially growth factors (as low as 0.0005%) can have a negative impact on differentiation potential of stem cells [57]. Patient sera contain natural source of endotoxin free-HGF, which can be used for differentiating hMSCs towards generation of autologous hepatocytes and can be applied clinically in future.

In conclusion, our findings report that hMSCs can effectively trans-differentiate towards functional hepatocytes *in vitro* using a novel hepatogenic conditioned culture system containing sera from cardiac-failure-associated congestive/ischemic liver patients. These hMSCs-derived hepatocytes might be clinically relevant for autologous cellular therapy during end-stage liver failure as well as for *in vitro* toxicity testing.

Supporting Information

Figure S1 Correlation between serum HGF levels and total bilirubin level in cardiac failure-associated jaundiced patients and volunteer donors.

Table S1 Liver function abnormality and serum HGF levels in heart failure patients.

Table S2 The genes and primer sequences used in real-time qPCR.

Acknowledgments

We acknowledge Prof. Jonathan M. G. Higgins, Division of Rheumatology, Immunology and Allergy, Brigham and Women's Hospital, Harvard Medical School, Boston, Massachusetts, United States of America for kindly gifting us E-Cadherin antibody.

Author Contributions

Conceived and designed the experiments: DKB SG KMC RSV. Performed the experiments: DKB SM SG. Analyzed the data: DKB SM SG RSV. Contributed reagents/materials/analysis tools: SG KMC RSV. Wrote the paper: DKB SM SG RSV.

References

1. Williams R (2006) Global challenges in liver disease. Hepatology 44: 521–526.
2. van Deursen VM, Damman K, Hillege HL, van Beek AP, van Veldhuisen DJ, et al. (2010) Abnormal liver function in relation to hemodynamic profile in heart failure patients. J Card Fail 16: 84–90.
3. Maleki M, Vakilian F, Amin A (2011) Liver diseases in heart failure. Heart Asia 3: 143–149.
4. Lau GT, Tan HC, Kritharides L (2002) Type of liver dysfunction in heart failure and its relation to the severity of tricuspid regurgitation. Am J Cardiol 90: 1405–1409.
5. Giallourakis CC, Rosenberg PM, Friedman LS (2002) The liver in heart failure. Clin Liver Dis 6: 947–967.
6. Schuppan D, Afdhal NH (2008) Liver cirrhosis. Lancet 371: 838–851.
7. Gallegos-Orozco JF, Vargas HE (2009) Liver transplantation: from child to MELD. Med Clin North Am 93: 931–950.
8. Dalgetty DM, Medine CN, Iredale JP, Hay DC (2009) Progress and future challenges in stem cell-derived liver technologies. Am J Physiol Gastrointest Liver Physiol 297: G241–G248.
9. Zhao Q, Ren H, Zhu D, Han Z (2009) Stem/progenitor cells in liver injury repair and regeneration. Biol Cell 101: 557–571.
10. Zhao Y, Xu Y, Zhang B, Wu X, Xu F, et al. (2010) *In vivo* generation of thick, vascularized hepatic tissue from collagen hydrogel-based hepatic units. Tissue Eng Part C Methods 16: 653–659.
11. Herrera MB, Bruno S, Buttiglieri S, Tetta C, Gatti S, et al. (2006) Isolation and Characterization of a Stem Cell Population from Adult Human Liver. Stem Cells 24: 2840–2850.
12. Czyz J, Wiese C, Rolletschek A, Blyszczuk P, Cross M, et al. (2003) Potential of embryonic and adult stem cells *in vitro*. Biol Chem 384: 1391–1409.
13. Asgari S, Pournasr B, Salekdeh GH, Ghodsizadeh A, Ott M, et al. (2010) Induced pluripotent stem cells: A new era for hepatology. J Hepatol 53: 738–751.
14. Petersen BE, Bowen WC, Patrene KD, Mars WM, Sullivan AK, et al. (1999) Bone marrow as a potential source of hepatic oval cells. Science 284: 1168–1170.
15. Houlihan DD, Newsome PN (2008) Critical review of clinical trials of bone marrow stem cells in liver disease. Gastroenterology 135: 438–450.
16. Gilchrist ES, Plevris JN (2010) Bone marrow-derived stem cells in liver repair: 10 years down the line. Liver Transpl 16: 118–129.
17. Terai S, Ishikawa T, Omori K, Aoyama K, Marumoto Y, et al. (2006) Improved liver function in patients with liver cirrhosis after autologous bone marrow cell infusion therapy. Stem Cells 24: 2292–2298.
18. Mohamadnejad M, Alimoghaddam K, Mohyeddin-Bonab M, Bagheri M, Bashtar M, et al. (2007) Phase 1 trial of autologous bone marrow mesenchymal stem cell transplantation in patients with decompensated liver cirrhosis. Arch Iran Med 10: 459–466.
19. Kharaziha P, Hellstrom PM, Noorinayer B, Farzaneh F, Aghajani K, et al. (2009) Improvement of liver function in liver cirrhosis patients after autologous mesenchymal stem cell injection: a phase I-II clinical trial. Eur J Gastroenterol Hepatol 21: 1199–1205.
20. Schwartz RE, Reyes M, Koodie L, Jiang Y, Blackstad M, et al. (2002) Multipotent adult progenitor cells from bone marrow differentiate into functional hepatocyte-like cells. J Clin Invest 109: 1291–1302.
21. Lee KD, Kuo TKC, Whang-Peng J, Chung YF, Lin CT, et al. (2004) *In vitro* hepatic differentiation of human mesenchymal stem cells. Hepatology 40: 1275–1284.
22. Snykers S, Vanhaecke T, Papeleu P, Luttun A, Jiang Y, et al. (2006) Sequential exposure to cytokines reflecting embryogenesis: The key for *in vitro* differentiation of adult bone marrow stem cells into functional hepatocyte-like cells. Toxicol Sci 94: 330–341.
23. Wang PP, Wang JH, Yan ZP, Hu MY, Lau GK, et al. (2004) Expression of hepatocyte-like phenotypes in bone marrow stromal cells after HGF induction. Biochem Biophys Res Commun 320: 712–716.
24. Avital I, Inderbitzin D, Aoki T, Tyan DB, Cohen AH, et al. (2001) Isolation, Characterization, and Transplantation of Bone Marrow-Derived Hepatocyte Stem Cells. Biochem Biophys Res Commun 288: 156–164.
25. Yamazaki S, Miki K, Hasegawa K, Sata M, Takayama T, et al. (2003) Sera from liver failure patients and a demethylating agent stimulate transdifferentiation of murine bone marrow cells into hepatocytes in coculture with nonparenchymal liver cells. J Hepatol 39: 17–23.
26. Cai YF, Zhen ZJ, Min J, Fang TL, Chu ZH, et al. (2004) Selection, proliferation and differentiation of bone marrow-derived liver stem cells with a culture system containing cholestatic serum *in vitro*. World J Gastroenterol 10: 3308–3312.
27. Xu J, Deng X, Demetriou AA, Farkas DL, Hui T, et al. (2008) Factors released from cholestatic rat livers possibly involved in inducing bone marrow hepatic stem cell priming. Stem Cells Dev 17: 143–155.
28. Yang Y, Qu B, Huo JH, Wu SL, Zhang MY, et al. (2009) Serum from radiofrequency-injured livers induces differentiation of bone marrow stem cells into hepatocyte-like cells. J Surg Res 155: 18–24.
29. Shiota G, Okano J, Kawasaki H, Kawamoto T, Nakamura T (1995) Serum hepatocyte growth factor levels in liver diseases: clinical implications. Hepatology 21: 106–112.
30. Yamagamim H, Moriyama M, Matsumura H, Aoki H, Shimizu T, et al. (2002) Serum concentrations of human hepatocyte growth factor is a useful indicator for predicting the occurrence of hepatocellular carcinomas in C-viral chronic liver diseases. Cancer 95: 824–834.
31. Matsumori A, Furukawa Y, Hashimoto T, Ono K, Shioi T, et al. (1996) Increased circulating hepatocyte growth factor in the early stage of acute myocardial infarction. Biochem Biophys Res Commun 221: 391–395.
32. Nakamura Y, Morishita R, Nakamura S, Aoki M, Moriguchi A, et al. (1996) A vascular modulator, hepatocyte growth factor, is associated with systolic pressure. Hypertension 28: 409–413.
33. Zhu Y, Hojo Y, Ikeda U, Shimada K (2000) Production of hepatocyte growth factor during acute myocardial infarction. Heart 83: 450–455.
34. Ueno S, Ikeda U, Hojo Y, Arakawa H, Nonaka M, et al. (2001) Serum hepatocyte growth factor levels are increased in patients with congestive heart failure. J Card Fail 7: 329–334.
35. Rickham PP (1964) Human experimentation. Code of ethics of the world medical association. Declaration of helsinki. Br Med J 2: 177.
36. Chang CY, Schiano TD (2007) Drug hepatotoxicity. Aliment Pharmacol Ther 25: 1135–1151.
37. Snykers S, De Kock J, Tamara V, Rogiers V (2011) Hepatic differentiation of mesenchymal stem cells: *in vitro* strategies. Methods Mol Biol 698:305–314.
38. Rius C, Smith JD, Almendro N, Langa C, Botella LM, et al. (1998) Cloning of the promoter region of human endoglin, the target gene for hereditary hemorrhagic telangiectasia type 1. Blood 92: 4677–4690.
39. Poelzl G, Ess M, Mussner-Seeber C, Pachinger O, Frick M, et al. (2012) Liver dysfunction in chronic heart failure: prevalence, characteristics and prognostic significance. Eur J Clin Invest 42: 153–163.
40. Allen LA, Felker GM, Pocock S, McMurray JJV, Pfeffer MA, et al. (2009) Liver function abnormalities and outcome in patients with chronic heart failure: data from the Candesartan in heart failure: assessment of reduction in mortality and morbidity (CHARM) program. Eur J Heart Fail 11: 170–177.
41. Szygula-Jurkiewicz B, Wojnicz R, Lekston A, Duszanska A, Spinczyk B, et al. (2007) Effect of elevated bilirubin levels on the long-term outcome in patients with chronic heart failure due to hypertension. Pol Arch Med Wewn 117: 227–233.
42. Sherlock S (1951) The liver in heart failure; relation of anatomical, functional, and circulatory changes. Br Heart J 13: 273–293.
43. Lin JP, O'Donnell CJ, Schwaiger JP, Cupples LA, Lingenhel A, et al. (2006) Association between the UGT1A1*28 allele, bilirubin Levels, and coronary heart disease in the Framingham heart study. Circulation 114: 1476–1481.
44. Stocker R (2004) Antioxidant activities of bile pigments. Antioxid Redox Signal 6: 841–849.
45. Ueno S, Tanabe G, Kawaida K, Hamanoue M, Mitsue S, et al. (1997) Serum hepatocyte growth factor (HGF) levels predict the outcomes in hepatectomized patients with postoperative hyperbilirubinemia. Int Hepatol Commun 6: 294–299.
46. Nakamura T, Nishizawa T, Hagiya M, Seki T, Shimonishi M, et al. (1989) Molecular cloning and expression of human hepatocyte growth factor. Nature 342: 440–443.
47. Lamblin N, Susen S, Dagorn J, Mouquet F, Jude B, et al. (2005) Prognostic significance of circulating levels of angiogenic cytokines in patients with congestive heart failure. Am Heart J 150: 137–143.

48. Rychli K, Richter B, Hohensinner PJ, Mahdy Ali K, Neuhold S, et al. (2011) Hepatocyte growth factor is a strong predictor of mortality in patients with advanced heart failure. Heart 97: 1158–1163.
49. Funakoshi H, Nakamura T (2003) Hepatocyte growth factor: from diagnosis to clinical applications. Clin Chim Acta 327: 1–23.
50. Min J, Shang CZ, Chen YJ, Zhang L, Liu L, et al. (2007) Selective enrichment of hepatocytes from mouse embryonic stem cells with a culture system containing cholestatic serum. Acta Pharmacol Sin 28: 1931–1937.
51. Li Z, Mizuno S, Nakamura T (2007) Antinecrotic and antiapoptotic effects of hepatocyte growth factor on cholestatic hepatitis in a mouse model of bile-obstructive diseases. Am J Physiol Gastrointest Liver Physiol 292: G639–G646.
52. Janes CH, Dickson ER, Okazaki R, Bonde S, McDonagh AF, et al. (1995) Role of hyperbilirubinemia in the impairment of osteoblast proliferation associated with cholestatic jaundice. J Clin Invest 95: 2581–2586.
53. Ruiz-Gaspà S, Martinez-Ferrer A, Guañabens N, Dubreuil M, Peris P, et al. (2011) Effects of bilirubin and sera from jaundiced patients on osteoblasts: Contribution to the development of osteoporosis in liver diseases. Hepatology 54: 2104–2113.
54. Ong SY, Dai H, Leong KW (2006) Hepatic differentiation potential of commercially available human mesenchymal stem cells. Tissue Eng 12: 3477–3485.
55. Si-Tayeb K, Noto FK, Nagaoka M, Li J, Battle MA, et al. (2010) Highly efficient generation of human hepatocyte-like cells from induced pluripotent stem cells. Hepatology 51: 297–305.
56. Ochiya T, Yamamoto Y, Banas A (2010) Commitment of stem cells into functional hepatocytes. Differentiation 79: 65–73.
57. Sivasubramaniyan K, Atluri RR, Sarda K, Arvind M, Balaji V et al. (2008) Endotoxin-induced silencing of mesoderm induction and functional differentiation: role of HMGB1 in pluripotency and infection. Regen Med 3:23–31.

Low-Power Laser Irradiation Suppresses Inflammatory Response of Human Adipose-Derived Stem Cells by Modulating Intracellular Cyclic AMP Level and NF-κB Activity

Jyun-Yi Wu[1]◑, Chia-Hsin Chen[2,3,4]◑, Chau-Zen Wang[5,6,7], Mei-Ling Ho[5,6,7], Ming-Long Yeh[1]*, Yan-Hsiung Wang[5,8]*

1 Institute of Biomedical Engineering, National Cheng Kung University, Tainan, Taiwan, Republic of China, **2** Department of Physical Medicine and Rehabilitation, Kaohsiung Medical University Hospital, Kaohsiung, Taiwan, Republic of China, **3** Department of Physical Medicine and Rehabilitation, Faculty of Medicine, College of Medicine, Kaohsiung Medical University, Kaohsiung, Taiwan, Republic of China, **4** Department of Physical Medicine and Rehabilitation, Kaohsiung Municipal Ta-Tung Hospital, Kaohsiung, Taiwan, Republic of China, **5** Orthopaedic Research Center, College of Medicine, Kaohsiung Medical University, Kaohsiung, Taiwan, Republic of China, **6** Department of Physiology, College of Medicine, Kaohsiung Medical University, Kaohsiung, Taiwan, Republic of China, **7** Graduate Institute of Medicine, College of Medicine, Kaohsiung Medical University, Kaohsiung, Taiwan, Republic of China, **8** School of Dentistry, College of Dental Medicine, Kaohsiung Medical University, Kaohsiung, Taiwan, Republic of China

Abstract

Mesenchymal stem cell (MSC)-based tissue regeneration is a promising therapeutic strategy for treating damaged tissues. However, the inflammatory microenvironment that exists at a local injury site might restrict reconstruction. Low-power laser irradiation (LPLI) has been widely applied to retard the inflammatory reaction. The purpose of this study was to investigate the anti-inflammatory effect of LPLI on human adipose-derived stem cells (hADSCs) in an inflammatory environment. We showed that the hADSCs expressed Toll-like Receptors (TLR) 1, TLR2, TLR3, TLR4, and TLR6 and that lipopolysaccharide (LPS) significantly induced the production of pro-inflammatory cytokines (Cyclooxygenase-2 (Cox-2), Interleukin-1β (IL-1β), Interleukin-6 (IL-6), and Interleukin-8 (IL-8)). LPLI markedly inhibited LPS-induced, pro-inflammatory cytokine expression at an optimal dose of 8 J/cm^2. The inhibitory effect triggered by LPLI might occur through an increase in the intracellular level of cyclic AMP (cAMP), which acts to down-regulate nuclear factor kappa B (NF-κB) transcriptional activity. These data collectively provide insight for further investigations of the potential application of anti-inflammatory treatment followed by stem cell therapy.

Editor: Fayyaz S. Sutterwala, University of Iowa Carver College of Medicine, United States of America

Funding: This study was supported by a grant from the National Health Research Institutes (NHRI-EX101-9914EC), a grant from the Kaohsiung Municipal Ta-Tung Hospital (KMTTH-99-013), and a grant from the Kaohsiung Medical University (KMU-Q099018) of Taiwan. The funders had no role in study design, data collection and analysis, decision to publish, or preparation of the manuscript.

Competing Interests: The authors have declared that no competing interests exist.

* E-mail: mlyeh@mail.ncku.edu.tw (MLY); yhwang@kmu.edu.tw (YHW)

◑ These authors contributed equally to this work.

Introduction

Adult stem cells isolated from fat tissue, or adipose-derived stem cells (ADSCs), are capable of differentiating into bone, cartilage, fat, and muscle in vitro [1] and are a potential source of mesenchymal stem cells (MSCs). The advantages of ADSCs comparing to bone marrow and embryotic stem cells include the fact that they are less invasive when used in operations, can be safely transplanted to an autologous or an allogeneic host, and exhibit increased stem cell activities [2]. A variety of studies involving animal models, preclinical trials, and clinical trials have suggested that MSC-based tissue regeneration provides beneficial effects with respect to replacing damaged tissues and restoring organ function [3–6]. Although the mechanism underlying this type of regeneration is still not well understood, stem cell regeneration has been postulated to relate to the survival rate of engrafted cells [7] and the microenvironment of the local injury site. During tissue repair, inflammation represents an important phase in the healing process resulting in an influx of neutrophils/macrophages and release of pro-inflammatory cytokines and growth factors to remodel injury sites [8]. The existence of inflammatory mediators is likely restrictive regarding the applicability of stem cell-based implantation [9].

Lipopolysaccharide (LPS), the ligand of Toll-like receptor (TLR) 4, is a common model system that has been used in several recent studies to investigate inflammatory reactions. Toll-like receptors (TLRs) are characterized as transmembrane proteins composed of a globular intracellular region and an extracellular domain that contains numerous leucine-rich repeats (LRR) [10]. Ligation through TLRs leads to the recruitment of TIR-domain-containing adaptor proteins (MyD88, TRIF, and TIRAP) that trigger the activation of a nuclear factor kappa B (NF-κB) signaling cascade

[11,12]. As a result of this activation, transcription factor NF-κB induces the secretion of pro-inflammatory cytokines, including TNF-α, IL-1β, IFN-γ, and IL-6, thereby promoting the immune response.

Over the last few decades, low-power laser irradiation (LPLI) has been applied clinically for treating musculoskeletal aches and pains, wound healing, chronic inflammation, and acute inflammation. Several studies have suggested benefits of using LPLI. For example, Bortone et al. noted that LPLI induced an anti-inflammatory effect through the regulation of transcription factors linked to Cox-2 and through regulating the expression of pro-inflammatory mRNAs [13]. Additionally, Aimbire et al. reported that LPLI reduced neutrophil influx and IL-1β mRNA expression [14], and Sakurai et al. showed that LPLI inhibited PGE2 production and led to a reduction of Cox-2 mRNA levels [15]. However, few studies have been performed to investigate the effect of LPLI on the LPS-induced inflammatory response of ADSCs. The purpose of this study was to evaluate the effect of LPLI on the LPS-induced inflammation response of hADSCs and to provide basic knowledge related to the clinical application of stem cell therapy.

Materials and Methods

Cell Culture

The hADSCs used in this research was purchased from Cellular Engineering Technologies (CET, Coralville, IA, USA). The hADSCs were cultured and expanded in Keratinocyte serum-free medium (KSFM, GIBCO-BRL, Rockville, MD, USA) containing the supplements bovine pituitary extract (25 mg) and Human Recombinant EGF protein (2.5 μg) as well as 0.2 mM N-acetyl-Cysteine (Sigma-Aldrich, Saint Louis, MO, USA), 0.2 mM L-Ascorbic acid 2-phosphate magnesium (Sigma-Aldrich, Saint Louis, MO, USA), 5% fetal bovine serum, and 1% penicillin/streptomycin (GIBCO-BRL, Rockville, MD, USA) and a mixture of Dulbecco's modified Eagle's medium-low glucose (GIBCO-BRL, Rockville, MD, USA) containing 0.06 μg/ml insulin (Sigma-Aldrich, Saint Louis, MO, USA), 5% fetal bovine serum, and 1% penicillin/streptomycin in a ratio of 1:1. The cells were cultured in a humidified 5% CO_2 atmosphere in a 37°C incubator, and the medium was changed every other day.

The U937 human leukemic monocyte lymphoma cell line was obtained from the American Type Culture Collection (a kind gift from Dr. Chi-Ming Chiu, Department of Biotechnology, Ming Chuan University, Taipei, Taiwan). Growing U937 cells were cultured using RPMI medium 1640 (GIBCO-BRL, Rockville, MD, USA) with L-glutamine, 10% fetal bovine serum and 1% penicillin/streptomycin; the cells were maintained at 37°C in an atmosphere of 5% CO_2. Fresh medium was added two times a week, and the cells were kept at a density between 2×10^5 and 8×10^5 cells/ml.

Low-power Laser Irradiation

A red gallium-aluminum-arsenide (GaAlAs) laser (wavelength 660 nm) (TRANSVERSE IND. CO., LTD., Taipei, Taiwan) was used as a light source follow our previous study [16]. The laser had a maximum power of 70 mW, and the distance between the laser source and the bottom of the plate could be adjusted to match the intended target size. The power density applied was 15.17 mW/cm^2, and the cells were irradiated for 264 s and 528 s to achieve energies of 4 J/cm^2 and 8 J/cm^2, respectively. All irradiation experiments were performed on a clean bench at room temperature. The control groups were processed under the same conditions, except without laser irradiation.

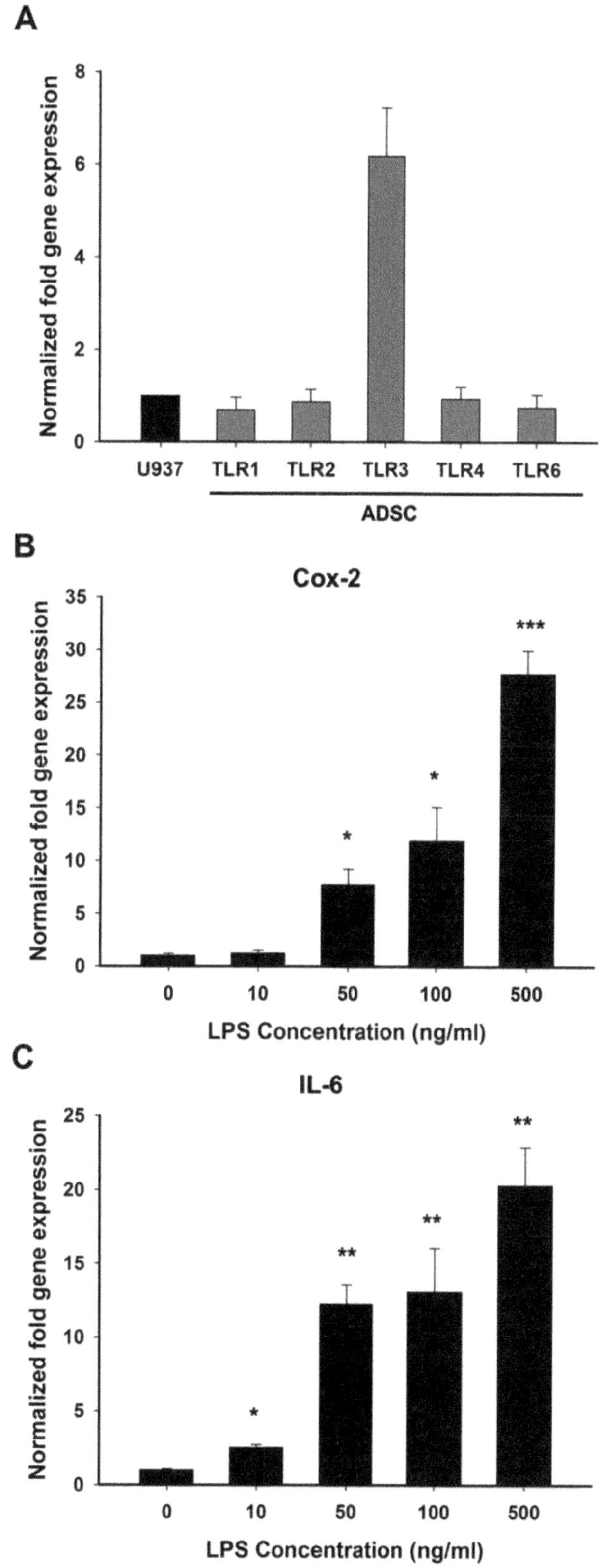

Figure 1. hADSCs expresses TLRs and stimulated by LPS. (**A**) The mRNA expression levels of all TLRs in the hADSCs were compared to the expression levels in U937 cells, as determined by real-time RT-PCR analysis; TLR3 showed a higher mRNA expression level in hADSCs than in U937 cells. (**B, C**) Determination of the optimal LPS treatment concentration by evaluating the expression of *Cox-2* and *IL-6* mRNA

levels. The data are shown as the mean ± SD (n = 3). The following statistical levels were applied: *p<0.05, **p<0.01, and ***p<0.001 compared to the control.

Chemical Reagent Treatment

LPS (Sigma-Aldrich, Saint Louis, MO, USA) was dissolved in 1X phosphate-buffered saline (PBS, Bio-Tech, Santa Cruz, CA, USA) to a final concentration of 10 mg/ml and then stored at 4°C. The LPS concentrations used in this study were 10, 50, 100, and 500 ng/ml; LPS was freshly diluted in culture medium immediately before being used. Forskolin (Sigma-Aldrich, Saint Louis, MO, USA), an adenylyl cyclase activator, was used as a cAMP-elevating agent. Forskolin was dissolved in DMSO (Sigma-Aldrich, Saint Louis, MO, USA) to a concentration of 60 μM, and cultures were incubated with this solution for 20 minutes. SQ22536 (Sigma-Aldrich, Saint Louis, MO, USA), an adenylyl cyclase inhibitor, was used to block the activity of adenylyl cyclase. SQ22536 was dissolved in DMSO to a final concentration of 100 μM, and the cultures were incubated with this solution for 15 minutes.

Real-time Reverse Transcription-polymerase Chain Reaction (RT-PCR)

Cells were harvested and rinsed twice with 1X PBS. Total RNA was extracted using TRIzol (GIBCO-BRL, Rockville, MD, USA), and the concentration of the obtained RNA was quantified by measuring the absorbance with a spectrophotometer (ND-1000, NanoDrop, Wilmington, DE, USA) at 260 nm and 280 nm. First-strand cDNA was reverse-transcribed from 1 μg of RNA using Moloney murine leukemia virus reverse transcriptase and oligo-dT primers. Quantitative real-time PCR was performed in a Bio-Rad iQ5 real-time detection system (Bio-Rad Laboratories Inc, Hercules, CA), and the reactions were carried out in a 12.5 μl mixture containing cDNA, forward and reverse primers for each gene and SYBR® Green Real-time PCR Master Mix (Toyobo CO.,LTD. OSAKA JANPAN). Primer pairs for the following human genes were used: *GAPDH* (5′-GAA GGT GAA GGT CGG AGT CA-3′ forward and 5′-GAA GAT GGT GAT GGG ATT TC-3′ reverse), *TLR1* (5′-GTT CCT GGC AAG AGC ATT GT-3′ forward and 5′-ATG GGT TCC AGC AAG ATC AG-3′ reverse), *TLR2* (5′-GCA TGT GCT GTG CTC TGT TC-3′ forward and 5′-TTC TCC ACC CAG TAG GCA TC-3′ reverse), *TLR3* (5′-AGT GCC CCC TTT GAACTC TT-3′ forward and 5′-TCC CAG ACC CAA TCC TTA TC-3′ reverse),

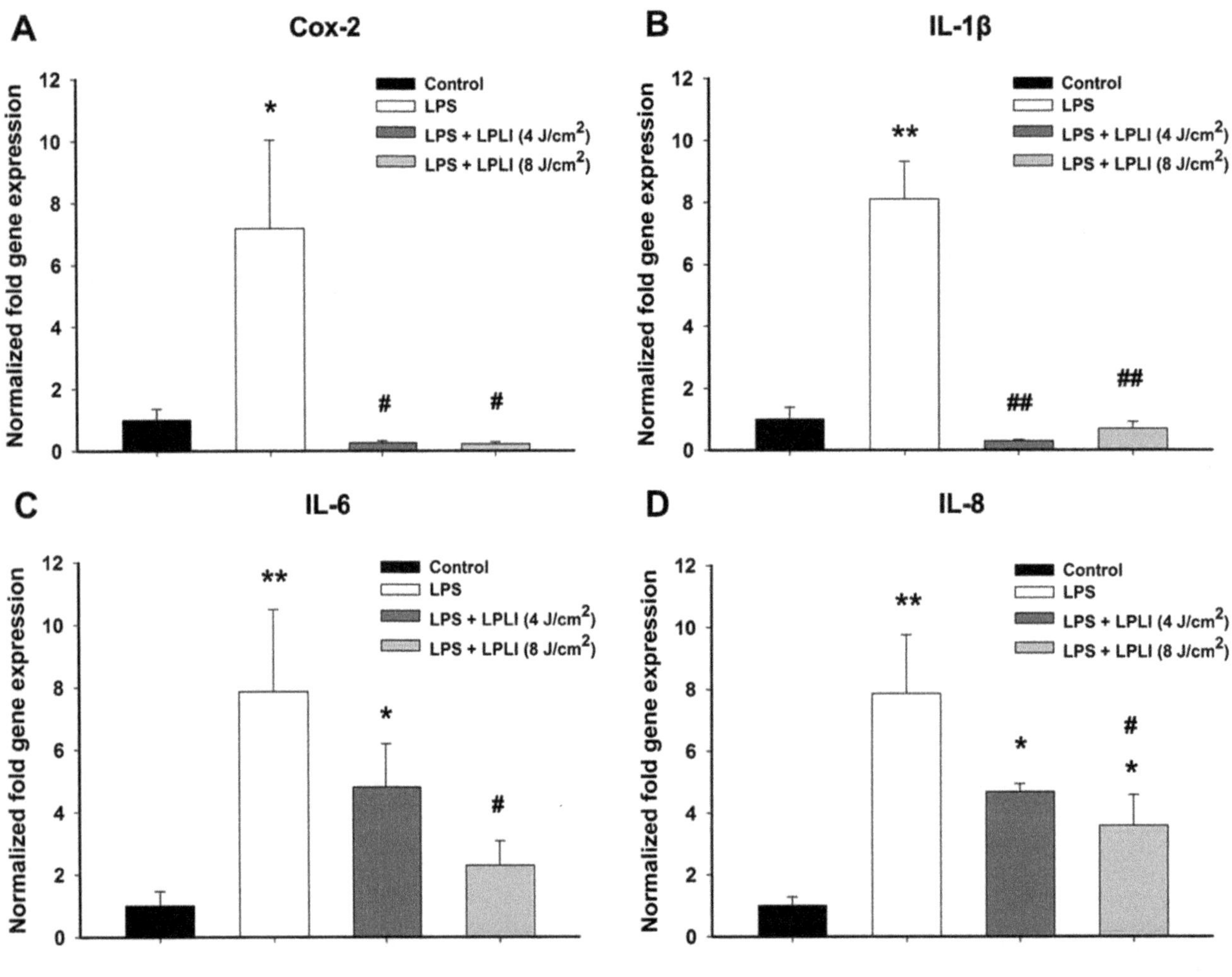

Figure 2. LPLI suppresses gene expression of pro-inflammatory cytokines by real-time RT-PCR. (**A**) *Cox-2.* (**B**) *IL-1β.* (**C**) *IL-6.* (**D**) *IL-8.* The results were analyzed using the $2^{-\Delta CT}$ method based on the control. The data are shown as the mean ± SD (n = 6). The following statistical levels were applied: *p<0.05 and **p<0.01 compared to the control and #p<0.05 and ##p<0.01 compared to LPS (50 ng/ml).

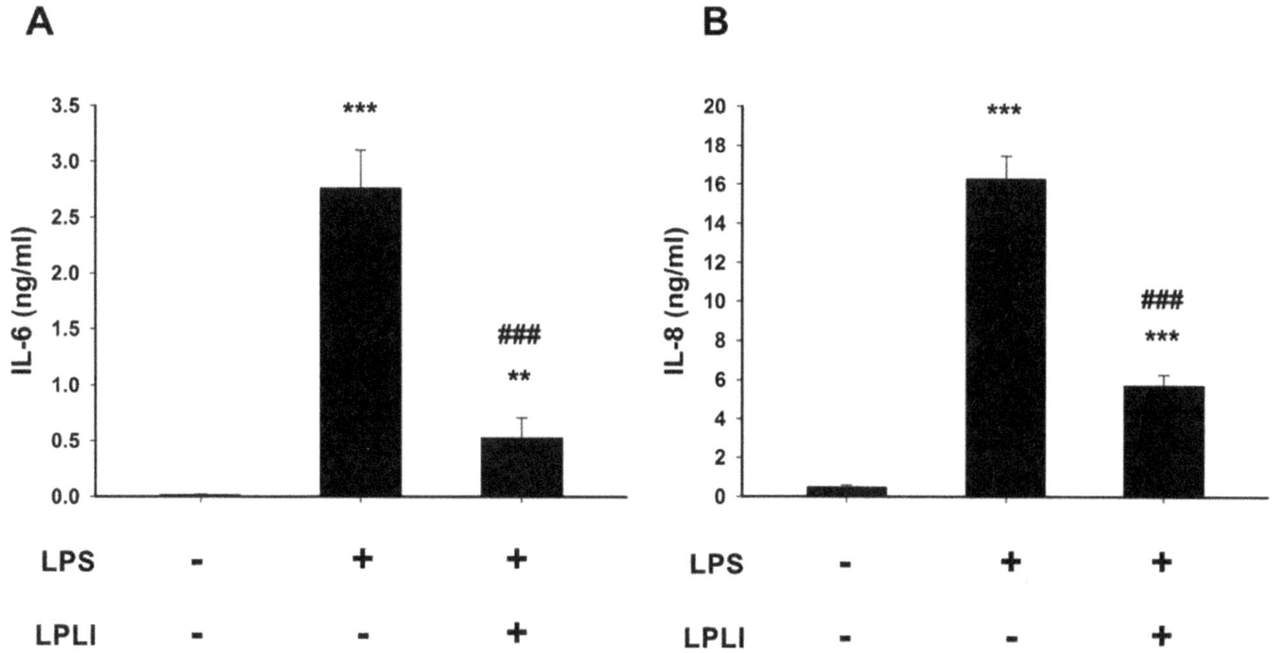

Figure 3. LPLI inhibits protein secretion of IL-6 and IL-8. The levels of pro-inflammatory proteins in LPS-stimulated hADSC culture media at 24 hours were determined by ELISA. (**A**) IL-6. (**B**) IL-8. The data are shown as the mean ± SD (n = 8). The following statistical levels were applied: **p<0.01 and ***p<0.001 compared to the control and ###p<0.001 compared to LPS (50 ng/ml).

TLR4 (5′-AAG CCG AAA GGT GAT TGT TG-3′ forward and 5′-TCC TCC CAC TCC AGG TAA GT-3′ reverse), *TLR6* (5′-AGT AGC TGG GCT TGC ATT GT-3′ forward and 5′-TTA TTG GAG GGC CTT GAG TG-3′ reverse), *Cox-2* (5′-TGA GCA TCT ACG GTT TGC TG-3′ forward and 5′-TGC TTG TCT GGA ACA ACT GC-3′ reverse), *IL-1β* (5′-GAC CTT CCA GGA GAA TGA CC-3′ forward and 5′-CAA AGG ACA TGG AGA ACA CC-3′ reverse), *IL-6* (5′-CTA GAG TAC CTC CAG AAC AG-3′ forward and 5′-TGA CCA GAA GAA GGA ATG C-3′ reverse) and *IL-8* (5′-TGC AGC TCT GTG TGA AGG TG-3′ forward and 5′-AAT TTC TGT GTT GGC GCA GT-3′ reverse). Relative mRNA expression levels were calculated from the threshold cycle (Ct) value of each PCR product and normalized to the housekeeping gene GAPDH using the comparative Ct method.

Enzyme-linked Immunosorbent Assay (ELISA)

The supernatants of the cell cultures were collected to evaluate the concentrations of IL-6 and IL-8 using the Human IL-6 ELISA Ready-SET-Go Kit (Bioscience, Inc., Allentown, PA, USA) and the Human IL-8 ELISA Ready-SET-Go Kit (Bioscience, Inc., Allentown, PA, USA), respectively. Both kits were used according to the manufacturer's instructions. The levels of IL-6 and IL-8 proteins were quantified at 450 nm using an ELISA plate reader. In addition, a Cyclic AMP EIA Kit (Cayman Chemical Company, Ann Arbor, MI, USA) was used to measure the generation of cyclic adenosine monophosphate (cAMP) within cells by collecting cell culture extracts. All procedures were based on the manufacturer's instructions, and the plates were read at a wavelength of 450 nm.

Western Blotting

Cells were washed twice in PBS and lysed using CelLytic™ M cell lysis reagent (Sigma-Aldrich, Saint Louis, MO, USA) with protease inhibitor (Complete Protease Inhibitor Cocktail Tablets; Roche Diagnostics Ltd. Burgess Hill, United Kingdom). Protein concentrations were determined using the BCA protein assay kit (Novagen, Darmstadt, Germany) based on the manufacturer's instructions. Cell lysate containing 40 μg of proteins were analyzed by 8% SDS-Polyacrylamide gel electrophoresis. Transferred membranes were blocked using 5% non-fat milk and incubated overnight with antibodies against Phospho-NF-κB p65 (Ser536) (dilution 1:1,000, Cell Signaling Technology, Inc., Beverly, MA, USA), NF-κB p65 (dilution 1:6,000, Epitomics, Inc., Burlingame, CA, USA), IκBα Phospho (dilution 1:2,000, Epitomics, Inc., Burlingame, CA, USA) and IκBα (dilution 1:2,000, Epitomics, Inc., Burlingame, CA, USA). The same membrane was probed with anti-β-actin (dilution 1:4,000) as a loading control. Finally, membranes were detected using the Western Lighting™ Chemiluminescence Reagent Plus (Perkin-Elmer, Boston, MA, USA). Blotted bands were digitally measured by using the UVP AutoChemi™ Image and Analysis System (UVP, Upland, CA, USA).

Immunofluorescence staining

Cells were fixed with 4% paraformaldehyde (PANREAC QUIMICA, Castellar del Vallès, Spain) for 1 hour. After being washed, cells were permeabilized with 0.1% Triton X-100 for 10 min, blocked in 1% BSA for 1 hour, and immunostained with NF-κB p65 (Epitomics, Inc., Burlingame, CA, USA) at 4°C overnight. Cells were then incubated with Alexa Fluor 594-labeled goat secondary antibodies (Invitrogen, Carlsbad, CA, USA) for 1 hour. Finally, cell nuclei were stained with DAPI (Thermo SCIENTIFIC, Madison, WI, USA) and mounting on a coverslip with 50% glycerol in PBS. Fluorescence images were visualized using Nikon TE300 microscope and created by Image-Pro Plus 5.0 (Media-Cybernetics, Silver Spring, MD, USA).

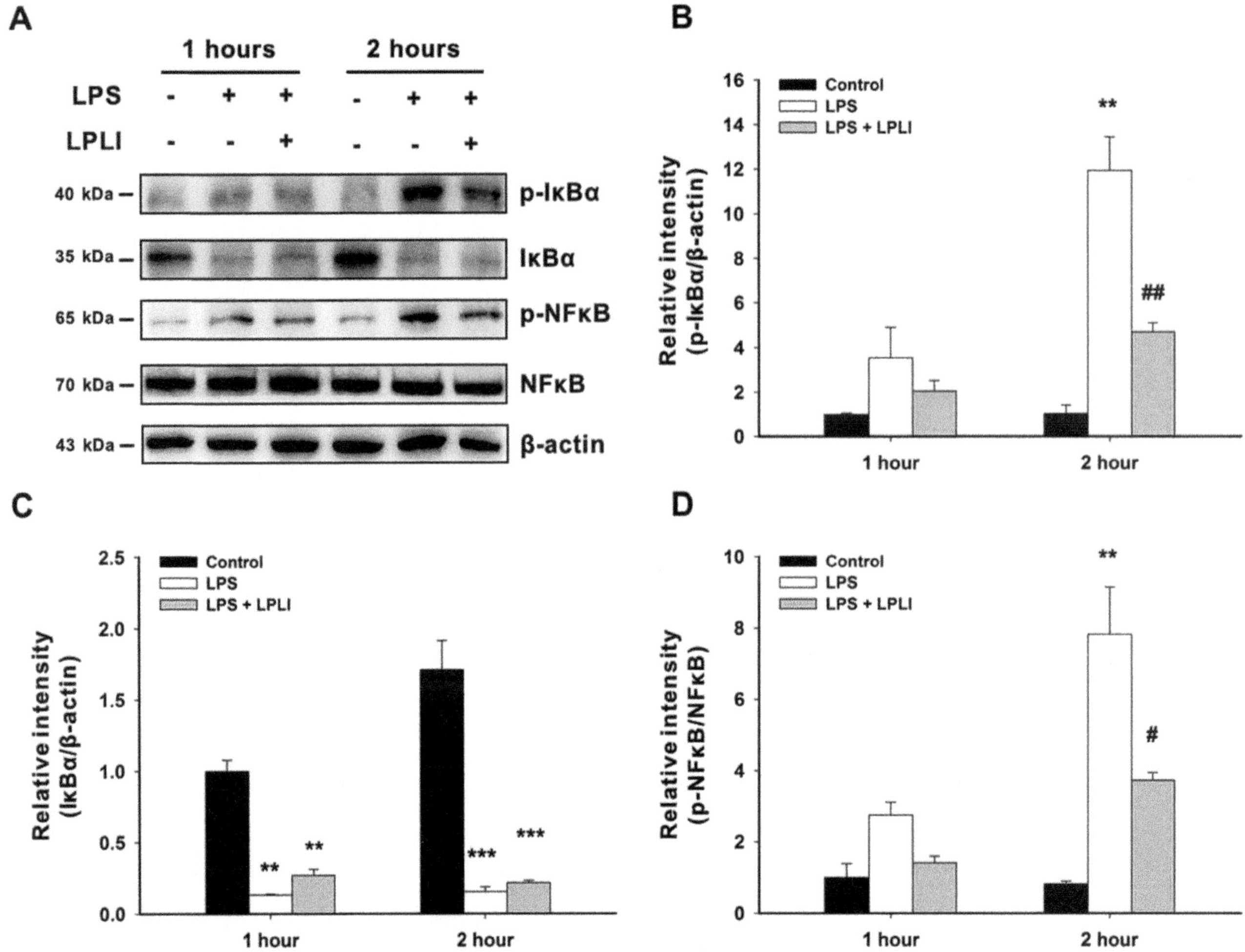

Figure 4. LPLI decreases IκBα degradation and NFκB activation. (**A**) hADSCs were treated with or without LPS (50 ng/ml), immediately followed by LPLI (8 J/cm^2). After 1 hour and 2 hours, the protein expression of p-IκBα, IκBα, p-NFκB and NFκB was determined using Western blot analysis, and β-actin served as loading control. The blots were quantified, and the results were expressed as ratios compared to LPS (0 ng/ml) at 1 hour, which was defined as 1. (**B**) p-IκBα vs. β-actin. (**C**) IκBα vs. β-actin. (**D**) p-NFκB vs. NFκB. The data are shown as the mean ± SD of 3 independent experiments. The following statistical levels were applied: **p<0.01 and ***p<0.001 compared to the control and #p<0.05 and ##p<0.01 compared to LPS (50 ng/ml).

Dual-luciferase Assay

Cells were co-transfected with the pGL 4.32[luc2P/NF-κB-RE/Hygro] vector (1 μg) (Promega, Madison, WI, USA) and pTK-Renilla vector (20 ng) (Promega, Madison, WI, USA). Electrotransfection was performed using a MicroPorator (Digital Bio Technology, South Korea) based on the manufacturer's instructions. After 5 hours, luciferase activity was initiated using a Dual-Luciferase® Reporter Assay System (Promega, Madison, WI, USA) and measured with TopCount Microplate Scintillation and Luminescence Counters (Packard Biosciences, Waltham, MA, USA).

Statistical Analysis

SPSS version 17.0 was used for statistical analysis. The results are expressed as the mean ± standard deviation. Statistically significant differences were determined using analysis of variance (ANOVA) followed by a post hoc Tukey's test for multiple comparisons. A p-value of less than 0.05 was considered statistically significant.

Results

Determination of the Activity and Expression Pattern of TLRs in hADSCs

To understand whether hADSCs can detect multiple pathogen-associated molecular patterns (PAMPs) via TLRs, we first had to ascertain the expression pattern of TLRs in hADSCs. Primers specific to TLR1, TLR2, TLR3, TLR4, and TLR6 were designed to analyze the basal mRNA expression levels of TLRs in hADSCs. The TLR expression levels in the hADSCs were compared to those of U937 cells, which express various TLRs [17]. The mRNA expression levels of all TLRs in the hADSCs except for TLR3 were similar to the expression levels in U937 cells, as determined by real-time RT-PCR analysis; TLR3 showed a higher mRNA expression level in hADSCs than in U937 cells (Fig. 1A).

Next, we analyzed the activity of TLRs in hADSCs. LPS, the ligand of TLR4, was used as a bacterial model in this study. hADSCs were treated with different concentrations of LPS, and the expression of *Cox-2* and *IL-6* mRNA was then evaluated. A dose-dependent increase in *Cox-2* and *IL-6* mRNA expression was

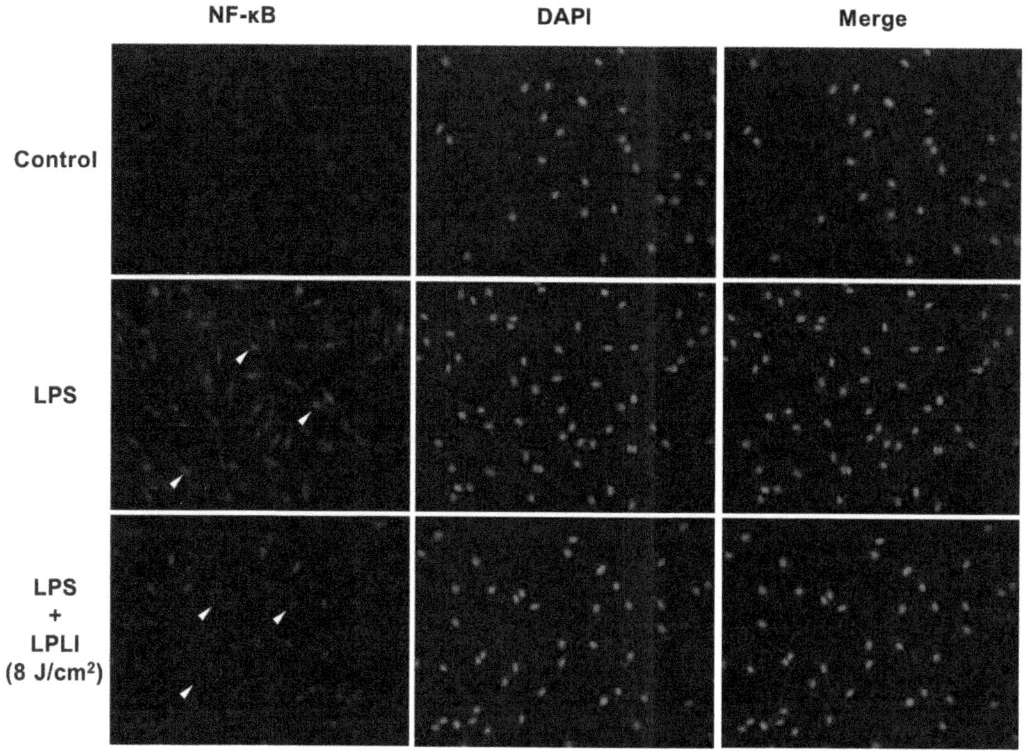

Figure 5. LPLI decrease nuclear translocation of NF-κB. hADSCs were pre-treated with or without LPS (50 ng/ml), followed by LPLI (8 J/cm^2), and the localization of the NF-κB was determined by fluorescence. The images in the upper panel show the cytoplasmic localization of NF-κB in the control cells. The image in the middle panel shows the nuclear translocation of NF-κB in cells treated with LPS. The image in the lower panel shows that LPLI treatment blocked the nuclear translocation of NF-κB caused by LPS stimulation.

observed 24 hours after LPS treatment (Fig. 1B, C). The levels of *Cox-2* and *IL-6* mRNA expression were significantly elevated when a concentration of LPS higher than 50 ng/ml was used ($p<0.05$ in *Cox-2* and $p<0.01$ in *IL-6*). These results indicated that hADSCs express TLRs and exhibit a functional TLR4. Additionally, treatment with LPS at a concentration of 50 ng/ml is sufficient to stimulate the production of pro-inflammatory mediators to initiate the microglial inflammation process.

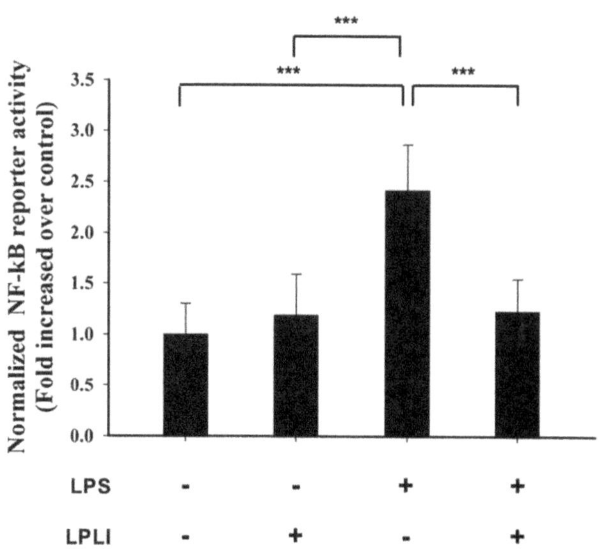

Figure 6. LPLI decreases the transcription activity of NFκB. hADSCs were transfected with the vectors pGL-NFκB and pTK-Renilla (control vector), and the cells were then incubated with 50 ng/ml LPS with or without LPLI. Luciferase activity was assayed 5 hours after LPLI treatment. The data are expressed as the mean ± SD of 3 independent experiments. The statistical level applied was ***$p<0.001$.

LPLI Decreases LPS-induced Pro-inflammatory Cytokine Gene Expression

hADSCs were treated with or without LPS (50 ng/ml), immediately followed by laser irradiation at energy densities of 4 J/cm^2 or 8 J/cm^2 to examine the effect of LPLI on the LPS-induced inflammatory response in hADSCs. The mRNA expression levels of pro-inflammatory cytokines, such as *Cox-2*, *IL-1β*, *IL-6*, and *IL-8*, were analyzed after 24 hours using real-time RT-PCR. LPS treatment strongly induced the expression of pro-inflammatory genes. However, when hADSCs were treated with a combination of LPLI and LPS, the mRNA levels of *Cox-2* and *IL-1β* were dramatically decreased, regardless of the LPLI energy density applied to the hADSCs (Fig. 2A, B). LPLI significantly reduced the levels of *IL-6* and *IL-8* mRNA expression in a dose-dependent manner (Fig. 2C, D); the *IL-6* and *IL-8* mRNA expression of group treated with energy density of 8 J/cm^2 was significantly lower than the group that was only treated with LPS.

To further confirm the effects of LPLI on the suppression of LPS-induced pro-inflammatory cytokines in hADSCs, the amounts of IL-6 and IL-8 secreted into the culture medium were measured using specific ELISAs. The concentrations of IL-6 and IL-8 showed similar trends of significant increases in the culture medium after 24 hours of LPS treatment. Co-treatment with LPLI

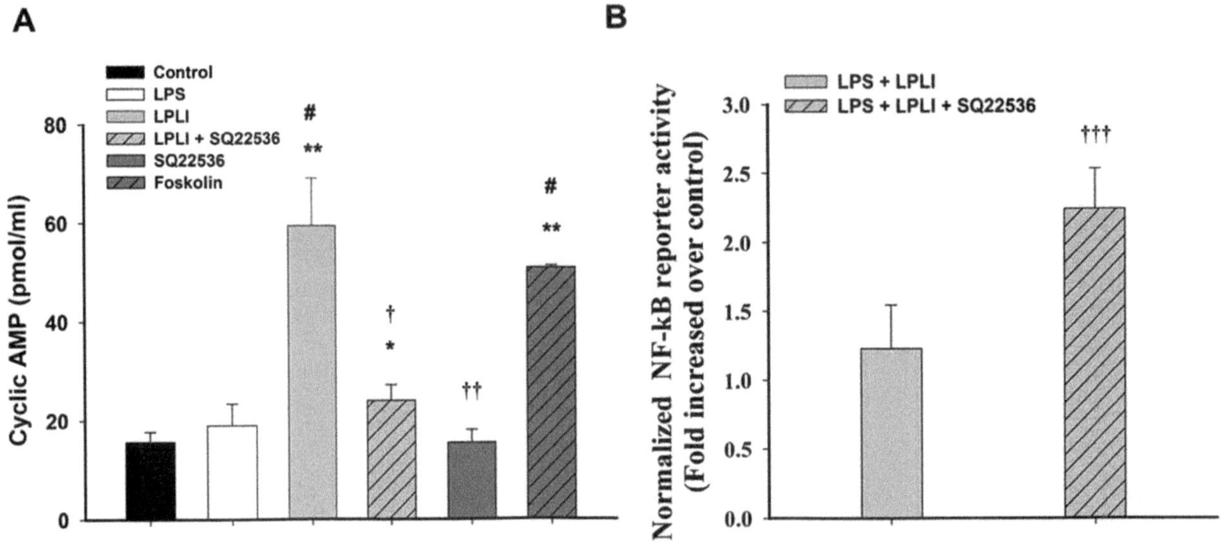

Figure 7. LPLI increases intracellular cAMP level to suppress NF-κB transcriptional activity. (**A**) Intracellular cAMP levels were measured by ELISA in hADSCs treated with LPS, LPLI, SQ22536, and forskolin. (**B**) NF-κB transcriptional activity was assayed 5 hours after LPLI treatment only and combined pretreatment of SQ22536. The data are expressed as the mean ± SD (n = 6). The following statistical levels were applied: *$p<0.05$ and **$p<0.01$ compared to the control, #$p<0.05$ compared to LPS (50 ng/ml), and †$p<0.05$, ††$p<0.01$, and †††$p<0.001$ compared to LPLI (8 J/cm^2).

(8 J/cm^2) and LPS significantly suppressed IL-6 and IL-8 production (Fig. 3A, B). These results showed that LPLI significantly reduced the expression of LPS-induced pro-inflammatory genes at both the transcriptional and translational levels. However, the strength of the suppression effect generated by LPLI varied among different pro-inflammatory genes; this difference indicated that the regulatory mechanisms induced by LPLI affecting these genes might differ depending on the gene. In addition, we also used the lactate dehydrogenase (LDH) assay to determine whether LPLI induced cytotoxic effects on hADSCs. It showed that LPLI treatments did not induce significant cytotoxic effects on hADSCs (Fig. S1). The data indicated that LPLI reduced LPS-induced inflammatory response was not caused by harmful effects of LPLI.

LPLI Downregulates LPS-induced NF-κB Activation

It has been shown that NF-κB is an important regulator in the LPS-induced inflammatory response [18], Thus, we investigated whether LPLI decreases the expression of LPS-induced pro-inflammatory cytokines by regulating NF-κB activity.

hADSCs were treated with or without LPS (50 ng/ml), immediately followed by LPLI (8 J/cm^2). Treatment with LPS induced a slight increase in IκBα phosphorylation at 1 hour post-incubation and strongly increased IκBα phosphorylation at 2 hours post-incubation. In addition to the increases in p-IkBα expression, IKBα degradation was observed both 1 hour and 2 hours post-incubation (Fig. 4A–C). Treatment with LPS induced a 2.75-fold increase in the phosphorylation of NF-κB (p-NF-κB/NF-κB) at 1 hour post-incubation and a 7.82-fold increase at 2 hours post-incubation, which is linked to enhancement of NF-κB activation (Fig. 4A,D). However, when hADSCs were co-treated with LPS and LPLI, the expression of p-IκBα and p-NF-κB was inhibited at these two time points (Fig. 4A, B and D). Although our data showed that co-treatment with LPS and LPLI induced a 1.41-fold increase in NF-κB phosphorylation at 1 hour post-incubation and a 3.72-fold increase at 2 hours post-incubation, these increases

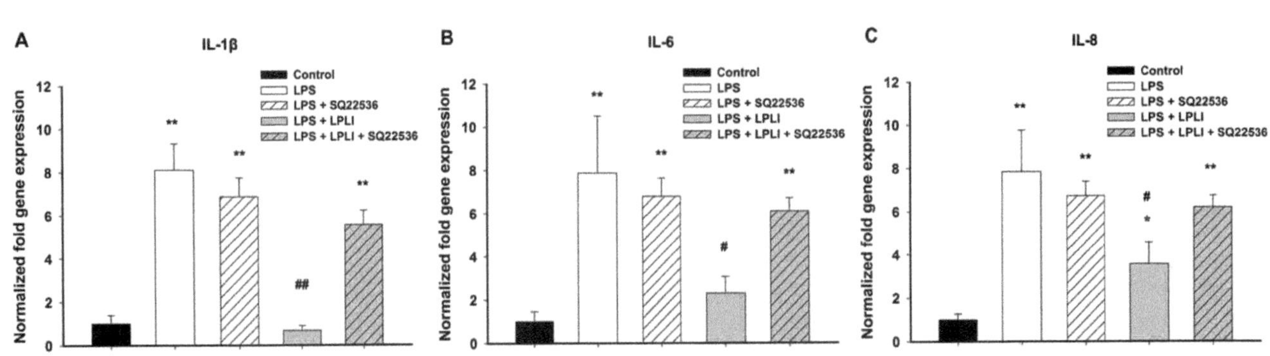

Figure 8. SQ22536 blocks the anti-inflammatory response of LPLI by real-time RT-PCR. (**A**) *IL-1β*. (**B**) *IL-6*. (**C**) *IL-8*. The results were analyzed by the $2^{-\Delta CT}$ method based on the control. The data are shown as the mean ± SD (n = 6). The following statistical levels were applied: *$p<0.05$ and **$p<0.01$ compared to the control and #$p<0.05$ and ##$p<0.01$ compared to LPS (50 ng/ml).

are much lower than those seen following LPS treatment alone (Fig. 4D).

Phosphorylation and nuclear translocation of NF-κB are required for NF-κB activation. We monitored the cellular distribution of NF-κB using fluorescence microscopy and found that LPLI blocked the nuclear accumulation of NF-κB resulting from LPS stimulation (Fig. 5). The transcriptional activity of NF-κB was determined by measuring relative luciferase activity. Treatment of hADSCs with LPLI alone did not affect luciferase activity, whereas treatment with LPS caused a 2.5-fold increase in luciferase activity. Luciferase activity was significantly inhibited following co-treatment of hADSCs with LPLI and LPS (Fig. 6). Together, these results indicated that LPLI suppressed pro-inflammatory cytokine expression through inhibition of NF-κB activation.

LPLI Modulated NF-κB via the cAMP Signaling Pathway

cAMP acts not only as a modulator of immune function but also as a regulator of LPLI signaling [19]. Thus, we determined the cAMP levels associated with LPLI-mediated anti-inflammatory activity in hADSCs. Treatment of hADSCs with either LPS or SQ22536, an adenylyl cyclase inhibitor, did not alter the level of cAMP, whereas the adenylyl cyclase activator forskolin significantly increased cAMP accumulation (Fig. 7A). A similar result was obtained by treating hADSCs with LPLI, which resulted in significant elevation of the cAMP concentration. This LPLI-mediated cAMP accumulation was blocked by pretreatment with SQ22536 (Fig. 7A). To investigate whether LPLI-mediated cAMP accumulation is the crucial mechanism involved in decreasing pro-inflammatory cytokine expression in hADSCs, we first determined whether cAMP signaling regulates the NF-κB activity using luciferase reporter assay. We found that SQ22536 pretreatment blocked the inhibitory effect of LPLI-mediated NF-κB transcriptional activity in LPS treated hADSCs (Fig. 7B). Next, we measured the mRNA expression levels of pro-inflammatory cytokines (IL-1, IL-6, and IL-8) in hADSCs treated with LPS or LPLI with or without SQ22536 pretreatment. Following pretreatment with SQ22536, the LPLI-mediated inhibition of pro-inflammatory cytokine expression was completely reversed. However, SQ22536 did not affect the mRNA expression of pro-inflammatory cytokines (IL-1, IL-6, and IL-8) induced by LPS (Fig. 8A–C). These results indicated that the LPLI-mediated anti-inflammatory activity observed in hADSCs might occur through the cAMP signaling pathway.

Discussion

The results of the present study show for the first time that LPLI (660 nm) suppresses the inflammatory reaction induced by LPS in hADSCs. We first confirmed that hADSCs exhibit a similar expression profile of TLRs compared with U937 cells, a human leukemic monocyte lymphoma cell line. Our results correspond with those of other recent studies. Cho et al. reported that hADSCs showed intense expression of TLR2, TLR3, TLR4, and TLR6 and lower expression of TLR1, TLR5 and TLR9 [20]. Lombardo et al. showed that hADSCs express TLR1 through TLR10, with the exception of TLR8 [21]. Together, these data indicated that hADSCs express TLRs and are able to respond to various TLR ligands associated with inflammation.

In the past few years, a number of findings have indicated that LPLI suppresses the inflammatory reaction both in vitro and in vivo. Correa et al. used LPS to induce peritonitis in mice and found that LPLI (GaAs laser, 904 nm) can reduce inflammatory cell migration in a dose-dependent manner, with an energy dose of 3 J/cm^2 being the most effective [22]. Pires et al. found that LPLI (780 nm), at an energy dose of 7.7 J/cm^2, decreased the expression of IL-6, COX-2, and TGF-β associated with collagenase-induced tendinitis [23]. Boschi et al. showed that LPLI (InGaAlP laser, 660 nm) significantly reduced the expression of NO, IL-6, MCP-1, IL-10, and TNFα [24]. In LPS challenged hADSC, we obtained similar results to those observed in the aforementioned studies, finding that LPLI significantly inhibited the mRNA expression of pro-inflammatory cytokines (Cox-2, IL-1, IL-6, and IL-8) and secreted proteins (IL-6 and IL-8), leading us to conclude that LPLI has an anti-inflammatory effect in hADSCs.

It is well known that NF-kB is a crucial transcription factor involved in the regulation of inflammation. Our data showed that LPLI suppressed the expression of pro-inflammatory cytokine genes. Therefore, we focused on the relationship between LPLI and the activity of NF-κB. The classic NF-κB is a heterodimer composed of class I (p50, p52) and class II proteins (Rel-A/p65, Rel-B, and c-Rel). In the cytoplasm of unstimulated cells, NF-κB forms a complex with inhibitors of NF-κB (IκBs). After stimulation by upstream signals, the IκB kinase (IKK) complex phosphorylates IκBs, leading to proteasome-mediated degradation and dissociation of IκBα and NF-κB [25]. Next, nuclear translocation of NF-κB occurs and induces gene transcription; in this process, the phosphorylation state of NF-κB has been linked to enhancement of NF-κB activation. We observed that LPLI decreased the level of phosphorylated IκBα and maintained more inactive IκBα in the cytoplasm. LPLI also inhibited the LPS-induced expression of phospho-NF-κB and clearly decreased the amount of NF-κB that was translocated. Furthermore, LPLI significantly inhibited the transcriptional activity of NF-κB. Results similar to ours were also reported by Aimbire et al [26], who showed that a low-level diode laser (660 nm) at an energy density of 7.5 J/cm^2 reduced the mRNA levels of both Bcl-xL and A1, which are produced by NF-κB nuclear translocation, leading these authors to suggest that the anti-inflammatory effect of LPLI can be mediated by NF-κB. Rizzi et al. indicated that LPLI (904 nm) at an energy density of 5 J/cm^2 was able to block the degradation of IkBα, which is consistent with our findings [27]. Together, these data suggest that LPLI inhibits the inflammatory process via inhibition of NF-κB transcriptional activity.

Several studies have shown that the intracellular levels of cAMP are increased by LPLI [28,29]. cAMP is an important second messenger in many biological processes, such as cell proliferation, differentiation, apoptosis, and inflammation, that is produced by the activation of adenylyl cyclases and converted from ATP [30]. Recent data have shown that elevation of intracellular cAMP levels inhibits the transcriptional activity of NF-κB [31,32]. Possible mechanisms underlying the role of cAMP in regulating NF-κB activity include the ability of cAMP to manage Iκκ activity and IκB degradation as well as to change the composition of NF-κB dimers and thereby block transcription [33]. In the present study, we observed that the level of cAMP increased by approximately 3–4 fold following LPLI treatment. This finding indicated that LPLI can act to stimulate the level of cAMP. To confirm that the inhibition of inflammation by LPLI occurred via cAMP, we treated hADSCs with the adenylyl cyclase inhibitor SQ22536 and observed that the inhibitory effect by LPLI was dramatically reduced. Overall, our results indicated that cAMP is an important mediator involved in the inhibitory effect of LPLI on LPS-induced inflammation in hADSCs.

In conclusion, our results show that hADSCs express TLR1, TLR2, TLR3, TLR4, and TLR6. Following LPS (50 ng/ml) stimulation, hADSCs exhibited a significant increase in the production of pro-inflammatory mediators (Cox-2, Il-1β, IL-6, and IL-8). The major findings of the present study are that low power laser irradiation markedly inhibited the LPS-mediated inflammatory response in hADSCs at an optimal dose of 8 J/cm^2. The inhibitory effect stimulated by LPLI might act via increasing the intracellular level of cAMP, resulting in down-regulation of NF-κB transcriptional activity. Our results indicate that LPLI treatment can potentially be applied in anti-inflammatory therapy followed by stem cell therapy.

Supporting Information

Figure S1 LPLI treatments did not induce cytotoxic effects on hADSCs. hADSCs were treated with LPLI at doses of 0 (control), 4, or 8 J/cm^2. LDH leakage was analyzed to evaluate cell cytotoxicity at 24 hours. There were no significant differences between the groups (n = 12).

Author Contributions

Conceived and designed the experiments: JYW CHC YHW MLY. Performed the experiments: JYW. Analyzed the data: JYW YHW MLY CHC. Contributed reagents/materials/analysis tools: MLH CZW MLY CHC. Wrote the paper: JYW CHC YHW.

References

1. Bunnell BA, Flaat M, Gagliardi C, Patel B, Ripoll C (2008) Adipose-derived stem cells: isolation, expansion and differentiation. Methods 45: 115–120.
2. Zhu Y, Liu T, Song K, Fan X, Ma X, et al. (2008) Adipose-derived stem cell: a better stem cell than BMSC. Cell Biochem Funct 26: 664–675.
3. Garcia-Gomez I, Elvira G, Zapata AG, Lamana ML, Ramirez M, et al. (2010) Mesenchymal stem cells: biological properties and clinical applications. Expert Opin Biol Ther 10: 1453–1468.
4. Crisostomo PR, Wang M, Markel TA, Lahm T, Abarbanell AM, et al. (2007) Stem cell mechanisms and paracrine effects: potential in cardiac surgery. Shock 28: 375–383.
5. Caplan AI (2007) Adult mesenchymal stem cells for tissue engineering versus regenerative medicine. J Cell Physiol 213: 341–347.
6. Behfar A, Perez-Terzic C, Faustino RS, Arrell DK, Hodgson DM, et al. (2007) Cardiopoietic programming of embryonic stem cells for tumor-free heart repair. J Exp Med 204: 405–420.
7. Uemura R, Xu M, Ahmad N, Ashraf M (2006) Bone marrow stem cells prevent left ventricular remodeling of ischemic heart through paracrine signaling. Circ Res 98: 1414–1421.
8. Martin P, D'Souza D, Martin J, Grose R, Cooper L, et al. (2003) Wound healing in the PU.1 null mouse–tissue repair is not dependent on inflammatory cells. Curr Biol 13: 1122–1128.
9. Anderson JM, Rodriguez A, Chang DT (2008) Foreign body reaction to biomaterials. Semin Immunol 20: 86–100.
10. Doyle SL, O'Neill LA (2006) Toll-like receptors: from the discovery of NFkappaB to new insights into transcriptional regulations in innate immunity. Biochem Pharmacol 72: 1102–1113.
11. Horng T, Barton GM, Medzhitov R (2001) TIRAP: an adapter molecule in the Toll signaling pathway. Nat Immunol 2: 835–841.
12. O'Neill LA, Bowie AG (2007) The family of five: TIR-domain-containing adaptors in Toll-like receptor signalling. Nat Rev Immunol 7: 353–364.
13. Bortone F, Santos HA, Albertini R, Pesquero JB, Costa MS, et al. (2008) Low level laser therapy modulates kinin receptors mRNA expression in the subplantar muscle of rat paw subjected to carrageenan-induced inflammation. Int Immunopharmacol 8: 206–210.
14. Aimbire F, Ligeiro de Oliveira AP, Albertini R, Correa JC, Ladeira de Campos CB, et al. (2008) Low level laser therapy (LLLT) decreases pulmonary microvascular leakage, neutrophil influx and IL-1beta levels in airway and lung from rat subjected to LPS-induced inflammation. Inflammation 31: 189–197.
15. Sakurai Y, Yamaguchi M, Abiko Y (2000) Inhibitory effect of low-level laser irradiation on LPS-stimulated prostaglandin E2 production and cyclooxygenase-2 in human gingival fibroblasts. Eur J Oral Sci 108: 29–34.
16. Wu JY, Wang YH, Wang GJ, Ho ML, Wang CZ, et al. (2012) Low-power GaAlAs laser irradiation promotes the proliferation and osteogenic differentiation of stem cells via IGF1 and BMP2. PLoS One 7: e44027.
17. Xiong F, Wang XB, Zhang JH, Liu W, Sun S, et al. (2007) Expression and role of toll-like receptors in U937 cells. Zhongguo Shi Yan Xue Ye Xue Za Zhi 15: 449–453.
18. Sen R, Baltimore D (1986) Multiple nuclear factors interact with the immunoglobulin enhancer sequences. Cell 46: 705–716.
19. de Lima FM, Moreira LM, Villaverde AB, Albertini R, Castro-Faria-Neto HC, et al. (2011) Low-level laser therapy (LLLT) acts as cAMP-elevating agent in acute respiratory distress syndrome. Lasers Med Sci 26: 389–400.
20. Hwa Cho H, Bae YC, Jung JS (2006) Role of toll-like receptors on human adipose-derived stromal cells. Stem Cells 24: 2744–2752.
21. Lombardo E, DelaRosa O, Mancheno-Corvo P, Menta R, Ramirez C, et al. (2009) Toll-like receptor-mediated signaling in human adipose-derived stem cells: implications for immunogenicity and immunosuppressive potential. Tissue Eng Part A 15: 1579–1589.
22. Correa F, Lopes Martins RA, Correa JC, Iversen VV, Joenson J, et al. (2007) Low-level laser therapy (GaAs lambda = 904 nm) reduces inflammatory cell migration in mice with lipopolysaccharide-induced peritonitis. Photomed Laser Surg 25: 245–249.
23. Pires D, Xavier M, Araujo T, Silva JA, Jr., Aimbire F, et al. (2011) Low-level laser therapy (LLLT; 780 nm) acts differently on mRNA expression of anti- and pro-inflammatory mediators in an experimental model of collagenase-induced tendinitis in rat. Lasers Med Sci 26: 85–94.
24. Boschi ES, Leite CE, Saciura VC, Caberlon E, Lunardelli A, et al. (2008) Anti-Inflammatory effects of low-level laser therapy (660 nm) in the early phase in carrageenan-induced pleurisy in rat. Lasers Surg Med 40: 500–508.
25. Akira S, Takeda K (2004) Toll-like receptor signalling. Nat Rev Immunol 4: 499–511.
26. Aimbire F, Santos FV, Albertini R, Castro-Faria-Neto HC, Mittmann J, et al. (2008) Low-level laser therapy decreases levels of lung neutrophils anti-apoptotic factors by a NF-kappaB dependent mechanism. Int Immunopharmacol 8: 603–605.
27. Rizzi CF, Mauriz JL, Freitas Correa DS, Moreira AJ, Zettler CG, et al. (2006) Effects of low-level laser therapy (LLLT) on the nuclear factor (NF)-kappaB signaling pathway in traumatized muscle. Lasers Surg Med 38: 704–713.
28. Zungu IL, Hawkins Evans D, Abrahamse H (2009) Mitochondrial responses of normal and injured human skin fibroblasts following low level laser irradiation–an in vitro study. Photochem Photobiol 85: 987–996.
29. Karu TI, Lobko VV, Lukpanova GG, Parkhomenko IM, Chirkov I (1985) Effect of irradiation with monochromatic visible light on the cAMP content in mammalian cells. Dokl Akad Nauk SSSR 281: 1242–1244.
30. Gether U (2000) Uncovering molecular mechanisms involved in activation of G protein-coupled receptors. Endocr Rev 21: 90–113.
31. Zhong H, SuYang H, Erdjument-Bromage H, Tempst P, Ghosh S (1997) The transcriptional activity of NF-kappaB is regulated by the IkappaB-associated PKAc subunit through a cyclic AMP-independent mechanism. Cell 89: 413–424.
32. Parry GC, Mackman N (1997) Role of cyclic AMP response element-binding protein in cyclic AMP inhibition of NF-kappaB-mediated transcription. J Immunol 159: 5450–5456.
33. Gerlo S, Kooijman R, Beck IM, Kolmus K, Spooren A, et al. (2011) Cyclic AMP: a selective modulator of NF-kappaB action. Cell Mol Life Sci 68: 3823–3841.

17

The Role of Indoleamine 2,3 Dioxygenase in Beneficial Effects of Stem Cells in Hind Limb Ischemia Reperfusion Injury

Mohamad Masoumy[1], Jack Yu[1], Jun Yao Liu[2], Nathan Yanasak[3], Christopher Middleton[4], Folami Lamoke[5], Mahmood S. Mozaffari[2], Babak Baban[1,2]*

1 Department of Surgery, Georgia Regents University, Augusta, Georgia, United States of America, **2** Department of Oral Biology, Georgia Regents University, Augusta, Georgia, United States of America, **3** Department of Radiology and Imaging, Georgia Regents University, Augusta, Georgia, United States of America, **4** Cancer Center, Georgia Regents University, Augusta, Georgia, United States of America, **5** Department of Ophthalmology, Georgia Regents University, Augusta, Georgia, United States of America

Abstract

Ischemia-Reperfusion (IR) injury of limb remains a significant clinical problem causing secondary complications and restricting clinical recovery, despite rapid restoration of blood flow and successful surgery. In an attempt to further improve post ischemic tissue repair, we investigated the effect of a local administration of bone marrow derived stem cells (BMDSCs) in the presence or absence of immune-regulatory enzyme, IDO, in a murine model. A whole limb warm ischemia-reperfusion model was developed using IDO sufficient (WT) and deficient (KO) mice with C57/BL6 background. Twenty-four hours after injury, 5×10^5 cells (5×10^5 cells/200 μL of PBS solution) BMDSCs (Sca1 + cells) were injected intramuscularly while the control group received just the vehicle buffer (PBS). Forty-eight to seventy-two hours after limb BMDSC injection, recovery status including the ratio of intrinsic paw function between affected and normal paws, general mobility, and inflammatory responses were measured using video micrometery, flow cytometry, and immunohistochemistry techniques. Additionally, MRI/MRA studies were performed to further study the inflammatory response between groups and to confirm reconstitution of blood flow after ischemia. For the first time, our data, showed that IDO may potentially represent a partial role in triggering the beneficial effects of BMDSCs in faster recovery and protection against structural changes and cellular damage in a hind limb IR injury setting ($P=0.00058$).

Editor: Michael Platten, University Hospital of Heidelberg, Germany

Funding: Institutional Seed Money. The funders had no role in study design, data collection and analysis, decision to publish, or preparation of the manuscript.

Competing Interests: The authors have declared that no competing interests exist.

* E-mail: bbaban@gru.edu

Introduction

Ischemia-Reperfusion injury is seen in many clinical scenarios from stroke, myocardial infarction, traumatic limb injury, embolic events and organ transplantation. According to statistics from the CDC, nearly 2 million people a year in the US are affected by the above conditions [1]. Therefore, there is no shortage of opportunity to promote the health and lives of patients under each circumstance by improving our understanding of the mechanisms by which IR injury occurs, so that modulation and attenuation of injury can be achieved for therapeutic purposes. IR injury begins with the initial ischemic event leading to decreased oxygen and nutrient delivery. The resulting metabolic derangements in affected tissues lead to a buildup of acidic metabolites, dysfunction of organelles and build of free radicals, which include reactive oxygen and nitrogen species. Additionally, leukocytes become activated releasing cytokines and complement mediators, and inflammatory cascades are activated as a result of cell necrosis.

Moreover, these structural changes markedly increase mitochondrial dysfunction secondary to the ischemic injury, leading to increased membrane permeability, influxes of calcium, uncoupling of the electron transport chain and release of ROS [2]. Collectively, the response to IR injury leads to innate and adaptive immune activation, microvascular dysfunction, severe inflammation and ultimately, cell death.

Stem cells possess the capacity to self-renew and develop into functionally specialized cells. Implanted stem cells can be integrated into various host organs, survive, and reverse different ischemic deficits [3]. Adult stem cells and bone marrow stromal cells have now shown promise in experimental models of ischemic diseases including stroke, cardiac infarction, ischemic retinopathy, and limb ischemic injury. The plasticity of stem cells and the potential to manipulate their response depending on the milieu to which they are exposed has led to stem cell therapy (SCT) being an attractive field of investigation for its many potential clinical applications including ischemia-reperfusion (IR) injury. Numerous studies have shown stem cells to not only promote angiogenesis, but also to exert anti-inflammatory effects via paracrine and autocrine signaling [4–8]. The crosstalk that occurs between the stem cells and the microenvironment is the key to their observed response. The cytokines, chemokines and growth factors released and the receptors expressed by stem cells during this time, alters the milieu ultimately attenuating the inflammatory response and damage from IR injury, while helping promote and enhance tissue repair and recovery [9–12]. Our previous studies have shown that the immunomodulator leflunomide, which pharmacologically recruits bone marrow derived stem cells (BMDSC), in addition to inhibiting the proliferation of lymphocytes, helps protect renal tissue subjected to IR injury by up-regulating interleukin (IL)-10, down regulating IL-17 and IL-23, while persevering the mitochondrial membrane potential (ψm) and reducing overall cell death [13,14].

Indoleamine 2,3-dioxygenase (IDO) has emerged as a pivotal modulator/regulator of the immune response [13,28,34]. IDO is a heme-containing cytosolic enzyme that is the rate limiting catalyst to the metabolism of the essential amino acid tryptophan within the kynurenine pathway [15,16]. These genes contain interferon (IFN) response elements and therefore, IFNs are powerful inducers of IDO expression [23]. Over the last two decades, IDO has been investigated in the fields of immunology and microbiology to help develop a better understanding of its role within eukaryotic cells beyond its role in tryptophan metabolism. More recently, studies have shown IDO expression to be important in the ability for tumor cells to be immune tolerant [17–19]. Additionally, IDO helps in activation of t-regulatory cells, resulting in containment of hyperinflammatory responses and inhibition of t-helper 17 cells recruitment [14]. Importantly, stem cells express the complete kynurenine pathway indicative of its functional role(s) in stem cell biology. Up regulation of IDO expression using interferon-γ within mesenchymal stem cells (MSCs) have shown to reduce inflammatory conditions [8,20–22]. However, it remains to be established whether the status of stem cells with respect to IDO is a determinant of the local inflammatory response and prevention of tissue injury in the setting of IR injury. Therefore, developing a more profound understanding of the specific mechanisms by which SCT and IDO deliver their effects and how they are related will contribute to exploring the translational potential for more targeted and effective therapies in IR injury and other inflammatory conditions.

IDO is an important immune-modulator with the ability to help confer tolerance and suppression depending on the environmental milieu, and the therapeutic effects of stem cells in ischemic models. This article is to investigate whether the beneficial function of stem cells in the treatment of an ischemia reperfusion model is an IDO-dependent mechanism. To the best of our knowledge, this is the first study focused on the interaction of IDO and stem cells in a hind limb ischemia reperfusion model, which can be potentially translated into an effective therapeutic protocol in the treatment of hind limb ischemia reperfusion.

Materials and Methods

Animals

All animal procedures were performed in accordance with the approval of the Institutional Animal Care and Use Committee of Georgia Regents University. Male IDO sufficient wild-type (WT) and IDO deficient C57BL/6 mice (n = 5/group) were obtained (Jackson Laboratories, Bar Harbor, ME, USA).

Bone Marrow Derived Stem Cells Isolation

Bone marrow derived stem cell (BMDSCs) was isolated from the femur and tibia of IDO WTC57BL/6 mice 10–14 weeks of age. Briefly, marrow was flushed out of the bones by inserting a 27 g connected to a 3 ml syringe (filled with medium, RPMI-1640 plus FBS 10%) into the cavity and washing out the central core of cells into a Petri dish. To homogenize, any clumps was broken up by pipetting or with the plunger of the syringe followed by harvesting the cell suspension into a 15 ml tube through 70 mm nylon mesh to remove any smaller clumps of cells. Cell suspension then was span down (1500 rpm for 5 minutes) and re-suspended in medium. Using the magnetic-activated cell sorting (MACS) technique following manufacturer's instructions (Miltenyi Biotech, Auburn, CA), BMDSCs were separated out from total bone marrow cells by targeting Sca1 (marker for murine stem cells).

Induction of Ischemia and Treatment

To induce the whole limb warm ischemia-reperfusion model, both the arterial inflow and venous outflow were occluded via external constriction for 90 minutes using an elastic lasso device. Twenty-four hours after inducing limb IR injury, 5×10^5 BMDSCs were injected intramuscularly. Forty-eight to seventy-two hours after limb BMDSC injection, recovery status including ratio of toe spread between affected and normal paws, general mobility, and inflammatory responses were measured using video micrometer, and flow cytometry techniques. Additionally, MRI/MRA studies were performed to further study the inflammatory response between groups and to confirm reconstitution of blood flow after ischemia.

Analytical Flow Cytometry

Soft tissues from the site of injury were sieved through a cell strainer (BD Biosciences, San Diego, CA), followed by centrifugation (1,500 rpm, 10 min) to prepare single-cell suspensions. Phenotypic and intracellular analyses of cells were performed as described previously [28,38]. Briefly, samples were treated with reagents, and stained with fluorochrome-conjugated antibodies of interest based on manufacturers' instructions. Antibodies against Sca1 (marker of stem cells), Annexin V, and 7-aminoactinomycin D (7AAD; for apoptosis/necrosis), and FoxP3 (for Tregs) were obtained from BD Biosciences (San Diego, CA). Further, IL-23, IL-17, and IL-10 antibodies were purchased from eBioScience (San Diego, CA). Next, cell samples were run through a four-color flow cytometer (FACS Calibur, BD Biosciences, San Diego, CA), and data were collected using CellQuest software. Samples were double-stained with control IgG, and cell markers were used to assess any spillover signal of fluorochromes; proper compensation was set to ensure the median fluorescence intensities of negative and positive cells were identical and were both gated populations. Gating was used to exclude dead cells and debris using forward and side scatterplots. As a gating strategy for each sample, isotype-matched controls were analyzed to set the appropriate gates. For each marker, samples were analyzed in duplicate measurements. To minimize false-positive events, the number of double positive events detected with the isotype controls were subtracted from the number of double-positive cells stained with corresponding antibodies (non-isotype control), respectively. Cells expressing a specific marker were reported as a percentage of the number of gated events.

Immunohistochemical Analysis

Formalin-fixed paraffin-embedded soft tissues were cut in 4 μm sections and processed for immunohistochemical assessment of IDO, IL-17, IL-23, IL-10 and FOXP3 according to previously described protocols [28,38]. All antibodies for immunohistochemistry analysis were purchased from Santa Cruz Biotechnology, USA.

Video Micrometer/Toe Spread

A 60 sec video recording of the animal was created in a clear plastic cage with all beddings removed to allow for an unimpeded view of the extremities. At 5 random time points every 5–

10 seconds, the video frame was frozen and the picture magnified and examined for quality adequacy. We used a modified version of walking track analysis to measure the functional outcome of stem cell (SC) injection in ischemia-reperfusion (IR) injury, as described by Lee *et al.* [40]. Briefly, the toe spread was measured in the IR limb (Ti) and control contralateral limb (Tc). The ratio of the toe spread of the paw of the IR limb (Ti) to the control limb (Tc) was then calculated by Ti/Tc. As shown, 100% indicated equal width and, thus normal intrinsic function. If both the right and left hind paws could be seen, then it was deemed acceptable and the width of the hind paws were measured using a micrometer and the ratio of the smaller one to the larger one was calculated. The individual performing this part of the outcome measure was blinded in regards to the identity and treatment history of the animal in the video being analyzed. Standard descriptive statistics, including mean and standard deviations, were then calculated and the inferential statistics were done using two-tailed student t-test. Alpha was set at 0.05 and with power at 80%.

MRI/MRA Imaging Acquisition

MRI images were acquired for physiological characterization of mouse thigh and calf muscle during and after the application of a tourniquet for 90 minutes. Additionally, MRA studies were performed to confirm the occlusion and reconstitution of blood flow to validate our model. Adult male wild-type and IDO knockout C57/BL6 background mice were imaged using MRI following the guidelines of the Animal Care and Use Committee and the Animal Health and Care Section of GRU.

Initially, each mouse underwent 3 imaging sessions. They were imaged while the tourniquet was on, immediately after the tourniquet was released and 48–72 hours after injury. The purpose of performing the MRA imaging sequences was to confirm that tourniquet application occluded blood flow and that the blood flow reconstituted once the tourniquet was released. Once these initial parameters were confirmed, subsequent experimental groups were imaged 48–72 hours after their treatment was administered. There were a total of 4 experimental groups imaged: WT mice treated with stem cells (n = 4) or PBS (n = 3), and the IDO KO mice treated with stem cells (n = 4) and PBS (n = 4). The animals were anesthetized with a mixture of medical air, oxygen (1:1), and 2.5% isoflurane. After positioning the animal on a cradle to maintain a constant body temperature (37.8°C), the animal was maintained on ~1.25% isoflurane throughout the MRI session. For the purposes of constraining motion, the abdomen of each mouse was secured to the cradle using medical tape.

Images were acquired on a 7.0T 20 cm bore BioSpec MRI spectrometer (Bruker Instruments, Billerica, MA, USA). A standard transmit/receive volume coil (35 mm i.d.) was used for imaging. ECG and respiratory signals were monitored by a physiological monitoring system (SA Instruments, INC., Stony Brook, NY). For investigation of tissue edema, a 2D RARE sequence was used to acquire T2-weighted axial images (TE_{eff}/TR = 12/2890msec; RARE Factor = 8; # of slices = 24; slice thickness = 1 mm; slice gap = 0.5 mm; # of averages = 5; Matrix = 256×256; FOV = 32.0×32.0 mm^2). The imaging volume was set to begin just superior to the hip joint at the pelvis. For all mice, the stack of imaging slices subtended a volume reaching to the bottom of the calf. For visualization of reperfusion of major leg vessels post-tourniquet, images were collected in a subset of animals during application of and immediately subsequent to removal of a tourniquet using a 2D time-of-flight MR Angiography (2D TOF MRA) sequence (TE/TR = 4/19.5 msec; # of slices = 90; slice thickness = 0.3 mm; slice gap = 0.1 mm; # of averages = 4; Matrix = 352×320; FOV = 32.0×32.0 mm^2). In the case of imaging during application of the tourniquet and shortly thereafter, the mouse cradle was moved out of the scanner. The tourniquet was removed with the animal attached to the cradle so that it could be relocated to a similar position for scanning post-tourniquet.

MRI/MRA Imaging Processing

The mice from each of the 4 groups underwent T2 weighted imaging of the injured and non-injured limbs. Using Image J to process the images taken [23], measurements of the soft tissue intensity were taken from the anterior lateral, anterior medial and posterior thigh in each leg in 3 consecutive image slices distal to the level of the injury. Given that T2 weighted imaging enhances with increases in soft tissue inflammation, such measurements were used to help correlate clinical findings of response to injury and therapy. The average intensity measured from each leg was then used to create a ratio comparing the intensity of the injured leg to the contra-lateral leg. Once each ratio was obtained, the measured ratios were compared to one another to evaluate the response to the different treatments and then plotted onto representative graphs.

MRA images were processed for display and qualitative inspection of vessel clamping during application of the tourniquet, using Image J [23]. Images were loaded into the software, and vessels were manually segmented into groups including left leg, right leg, and tail. Each of these groups was stacked into separate maximum intensity projection images (MIPs). After each group was assigned a color (red, green, or blue), the MIPs were summed together into a composite image for easy inspection of vessels.

Statistical Analysis

All data are presented as mean ± SEM. Laser Doppler measurements were analyzed using SPSS 17.0 repeated measures ANOVA and LSD post-hoc test. In all cases, p-value ≤0.05 was considered to be significant.

Results

Stem Cells Insert their Beneficial Effects in an IDO Dependent Fashion

One of the first parameters measured the response each treatment group had from a clinical standpoint. After the IR injury was induced, the mice were observed and videoed to follow their recovery. Forty-eight hours after treatment (72 hours after the initial injury), each mouse, in a blinded manner, was observed looking at their mobility and intrinsic paw function. As described in above, the intrinsic function of each mouse was observed and measured. In the SC treated wild type, the average Ti/Tc was 83.2%. In the wild type with the same IR but no SC, the average Ti/Tc = 57.4%. The improvement by SC in wild type, as measured by toe spread is thus [wild type Ti/Tc with SC]–[wild type Ti/Tc without SC], or 35.8%, and when normalized to the wild type Ti/Tc without SC, by the equation: [Δ Ti/Tc]/[Ti/Tc without SC], or [wild type Ti/Tc with SC – wild type Ti/Tc without SC]/[wild type Ti/Tc without SC] = 35.8/57.4, or 62.4%. In IDO knockout (IDO ko) mice, the same IR injury without SC caused Ti/Tc to decrease to 37.2%. SC was able to increase it in IDO ko group with the same IR injury to 59.4%. [IDO ko Ti/Tc with SC] – [IDO ko Ti/Tc without SC] = 59.4%–37.2% = 22.2%. Using the same equation to calculate the normalized magnitude of improvement, [IDO ko Ti/Tc with SC–IDO ko Ti/Tc without SC]/[IDO ko Ti/Tc without SC] = 22.2/37.2 = 59.7%. The improvement by SC was greater in the wild type (35.8%) compared to the IDO ko group (22.2%)

and, though less strong, this still holds when normalized to the controls, 62.4% versus 59.7%. As Figure 1 shows, the WT mice treated with stem cells demonstrated the greatest recovery with 83.2% recovery of intrinsic function of the injured paw to the non-injured paw (*p-value* = 0.0027), compared to the WT group treated with PBS, or a 45% increase in recovery was seen showing the efficacy of stem cell therapy alone in the presence of an environment where IDO expression is present. Additionally, when the WT group treated with SC was compared with its KO counterpart, a 40% increase in recovery (*p-value* = 0.04) is seen.

Enhancement of Soft Tissues in the Injured Legs

As Figure 2a shows, enhancement of the soft tissue can be observed within the injured legs compared to the non-injured legs. Image J was used to calculate the enhancement seen in the soft tissues within the injured leg. As shown in Figure 2b, WT mice treated with stem cells showed the lowest ratio of soft tissue enhancement, comparing the injured to non-injured leg, indicating the stem cell therapy in the WT mice had the greatest ability to recover.

Stem Cells Enhance IDO Expression in Tissue Microenvironment

As demonstrated in Figure 3 (panel a), immunohistochemical assessment indicated the fact that IDO expression was increased in WT mice treated with SCs after injury compared to other groups. Indeed, as shown in Figure 3, enhancement of IDO expression is mediated by SCs as evidenced by both higher expression of IDO in treated WT compared to non-treated WT, as well as by minimum expression of IDO in IDO KO animals treated with SCs compared to no expression of IDO in non-treated IDO KO animals (as SCs were the only source of IDO in these KO recipient animals). These results suggest that SCs are inserting their regulatory effects into the tissue microenvironment through a variety of mechanisms of which, enhancement of IDO may potentially be a crucial determinant of recovery in a hind limb IR injury.

Stem Cells Attenuate Inflammation with their Most Profound Effects in the Presence of IDO

Immunohistochemical analysis. Consistent with our clinical observation and imaging results, histological and immunohistocheical analysis demonstrated the significant reduction in the expression level of IL-17, IL-23 in WT mice treated with SCs compared to the other experimental groups (Figure 3, panels C and D). Meanwhile, the expression level of FOXP3 (Treg marker) and IL-10 were increased in WT mice treated with SCs compared to the other experimental groups (Figure 3, panels B and E). These results suggest that IDO may be an effective factor in the efficiency and beneficial impact of SCs in a hind limb IRI setting, which may imply a therapeutic application based on SCs and enhancement of IDO.

Analytical flow cytometry. As demonstrated in Figure 4, in accordance with our histological data and consistent with our clinical observation, flow cytometry analysis showed WT mice treated with stem cells attenuate the injury-induced inflammatory response (*p-value*<0.04). Figure 4 shows that the expression of cytokines IL-17, well-established marker for inflammation derived from T-helper-17 (Th-17) cells, and IL-23, a powerful pro-inflammatory cytokine, was lower in the tissue from the WT mice treated with stem cells than any other group, while in IDO KO group, both IL-17 and IL-23 expression were the highest. More importantly, the regulatory T cells (FOXP3+T regs), anti-inflammatory index, were detected 8.75 times higher in the WT mice treated with SC compared to the KO group treated with SC. Moreover, IL-10, an anti-inflammatory cytokine, was 56% higher in the WT group treated with SC when compared to its KO counterpart at 12.8% (compared to 8.2%) (Figure 4c). These findings reaffirm the important anti-inflammatory effects of stem

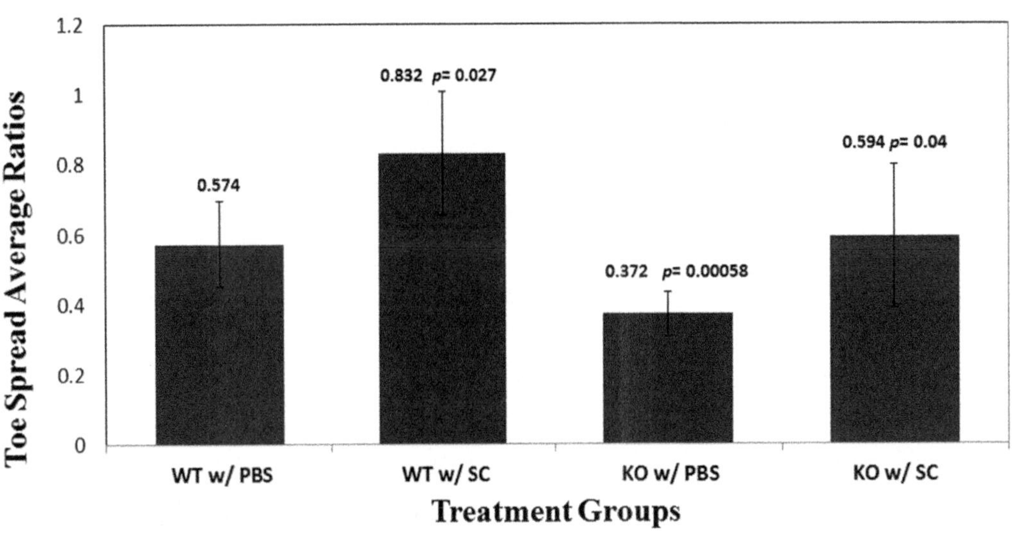

Figure 1. Treatment Effect on Toe Spread Ratio Averages (48–72 hours after treatment). The outcome of stem cell (SC) therapy indicates that IDO may improve recovery. IDO-KO mice treated with SC demonstrated an accelerated recovery compared with IDO-KO treated with PBS (p-value <0.05). However, the WT mice treated with SC showed the greatest recovery of intrinsic paw function when expressed as a ratio comparing it to the non-injured paw (p-value = 0.027). Functional recovery from ischemia-reperfusion (IR) injury in the different treatment groups was measured, using a modified version of walking track analysis. For each subject, toe spread was measured in the IR limb (Ti) and control contralateral limb (Tc). The ratio of the toe spread in the injured limb (Ti) to the control limb (Tc) was then calculated by Ti/Tc. A ratio of 1 indicates 100% recovery or equal width and thus normal intrinsic function. When comparing the WT group treated with stem cells to those treated with PBS, a 45% increase in recovery was seen demonstrating the efficacy of stem cell therapy alone in the presence of an environment where IDO expression is present.

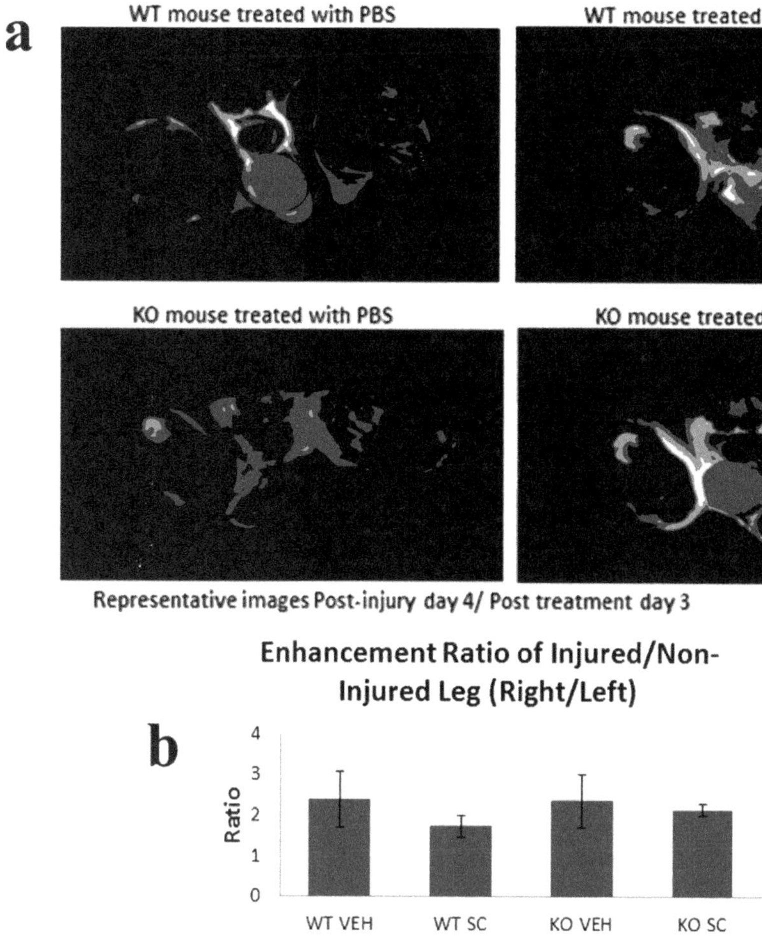

Figure 2. T2 weighted images. a) Representative T2 weighted imaging cuts where the enhancement of the soft tissue can be observed within the injured legs and compared to the non-injured legs and, **b)** The average ratio of enhancement measured between the injured leg over the non-injured leg are shown in this representative graph for each of the 4 groups.

cell therapy and the critical role IDO plays in the efficacy of stem cell therapy.

Stem Cells Reduces Cell Death with their Most Profound Effects in the Presence of IDO

As shown in Figure 5a, while an IR insult resulted in a marked increase in cell death (both apoptosis and necrosis) in the ischemic-reperfused limbs, treatment with SCs resulted in a marked decrease in cell death in the ischemic-reperfused limbs with a higher rate of reduction in the presence of IDO compared to IDO KO mice. Using flow cytometry analysis, markers for necrosis (7-aminoactinomycin D, 7-AAD) and apoptosis (Caspase 3) were measured and consistent with all results shown above. The lowest levels of both markers were seen in the WT group treated with SC by a considerable margin when compared to the corresponding groups and highest in the two KO groups. When analyzing data for Caspase 3, the WT group treated with stem cells showed an average of 2.4% apoptotic expression compared to 5.4% in the vehicle treated WT group and 9% in the KO group treated with stem cells. The studies of necrosis showed similar trends with an average of 4% necrosis in the WT treated with stem cells group, while the vehicle treated WT mice showed an average of 8.8% cell necrosis and the vehicle treated KO group up to 10.2% average cell necrosis. Importantly, as is showed in Fig. 5b, CD45 (a marker to differentiate the hematopoietic cells from non-hematopoietic ones), was used to show that the effect of SCs cell on reducing apoptosis/necrosis was mainly on non-hematopoietic cells, leading to restoring tissue viability, healing and protection. These findings suggest that the adverse effects of ischemia reperfusion can be reduced by insertion of SCs into an IDO sufficient micro-environment, resulting in less cell death and faster recovery, supporting our hypothesis that IDO may play a critical role in enhancing the beneficial effects of stem cell therapy.

Discussion

We demonstrated that intramascular injections of BMDSC decrease the recovery time after IR injury in hind limb ischemia reperfusion injury. Most importantly, our results suggest that IDO modulatory feature may play a potential role in beneficial effects of BMDSCs in an ischemia reperfusion setting of hind limb injury, manifested by marked anti-inflammatory effect of these cells and reduction of apoptosis/necrosis inside the tissue microenvironment.

Ischemia reperfusion injury results from the culmination of events starting with the initial ischemic insult which leads to metabolic, cellular, immunologic and environmental changes and ultimately inflammation and cell death [2,24,25]. The immunologic response, including the release of inflammatory mediators

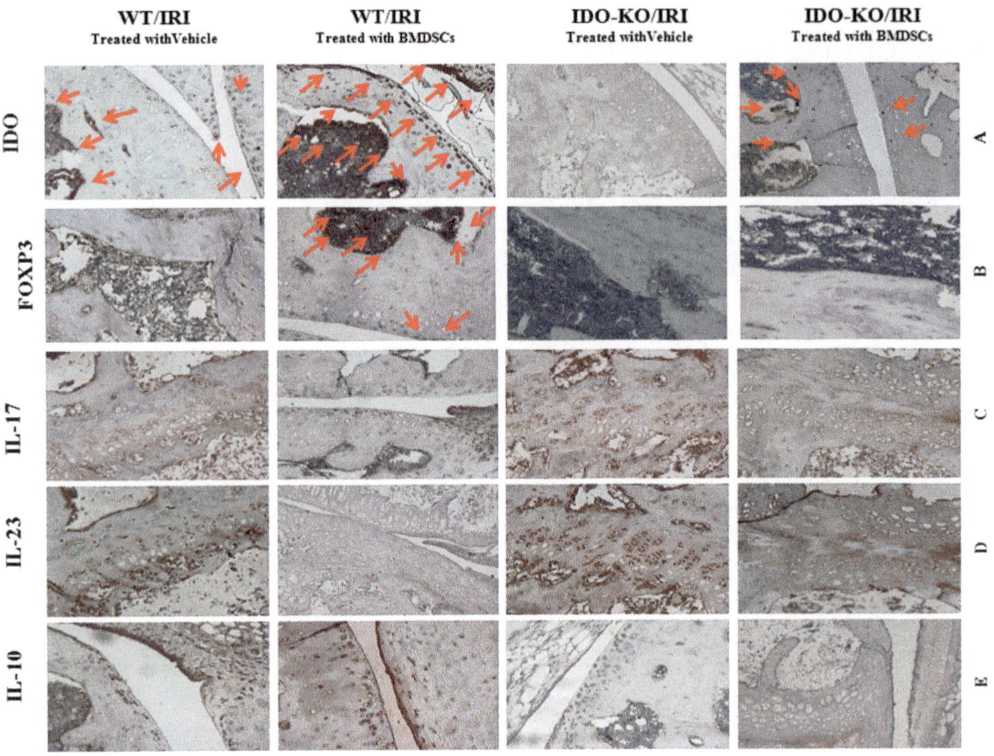

Figure 3. BMDScs can enhance IDO and regulatory T cells while reducing inflammatory cytokines in the hind limb IR injury. Immunohistochemical analysis of paraffin embedded tissues from murine model with IRI of hind limb showed that treating the animals with BMDSCs in an IDO sufficient microenvironment first: increased IDO and FOXP3 expression (panels A and B, red arrows), while decreased the inflammatory cytokines, IL-17 and IL-23 (panels C and D). Anti inflammatory cytokine, IL-10, was increased as demonstrated in panel E. All together, these analysis suggest a potential therapeutic role for BMDSCs, re-enforced by possible IDO dependent mechanisms. All pictures are 400X magnification.

and recruitment of cells, can be a key to modulating the response, and ultimately the degree of damage and cell death seen as the result of a given insult. The immunologic significance of IDO has been shown over the years from its involvement in host defense against pathogens, tumor immune evading properties and embryo survival from the mother's immune responses to name a few [17–20,26–34]. Our goal was to better understand IDO's role in the immunologic response seen with IR injury as it relates to stem cell therapy. This knowledge can be used to favorably decrease the injury which is extremely valuable given the myriad of clinical conditions by which such therapeutic potentials can be used. Stem cell therapy has been described to lessen the inflammatory response and injury that results from IR through several mechanisms including paracrine and autocrine signaling, neutralization of ROS, promotion of angiogenesis, and modulation of the innate and adaptive host responses as described previously [4,5,9,11,12,35,36]. Our results bring to the forefront the novel idea that IDO, a known modulator of the immune response pathways, plays a critical role in the anti-inflammatory effects seen with stem cell therapy in the setting of IR injury. One approach that is readily translatable into clinical use is the up-regulation of IDO by small stable compounds such as CpG oligodeoxynucleotides (CpG-ODNs) alone or in conjunction with BMDSCs. CpG-ODN has been shown since 2005 to increase IDO-dependent T-regulatory cell function [28,38,39]. Any ischemic event, such as organ transplantation, revascularization after myocardial infarction, stroke, limb re-vascularization, and micro-vascular tissue transfer, CpG-ODNs can be given IV after blood flow is restored to reduce the effects of IR injury induced inflammation.

Stem cell use for clinical application has been researched extensively, including for IR injury. However, the ability to control stem cell response with consistent results has been unsuccessful, in part, because of an incomplete understanding of how they confer their effects. In the setting of IR injury using a murine model, our experiments showed injury to the hind limb improved more rapidly and returned to more normal intrinsic function in both the WT and KO groups when compared to their vehicle treated counterpart. Using the non-injured leg as a relative control and indication of both baseline function, and tissue properties, clinically observed intrinsic function and tissue inflammation could be compared across all groups. We consistently found that stem cell therapy, especially in the presence of IDO, greatly improved both the intrinsic function observed and attenuated the inflammatory response, and therefore lessened the soft tissue edema seen in the T2 weighted studies. These observations are congruent with previous studies citing the anti-inflammatory properties of BMDSCs.

In order to further investigate the anti-inflammatory effects of BMDSC therapy appreciated in our MRI studies, we also performed a variety of IHC and flow cytometry studies. The inflammatory response as indicated by the expression of IL-17 and the anti-inflammatory markers IL-10 and FoxP3+ confirmed the anti-inflammatory properties of stem cell therapy, especially in the WT group. The attenuated inflammatory response as indicated by decreased expression of IL-17 and increased IL-10 and FoxP3+ also seems to have a direct effect on the end result of cell death as indicated by the percent of cells undergoing apoptosis and necrosis. Our data showed that most of the IL-10 expressing cells were non-CD4 cells, suggesting that other type of cells rather than

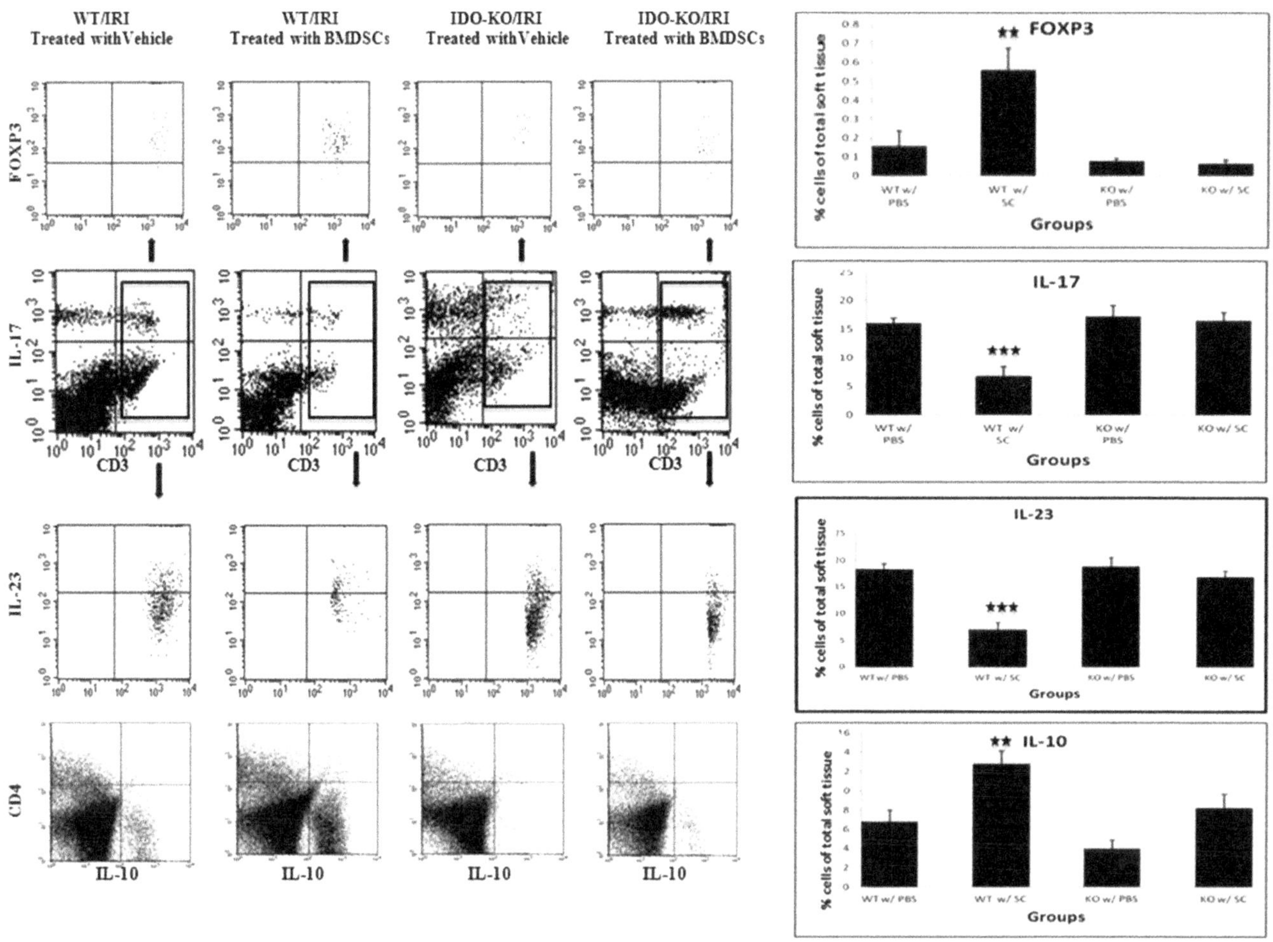

Figure 4. Regulatory effects of SCs may be partially associated with IDO presence in tissue microenvironment. Dot Plots are representatives of flow cytometric analysis of live single cell preparations from murine hind limb tissues with IRI showed that in the presence of IDO, injection of BMDSCs decreased CD3+ cells (T cells), IL-17 and IL-23 expression (IL-23 is based on gating of IL-17+ cells). Further, applying BMDCs could increase both CD3+ FOXP3+ cells (Tregs, gated on CD3+ cells) and IL-10 expression in an IDO sufficient microenvironment. Bargraphs on the right side of each group are reflecting the accumulative analysis of five animal per experimental group. Statistical analysis was performed using Kruskal-Wallis test with Dunn's multiple-comparison post- hoc test. * = $p<0.05$, ** = $p<0.01$, *** = $p<0.001$.

CD4+ cells, including (not limited to) bone marrow derived stem cells, CD8+ lymphocytes, as well as dendritic cells may be the potential source of IL-10. Several reports previously have demonstrated the production of IL-10 by SCs, monocytes and non-CD4 lymphocytes [41–43]. However, these findings provide additional evidence that SCs may alter cytokine production profile from a pro-inflammatory (e.g., IL-17) to an anti-inflammatory (e.g., IL-10) status in a paracrine/autocrine fashion.

The observations of improved clinical recovery decreased tissue edema, attenuated markers of inflammation and decreased overall cell death seen with BMDSCs and especially in the WT group which bring us to our last point about the critical role IDO plays in this process.

In studies from both our team and others, IDO has shown to be a vital enzyme used by cells to modulate the host immune response depending on the environmental ques. In instances of pathogenic insults, IDO has been shown to help up-regulate the immune response in a host protective manner. During pregnancy, IDO expression seen in trophoblasts surrounding the embryo have been shown to be crucial to maintaining the embryo and protecting it from T cell mediated rejection [37]. As cited earlier, many other experiments have shown the duality of IDO which is dependent on the timing of expression. The cells which express it and the micro environmental conditions at the time of expression help determine the effects its expression has on immune response. In the experiments performed during this study, we have repeatedly shown that the effects of SCT therapy have an impressive anti-inflammatory effect with down-regulation of inflammatory mediators including IL-17, IL-23 and increased expression of anti-inflammatory index IL-10 and FoxP3+. Additionally, the percent of cells (non-hematopoitic cells) undergoing apoptosis and necrosis at 48–72 hours after BMDSC treatment in the WT group were less than half of any other group and correlate well with the clinical findings of improved recovery, mobility and intrinsic function. Additionally, these findings were supported with MRI T2 weighted studies showing less edema and inflammation within the soft tissues of the stem cell treatment mice compared to the non-treated mice and a significant difference ($p<0.05$, $n=4$, using 2 tailed t-test) between the IDO WT and KO mice treated. The constellation of these findings suggests the protective effects of SCT are critically dependent on the expression of IDO to induce its anti-inflammatory effects. Without IDO present in the environment, inflammation induced by the IR injury is more

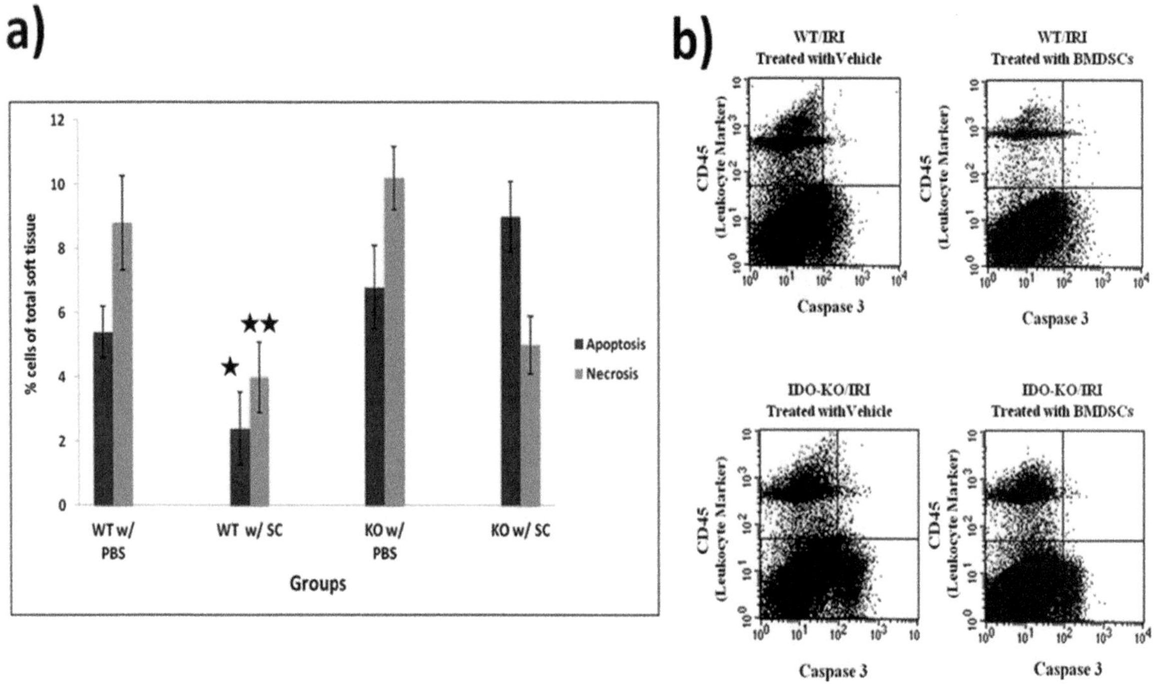

Figure 5. SCs treatment reduces cell death in an IDO dependent manner. Flow cytometry analysis showed (5a) that stem Cells reduce cell death (apoptosis/necrosis) with their most profound effects in the presence of IDO. Further, flow cytometric analysis demonstrated that a majority of reduction in cell death occurred among parenchymal cells (CD45 negative) and not infiltrated leukocytes (CD45+), indicating the beneficial impact of SCs in reducing cell death and necrotic responses, leading to less inflammation, more protection and faster recovery. Statistical analysis was performed using Kruskal-Wallis test with Dunn's multiple-comparison post- hoc test. * = $p<0.05$, ** = $p<0.01$, *** = $p<0.001$.

pronounced and so is necrosis and apoptosis of the tissues, which leads to a longer recovery time seen clinically.

The future of IDO for clinically relevant therapeutic interventions and the responsible mechanisms need to be investigated in more detailed and comprehensive studies. Antibiotics cannot be developed before we understand the properties of the pathogen and how it attacks the host. Similarly, the most important step toward being able to tap into IDO's therapeutic potential is by understanding the mechanisms and principles, which it modulates and interacts with the immune system during various clinical scenarios. Teasing out the mechanisms of cross talk between different intra and intercellular pathways during inflammatory conditions like IR and relating them in response to new ideas and findings such as IDO will allow us to more appropriately and effectively apply these concepts towards new therapies. These studies were done to help define the importance of IDO's presence during stem cell therapy for IR injury, but its translation is not limited to SCT alone, but all incidents were IDO and inflammation are involved. IDO has been linked to a tumors ability to remain immune tolerant and so modulation of IDO expression could potentially help therapeutic measures like chemotherapy and radiation be more effective in our ability to fight oncological disease. Ultimately, allograft transplantation fails in the long term due to a slow, T-cell mediated rejection of the organ. It has been discussed that IDO inducers and enhancement within the tissue can help protect the organ from such rejection much in the same way trophoblasts protect the embryo from the mother's immune system [28,38,39]. With continued efforts to more profoundly understand the functions of IDO and how it influences states of inflammation and immune response, the potential clinical implications are vast and deserve further intense investigation.

In conclusion, the beneficial anti-inflammatory effects of SCs may be partially actuated through an IDO dependent mechanism, which not only shortens the recovery time, but also confers significant protection against cell death in the ischemic re-perfused limb. These effects were associated with increased recruitment and mobilization of Tregs as well as regulatory cytokines such as IL-10. Collectively, the results suggest that the immune-modulatory role of IDO in SCs may offer a novel approach to protecting the hind limb against acute injury secondary to an IR insult. However, the novel demonstration of the prominent impact of IDO on mobilization and function of stem cells is not only of clinical relevance and significance for the hind limb ischemia reperfusion injury, but for other immune and inflammatory disorders, including organ transplant and cancer.

Acknowledgments

This study was supported by institutional funds (provided to BB) and the Milford B. Hatcher Endowment (provided to JY). The authors are grateful to Mrs. Michelle Barnes for her administrative support.

Author Contributions

Conceived and designed the experiments: M. Masoumy JYL JY M. Mozaffari CM NY BB. Analyzed the data: M. Masoumy JYL JY CM NY FL BB. Contributed reagents/materials/analysis tools: JY CM NY FL BB. Wrote the first draft of the manuscript: M. Masoumy JY BB. Contributed to the writing of the manuscript: M. Masoumy JY NY CM BB. Criteria for authorship read and met: M. Masoumy JY JYL NY CM FL M. Mozaffari BB. Agree with manuscript results and conclusions: M. Masoumy JY JYL NY CM FL M. Mozaffari BB.

References

1. Centers for Disease Control and Prevention website. Available:http://www.cdc.gov/nchs/fastats. Accessed 2014 April 1st.
2. Eltzschig HK, Eckle T (2011) Ischemia and reperfusion–from mechanism to translation. Nat Med 17(11): 1391–401.
3. Cogle CR, Guthrie SM, Sanders RC, Allen WL, Scott EW et al. (2003) An overview of stem cell research and regulatory issues. Mayo Clin Proc 78(8): 993–1003.
4. Castillo M, Liu K, Bonilla L, Rameshwar P (2007) The immune properties of mesenchymal stem cells. Int J Biomed Sci 3(2): 76–80.
5. Dazzi F, Lopes L, Weng L (2012) Mesenchmal stromal cells: a key player in 'innate tolerance'? Immunology 137(3): 206–13.
6. Di Nicola M, Carlo-Stella C, Magni M, Milanesi M, Longoni PD et al. (2002) Human bone marrow stromal cells suppress T-lymphocyte proliferation induced by cellular or nonspecific mitogenic stimuli. Blood 99(10): 3838–43.
7. Le Blanc K, Pittenger M (2005) Mesenchymal stem cells: progress toward promise. Cytotherapy 7(1): 36–45.
8. Shi Y, Su J, Roberts AI, Shou P, Rabson AB et al. (2012) How mesenchymal stem cells interact with tissue immune responses. Trends Immunol 33(3): 136–43.
9. Gnecchi M, Zhang Z, Ni A, Dzau VJ (2008) Paracrine mechanisms in adult stem cell signaling and therapy. Circ Res 103(11): 1204–19.
10. Wang H, Wu M, Liu Y (2012) Are mesenchymal stem cells major sources of safe signals in immune system? Cell Immunol 272(2): 112–6.
11. Yang SH, Park MJ, Yoon IH, Kim SY, Hong SH et al. (2009) Soluble mediators from mesenchymal stem cells suppress T cell proliferation by inducing IL-10. Exp Mol Med 41(5): 315–24.
12. Salem HK, Thiemermann C (2010) Mesenchymal stromal cells: current understanding and clinical status. Stem Cells 28(3): 585–96.
13. Baban B, Liu JY, Mozaffari MS (2012) Aryl hydrocarbon receptor agonist, leflunomide, protects the ischemic-reperfused kidney: role of Tregs and stem cells. Am J Physiol Regul Integr Comp Physiol 303(11): R1136–46.
14. Casado FL, Singh KP, Gasiewicz TA (2011) Aryl hydrocarbon receptor activation in hematopoietic stem/progenitor cells alters cell function and pathway-specific gene modulation reflecting changes in cellular trafficking and migration. Mol Pharmacoln 80(4): 673–82.
15. Ball HJ, Yuasa HJ, Austin CJ, Weiser S, Hunt NH (2009) Indoleamine 2,3-dioxygenase-2; a new enzyme in the kynurenine pathway. Int J Biochem Cell Biol 41(3): 467–71.
16. Sugimoto H, Oda S, Otsuki T, Hino T, Yoshida T et al. (2006) Crystal structure of human indoleamine 2,3-dioxygenase: catalytic mechanism of O2 incorporation by a heme-containing dioxygenase. Proc Natl Acad Sci U S A 103(8): 2611–6.
17. Munn DH (2006) Indoleamine 2,3-dioxygenase, tumor-induced tolerance and counter-regulation. Curr Opin Immunol 18(2): 220–5.
18. Munn DH, Sharma MD, Hou D, Baban B, Lee JR et al. (2004) Expression of indoleamine 2,3-dioxygenase by plasmacytoid dendritic cells in tumor-draining lymph nodes. J Clin Invest 114(2): 280–90.
19. Uyttenhove C, Pilotte L, Theate I, Stroobant V, Colau D et al. (2003) Evidence for a tumoral immune resistance mechanism based on tryptophan degradation by indoleamine 2,3-dioxygenase. Nat Med 9(10): 1269–74.
20. Croitoru-Lamoury J, Lamoury FM, Caristo M, Suzuki K, Walker D et al. (2011) Interferon-gamma regulates the proliferation and differentiation of mesenchymal stem cells via activation of indoleamine 2,3 dioxygenase (IDO). PLoS One 6(2): e14698.
21. Matysiak M, Stasiolek M, Orlowski W, Jurewicz A, Janczar S et al. (2008) Stem cells ameliorate EAE via an indoleamine 2,3-dioxygenase (IDO) mechanism. J Neuroimmunol 193(1–2): 12–23.
22. Opitz CA, Litzenburger UM, Lutz C, Lanz TV, Tritschler I et al. (2009) Toll-like receptor engagement enhances the immunosuppressive properties of human bone marrow-derived mesenchymal stem cells by inducing indoleamine-2,3-dioxygenase-1 via interferon-beta and protein kinase R. Stem Cells 27(4): 909–19.
23. Abramoff MD, Magelhaes PJ, Ram SJ (2004) Image Processing with ImageJ. Biophotonics International 11: 36–42.
24. Blaisdell FW (2002) The pathophysiology of skeletal muscle ischemia and the reperfusion syndrome: a review. Cardiovasc Surg 10(6): 620–30.
25. Boros P, Bromberg JS (2006) New cellular and molecular immune pathways in ischemia/reperfusion injury. Am J Transplant 6(4): 652–8.
26. Metz R, Duhadaway JB, Kamasani U, Laury-Kleintop L, Muller AJ et al. (2007) Novel tryptophan catabolic enzyme IDO2 is the preferred biochemical target of the antitumor indoleamine 2,3-dioxygenase inhibitory compound D-1-methyl-tryptophan. Cancer Res 67(15): 7082–7.
27. Munn DH, Zhou M, Attwood JT, Bondarev I, Conway SJ et al. (1998) Prevention of allogeneic fetal rejection by tryptophan catabolism. Science 281(5380): 1191–3.
28. Baban B, Chandler PR, Sharma MD, Pihkala J, Koni PA et al. (2009) IDO activates regulatory T cells and blocks their conversion into Th17-like T cells. J Immunol 183(4): 2475–83.
29. Daubener W, Spors B, Hucke C, Adam R, Stins M et al. (2001) Restriction of Toxoplasma gondii growth in human brain microvascular endothelial cells by activation of indoleamine 2,3-dioxygenase. Infect Immun 69(10): 6527–31.
30. Hwu P, Du MX, Lapointe R, Do M, Taylor MW et al. (2000) Indoleamine 2,3-dioxygenase production by human dendritic cells results in the inhibition of T cell proliferation. J Immunol 164(7): 3596–9.
31. Kaufmann SH (1993) Immunity to intracellular bacteria. Annu Rev Immunol 11: 129–63.
32. Mellor AL, Munn D, Chandler P, Keskin D, Johnson T et al. (2003) Tryptophan catabolism and T cell responses. Adv Exp Med Biol 527: 27–35.
33. Pfefferkorn ER (1984) Interferon gamma blocks the growth of Toxoplasma gondii in human fibroblasts by inducing the host cells to degrade tryptophan. Proc Natl Acad Sci U S A 81(3): 908–12.
34. Sharma MD, Baban B, Chandler P, Hou DY, Singh N et al. (2007) Plasmacytoid dendritic cells from mouse tumor-draining lymph nodes directly activate mature Tregs via indoleamine 2,3-dioxygenase. J Clin Invest 117(9): 2570–82.
35. Bartholomew A, Sturgeon C, Siatskas M, Ferrer K, McIntosh K et al. (2002) Mesenchymal stem cells suppress lymphocyte proliferation in vitro and prolong skin graft survival in vivo. Exp Hematol 30(1): 42–8.
36. Giordano A, Galderisi U, Marino IR (2007) From the laboratory bench to the patient's bedside: an update on clinical trials with mesenchymal stem cells. J Cell Physiol 211(1): 27–35.
37. Baban B, Chandler P, McCool D, Marshall B, Munn DH et al. (2004) Indoleamine 2,3-dioxygenase expression is restricted to fetal trophoblast giant cells during murine gestation and is maternal genome specific. J Reprod Immunol 61(2): 67–77.
38. Baban B, Chandler PR, Johnson BA, 3rd, Huang L, Li M et al. (2011) Physiologic control of IDO competence in splenic dendritic cells. J Immunol 187(5): 2329–35.
39. Mellor AL, Baban B, Chandler PR, Manlapat A, Kahler DJ et al. (2005) Cutting edge: CpG oligonucleotides induce splenic CD19+ dendritic cells to acquire potent indoleamine 2,3-dioxygenase-dependent T cell regulatory functions via IFN Type 1 signaling. J Immunol 175(9): 5601–5.
40. Lee JY, Giusti G, Wang H, Friedrich PF, Bishop AT et al. (2013) Functional evaluation in the rat sciatic nerve defect model: a comparison of the sciatic functional index, ankle angles, and isometric tetanic force. Plast Reconstr Surg 132(5): 1173–80.
41. Harizi H1, Gualde N (2006) Pivotal role of PGE2 and IL-10 in the cross-regulation of dendritic cell-derived inflammatory mediators. Cell Mol Immunol 3(4): 271–7.
42. Milwid JM1, Ichimura T, Li M, Jiao Y, Lee J et al. (2012) Secreted factors from bone marrow stromal cells upregulate IL-10 and reverse acute kidney injury. Stem Cells Int. 2012: 392050.
43. Ming Li and Susumu Ikehara (2013) Immunomodulatory Properties and Therapeutic Application of Bone Marrow Derived-Mesenchymal Stem Cells. J Bone Marrow Res 1: 4.

18

Leukemic Stem Cell Frequency: A Strong Biomarker for Clinical Outcome in Acute Myeloid Leukemia

Monique Terwijn[1], Wendelien Zeijlemaker[1], Angèle Kelder[1], Arjo P. Rutten[1], Alexander N. Snel[1], Willemijn J. Scholten[1], Thomas Pabst[2], Gregor Verhoef[3], Bob Löwenberg[4], Sonja Zweegman[1], Gert J. Ossenkoppele[1], Gerrit J. Schuurhuis[1]*

1 Department of Hematology, VU University Medical Center, Amsterdam, The Netherlands, **2** Department of Medical Oncology, Inselspital, Bern University Hospital, University of Bern, Bern, Switzerland, **3** Department of Hematology, University Hospital Leuven, Leuven, Belgium, **4** Department of Hematology, Erasmus University Medical Center, Rotterdam, The Netherlands

Abstract

Introduction: Treatment failure in acute myeloid leukemia is probably caused by the presence of leukemia initiating cells, also referred to as leukemic stem cells, at diagnosis and their persistence after therapy. Specific identification of leukemia stem cells and their discrimination from normal hematopoietic stem cells would greatly contribute to risk stratification and could predict possible relapses.

Results: For identification of leukemic stem cells, we developed flow cytometric methods using leukemic stem cell associated markers and newly-defined (light scatter) aberrancies. The nature of the putative leukemic stem cells and normal hematopoietic stem cells, present in the same patient's bone marrow, was demonstrated in eight patients by the presence or absence of molecular aberrancies and/or leukemic engraftment in NOD-SCID IL-2Rγ-/- mice. At diagnosis (n = 88), the frequency of the thus defined neoplastic part of CD34+CD38- putative stem cell compartment had a strong prognostic impact, while the neoplastic parts of the CD34+CD38+ and CD34- putative stem cell compartments had no prognostic impact at all. After different courses of therapy, higher percentages of neoplastic CD34+CD38- cells in complete remission strongly correlated with shorter patient survival (n = 91). Moreover, combining neoplastic CD34+CD38- frequencies with frequencies of minimal residual disease cells (n = 91), which reflect the total neoplastic burden, revealed four patient groups with different survival.

Conclusion and Perspective: Discrimination between putative leukemia stem cells and normal hematopoietic stem cells in this large-scale study allowed to demonstrate the clinical importance of putative CD34+CD38- leukemia stem cells in AML. Moreover, it offers new opportunities for the development of therapies directed against leukemia stem cells, that would spare normal hematopoietic stem cells, and, moreover, enables *in vivo* and *ex vivo* screening for potential efficacy and toxicity of new therapies.

Editor: Kevin D Bunting, Emory University, United States of America

Funding: This work was supported by Netherlands Cancer Foundation KWF grant 2006-3695. The funders had no role in study design, data collection and analysis, decision to publish, or preparation of the manuscript.

Competing Interests: The authors have declared that no competing interests exist.

* Email: gj.schuurhuis@vumc.nl

Introduction

There is increasing evidence that the development of solid and hematological tumors depends on the presence of small populations of cells known as tumor-initiating cells or tumor stem cells [1]. The first proof of the stem cell concept came from studies by John Dick and colleagues in acute myeloid leukemia (AML) [2,3]. In AML, these cells are referred to as leukemia-initiating cells or leukemic stem cells (LSCs) [4]. Since the first studies on LSCs, cell compartments defined by immunophenotype (CD34/CD38 expression) and function (side population, SP, and aldehyde dehydrogenase [ALDH] activity) have been reported to contain LSCs [5–9]. The first LSC compartment that was described had the CD34+CD38- immunophenotype [2,3]. Although immune reactivity of the CD38 antibody used in earlier studies likely caused the lack of engraftment of CD38+ cells [10], the CD34+CD38- compartment still seemed to be the most robust compartment in CD34-positive (CD34+) patients, since it was found to be the predominant compartment containing leukemia-initiating cells in less immunocompromised mouse models [3]. On the other hand, in more severely immune-compromised mouse models, CD34+CD38+ and CD34-negative compartments were also found to contain leukemia initiating cells [5,7–9,11].

In bone marrow (BM) of AML patients, leukemic and normal cells are present within one compartment. The CD34+CD38- compartment in particular, was shown to contain both CD34+

CD38- LSCs and normal hematopoietic stem cells (HSCs) [10,12]. Previous papers reported only on the role of the size of the total CD34+CD38-stem cell compartment and only at diagnosis [13–17]. However, for proper identification of LSCs, with the aim to establish the prognostic impact of their frequencies at diagnosis and to follow their fate during and after therapy, it is important to identify distinguishing features between LSCs and HSCs. We previously have found that in a substantial number of AML cases, CD34+CD38- LSCs are characterized by the expression of C-type lectin-like molecule-1 (CLL-1) and aberrant expression of several lineage markers [18,19]. Since these markers are absent on HSCs in regenerating BM after chemotherapy [18,19], for the current study we chose to use these markers to distinguish between LSCs and HSCs in both diagnosis and post-diagnosis samples. Other LSC markers have also been described for AML diagnosis (reviewed in ref [20]), but since little is known about their behavior during and after therapy, their suitability for LSC tracking remains to be established. However, despite the usefulness of CLL-1 and lineage markers, in at least 25% of the AML patients, aberrant marker expression on CD34+/CD38- cells is absent or too weak, and therefore there is a need to identify other discriminative parameters [6,19]. In the first part of this paper, we will describe such additional parameters for CD34+CD38- compartment. Now being able to discriminate neoplastic from normal CD34+CD38- cells, in the second part we will focus on the prognostic role of the CD34+CD38- stem cell compartment. This will be compared with the other CD34/CD38 defined compartments.

Moreover, another important aspect of cancer stem cells is their putative therapy resistance [21], with a subsequent ability to cause re-growth. The present study is the first to address the resistance to therapy by showing the prognostic impact of AML stem cells post-therapy in a large patient group. The results demonstrate that defining residual LSCs post-therapy has important prognostic impact. Moreover, we show that it adds important prognostic impact to a well-established immunophenotypical MRD approach, known to identify and quantify the bulk of neoplastic cells.

Patients, Material and Methods

More detailed information can be found in the supporting text (Text S1).

Patient treatment and sampling

Patients between 18 and 60 years of age with AML, except those with FAB M3 and previously untreated RAEB and RAEB-t patients (with IPSS ≥ 1.5), were included in this study. Detailed information regarding treatment can be found at http://www.hovon.nl. The HOVON/SAKK 42a and 92 studies were reviewed and approved by an institutional review board (METc) of the Erasmus MC Rotterdam for the total study (number 2000-220 for Hovon 42a) and 2008/216 for Hovon 92). In addition, the VU Amsterdam review board approved both studies with METc number 2001/50 (LUV) and 2008/292 (LUV), respectively. Patients provided their written informed consent to participate in this study. In 250 CD34-positive patients, we used a specific gating strategy to identify normal and neoplastic cells in the CD34+CD38-, CD34+CD38+ and the CD34- compartment at diagnosis. Patient details are in Table S1 (patient groups 1 and 2). Additionally, in the last part of the results, an extra patient group (n = 23) was included (details are in Table S1, patient group 3) to study combination of LSC and minimal residual disease (MRD) data after second cycle of therapy.

Flow cytometry

Methods. LSC characterisation at diagnosis and LSC monitoring during follow up was performed on fresh patient samples. Purified white blood cells were obtained from BM or PB using lysing solution (Pharm lyse, Becton Dickinson, BD, San Jose, CA, U.S.A.) to eradicate red blood cells. After washing with PBS containing 0.1% human serum albumin (HSA), cells were re-suspended in PBS containing 0.1% HSA, incubated with monoclonal antibody combinations (mAbs) for 15 minutes at room temperature and washed with PBS containing 0.1% HSA. Details on antibodies (sources, clones, fluorochromes) are written in the supporting text (Text S1). Samples were analysed using a 4-color approach on a FACSCalibur from Becton Dickinson (BD, San Jose, CA, USA) using CellQuest and Infinicyte software. Cell sorting was performed using FACSAria (BD) with FACSDiva analysis software. More details are in the supporting text (Text S1).

Marker selection for the CD34+CD38- LSC. In a previous paper [19], we have shown that particular lineage markers, i.e. CD2, CD7, CD11b, CD19, CD22, CD56, can be positive on CD34+CD38- LSC, while always negative on CD34+CD38- HSC. This is true for CLL-1 too [18]. All markers have been shown to be negative on HSCs in normal bone marrow, but also on normal stem cells present at diagnosis and at follow up [18,19]. In addition, the absence of CD13, CD33 and HLA-DR was considered aberrant, since these markers are always expressed on normal HSCs [19,22]. For each new AML case one or more of these markers may be aberrantly expressed; in a next case this may be completely different. Therefore, for each new AML case, the choice for a marker was based on screening for presence (>10% positivity) of all markers. Even more heterogeneity was seen, since part of the CD34+CD38- population might be negative for these markers, and all or not covered by another marker.

Marker selection for the more mature CD34+CD38+ and CD34- populations. For CD34+CD38+ and CD34- leukemic cells, identification had to be done using lineage markers: it is only lineage markers that are absent (or present at very low frequencies) on these more mature populations [23,24]. This property offers the basis for detection of residual leukemic cells (all leukemic cells) in MRD approaches. CLL-1, however, cannot be used as a neoplastic marker on CD34+CD38+ and CD34- cells since it is present on part of these populations in normal bone marrow [25]. For LSC identification/quantification at diagnosis and follow up, only AML samples were used that, at diagnosis, showed>1% CD34 expression relative to the total WBC count. This was done because we have previously shown that the CD34+ population in AML with <1% CD34 is of normal origin [26]. CD34 negative AML samples, together with normal bone marrow, were used as a control for FSC/SSC values in HSC (see results section I.*A*). For details on the gating strategy to define the CD34+CD38- compartment, see Figure 1, part I. The details how to define scatter characteristics of marker positive and marker negative populations are outlined under Results. Stem cell numbers (LSC and HSC) were defined as a percentage of total white blood cells (WBC).

Engrafting studies

Animal experiments were performed after approval of the animal ethical committee of the VU University, Amsterdam, The Netherlands under DEC-number: KNO06-02. NOD/SCID IL-2Rγ -/- mice were obtained from the Jackson laboratory (Bar Harbor, ME, USA). At the age of 8-10 weeks, the mice were irradiated sub-lethally with a dose of 350 cGy, 24 hours prior to transplantation of the human AML cells. Details of irradiation, anesthesia, intravenous and intra-femoral injection are outlined in

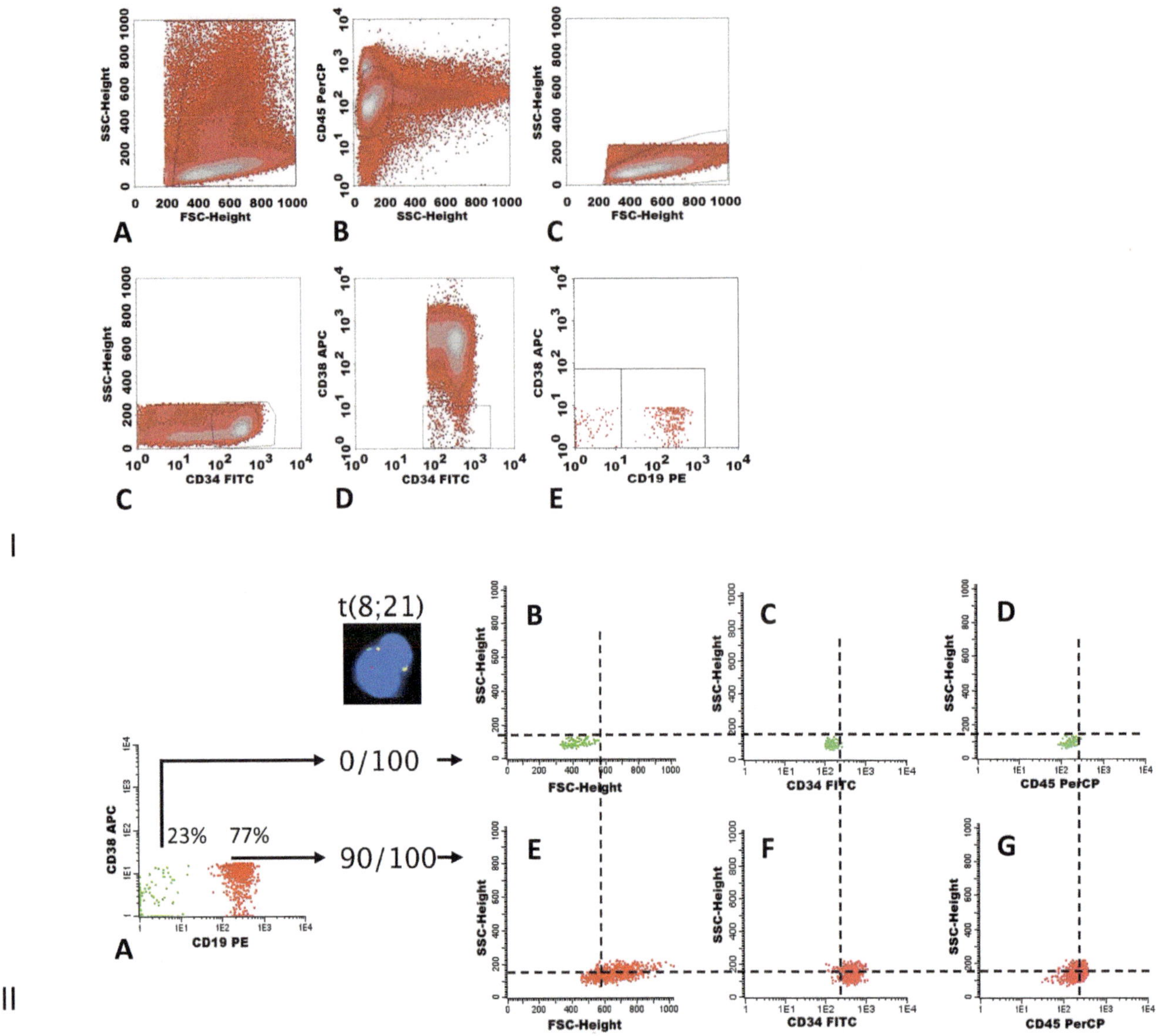

Figure 1. Gating strategy for the CD34+CD38- compartment and identification of pLSCs and HSCs in this compartment. I. Gating of CD34+CD38- AML cells. Cells were labeled with antibody-fluorochrome combinations as described in Patients, Materials and Methods. Remaining erythrocytes, debris and dead cells are largely excluded in an FSC/SSC plot (A). $CD45^{dim}/SSC^{dim}$ blast cells (B) were gated to homogeneity in FSC/SSC plot (C). CD34 positive cells are gated (D) and the CD38- stem cells are gated within this fraction (E). The CD38-negative fraction in D may contain two stem cell populations differing in CD34 expression (details in text). F. Within the CD34+CD38- gate, CD38 is plotted against an aberrant marker (in this case CD19) to indicate presence of putative LSCs (pLSCs) and HSCs. **II. Identification of pLSCs and HSCs.** This patient (nr 317) was diagnosed with t(8;21). Primary gating was as in I. Sorted CD34+CD38-/CD19+ cells (A, in red) were t(8;21) positive; sorted CD34+CD38-/CD19- cells (A, in green) were t(8;21) negative. These two populations were backgated in FSC/SSC (B,E), CD34/SSC (C,F) and CD45/SSC (D,G) plots. The CD19- cells are shown in the upper panels and the CD19+ cells in the lower panels. Dotted vertical lines (B–D) show that normal CD19- cells are FSC^{low} (B), $CD34^{low}$ (C) and slightly lower in CD45 (D), compared to CD19+ cells (E–G). The dotted horizontal line shows that SSC of the normal stem cells (in green) was slightly lower than that of neoplastic stem cells (in red). FISH data are from an example published previously [19]. Similar results were found in an additional series of 7 patients (Tables 1 and 2). FSC and SSC of CD19+ pLSC were factor 1.71 and 1.77 higher than lymphocyte present in the same samples. FSC and SSC of the CD19 negative cells were only 1.08 and 1.20 times lower than lymphocytes.

the supporting text (Text S1) under "Patients". Human leukemic and multi-lineage engraftment was determined based on positivity for CD45-PercP, the presence/absence of CD19-positive B-cells, CD13 and/or CD33-positive cells and, in three cases, by the presence/absence of aberrant marker expression.

Engraftment was defined as a clear clustered population in CD45 expression in a minimum of 200,000 acquired mice marrow cells. The minimum percentage of human engraftment that was detected was 0.1%, which shows up as a cluster of 200 cells on the scatterplot. Human leukemic engraftment was determined based on positivity for CD45-PercP, the absence of CD19-positive B cells and the presence of CD13 and/or CD33 positive cells: when CD13 was aberrantly absent on the AML that was injected, CD33 was used and when CD33 was absent, CD13 was used. Human multi-lineage engraftment was identified when CD45 positive cells consisted of both CD19 positive B-cells and myeloid cells with both CD13 and CD33 present (identified as monocytes and/or

Table 1. Molecular status of diagnosis total AML and sorted stem cell fractions from the same BM.

Patient nr	Molecular status of AML blasts[§]	Aberrant marker expressed onCD34+CD38- cells[*]	CD34+CD38- aberrant marker negative		CD34+CD38- aberrant marker positive	
			CD34+CD38- aberrant marker-	% aberrant marker- of CD34+CD38-[$]	CD34+CD38- aberrant marker+	% aberrant marker+ of CD34+CD38-
317[¥]	t(8:21) pos	CD19	t(8;21) neg	23	t(8;21) pos	77
808	50% FLT3-ITD[#]	CD33-	wt	67	45% ITD[#]	33
	NPM1 mut	CD33-	wt	67	NPM1 mut	33
945	NPM1 mut	CLL-1	wt	47	NPM1 mut	53
951	42% FLT3-ITD	CLL-1	wt	46	60% ITD	54
966	43% FLT3-ITD	CD7	wt	41	41% ITD	59
1263	60% FLT3-ITD	CD33-	wt	2.5	80% ITD	97.5
575	50% FLT3-ITD	CD33-	wt	58	50% ITD	42
	NPM1 mut	CD33-	wt	58	NPM1 mut	42
670	NPM1 mut	CLL-1	wt	74	NPM1 mut	26

[§]signal originating from the bulk of the leukemic blasts has been used for reference. Sorted lymphocytes from the same BM showed no molecular aberrancies.
[*]Aberrant markers with the highest coverage of the CD34+CD38- compartment were used for sorting HSC (aberrant marker negative) and pLSC (aberrant marker positive).
[¥]Shown in Figure 1.
[#]For FLT3-ITD+ cases, the percentage of ITD of the total signal is indicated.
[$]Median frequency of HSCs (46.5% of CD34+CD38- compartment) was higher than the median in a larger group of cases (median 18%), since cases in this table had to be selected on clear availability of both HSCs and pLSCs.
Abbreviations: wt, wild type: no FLT3-ITD or NPM1 peak present; mut: mutation of NPM1 (not quantitative), neg: negative, pos: positive, BM: bone marrow.

granulocytes in a FSC/SSC plot), and with absence of aberrant markers.

FISH, FLT3-ITD and NPM1 analysis

For the FISH analysis, cytospins were prepared with FACS-sorted cells. LSI AML1/ETO dual color for t(8;21) probe (Vysis, Abbott molecular, Illinois, U.S.A.) was applied to the denatured cells and incubated as previously described [17]. Genomic DNA from sorted cell populations was analyzed for the presence of an FLT3-ITD as described before [27]. Mutations in NPM1 exon 12 were analyzed by PCR using genomic DNA that had been isolated from sorted cell fractions (see Text S1).

Survival analysis

Statistical analysis of the stem cell data at follow-up was carried out using the SPSS 20.0 software program. The supplement provides details regarding the definition of overall survival (OS), event-free survival (EFS), relapse-free survival (RFS), Kaplan-Meier analyses, and Cox regression analyses for both univariate and multivariate analyses. P-values below 0.05 were considered significant.

Results

In the absence of formal proof of leukemia initiating ability in the stem cell compartments of most of the samples studied, the different stem cell compartments are referred to as *putative* LSCs (abbreviated as pLSC).

Since the aim of the study was to define the prognostic impact of the neoplastic CD34+CD38- compartment, and to compare with the neoplastic CD34+CD38+ and CD34 negative compartments, first, properties will be described that allow to discriminate between the neoplastic and the normal compartment within the CD34+CD38- compartment (Section I.A–D) and the CD34+ CD38+ and the CD34- compartments (Section I.E). Proof of the resulting concept was obtained using either molecular biological and/or murine engraftment experiments.

In Section II. the findings were used to assess the prognostic impact of the number of pLSCs at diagnosis and post-therapy for the CD34+CD38- compartment (II.A,B) and the CD34+CD38+ and CD34- compartment (II.C,D). Finally, the prognostic impact of the combination of follow up CD34+CD38- pLSC frequency, with MRD data will be described (II.E).

I. Discrimination between leukemic and normal stem cell compartments

A. CD34+CD38-: discrimination of pLSCs and HSCs can be made based on aberrant marker expression and scatter properties. We have previously shown that the expression of aberrant markers on CD34+CD38- cells indicates the leukemic nature of stem cells [18,19]. Figure 1.II illustrates the difference between CD34+CD38- pLSCs and HSCs based on aberrant marker expression and molecular aberrancies [in this example, t(8;21)]. In addition to Figure 1, Table 1 shows 7 other patients with aberrant marker positive cells, all with molecular aberrancies indicating that these are in fact neoplastic cells. The figure also shows that marker-positive pLSCs, compared to HSCs, are further characterized by a tight clustered cell population with higher forward scatter (FSC, reflecting cell size) and higher sideward scatter (SSC, reflecting granularity). This phenomenon was found in the other patients of Table 1 too (shown in Table 2). In Table 2, FSC and SSC of lymphocytes were used as internal controls to define the FSC/SSC position of pLSC and HSC. pLSC and HSC may also differ in CD34 and CD45 expression (in the case of Figure 1, CD34 expression of pLSC was higher than HSC). Such differences were not consistent but helped to define clusters of cells especially in cases with very low numbers of cells and/or small differences in aberrant marker expression and/or FSC/SSC.

Table 2. FSC/SSC position relative to lymphocytes.

Patient nr	Aberrant marker negative CD34+CD38-	Aberrant marker positive CD34+CD38-
	FSC/SSC position*	FSC/SSC position*
317[¥]	1.08/1.20	1.71/1.77
808	1.27/1.17	1.57/1.80
945	1.38/1.52	1.97/3.25
951	0.93/0.77	1.46/1.39
966	1.15/1.0	1.53/1.99
1263	1.04/1.22	1.39/2.08
575	1.06/1.02	1.34/1.77
670	1.21/1.07	1.52/1.73

¥Shown in Figure 1.
* FSC/SSC values are based on the position of cells in FSC/SSC relative to the FSC/SSC position of normal lymphocytes in the same BM sample.
The molecular status of the sorted cell fractions is summarized in Table 1.

Table S2 shows the FSC/SSC position relative to lymphocytes of CD34+CD38- HSCs present in normal BM, as well as CD34+ CD38- present in CD34 negative AML, which have previously been shown to be normal [26]. FSC/SSC of HSCs present in these controls are all lower than pLSC present in CD34+ AML (Table S2). Moreover, HSC present in CD34+ AML also have low FSC/SSC and altogether there is not even overlap in FSC/SSC between pLSC in CD34+ AML and the HSCs present in these three bone marrow sources. This allows accurate definition of the nature of a population especially in those cases with only a single population present.

B. CD34+CD38-: is there a role of FSC/SSC to define pLSC and HSC in aberrant marker-negative part of CD34+CD38- compartment?. Aberrant markers may cover only part of neoplastic CD34+CD38- cells [19]. We therefore investigated whether scatter differences may help to discriminate pLSCs from HSCs in aberrant marker-negative cells. This was indeed seen as illustrated for patient sample 456 in Figure 2.I. The investigation of expression of only aberrant markers would have led to an over-estimation of HSC numbers and an under-estimation of pLSC numbers. Figure 2.II provides more examples.

In approximately 25% of AML cases, there is no clear aberrant marker expression in the CD34+CD38- compartment, and in even more cases, marker expression is weak and overlaps with the marker negative cells [19]. In order to nevertheless discriminate between pLSCs and HSCs, we investigated whether scatter might replace marker expression in this respect. Figure 3 (A–D) shows that, even without clear aberrant marker expression (CD19 covers only a very low frequency CD34+CD38- population), differences in scatter can be used. Since, in this particular AML case, aberrant expression of CD7 was found (Figure 3 E–F), the validity of the scatter approach could be demonstrated (compare Figure 3 E–F with D). This shows that the scatter approach allows to discriminate pLSC and HSC in the absence of aberrant markers.

C. CD34+CD38-: discrimination between HSCs and pLSCs in a large patient group using marker and scatter differences. In cases with differences seen in FSC and SSC between pLSC and HSC, pLSC always had higher FSC/SSC than HSC. However, at this point it should be emphasized that differences in scatter were found in part of the patients. In 250 diagnosis AML cases studied, the combination of marker expression and FSC/SSC differences allowed accurate identification of both pLSCs and HSCs in 117/250 cases (47%), (outlined in detail in Table S3 and summarized in Table S4. For reasons mentioned earlier, differences in expression of CD34 and CD45 were only used for fine-tuning. As might be expected, the pLSC compartment made up the majority of the total CD34+CD38- compartment (median of 82%, ranging from 0%–100%). Apart from these 117 cases, there was an extra group of 102 patients (41%) with only marker expression available, while 31 patients (12%) had no marker and scatter properties usable for pLSC detection (summarized in Table S4).

D. CD34+CD38-: multilineage and leukemic engraftment of CD34+CD38- HSCs and CD34+CD38- pLSCs in NOD/SCID IL-2R $\gamma^{-/-}$ mice. To provide further proof of principle for the strategy based on marker expression, secondary gating and molecular profiles, murine engraftment experiments were performed. HSCs were sorted by marker negativity, and clustering as FSC/SSC^{low}, and/or distinct CD34/CD45 clustering, injected intrafemorally and evaluated for engraftment. In 5/6 cases, multilineage engraftment was found with no signs of leukemia (Figure 4, Table S5). In accordance with our *in vitro* gating strategy, AML engraftment with low cell numbers was observed in 2/6 cases for CD34+CD38- marker-positive samples and in 1/6 cases for the CD34+CD38- marker-negative cells, the latter being identified as neoplastic based on high FSC and aberrant CD34 expression (see legends of Table S5).

E. CD34+CD38+ and CD34-: discrimination between normal and leukemic CD34+CD38+ and CD34- cells and their leukemic engraftment in NOD/SCID IL-2R $\gamma^{-/-}$ mice. Although emphasis in this paper is on the CD34+ CD38- compartment, results described later for prognosis of CD34+CD38- pLSC will be compared with the other CD34/ CD38 defined compartments. To enable such, neoplastic CD34+ CD38+ and CD34- cells were identified by aberrant expression of markers used to define so-called Leukemia Associated (Immuno) Phenotypes established at diagnosis and used by others and us [23,24] for detection of minimal residual disease. Cells were sorted as described in Text S1. In agreement with recent reports, these cell compartments also had leukemia initiating potency since AML engraftment was found in 4/6 cases for the CD34+CD38+ compartment and in 2/6 cases using CD34- compartments (referred to in the legends of Table S5). In all but one case, this was accomplished with high cell numbers (100,000-1,000,000).

I

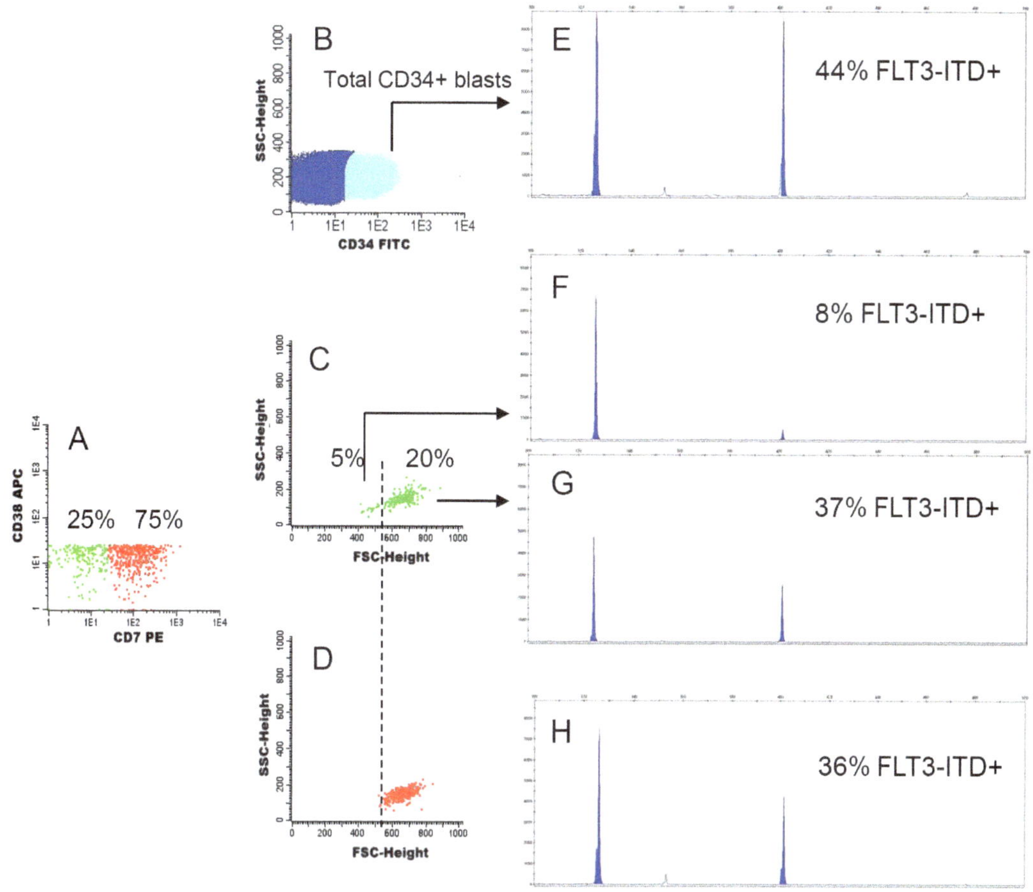

II

Patient nr	Aberrantly expressed marker	FLT-3 ITD (% of total) in cell fraction #				
		CD34+	Lymphocytes	CD34+CD38-/ aberrant marker+	CD34+CD38-/ aberrant marker-	
					FSC/SSC^{low} §	FSC/SSC^{high} §
598	CLL-1	54	0	50	0	38
1034	CLL-1	20	0	37	0	17
423	CLL-1	45	0	21	6	45
456*	CD7	44	0	36	8	37

Figure 2. Marker-negative cells may contain leukemic cells defined by aberrant scatter. I. Gating/sorting strategy and molecular analysis for patient 456. CD34+ CD38- cells (patient 456, figure 2) were gated as in figure 1. CD34+CD38- cells were either CD7-negative (green in A and C, 25% of CD34+CD38-) or CD7 positive (red in A and D, 75% of CD34+CD38-). CD7+ cells were FSC/SSC^{high} (D) and of neoplastic origin (H). CD7-negative cells were further subdivided into FSC/SSC^{low} (left of the broken line in C), and FSC/SSC^{high} (right of the broken line in C). The CD7-negative, but FSC/SSC^{high}, cells were neoplastic (G), while the CD7-negative FSC/SSC^{low} cells were essentially normal (F). CD34+ cells was the positive control (E). **II. Molecular analysis of stem cell subpopulations in four patients.** Sorting and analysis was done as in figure 2.I A for 3 additional patients. # FLT3-ITD (% of total signal: FLT3-ITD + wt) determined in cell populations sorted from CD34+ AML patients. * shown in figure 2.I A. § FSC/SSC^{low} and FSC/SSC^{high} defined as outlined in Table 2.

II. Prognostic role of putative stem cell compartments

To assess the clinical impact of our findings, we assessed the prognostic value of the size the three individual LSC compartments both at diagnosis and, where applicable, especially at clinical follow-up.

A. CD34+CD38- at diagnosis. To assess prognostic impact, different cut-off values were established to define patients with high pLSC count (above the cut-off level: referred to as pLSC+) and patients with low pLSC count (below the cut-off level: pLSC-). Patients who later turned out not to have achieved complete remission (non-CR patients) had median 6-fold higher pLSC count than patients who had achieved CR (see legends of Figure 5). Moreover, the CR patient group could be divided into two groups with significantly different prognoses at multiple cut-off points ranging from 0.002% to 1% (Figure 5A and Table S6A). Figure 5B shows that further splitting up the CR group resulted in the definition of three groups with large differences in relapse-free survival (RFS). The latter result was also found upon inclusion of non-CR patients (not shown) and replacement of RFS by EFS (Figure 5C). Multivariate analysis showed pLSC frequency to be an independent prognostic factor (Figure 5, legends).

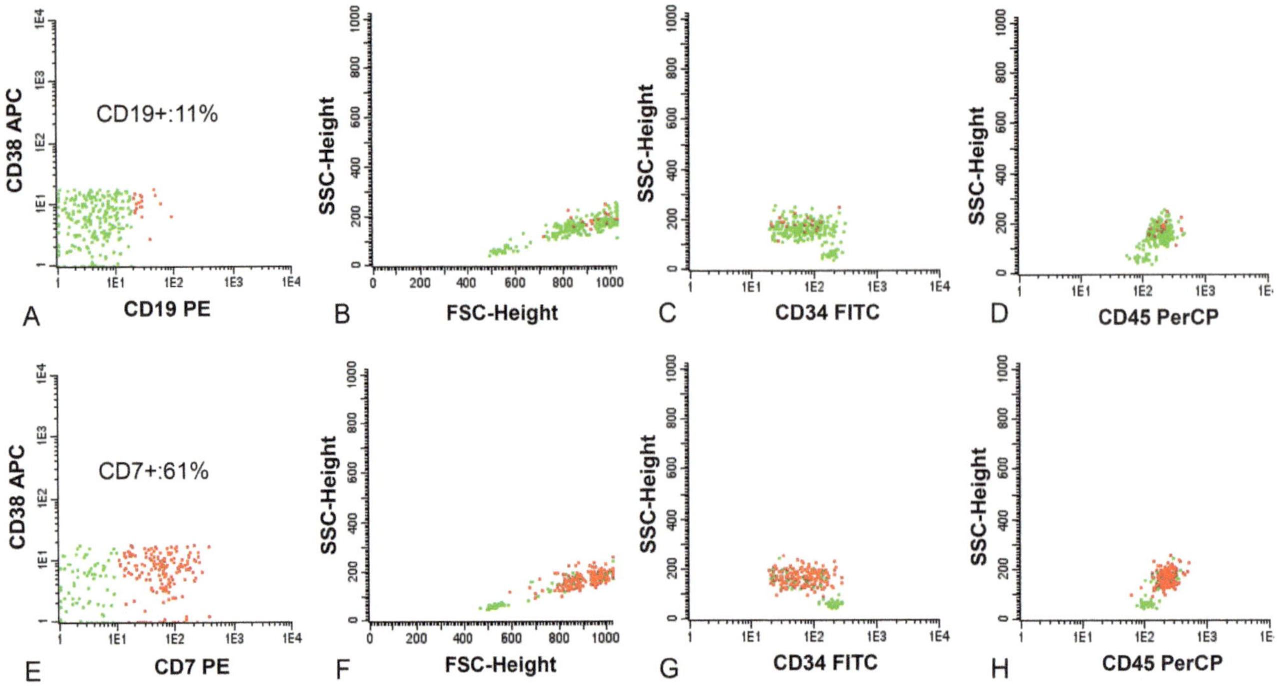

Figure 3. Marker negative pLSCs co-exist with marker positive pLSCs and are identified by scatter and CD34/CD45 expression patterns. CD34+ CD38- cells (patient 372) were identified as described in figure 1.I and gated for sub-compartments as described for figure 2.I. In the stem cell compartment of this AML case, only 11% could be identified as CD19+ (A). When back-gated in FSC/SSC (similar as performed in Figures 1 and 2), two different populations were identified based on the position in FSC/SSC: the small CD19+ fraction (events in red in A–D) is characterized by FSC/SSC^{high} (B), low CD34 expression (C) and high CD45 expression (D). CD19 negative cells (events in green in A–D), apart from a small FSC/SSC^{low}/CD34^{high}/CD45^{low} population of putative HSCs (A–D), contained a large population of cells that, similar to the CD19+ population in A–D, were FSC/SSC^{high} (B), CD34^{low} (C), and CD45^{high} (D). Apart from CD19, CD7 was an aberrant marker: 61% of the cells was CD7+ (E). Upon backgating, CD7+ cells (events in red in E–H), similar to CD19+ cells, were all FSC/SSC^{high} (F) CD34^{low} (G) and CD45^{high} (H). In contrast to CD19, CD7 negative cells (events in green in E–H) now completely consisted of a small FSC/SSC^{low} (F), CD34^{high}(G), CD45^{low} (H) fraction. CD7 thus covered the whole neoplastic CD34+CD38- population and shows perfect discrimination between HSC and pLSC. CD19 expression in the absence of both CD7 expression and other scatter and CD34/CD45 expression parameters would have under-estimated the pLSC in the CD34+CD38- compartment by a factor 5.5 (61%/11%), while the HSCs would have been over-estimated by a factor 2.3 (89%/39%). In this case CD7 was a good marker to compare with the poor CD19 marker; it can be seen, however, that in the absence of CD7 expression, but with the scatter and CD34/CD45 aberrancies present, these would have enabled a complete discrimination between putative HSC and LSC compartments. This patient was identified as NPM1-positieve and FLT3-ITD positive. Other molecular aberrancies were not detected.

B. CD34+CD38- during follow-up in CR. The gating approach shown in Figure 1 was used to follow the fate of pLSCs over time using sequential sampling (examples in Figure S1). Cox regression analysis showed a strong significant inverse correlation between the pLSC percentage after all therapy cycles and RFS and OS (for details, see legends to Table S7). Kaplan-Meier analyses showed markedly improved RFS and OS in pLSC-patients compared with pLSC+ patients (example in Figure 6, A–C), which applies for a large range of cut-off values (Table S7). Multivariate analyses with cut-off values showed that pLSC frequency after the first and second treatment cycles was an independent prognostic factor for RFS and OS (example with 0.0003% and 0.0001% shown in Table S8).

C. CD34+CD38+ and CD34- at diagnosis and at follow up. The CD34+CD38+ and CD34- compartments have been shown to contain leukemia initiating cells [5,11] (also in this paper). In order to assess whether these were clinically important, when present together with the leukemia initiating CD34+CD38- compartment, we assessed their prognostic impact.

At diagnosis, neoplastic CD34+CD38+ and CD34- blast cells were identified as described under I.B. For CD34+CD38+, there was no single cut-off value (tested in the range 2%–60%) that resulted in discrimination of two patient groups with different RFS (Figure 5D; Table S6B), even when non-CR patients were included in the latter (data not shown). Although only borderline significant, in contrast to CD34+CD38-, the neoplastic CD34+ CD38+ compartment was even smaller in non-CR patients than in CR patients (10.8% of WBC versus 23.7% of WBC; $p = 0.06$). Similar results were obtained for the CD34- compartment with cut-off points ranging from 0.1%–30% (Figure 5E; Table S6C), with the neoplastic compartment size, similar to CD34+CD38+, being even smaller (albeit non-significantly) in non-CR patients than in CR patients (0.87% of WBC versus 2.41% of WBC, $p = 0.48$).

At follow up, the neoplastic component of the CD34+CD38+ and CD34- compartment represented a considerable portion of the total neoplastic blast compartment. This reflects the total leukemic burden, or MRD, which, in turn, has previously been shown by many authors, including ourselves, to have prognostic impact (see also next paragraph) [23,24].

D. Post-diagnosis prognostic impact of CD34+CD38- LSC combined with MRD. The burden of leukemic stem cells after chemotherapy does not always reflect the total leukemic burden (known as MRD). It can thus be argued that the combination of pLSC frequency and MRD frequency may improve prognostic information at follow up. Because this combination would divide

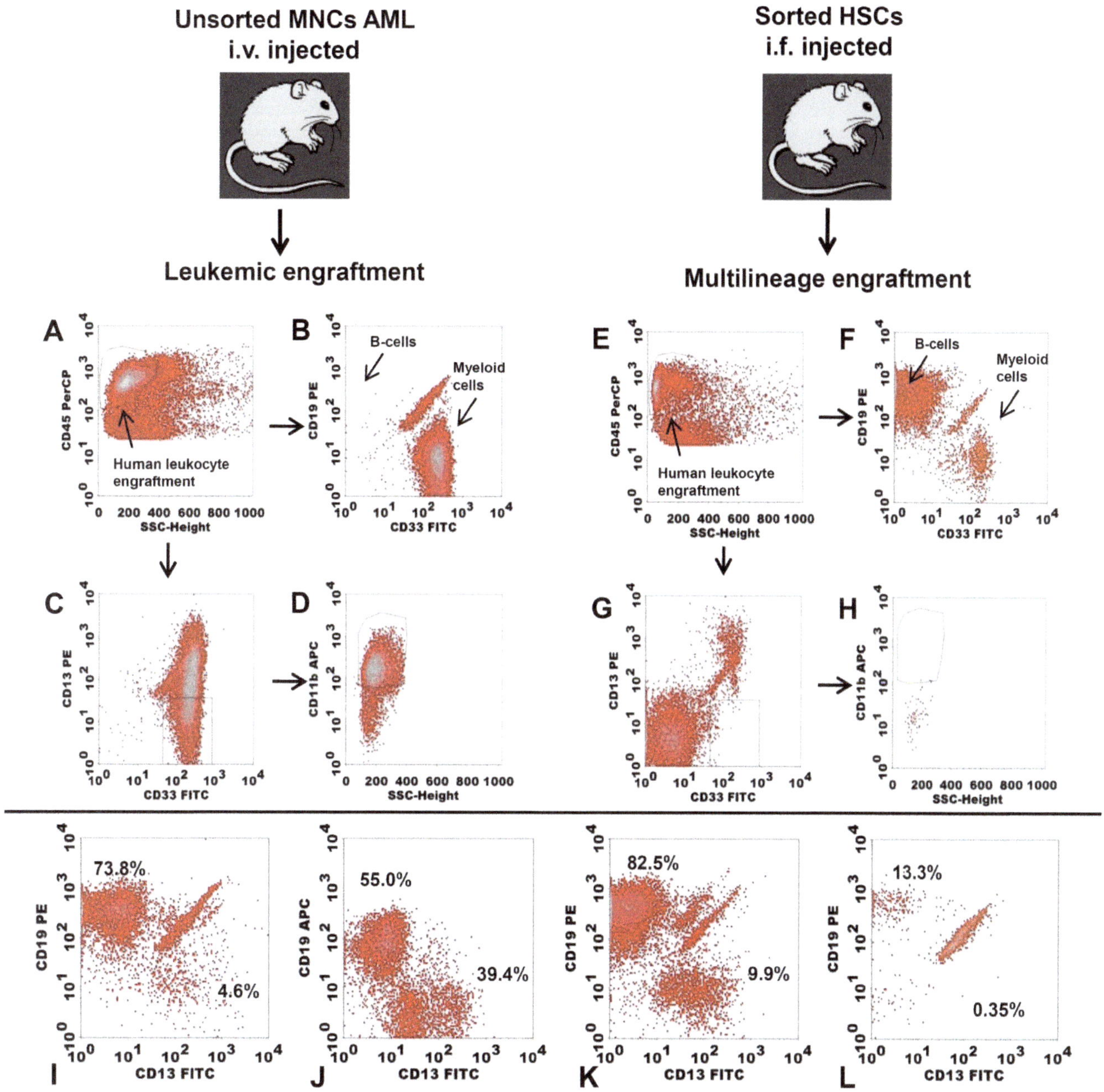

Figure 4. Multilineage engraftment of CD34/CD38 and scatter-defined putative HSCs. Unsorted mononuclear cells (MNCs) were injected intravenously and resulted in leukemic engraftment: cells were CD45+ (A) and of myeloid origin (B). In this case, the myeloid cells were positive for the diagnosis of leukemia-associated phenotype (LAP): partly CD33+CD13- (C) and CD11b+ (D). Sorted putative HSCs were injected intrafemorally (details, see Table S5). Engrafted CD45+ cells (E), contained both B-cells and myeloid cells (F), and lacked LAP (G,H). Multilineage engraftment of the sorted subpopulations was seen for patient 598 (I), 661 (J), 423 (K), and 928 (L). B-cells and myeloid cells (percentage of CD45+ cells) are in the upper left and lower right corners of the plots, respectively. The AML cells of patients 598 (I) and 661 (J) had an aberrant phenotype at diagnosis that was present in the neoplastic engrafted cells, but absent in the normal cells (not shown).

the total patient group in four relatively small sub-groups defined by both a pLSC cut-off (Figure 6, Table 3) and a fixed MRD cut-off of 0.1 [24]), we increased the size of the present patient remission group by including 23 similarly treated young adults in remission (<65 years) (details in Table S1 "Patients 3"). In the total group (n = 91) pLSC frequency again had strong prognostic impact over a range of cut-off points (from 0 in 10^6 WBC up to 10 in 10^6 WBC. As an example, the cut-off point of 0.0001% is shown in Figure 7A. For MRD, the cut-off point of 0.1% had the expected prognostic impact (Figure 7B) [24]. When combining MRD and pLSC in Figure 7C, four groups were identified. The main conclusions from this figure are: 1) within the total MRD-group (n = 64), pLSC+ patients (n = 31) have significantly poorer prognosis than pLSC- patients (n = 33; p = 0.01); 2) within the pLSC+ group, MRD- patients, although having a relatively poor prognosis, may do better than MRD+ patients (p = 0.04); the pLSC-/MRD- group had relatively good prognosis, while the pLSC+/MRD+ group had very poor prognosis.

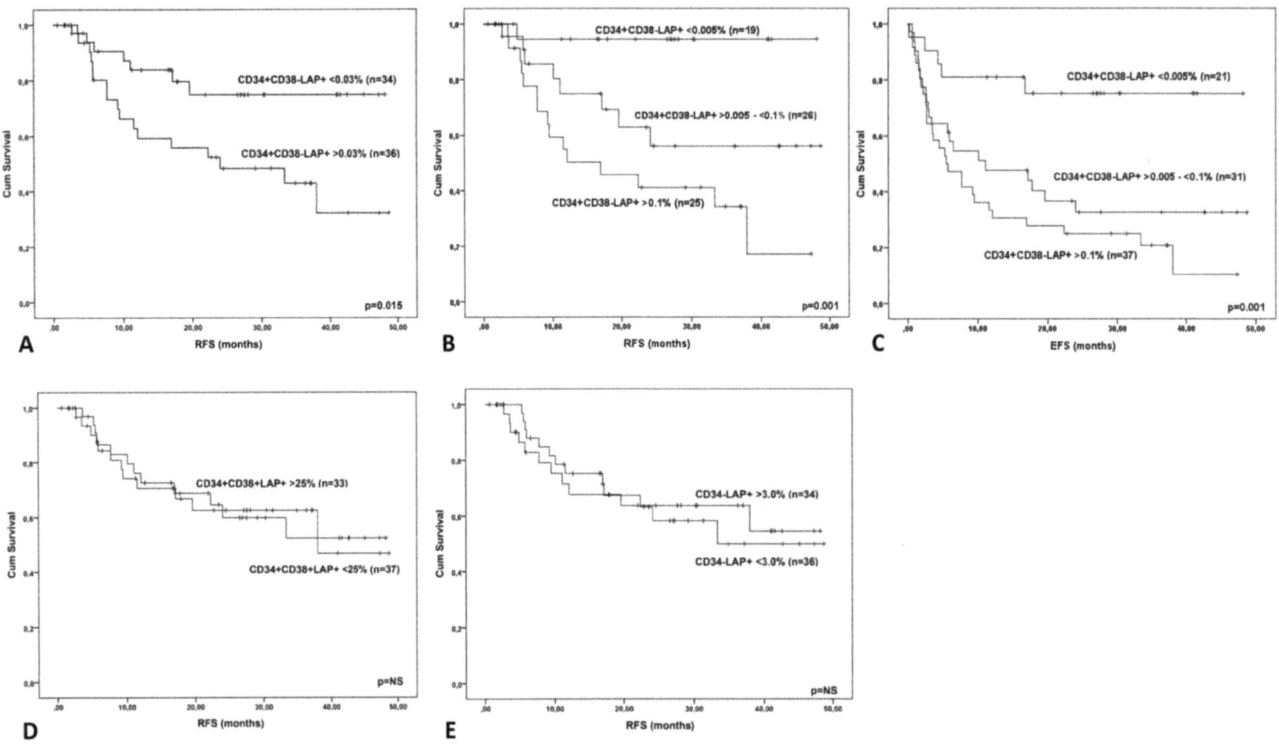

Figure 5. Prognostic value of frequencies of pLSC compartments at diagnosis. This figure shows the Kaplan-Meier analyses at diagnosis for the three compartments putatively containing pLSCs: CD34+CD38- (A,B,C), CD34+CD38+ (D) and CD34- (E). Of the 117 patients shown in Table S3, for Figures 5 and 6, 88 patients were chosen who had at least one follow-up time point. Of these, 70 entered Complete Remission, of whom 53 after the first course, 13 after the second course and 4 at later stages. Eighteen never reached CR. The size (median values) of the CD34+CD38- compartment at diagnosis was significantly (six-fold) higher in patients who did not enter CR (n = 18) compared with patients who did (n = 70): 0.225% of WBC versus 0.036% of WBC (p = 0.041). For CD34+CD38+ and CD34-, there were no significant differences (see text). Cut-off levels were defined to divide the total population into high stem cell frequencies (above cut-off) and low stem cell frequencies (below cut-off). A particular cut-off value was chosen (A, D, E) to ensure approximately equally numbers of patients in the resulting high and low stem cell frequency compartments. Results for other cut-offs for the three pLSC compartments are in Table S6. A–C: CD34+CD38-; D: CD34+CD38+; E: CD34-. A. RFS in remission patients (n = 70) with diagnosis CD34+CD38- cut-off of 0.03%; B. RFS in the same patient group (n = 70), but now with 2 cut-offs (0.005% and 0.1%); C. Event-free survival for all CR and non-CR patients (n = 88); D. RFS in remission patients (n = 70) with CD34+CD38+ cut-off of 25%; E. RFS in remission patients (n = 70) with CD34- cut-off of 3%. All relevant prognostic variables with statistical significance were investigated in a multivariate model. In this multivariate analysis it was found that risk group (according to the HOVON 102 trial) was an independent prognostic factor for OS at diagnosis (p = 0.001). For RFS, both risk group and CD34+CD38- leukemic stem cell load (using a 0.03% cut-off point) were independent prognostic factors at diagnosis (p = 0.017 and p = 0.011, respectively).

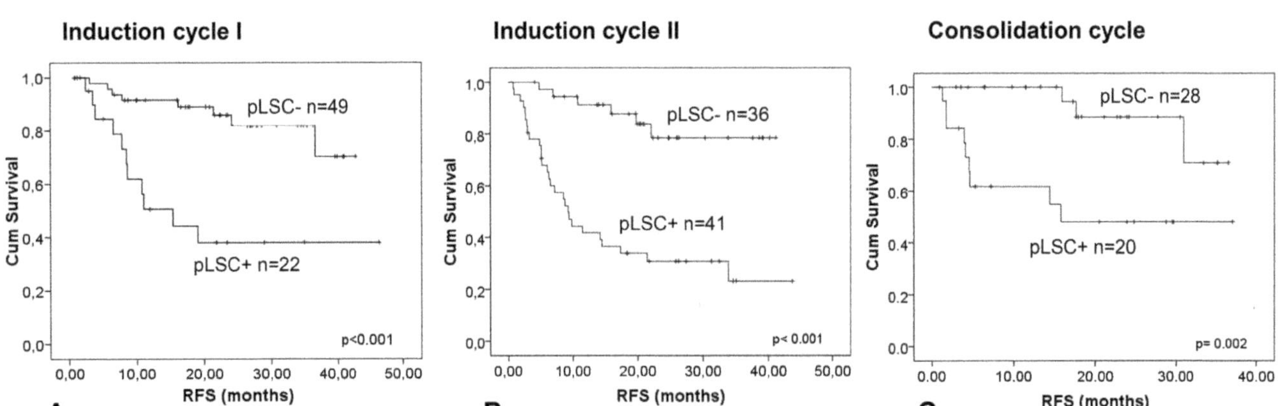

Figure 6. Prognostic value of frequencies of CD34+CD38- pLSC compartment at follow-up. This figure shows the Kaplan-Meier analyses for RFS for the CD34+CD38- pLSC compartment at follow up for three consecutive therapy cycles. The optimal cut-off levels were chosen to define pLSC + and pLSC- after 1st induction cycle (0.0003%,which is 3 pLSCs in 1,000,000 WBC) and after 2nd induction cycle and consolidation therapy 0.0001% (1 pLSC in 1,000,000 WBC). Results for other cut-offs are in Table S7. After the first induction cycle (B, 71 patients), second induction cycle (C, 77 patients), and after consolidation therapy (D, 48 patients), patients with high pLSC frequency (pLSC+) showed significantly more adverse performance compared with patients with low pLSC frequency (LSC-).

Table 3. Role of cytogenetically/molecularly defined risk groups in pLSC/MRD defined sub-groups.

Group sub-group (cytogen/mol)	N (nr per pLSC/MRD sub-group)	% per pLSC/MRD sub-group	40 months survival* (%)
pLSC-/MRD- (total)	**33**	**100**	**74**
good	8	24	69
intermediate	11	33	80
poor	10	30	80
very poor	4	12	50
pLSC-/MRD+ (total)	**9**	**100**	**56**
good	5	56	60
intermediate	3	33	67
poor	0	0	-
very poor	1	11	0
pLSC+/MRD- (total)	**31**	**100**	**45**
good	8	26	47
intermediate	7	23	57
poor	12	39	47
very poor	4	13	0
pLSC+/MRD+ (total)	**18**	**100**	**16**
good	4	22	25
Intermediate	2	11	0
Poor	7	39	15
Very poor	5	28	0

* 40 months was chosen as most survival curves had reached a plateau.
MRD cut-off: 0.1% of WBC, pLSC cut-off: 0.0001% of WBC.

When including the distribution of cytogenetically/molecularly good, intermediate, poor and very poor patients, these were all represented in the four LSC/MRD defined subgroups (Table 3). This shows that, the pLSC/MRD prognostic impact is across cytogenetic risk groups, although in the LSC+/MRD+ group poor and very poor cytogenetic/molecular risk groups are prevalent.

When including the post-diagnosis prognostic parameter "cycle after which CR is reached" in Figure 7, it turned out that there was no significant difference between the first three pLSC/MRD defined patient groups in number of cycles needed to reach CR: in the first group (pLSC-/MRD-) 29 after first cycle versus 4 after second cycle; in the third group (pLSC+/MRD-) this was 25 versus 6. The second group (pLSC-/MRD+) was too small (all patients in CR after one cycle). However, in the fourth group (pLSC+/MRD+) for 9 patients two cycles were needed, with one for the other 9 patients.

These data show that combining cytogenetic/molecular defined risk groups together and clinical parameters like cycles to CR, together with pLSC/MRD defined risk assessment may offer a very important new algorithm in risk assessment.

Discussion

One of the major challenges in the design of new therapies to eradicate leukemia stem cells is to achieve high therapeutic specificity. To this end, it is important to distinguish LSCs from the concomitantly present HSCs, and to assess whether this distinction is of prognostic value, since it would underline the clinical importance of LSCs. Consequently, it was necessary to identify parameters that allowed discrimination between these pLSCs and HSCs, preferably in all AML cases. Recent studies have shown that pLSCs may reside not only in CD34+CD38-, but also in CD34+CD38+ and CD34- compartments [5,11]. In the present paper, we first present methods to discriminate between the neoplastic and normal portions of these compartments, and we subsequently assessed the prognostic value of these putative stem cell compartments.

With regard to the prognostic impact, using a uniquely-designed multi-parameter flow cytometry protocol, we show for the first time that the CD34+CD38- pLSC load after different cycles of therapy was highly predictive of patient survival, independent of other prognostic parameters. In addition, following our own preliminary studies [13], we identified three patient groups at diagnosis defined by CD34+CD38- pLSC with very large differences in prognosis, again independent of other prognostic parameters. By comparison with literature [13–15], it can be appreciated that the prognostic impact using our new approach, in which the pLSC compartment within the total CD34+CD38- compartment was specifically used, is much higher than using the total CD34+CD38- compartment as done in the previous studies. This is likely due to the "contamination" of the leukemic CD34+ CD38- compartment with HSCs in the earlier studies (ranges of 0%–100% of pLSC, as seen in our study, also means that HSC range from 0%–100%), as seen in Table S3. In contrast, the CD34+CD38+ and CD34- compartments at diagnosis completely lacked prognostic impact, which strongly suggests that CD34+ CD38+ and CD34- pLSCs are of minor clinical importance, at least in AML cases where these compartments are accompanied by CD34+CD38- pLSCs (as by definition is the case in our current CD34 positive patient group). However, at follow-up, leukemic

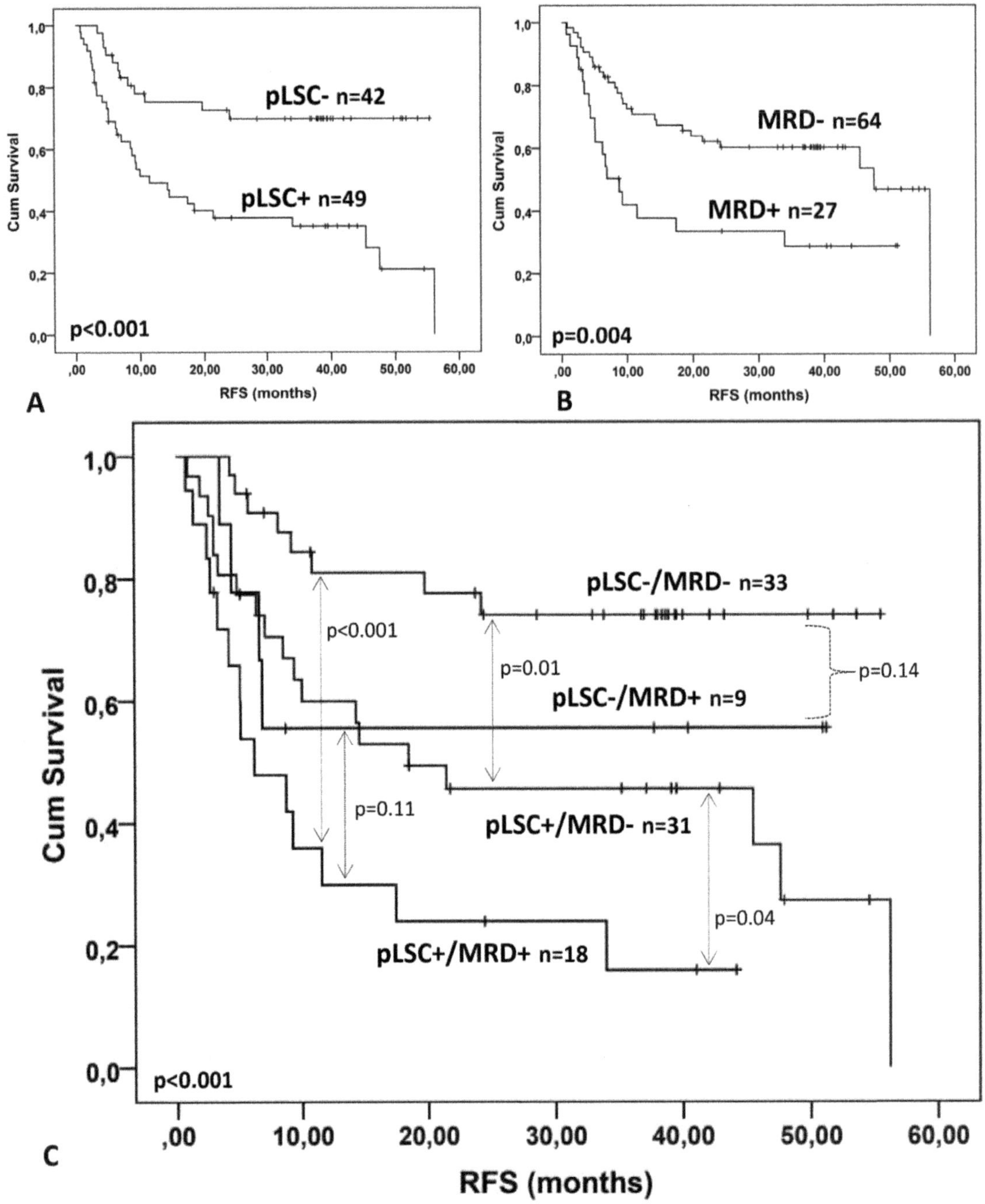

Figure 7. Prognostic value of combined p-LSC and MRD. (A) Kaplan-Meier analyses after cycle II for RFS for the pLSC data as shown in Figure 6, with an additional 23 patients (Table S1). The pLSC cut-off used is 0.0001%. (B) Kaplan Meier analysis of MRD data (cut-off 0.1%) obtained for the same patient group as in A (n = 91). (C) Combined pLSC and MRD (n = 91) data resulted in 4 patient groups: pLSC-/MRD-, pLSC-/MRD+, pLSC+/MRD- and pLSC+/MRD+.

CD34+CD38+ and CD34- compartments represent considerable portions of the total leukemic burden and thus reflect MRD cell frequency rather than pLSC frequency. Here, CD34+CD38+ and CD34- cells probably originate from (limited) differentiation of the CD34+CD38- pLSCs, a process that has been shown to occur *in vivo* by Goardon and colleagues [28].

The results are compatible with the following model: what the paper shows is that in CD34 positive AML cases, it is the percentage of the CD34+CD38- population at diagnosis that strongly correlates with clinical outcome and not the percentage of CD34+CD38+ or CD34- cells. This does not mean that CD34+ CD38+ and CD34- cells do not contain leukemia initiating ability; it simply strongly suggests that, in the presence of CD34+CD38- cells, these CD34+CD38+ and CD34- leukemia initiating cells are either less therapy resistant and/or less malignant compared to CD34+CD38- cells. Likely, leukemia initiating ability in mouse models of CD34/CD38 defined sub-populations do not reflect clinical importance (see also next paragraph), since this ability is always assessed using purified populations, whereby the "competition" between these populations in outgrow and/or the relative therapy resistance cannot be taken into account. This model also implies that in CD34 negative AML (with only neoplastic CD34- populations present), it is the CD34- pLSC that takes over the leukemia initiating ability. Also in CD34 positive AML in the absence of CD34+CD38- cells, but with neoplastic CD34+CD38+ and CD34- populations present, the latter two populations may take over the leukemia initiating ability. As a logical consequence of the model these putatively less aggressive and/or less therapy resistant populations should define a better clinical outcome, which is indeed the case: in a separate cohort of 438 patients, survival of CD34 negative patients was significantly better than survival of CD34 positive patients (unpublished results).

The lack of correlation between prognosis and the size of the different CD34+CD38+ and CD34- compartments, while all compartments in purified form do engraft in a mouse model, may be explained as follows: the overall immune status of a patient group like that in our study, may best be represented by a less immune-restricted mouse model, where CD34+CD38- pLSCs are the predominant engrafting cells [29,30]. In line with that, in our earlier engrafting experiments using CD34+ cells (i.e., containing CD34+CD38-, CD34+CD38+ and CD34- cells) in the less immune-restricted NOD/SCID mice, it was only the size of the CD34+CD38- compartment at diagnosis that correlated with levels of engraftment [13]. More immune-restricted mouse models are useful to study LSC engraftment of probably less aggressive pLSC sub-populations [5,28]. In this respect, an important initial observation was made by Costello and co-workers, who found that, *in vitro*, CD34+CD38- cells were more therapy-resistant and less immunogenic than other compartments [21]. Moreover, further compelling evidence that the CD34+CD38- compartment is most important in the clinical setting comes from our clinical observations: in most cases which had low frequency mutations at diagnosis, which became predominant at relapse, these mutations were present and/or enriched in the neoplastic CD34+CD38- diagnosis compartment [31].

Survival and outgrowth of leukemia cells after therapy may depend on many factors and include the LSC load and likely specific LIC properties, but also the frequency of AML blast cells referred to as MRD. Many authors, including us, have shown that MRD is a strong independent prognostic factor [24,32,33]. In this paper we have shown that both CD34+CD38- pLSC frequency and MRD cell frequency are complementary, thereby defining a new post-diagnosis combination factor that offers strong prognostic information, even across cytogenetics/molecular defined risk groups. This will prospectively be validated in a new patient cohort of the HOVON/SAKK cooperative study group for which patient enrollment has already started.

With regard to our approach to distinguish between pLSCs and HSCs within the CD34+CD38- compartment, we have shown that aberrant expression of antigens (including CLL-1) and lineage markers, as well as additional aberrancies (including differences in CD34 and CD45 expression and differences in light scatter) allow unequivocal discrimination between pLSCs and HSCs. Moreover, this approach even allowed CD34+CD38- pLSC detection in patients lacking aberrant antigen expression. By applying the additional parameters, underestimation of pLSC numbers or overestimation of HSC numbers (often seen when using marker expression alone) is avoided. The present study thereby confirms the findings of others [29] that different CD34+CD38- pLSC markers expressed in individual AML cases may miss substantial portions of LSCs present (Figure 3). Although secondary parameters are thus very useful, it requires experience.

The number of accurately-identified pLSCs would increase with the use of additional AML markers. Such markers may include: CD123, CD96, CD44, CD47, CD25, CD32, CD33, TIM-3 [12,18,29,34–38]. In the present study, both lineage markers and CLL-1 were used to identify CD34+CD38- pLSCs and HSCs during follow-up, since these markers remained highly specific for these LSCs during and after therapy [18,19]. To date, other markers have not yet been tested and considerable efforts will be necessary to demonstrate their applicability both for pLSC quantification at diagnosis and especially at follow up, and as putative targets in antibody therapies. In particular, undesirable post-chemotherapy up-regulation on HSCs may occur for some of these markers [18,19]. Thus, our study implies that, for adequate pLSC (and HSC) tracking, a spectrum of markers and probably additional parameters should be screened for each individual AML case. Therefore, broad application of a single target antigen for diagnostic purposes, as well as for future LSC-directed antibody therapies, still seems unlikely.

Although FSC/SSC characteristics are often effective in discriminating between CD34+CD38- pLSCs and HSCs, the underlying cause of this difference is not yet known. An attractive option is that, similar to normal stem cells, CD34+CD38- AML sub-populations may exist with slightly different levels of differentiation [28]. In fact, we now have more formal proof for that: AML cases with high FSC and SSC of the CD34+CD38- cells all have expression of the differentiation marker CD45RA. It is known that this marker identifies GMP CD34+CD38+ progenitors in contrast to MEP and CMP CD34+CD38+ progenitors. We have found that CD45RA is completely absent on the real CD34+CD38- HSCs, but marked corresponding CD34+CD38- pLSC populations in roughly half of the AML cases. In the other half, the pLSC had FSC and SSC close to those of corresponding HSCs; these pLSCs were CD45RA negative. The CD45RA positive cases may indeed reflect more progenitor-like pLSCs compared to the CD45RA negative pLSCs that resemble HSCs [28].

In AML cases with no malignant CD34+CD38- compartments, the pLSCs will be located in the CD34+CD38+ and/or CD34- compartments. It is likely, however, that within these relatively large compartments, further compartmentalization will be necessary to identify the true LSC sub-compartment which likely occurs at low frequencies. One candidate fraction is the side population (SP), which does occur in low numbers (usually <1% of the WBC population) and contains both normal and AML cells at diagnosis [8,9]. Moreover, the SP has also been found to contain different CD34/CD38-defined sub-fractions [6], which may represent CD34/CD38/SP pLSCs. In the future, it would be interesting to examine the relationships between the functional phenotypes (based on SP or aldehyde dehydrogenase activity) [6–9,39,40] and the CD34/CD38 immunophenotype.

In conclusion, the present study offers tools for detecting concomitantly present pLSCs and HSCs, both at diagnosis and at disease follow-up, and provides the first proof that CD34+CD38- pLSCs are not only clinically important at diagnosis, but also at follow-up. No evidence was found to suggest that the CD34+ CD38+ and CD34- leukemic fractions contain clinically important LSCs, at least not in CD34 positive AML with leukemic CD34+ CD38- present. The combination with well-established MRD assessment likely opens a new field in prognostication in patients with AML. Ultimately, our findings may contribute to the development of new diagnostic tools and to novel, more selective therapies, including antibody-based therapies that would be highly effective against AML stem cells, while leaving the normal HSCs intact. Finally, these results may stimulate further research into the role of cancer stem cells in other cancers, such as solid tumors.

Supporting Information

Figure S1 pLSCs monitoring during sequential BM sampling. A patient (243) with AML positive for a t(8;21) showed CLL-1 expression in the CD34+CD38- compartment (A), while normal BM misses CLL-1+ CD34+CD38- cells [1]. In CR, CLL-1+ CD34+CD38- cells (B) were sorted and assessed for t(8;21): mainly neoplastic cells were present (E). Normal CD34+ CD38-CLL-1 negative cells were almost completely absent here. Shown in F-H are examples of sequential monitoring in three cases with increasing periods of complete remission until relapse (F, pt 253; G, pt 372) and in continuous remission (H, pt 572). Note the increase of pLSC frequency preceding relapse (F, G).

Table S1 Patient characteristics. *"Patients 1" are all CD34+ patients. "Patients 2" are all patients were accurate discrimination between HSC and pLSC was enabled using our extensive gating strategy. "Patients 3" are CR patients with MRD and pLSC data to enlarge the total patient group as shown in Figure 7.

Table S2 FSC and SSC position relative to lymphocytes. FSC, forward scatter; SSC, side scatter; HSC hematopoietic stem cells; pLSC, putative leukemia stem cell; NA, not applicable. * FSC and SSC values relative to those of lymphocytes present in the same sample.

Table S3 Gating details of 117 patients with a secondary gating strategy to define pLSC and HSC at diagnosis AML.

Table S4 Number of patients for different strategies in 250 CD34+ AML cases. *>20% aberrant marker expression was considered substantial to identify directly at least a substantial part of the pLSC population (179/250 patients; rows 3 and 4). In 102/179 patients (41% of all 250 CD34+ patients, row 3), pLSC frequencies may be under-estimated since additional gating strategy (with FSC/SSC etc, referred to in columns 3–7) was not possible, probably leaving part of marker negative pLSCs unidentified. In 77 of these 179 patients, an additional gating step could be performed (FSC/SSC etc, see row 4), allowing a more accurate assessment of both pLSC and HSC frequencies. #: <20% aberrant marker expression (71/250 cases) is shown in rows 5 and 6. In 31 cases (12%) only inadequate LSC assessment was possible (row 5). However, in 40 of these 71 cases HSCs could still be distinguished from pLSCs with the use of secondary parameters (row 6). Highly adequate LSC assessment, using both aberrant marker expression and secondary parameters was thus possible in 77+40 cases (47%). Columns show parameters/plots used to distinguish HSCs from pLSCs.

Table S5 Multi-lineage engraftment of marker negative FSC/SSC^{low} (CD34^{high}) CD34+CD38- cells present in AML. * in the missing mouse, engraftment could not be assessed since this mouse died before examination was possible. # In the missing mouse, no human engraftment was detected. In terms of leukemic engraftment our results also confirmed the observation of Bonnet's group that purified CD34+CD38+ and CD34- were able to engraft be it in our case after injection of high cell numbers. CD34+CD38-/CLL-1+ in pts 1 and 2 (40,000 and 130,000 cells, respectively) CD34+CD38-/CLL-1-/FSC ^{high} CD34^{low} in pt 1 (6,000 cells) CD34+CD38+ in pts 2, 4, 5, 6 (high cell numbers, 100,000-10^6 injected in pts 2, 4, 6 and 1,000 in pt 5) CD34- in pts 2 and 5 (high cell numbers injected:100,000-10^6).

Table S6 Cut-off values in the CD34+CD38-, CD34+CD38+ and CD34- cell compartment at diagnosis to identify patient groups with different survival. *p-values refer to significance of differences in RFS between patients above and patients below the indicated cut-offs.

Table S7 Relative risk of relapse determined for pLSC- and pLSC+ patients at follow up defined by different cut-off points. Not shown in the Table: for RFS and OS, without use of cut-offs, Cox regression analysis showed a strong significant inverse correlation between pLSC percentage and **RFS** after 1st cycle (n = 71, RR = 2.4, 95%CI:1.3–4.6, p = 0.008), 2nd cycle (n = 77, RR = 2.5, 95%CI:1.7–3.7, p<0.001) and consolidation cycle (n = 48, RR = 3.0, 95%CI:1.4–6.2, p = 0.004). For **OS**, these figures were RR = 1.8 (p = 0.04), RR = 2.7 (p<0.001) and RR = 2.0 (p = 0.07). Hereafter different cut offs were applied for risk on relapse. RR, relative risk of relapse using these different cut offs. Cut-offs of 0.0003% (3 in a million, 1st cycle) and 0.0001% (2nd and consolidation cycle) were used for relapse-free survival **(RFS)** in Kaplan-Meier analyses shown in Figure 6. With these cut-offs median overall survival (**OS**, not shown in Figure 6) was not reached (>42 months) for pLSC+ patients after 1st cycle, but more patients survived in the pLSC- group (p = 0.002). After 2nd cycle median **OS** in pLSC- group was>45 months versus 35 months in pLSC+ group (p<0.001). After consolidation cycle these figures were both>37 months (p = 0.05).

Table S8 Multivariate analysis# for impact of pLSC frequency on RFS. # in univariate analyses performed on 162 CD34+ patients, cytogenetic/molecular risk (p = 0.001, n = 162), number of chemotherapy cycles needed to achieve CR (p<0.001, n = 162), and WBC count at diagnosis (p = 0.002, n = 162) were significant. Other factors showed a trend: NPM1 mutation (p = 0.093, n = 140) and EVI-1 (p = 0.18, n = 140). n.r. not relevant since all evaluated patients achieved CR after first cycle RR, relative risk of relapse. * only the most optimal cut-offs (0.0003% after 1st cycle and 0.0001% after 2nd and consolidation cycle) are shown.

Text S1 Supporting text.

Acknowledgments

The authors would like to thank Hans Meel and Rolf Wouters for their help with the sorting and PCR experiments.

Author Contributions

Conceived and designed the experiments: MT AK ANS WZ APR SZ GJO GJS. Performed the experiments: MT AK ANS WZ APR WJS. Analyzed the data: MT AK ANS WZ APR WJS SZ GJO GJS. Contributed reagents/materials/analysis tools: SZ GJO TP GV BL GJS. Wrote the paper: MT AK ANS WZ APR WJS TP GV BL SZ GJO GJS.

References

1. Valent P, Bonnet D, De MR, Lapidot T, Copland M, et al. (2012) Cancer stem cell definitions and terminology: the devil is in the details. Nat Rev Cancer 12: 767–775.
2. Lapidot T, Sirard C, Vormoor J, Murdoch B, Hoang T, et al. (1994) A cell initiating human acute myeloid leukaemia after transplantation into SCID mice. Nature 367: 645–648.
3. Bonnet D, Dick JE (1997) Human acute myeloid leukemia is organized as a hierarchy that originates from a primitive hematopoietic cell. Nature medicine 3: 730–737.
4. Dick JE (2008) Stem cell concepts renew cancer research. Blood 112: 4793–4807.
5. Taussig DC, Vargaftig J, Miraki-Moud F, Griessinger E, Sharrock K, et al. (2010) Leukemia-initiating cells from some acute myeloid leukemia patients with mutated nucleophosmin reside in the CD34(-) fraction. Blood 115: 1976–1984.
6. Moshaver B, Van Rhenen A, Kelder A, Van der Pol M, Terwijn M, et al. (2008) Identification of a small subpopulation of candidate leukemia-initiating cells in the side population of patients with acute myeloid leukemia. Stem cells 26: 3059–3067.
7. Pearce DJ, Taussig D, Simpson C, Allen K, Rohatiner AZ, et al. (2012) Characterization of cells with a high aldehyde dehydrogenase activity from cord blood and acute myeloid leukemia samples. Stem cells 23: 752–760.
8. Wulf GG, Wang RY, Kuehnle I, Weidner D, Marini F, et al. (2001) A leukemic stem cell with intrinsic drug efflux capacity in acute myeloid leukemia. Blood 98: 1166–1173.
9. Feuring-Buske M, Hogge DE (2001) Hoechst 33342 efflux identifies a subpopulation of cytogenetically normal CD34(+)CD38(-) progenitor cells from patients with acute myeloid leukemia. Blood 9: 3882–3889.
10. Taussig DC, Miraki-Moud F, Anjos-Afonso F, Pearce DJ, Allen K, et al. (2008) Anti-CD38 antibody-mediated clearance of human repopulating cells masks the heterogeneity of leukemia-initiating cells. Blood 112: 568–575.
11. Sarry JE, Murphy K, Perry R, Sanchez P V, Secreto A, et al. (2011) Human acute myelogenous leukemia stem cells are rare and heterogeneous when assayed in NOD/SCID/IL2Rgammac-deficient mice. J Clin Invest 121: 384–395.
12. Majeti R, Chao MP, Alizadeh AA, Pang WW, Jaiswal S, et al. (2009) CD47 is an adverse prognostic factor and therapeutic antibody target on human acute myeloid leukemia stem cells. Cell 138: 286–299.
13. Van Rhenen A, Feller N, Kelder A, Westra AH, Rombouts E, et al. (2005) High stem cell frequency in acute myeloid leukemia at diagnosis predicts high minimal residual disease and poor survival. Clinical cancer research 11: 6520–6527.
14. Ran D, Schubert M, Taubert I, Eckstein V, Bellos F, et al. (2012) Heterogeneity of leukemia stem cell candidates at diagnosis of acute myeloid leukemia and their clinical significance. Experimental hematology 40: 155–165.
15. Witte KE, Ahlers J, Schafer I, Andre M, Kerst G, et al. (2011) High proportion of leukemic stem cells at diagnosis is correlated with unfavorable prognosis in childhood acute myeloid leukemia. Pediatr Hematol Oncol 28: 91–99.
16. Roshal M, Chien S, Othus M, Wood BL, Fang M, et al. (2013) The proportion of CD34(+)CD38(low or neg) myeloblasts, but not side population frequency, predicts initial response to induction therapy in patients with newly diagnosed acute myeloid leukemia. Leukemia 27: 728–731.
17. Vergez F, Green AS, Tamburini J, Sarry JE, Gaillard B, et al. (2011) High levels of CD34+CD38low/-CD123+ blasts are predictive of an adverse outcome in acute myeloid leukemia: a Groupe Ouest-Est des Leucemies Aigues et Maladies du Sang (GOELAMS) study. Haematologica 96: 1792–1798.
18. Van Rhenen A, Van Dongen GA, Kelder A, Rombouts EJ, Feller N, et al. (2007) The novel AML stem cell associated antigen CLL-1 aids in discrimination between normal and leukemic stem cells. Blood 110: 2659–2666.
19. Van Rhenen A, Moshaver B, Kelder A, Feller N, Nieuwint AW, et al. (2007) Aberrant marker expression patterns on the CD34+ CD38- stem cell compartment in acute myeloid leukemia allows to distinguish the malignant from the normal stem cell compartment both at diagnosis and in remission. Leukemia 21: 1700–1707.
20. Becker MW, Jordan CT (2011) Leukemia stem cells in 2010: current understanding and future directions. Blood Rev25: 75–81.
21. Costello RT, Mallet F, Gaugler B, Sainty D, Arnoulet C, et al. (2000) Human acute myeloid leukemia CD34+/CD38- progenitor cells have decreased sensitivity to chemotherapy and Fas-induced apoptosis, reduced immunogenicity, and impaired dendritic cell transformation capacities. Cancer Res 60: 4403–4411.
22. Taussig DC, Pearce DJ, Simpson C, Rohatiner AZ, Lister TA, et al. (2005) Hematopoietic stem cells express multiple myeloid markers: implications for the origin and targeted therapy of acute myeloid leukemia. Blood 106: 4086–4092.
23. Feller N, Van der Pol MA, van Stijn A, Weijers GW, Westra AH, et al. (2004) MRD parameters using immunophenotypic detection methods are highly reliable in predicting survival in acute myeloid leukaemia. Leukemia 18: 1380–1390.
24. Terwijn M, Van Putten WLJ, Kelder A, Van der Velden VHJ, Brooimans RA, et al. (2013) High Prognostic Impact of Flow Cytometric Minimal Residual Disease Detection in Acute Myeloid Leukemia: Data From the HOVON/SAKK AML 42A Study. Journal of clinical oncology 31: 3889–3897.
25. Bakker AB, Van den Oudenrijn S, Bakker AQ, Feller N, van Meijer M, et al. (2004) C-type lectin-like molecule-1: a novel myeloid cell surface marker associated with acute myeloid leukemia. Cancer Res 64: 8443–8450.
26. Van der Pol MA, Feller N, Roseboom M, Moshaver B, Westra G, et al. (2006) Assessment of the normal or leukemic nature of CD34+ cells in acute myeloid leukemia with low percentages of CD34 cells. Haematologica 88: 983–993.
27. Cloos J, Goemans BF, Hess CJ, Van Oostveen JW, Waisfisz Q, et al. (2006) Stability and prognostic influence of FLT3 mutations in paired initial and relapsed AML samples. Leukemia 20: 1217–1220.
28. Goardon N, Marchi E, Atzberger A, Quek L, Schuh A, et al. (2011) Coexistence of LMPP-like and GMP-like leukemia stem cells in acute myeloid leukemia. Cancer cell 19: 138–152.
29. Jan M, Chao MP, Cha AC, Alizadeh AA, Gentles AJ, et al. (2011) Prospective separation of normal and leukemic stem cells based on differential expression of TIM3, a human acute myeloid leukemia stem cell marker. Proc Natl Acad Sci USA 108: 5009–5014.
30. Ishikawa F, Yoshida S, Saito Y, Hijikata A, Kitamura H, et al. (2007) Chemotherapy-resistant human AML stem cells home to and engraft within the bone-marrow endosteal region. Nature biotechnology 25: 1315–1321.
31. Bachas C, Schuurhuis GJ, Assaraf YG, Kwidama ZJ, Kelder A, et al. (2012) The role of minor subpopulations within the leukemic blast compartment of AML patients at initial diagnosis in the development of relapse. Leukemia 26: 1313–1320.
32. Buccisano F, Maurillo L, Del Principe MI, Del PG, Sconocchia G, et al. (2012) Prognostic and therapeutic implications of minimal residual disease detection in acute myeloid leukemia. Blood 119: 332–341.
33. Freeman SD, Virgo P, Couzens S, Grimwade D, Russell N, et al. (2013) Prognostic relevance of treatment response measured by flow cytometric residual disease detection in older patients with acute myeloid leukemia. J Clin Oncol 31: 4123–4131.
34. Hosen N, Park CY, Tatsumi N, Oji Y, Sugiyama H, et al. (2007) CD96 is a leukemic stem cell-specific marker in human acute myeloid leukemia. Proc Natl Acad Sci USA 104: 11008–11013.
35. Jordan CT (2002) Unique molecular and cellular features of acute myelogenous leukemia stem cells. Leukemia 16: 559–562.
36. Jordan CT, Upchurch D, Szilvassy SJ, Guzman ML, Howard DS, et al. (2000) The interleukin-3 receptor alpha chain is a unique marker for human acute myelogenous leukemia stem cells. Leukemia 14: 1777–1784.
37. Jin L, Hope KJ, Zhai Q, Smadja-Joffe F, Dick JE (2006) Targeting of CD44 eradicates human acute myeloid leukemic stem cells. Nat med 12: 1167–1174.
38. Saito Y, Kitamura H, Hijikata A, Tomizawa-Murasawa M, Tanaka S, et al. (2010) Identification of therapeutic targets for quiescent, chemotherapy-resistant human leukemia stem cells. Sci Transl Med 2: 17ra9.
39. Ran D, Schubert M, Pietsch L, Taubert I, Wuchter P, et al. (2009) Aldehyde dehydrogenase activity among primary leukemia cells is associated with stem cell features and correlates with adverse clinical outcomes. Exp Hematol 37: 1423–1434.
40. Cheung AMS, Wan TSK, Leung JCK, Chan LYY, Huang H, et al. (2007) Aldehyde dehydrogenase activity in leukemic blasts defines a subgroup of acute myeloid leukemia with adverse prognosis and superior NOD/SCID engrafting potential. Leukemia 2: 1423–1430.

A Retrospective Review of the Prognostic Value of ALDH-1, Bmi-1 and Nanog Stem Cell Markers in Esophageal Squamous Cell Carcinoma

Cheng-Cheng Hwang[1], Shin Nieh[2], Chien-Hong Lai[3], Chien-Sheng Tsai[4], Liang-Che Chang[1], Chung-Ching Hua[5], Wen-Ying Chi[1], Hui-Ping Chien[1], Chih-Wei Wang[7], Siu-Cheung Chan[6], Tsan-Yu Hsieh[1], Jim-Ray Chen[1]*

1 Department of Pathology, Chang Gung Memorial Hospital, Keelung, Taiwan, **2** Department of Pathology and Graduate Institute of Pathology, National Defense Medical Center & Tri-Service General Hospital, Taipei, Taiwan, **3** Division of Hematology-Oncology, Department of Internal Medicine, Chang Gung Memorial Hospital, Keelung, Taiwan, **4** Department of Radiation Oncology, Chang Gung Memorial Hospital, Keelung, Taiwan, **5** Division of Pulmonary, Critical Care and Sleep Medicine, Chang Gung Memorial Hospital, Keelung, Taiwan, **6** Department of Diagnostic Radiology, Chang Gung Memorial Hospital, Keelung, Taiwan, **7** Department of Anatomic Pathology, Chang Gung Memorial Hospital, Tao-Yuan, Taiwan

Abstract

Stem cell markers are upregulated in various cancers and have potential as prognostic indicators. The objective of this study was to determine the expression of three stem cell markers, aldehyde dehydrogenase 1 (ALDH-1), B cell-specific Moloney murine leukemia virus integration site 1 (Bmi-1), and Nanog, in esophageal squamous cell carcinoma (ESCC) tissues. Immunohistochemistry was used to measure the expression of ALDH-1, Bmi-1, and Nanog in ESCC tissues from 41 patients who received pre-operative chemoradiation. We evaluated the relationship between expression of these markers, and clinicopathological features, tumor regression grade (TRG), and 5-year overall survival (OS). There were no significant associations of ALDH-1 or Bmi-1 expression with age, gender, clinical stage, and treatments ($p>0.05$). However, patients with Nanog-positive tumors were significantly older than those whose tumors were Nanog-negative ($p=0.033$). TRG after treatment was significantly associated with expression of ALDH-1 ($p=0.001$), Bmi-1 ($p=0.004$), and Nanog ($p<0.001$). Although OS was significantly better in patients with low TRGs ($p=0.001$), there were no significant correlations between ALDH-1, Bmi-1, or Nanog with OS. Expression of ALDH-1, Bmi-1, and Nanog correlated with TRG, but not OS. Further large studies are necessary to fully elucidate the prognostic value of these stem cell markers for ESCC patients.

Editor: Anthony W.I. Lo, The Chinese University of Hong Kong, Hong Kong

Funding: This study was supported by grant CMRPG290391 to C.-C. Hwang from Chang Gung Memorial Hospital, Keelung, Taiwan. The funders had no role in study design, data collection and analysis, decision to publish, or preparation of the manuscript.

Competing Interests: The authors have declared that no competing interests exist.

* Email: rrooyyaall@yahoo.com

Introduction

Esopahgeal adenocarcinoma (EAC) and esophageal squamous cell carcinoma (ESCC) are the 2 major histological types of esophageal cancer [1]. EAC is more common in Western countries, but ESCC is more common in East Asia [2]. In China, ESCC is the fourth-leading cause of cancer-associated death [3]. Despite improvements in surgery, chemotherapy, and radiotherapy, the 5-year overall survival remains poor, in part because patients typically have advanced-stage cancer at diagnosis [4]. Thus, in addition to prevention strategies, which focus on modification of risk factors associated with ESCC such as alcohol consumption and cigarette smoking [5], identification of prognostic markers of ESCC may also help to reduce the mortality associated with this cancer.

The levels of serum anti-p53 antibodies and C-reactive protein predict the response to therapy in patients with recurrent esophageal cancer [6]. In addition, the phosphorylation level of mammalian target of rapamycin (mTOR) is associated with response to chemoradiotherapy and overall disease-free survival in ESCC patients [7]. Furthermore, 2 mesenchymal markers of the epithelial mesenchymal transition (EMT), vimentin and fibronectin, are indicators of poor prognosis [4]. The EMT enhances cancer invasion and metastasis, in part because cells develop stem cell-like properties [8].

Cancer stem cells have the capacity of self-renewal, and are thought to be responsible for the generation of multiple cell lineages that are characteristic of many tumors [9–10]. Cancer stem cells have roles in tumorigenesis and resistance to therapy, so the identification of markers for cancer stem cell may provide important information regarding patient prognosis and response to therapy [11–12]. For example, previous research indicated that adenosine triphosphate-binding cassette superfamily G member 2 (ABCG2) expression was associated with ESCC patient survival [13] and Notch1 expression was associated with greater pathological grade and shorter OS in ESCC patients [14].

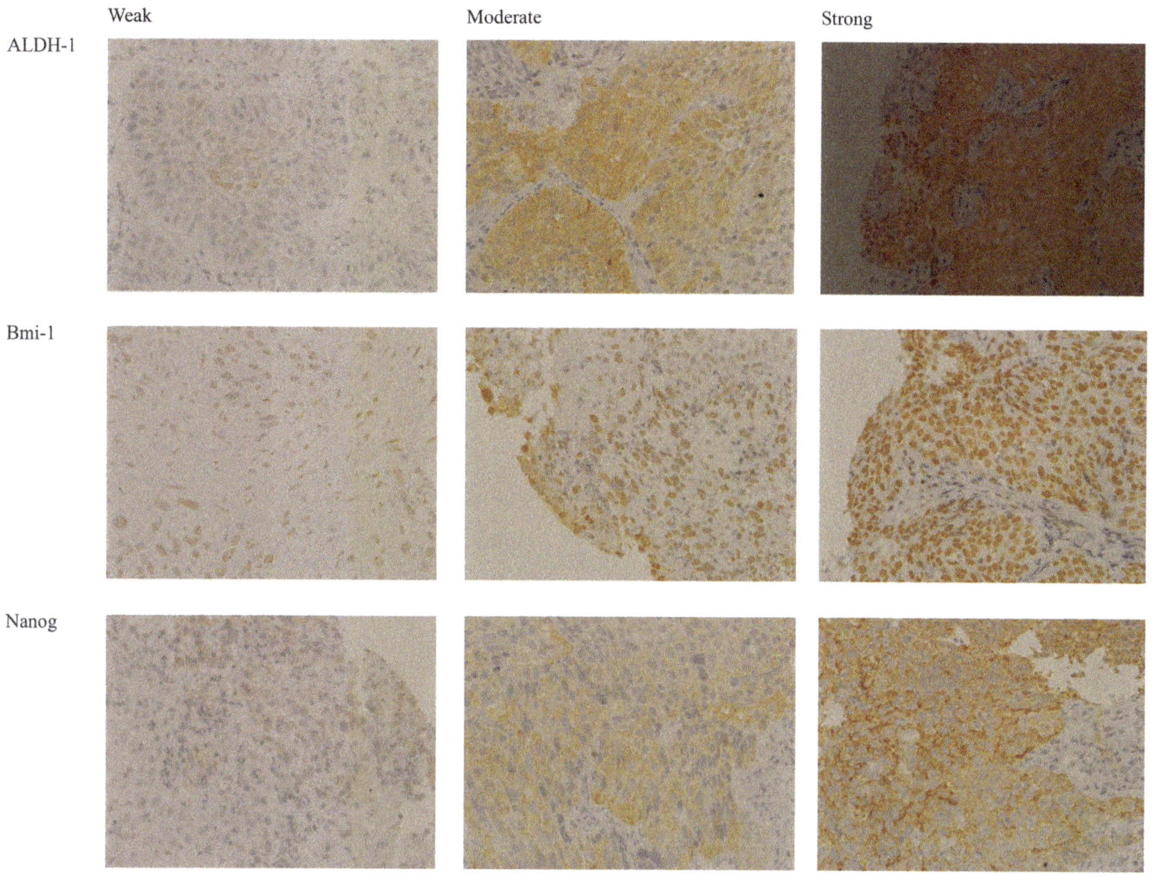

Figure 1. Representative images of cancer stem cell markers (top: ALDH1, middle: Bmi-1, bottom: Nanog) with different stain intensities (weak, moderate, strong).

Identification of markers associated with ESCC may be useful for determining prognosis and response to therapy. Previous research indicated that the expression of 3 cancer stem cell markers, aldehyde dehydrogenase 1 (ALDH-1), B cell-specific Moloney murine leukemia virus integration site 1 (Bmi-1), and Nanog, are associated with esophageal cancer [15–17]. ALDH-1 is one of a family of enzymes that catalyze the oxidation of aldehydes to carboxylic acids; Bmi-1 is a polycomb ring finger oncogene that regulates p16 and p19; and Nanog is a homeobox transcription factor that has elevated expression in stem cells and some solid tumors. The present study used immunohistochemistry to investigate the expression of these 3 genes in human ESCC tissues and examined the relationship of their expression with clinicopathological characteristics and prognosis.

Material and Methods

Sample and data collection

The records of 41 ESCC patients who underwent preoperative chemoradiation therapy and radical esophagectomy/curative surgery from 2000 to 2012 at Keelung Chang Gung Memorial Hospital of Chang Gung Medical Foundation were retrospectively reviewed. Patient information, including gender, age, and clinicopathologic characteristics, was obtained from the Taiwan Cancer Registry, a national database. ESCC tissue samples were obtained from all participants for immunohistochemical analysis. The enrolled patients included 39 males and 2 females, and the median age was 54 years (range, 37–78 years). Histological analysis indicated that 2 patients (4.9%) had well-differentiated tumors, 27 patients (65.9%) had moderately differentiated tumors, and 12 patients (29.3%) had poorly differentiated tumors. The median follow-up time was 13.0 months (range 0.3–57.4 months). The duration of overall survival (OS) began at the day of surgery and ended at last follow-up or death.

Diagnosis was confirmed by pathological examination of hematoxylin and eosin staining results as described previously [18]. There were 5 tumor regression grades (TRGs), ranging from complete tumor response (TRG I) to no regressive changes within the tumor (TRG V) as previously described [19]. Specifically, TRG I was characterized by the absence of residual cancer and fibrosis throughout the esophageal wall; TRG II by the presence of rare residual cancer cells scattered through the fibrosis; TRG III by an increased number of residual cancer cells with predominant fibrosis; TRG IV by residual cancer outgrowing fibrosis; and TRG V by no regressive changes [19]. This study was approved by the institutional review board of the Keelung Chang Gung Memorial Hospital of CGMF, and written informed consent was obtained from all participants.

Immunohistochemical staining

Tissues were embedded in paraffin and sectioned. After deparaffinization, they were hydrated and incubated with an epitope retrieval solution (10 mM citric acid, pH 6.0) at 95°C for 40 min. After incubation with 0.3% H_2O_2 to inactivate endogenous peroxidase, nonspecific binding was inhibited by incubation

Table 1. Relationship between expression of stem cell markers and characteristics of 41 patients with ESCC.

	ALDH-1			Bmi-1			Nanog		
	Positive (n = 24)	Negative (n = 17)	*P*-value	Positive (n = 36)	Negative (n = 5)	*P*-value	Positive (n = 37)	Negative (n = 4)	*P*-value
Age (years)[1]	55 (51, 64)	51 (46, 59)	0.255	54 (47, 62)	56 (53, 61)	0.282	55 (50, 62)	46 (41, 50)	0.033
Gender, n (%)[2]			1.000			1.000			1.000
Male	23 (95.8)	16 (94.1)		34 (94.4)	5 (100)		35 (94.6)	4 (100)	
Female	1 (4.2)	1 (5.9)		2 (5.6)	0 (0.0)		2 (5.4)	0 (0.0)	
Clinical stage, n(%)[2]			0.212			0.435			0.621
II	2 (8.3)	5 (29.4)		6 (16.7)	1 (20.0)		7 (18.9)	0 (0.0)	
III	14 (58.3)	9 (52.9)		19 (52.8)	4 (80.0)		21 (56.8)	2 (50.0)	
IV	8 (33.3)	3 (17.6)		11 (16.7)	0 (0.0)		9 (24.3)	2 (50.0)	

Each number indicates median and interquartile range unless otherwise specified.
P-values are based on [1]Mann-Whitney U test and [2]Fisher's exact test.

with 5% normal serum in PBS at room temperature for 30 min. The sections were next incubated with primary antibodies specific for ALDH-1, Bmi-1, and Nanog (Epitomics, Burlingame CA, USA) at 4°C overnight. The sections were next incubated with biotinylated secondary antibodies (Lab Vision, Kalamazoo, MI, USA) at room temperature for 10 min, and then with streptavidin peroxidase (Lab Vision) at room temperature. Visualization was done with DAB following the manufacturer's instructions (DaKo, Carpinteria, CA, USA). After sections were counterstained with hematoxylin, they were air-dried and observed under a microscope (E600, Nikon, Tokyo, Japan). All cells of each section were observed. The stain intensity was graded as negative (no expression), weak (fewer than one-third positive cells), moderate (one-third to two-thirds positive cells), or strong (more than two-thirds positive cells). Figure 1 shows representative images of the different stain intensities of each marker. Two pathologists who were blinded to the clinical data performed all histological evaluations. Inconsistent results were reviewed and discussed before making a final decision.

Statistical Analysis

Due to the small sample size, continuous data were compared using the Mann-Whitney U test, and categorical variables were compared by Fisher's exact test. Continuous variables are presented as medians and interquartile ranges, and categorical data as numbers and percentages. Correlations between stem cell marker expression and tumor regression grade were performed using Kendall's correlation coefficient. Kaplan-Meier analysis and the log-rank test were used for analysis of cumulative survival curves. All statistical assessments were two-sided, and a *p*-value less than 0.05 was considered significant. Statistical analyses were performed with SPSS 15.0 Statistics Software (SPSS Inc., Chicago, IL, USA).

Results

Association of stem cell markers with patient characteristics

Table 1 shows that there were no significant correlations between expression of ALDH-1 or Bmi-1 with age, gender, clinical stage ($p>0.05$ for all). However, patients with Nanog-positive tumors were significantly older than those whose tumors were Nanog-negative ($p = 0.033$). Additional analysis and calculation of Kendall's rank correlation coefficients indicated that there were several significant positive correlations in the expression of these markers (ALDH-1 and Bmi-1: $\tau = 0.34$, $p = 0.014$; ALDH-1 and Nanog: $\tau = 0.45$, $p = 0.001$; Bmi-1 and Nanog: $\tau = 0.28$, $p = 0.040$) (Table S1).

Association of stem cell markers with TRG and patient survival

TRG was classified based on complete tumor response to therapy [19]. Figure 2 shows that TRG after treatment was significantly associated with greater expression of ALDH-1 ($p = 0.001$), Bmi-1 ($p = 0.004$), and Nanog ($p<0.001$).

We also analyzed the association between the expression of the 3 stem cell markers with patient survival. By the end of the study, 15 patients had died from ESCC, 9 patients died from complications associated with ESCC, and 2 patients died from unknown causes. The median survival time was 14.8 months (95% CI, 8.8–20.8 months). Kaplan–Meier analysis and the log-rank tests show that there was no significant correlation between expression of any of the stem cell markers with OS ($p>0.05$).

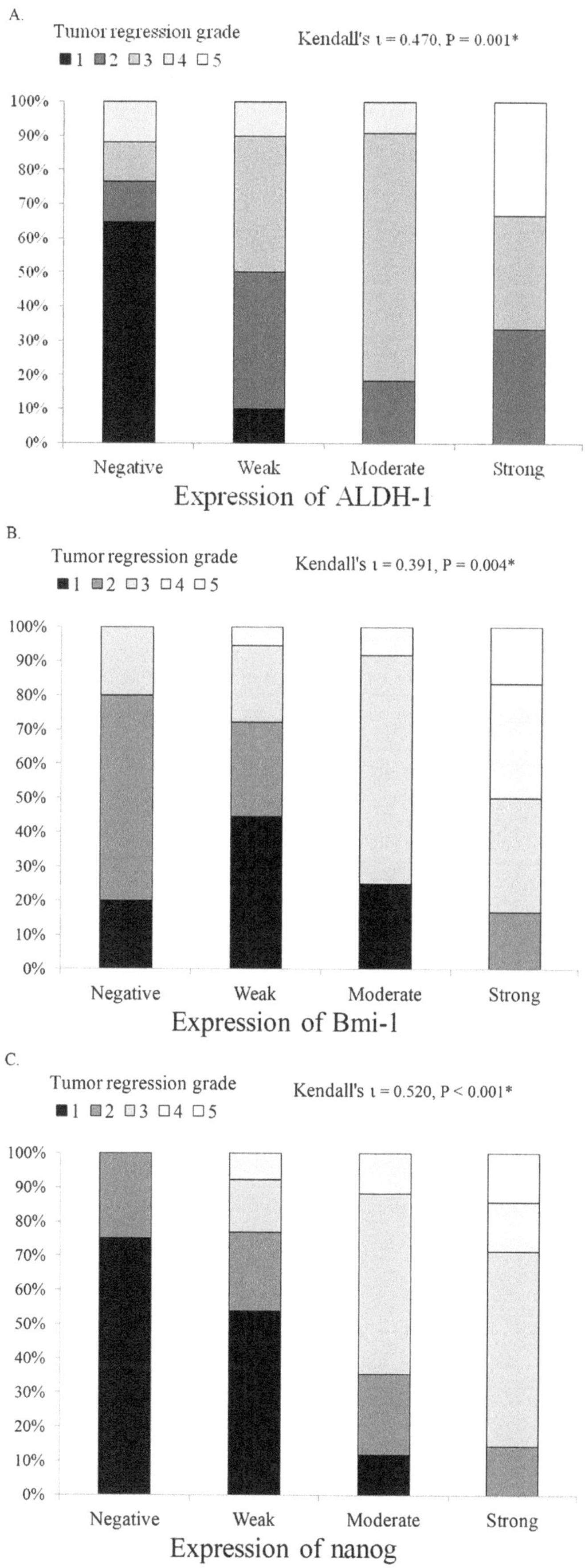

Figure 2. Relationship of expression of cancer stem cell markers (A: ALDH-1, B: Bmi-1, C: Nanog) with tumor regression grade. *$P<0.01$.

A.

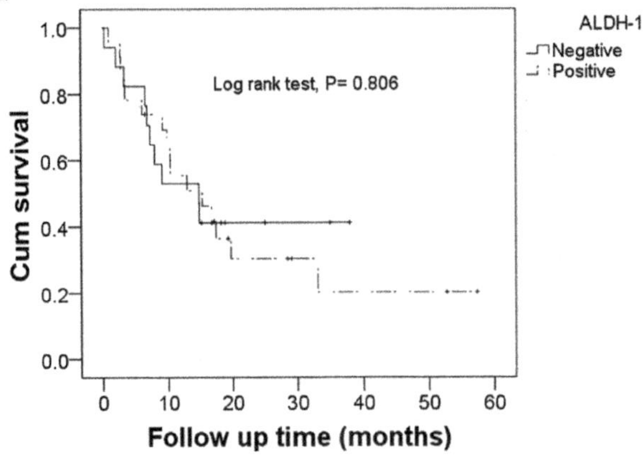
ALDH-1
Negative
Positive
Log rank test, P= 0.806
Cum survival
1.0
0.8
0.6
0.4
0.2
0.0
0
10
20
30
40
50
60
Follow up time (months)

B.

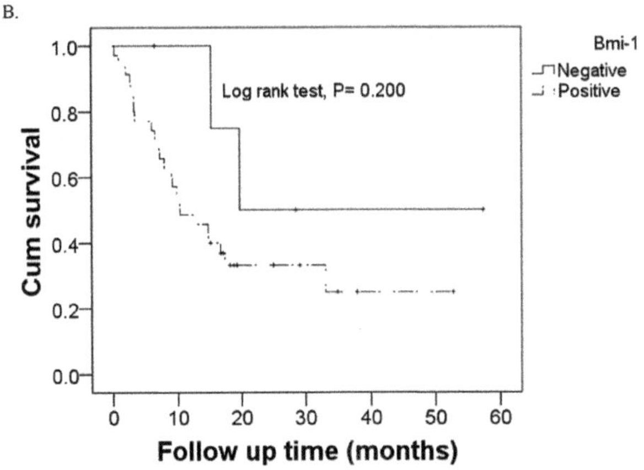
Bmi-1
Negative
Positive
Log rank test, P= 0.200
Cum survival
1.0
0.8
0.6
0.4
0.2
0.0
0
10
20
30
40
50
60
Follow up time (months)

C.

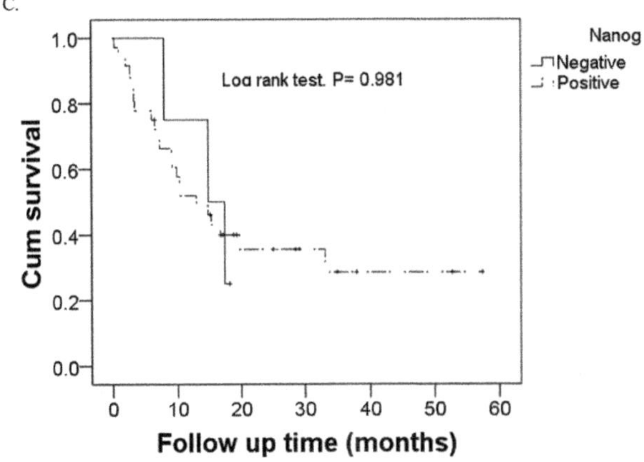
Nanog
Negative
Positive
Log rank test, P= 0.981
Cum survival
1.0
0.8
0.6
0.4
0.2
0.0
0
10
20
30
40
50
60
Follow up time (months)

D.

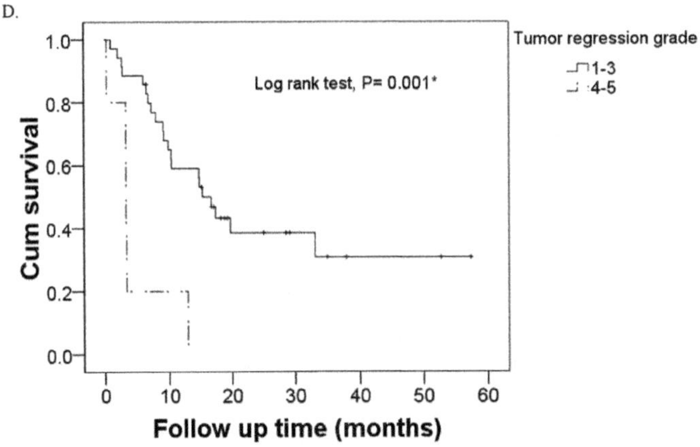
Tumor regression grade
1-3
4-5
Log rank test, P= 0.001*
Cum survival
1.0
0.8
0.6
0.4
0.2
0.0
0
10
20
30
40
50
60
Follow up time (months)

Figure 3. Kaplan-Meier analysis of the prognostic value of the expression of cancer stem cell markers (A: ALDH-1, B: Bmi-1, C: Nanog) and tumor regression grade (D) in prediction of overall survival. $^*P<0.001$.

However, OS was significantly better in patients TRG 1–3 than in patients with TRG 4–5 ($p = 0.001$, Figure 3).

Discussion

More than 50% of patients with esophageal cancer have metastases or unresectable tumors at diagnosis [20], and this largely explains their poor prognosis [4]. Therefore, identification of early diagnostic markers of esophageal cancer may be useful for improving prognosis and informing treatment decisions. Previous research reported that the expression of ALDH-1, Bmi-1, and Nanog were associated with ESCC and EAC [15–17], and this motivated our examination of the relationship between their expression and the clinicopathological features of patients with ESCC. Although Nanog expression was associated with age, there were no other significant correlations between expression of the 3 examined markers with age, gender, and clinical stage. However, elevated expression of each marker was associated with more advanced TRG after treatment. Although poor OS was associated with more advanced TRG, there were no significant correlations between expression of ALDH-1, Bmi-1, or Nanog with OS.

TNM staging, which considers tumor size, its spread to lymph nodes, and distant metastasis, is the most common staging system but a recent study highlighted several shortcomings of the 7th edition regarding its prognostic value for patients receiving neoadjuvant chemoradiotherapy [21]. Thus, Weider et al. [22] suggested that assessments of early tumor response to chemoradiotherapy predicted esophageal cancer patient survival. Similarly, Mandard et al. [19] observed that tumor regression was a significant predictor of disease-free survival of esophageal cancer patients, consistent with the results of the present study, which indicated that advanced TRG was associated with poor OS. However, identification of prognostic markers that can be detected by noninvasive methods would enable clinicians to continuously monitor response to therapy. Cancer stem cells are important for tumorigenesis and therapy resistance, so analysis of cancer stem cell markers in tumors may provide important information on patient prognosis and response to therapy [11–12].

ALDH-1 has a role in stem cell differentiation by regulating retinal oxidation to retinoic acid [23], and its expression has been used as a marker for normal and cancer stem cells [24]. In Chinese patients with ESCC, nuclear expression of ALDH-1 is associated with poor histological differentiation, lymph node metastasis, TNM stage, and poor 5-year OS [15]. In the present study, ALDH-1 expression was associated with advanced TRG, but not with clinical stage or OS.

Bmi-1, a polycomb-group protein, regulates gene expression by chromatin remodeling, and control of axial patterning, cell proliferation, and senescence [25–26]. Expression of Bmi-1 is upregulated in many solid and hematologic cancers [27]. In addition, previous research reported associations between Bmi-1 expression and ESCC and EAC [17,28–29]. Specifically, He et al. [28] reported an association between Bmi-1 overexpression and advanced pathological stage and poor OS and the presence of Bmi-1 autoantibodies correlated with tumor stage and lymph node metastasis [29]. However, Choy et al. [17] found no association between Bmi-1 expression and OS in patients with EAC or ESCC, consistent with the results of the present study.

The expression of Nanog, a homeobox transcription factor, correlates with cell differentiation [30] and resistance to therapy [31] in patients with breast cancer and with poor survival in patients with oral cancer [32]. Furthermore, a previous study reported that Nanog expression was associated with TNM stage and differentiation of ESCC [16]. The present study indicated that Nanog expression was associated with TRG, but not with clinical stage or OS.

The small sample size of the present study is a limitation. Thus, the findings of no associations between ALDH-1, Bmi-1, or Nanog expression with clinicopathological features or OS cannot be completely ruled out. A larger cohort study would also allow the analysis of the prognostic value of these markers on tumor subpopulations, as in Sudo et al. [4]. Alternatively, it remains possible that the selected stem cell markers may not be suitable for predicting the outcome of patients with ESCC. Therefore, larger studies are necessary. In addition, the present study did not perform survival analysis based on tumor node metastasis stage of ESCC.

Conclusions

Examination of the expression of 3 cancer stem cell markers, ALDH-1, Bmi-1, and Nanog, in patients with ESCC indicated correlations with TRG but not OS. Further large studies are necessary to fully elucidate the prognostic value of these stem cell markers for ESCC patients.

Supporting Information

Table S1 Correlations in the expression of ALDH-1, Bmi-1, and Nanog in 41 patients with ESCC.

Acknowledgments

We thank Dr. Marjet Heitzer who provided medical writing services on behalf of MedCom Asia, Inc.

Author Contributions

Conceived and designed the experiments: CCH JRC SN. Performed the experiments: CCH JRC. Analyzed the data: CHL CST. Contributed reagents/materials/analysis tools: LCC CCH. Contributed to the writing of the manuscript: WYC SCC. Acquired the data: CHL CST. Contributed to review of slides: HPC CWW TYH.

References

1. Kelsen DP, Daly JM, Kern SE, Levin B, Tepper JE, et al. (2008) Principles and practice of gastrointestinal oncology, 2nd ed. Philadelphia: Lippincott Williams & Wilkins.
2. Lin Y, Totsuka Y, He Y, Kikuchi S, Qiao Y, et al. (2013) Epidemiology of Esophageal Cancer in Japan and China. J Epidemiol 2013 Apr 27. [Epub ahead of print]
3. Jemal A, Bray F, Center MM, Ferlay J, Ward E, et al. (2011) Global cancer statistics. CA Cancer J Clin 61: 69–90.
4. Sudo T, Iwaya T, Nishida N, Sawada G, Takahashi Y, et al. (2012) Expression of mesenchymal markers vimentin and fibronectin: the clinical significance in esophageal squamous cell carcinoma. Ann Surg Oncol 2012 May 30. [Epub ahead of print]

5. Oze I, Matsuo K, Ito H, Wakai K, Nagata C, et al. (2012) Cigarette smoking and esophageal cancer risk: an evaluation based on a systematic review of epidemiologic evidence among the Japanese population. Jpn J Clin Oncol 42: 63–73.
6. Shimada H, Kitabayashi H, Nabeya Y, Okazumi S, Matsubara H, et al. (2003) Treatment response and prognosis of patients after recurrence of esophageal cancer. Surgery 133: 24–31.
7. Li SH, Huang EY, Lu HI, Huang WT, Yen CC, et al. (2012) Phosphorylated mammalian target of rapamycin expression is associated with the response to chemoradiotherapy in patients with esophageal squamous cell carcinoma. J Thorac Cardiovasc Surg 144: 1352–1359.
8. Mani SA, Guo W, Liao MJ, Eaton EN, Ayyanan A, et al. (2008) The epithelial-mesenchymal transition generates cells with properties of stem cells. Cell 133: 704–715.
9. Tu SM, Lin SH, Logothetis CJ (2002) Stem-cell origin of metastasis and heterogeneity in solid tumours. Lancet Oncol 3: 508–513.
10. Stingl J, Caldas C (2007) Molecular heterogeneity of breast carcinomas and the cancer stem cell hypothesis. Nat Rev Cancer 7: 791–799.
11. Clarke MF, Dick JE, Dirks PB, Eaves CJ, Jamieson CH, et al. (2006) Cancer stem cells perspectives on current status and future directions: AACR Workshop on cancer stem cells. Cancer Res 66: 9339–9344.
12. Eyler CE, Rich JN (2008) Survival of the fittest: cancer stem cells in therapeutic resistance and angiogenesis. J Clin Oncol 26: 2839–2845.
13. Hang D, Dong HC, Ning T, Dong B, Hou DL, et al. (2012) Prognostic value of the stem cell markers CD133 and ABCG2 expression in esophageal squamous cell carcinoma. Dis Esophagus 25: 638–644.
14. Liu J, Fan H, Ma Y, Liang D, Huang R, et al. (2013) Notch1 is a 5-fluorouracil resistant and poor survival marker in human esophagus squamous cell carcinomas. PLoS One 8:e56141.
15. Wang Y, Zhe H, Gao P, Zhang N, Li G, et al. (2012) Cancer stem cell marker ALDH1 expression is associated with lymph node metastasis and poor survival in esophageal squamous cell carcinoma: a study from high incidence area of northern China. Dis Esophagus 25: 560–565.
16. Du Y, Shi L, Wang T, Liu Z, Wang Z (2012) Nanog siRNA plus Cisplatin may enhance the sensitivity of chemotherapy in esophageal cancer. J Cancer Res Clin Oncol 138: 1759–1767.
17. Choy B, Bandla S, Xia Y, Tan D, Pennathur A, et al. (2012) Clinicopathologic characteristics of high expression of Bmi-1 in esophageal adenocarcinoma and squamous cell carcinoma. BMC Gastroenterol 12: 146.
18. Gore RM (1997) Esophageal cancer: clinical and pathologic features. Radiol Clin North Am 35: 243–263
19. Mandard AM, Dalibard F, Mandard JC, Marnay J, Henry-Amar M, et al. (1994) Pathologic assessment of tumor regression after preoperative chemoradiotherapy of esophageal carcinoma. Clinicopathologic correlations. Cancer 73: 2680–2686.
20. Enzinger PC, Mayer RJ (2003) Esophageal cancer. N Engl J Med 349: 2241–2252.
21. Nomura M, Shitara K, Kodaira T, Hatooka S, Mizota A, et al. (2012) Prognostic impact of the 6th and 7th American Joint Committee on Cancer TNM staging systems on esophageal cancer patients treated with chemoradiotherapy. Int J Radiat Oncol Biol Phys 82: 946–952.
22. Wieder HA, Brucher BL, Zimmermann F, Becker K, Lordick F, et al. (2004) Time course of tumor metabolic activity during chemoradiotherapy of esophageal squamous cell carcinoma and response to treatment. J Clin Oncol 22: 900–908.
23. Kastan MB, Schlaffer E, Russo JE, Colvin OM, Civin CI, et al. (1990) Direct demonstration of elevated aldehyde dehydrogenase in human hematopoietic progenitor cells. Blood 75: 1947–1950.
24. Ma I, Allan AL (2011) The role of human aldehyde dehydrogenase in normal and cancer stem cells. Stem Cell Rev 7: 292–306.
25. van der Lugt NM, Domen J, Linders K, van Roon M, Robanus-Maandag E, et al. (1994) Posterior transformation, neurological abnormalities, and severe hematopoietic defects in mice with a targeted deletion of the bmi-1 protooncogene. Genes Dev 8: 757–769.
26. Jacobs JJ, Kieboom K, Marino S, DePinho RA, van Lohuizen M (1999) The oncogene and Polycomb-group gene bmi-1 regulates cell proliferation and senescence through the ink4a locus. Nature 397: 164–168.
27. Cao L, Bombard J, Cintron K, Sheedy J, Weetall ML, et al. (2011) BMI1 as a novel target for drug discovery in cancer. J Cell Biochem 112: 2729–2741.
28. He XT, Cao XF, Ji L, Zhu B, Lv J, et al. (2009) Association between Bmi1 and clinicopathological status of esophageal squamous cell carcinoma. World J Gastroenterol 15: 2389–2394.
29. Liu WL, Guo XZ, Zhang LJ, Wang JY, Zhang G, et al. (2010) Prognostic relevance of Bmi-1 expression and autoantibodies in esophageal squamous cell carcinoma. BMC Cancer 10: 467.
30. Ben-Porath I, Thomson MW, Carey VJ, Ge R, Bell GW, et al. (2008) An embryonic stem cell-like gene expression signature in poorly differentiated aggressive human tumors. Nat Genet 40: 499–507.
31. Bourguignon LY, Peyrollier K, Xia W, Gilad E (2008) Hyaluronan-CD44 interaction activates stem cell marker Nanog, Stat-3-mediated MDR1 gene expression, and ankyrin-regulated multidrug efflux in breast and ovarian tumor cells. J Biol Chem 283: 17635–17651.
32. Chiou SH, Yu CC, Huang CY (2008) Positive correlations of Oct-4 and Nanog in oral cancer stem-like cells and high-grade oral squamous cell carcinoma. Clin Cancer Res 14: 4085–4095.

New Therapy of Skin Repair Combining Adipose-Derived Mesenchymal Stem Cells with Sodium Carboxymethylcellulose Scaffold in a Pre-Clinical Rat Model

Cristiano Rodrigues[1], Adriano M. de Assis[1,4], Dinara J. Moura[2], Graziele Halmenschlager[3], Jenifer Saffi[2], Léder Leal Xavier[5], Marilda da Cruz Fernandes[6], Márcia Rosângela Wink[1]*

1 Laboratórios de Pesquisa em Biologia Celular, Departamento de Ciências Básicas da Saúde, Universidade Federal de Ciências da Saúde de Porto Alegre, Porto Alegre/RS, Brazil, **2** Genética Toxicológica, Departamento de Ciências Básicas da Saúde, Universidade Federal de Ciências da Saúde de Porto Alegre, Porto Alegre/RS, Brazil, **3** Estresse Oxidativo e Poluição Atmosférica, Departamento de Ciências Básicas da Saúde, Universidade Federal de Ciências da Saúde de Porto Alegre, Porto Alegre/RS, Brazil, **4** Programa de Pós-Graduação em Ciências Biológicas: Bioquímica, Departamento de Bioquímica, Universidade Federal do Rio Grande do Sul, Porto Alegre/RS, Brazil, **5** Laboratório de Biologia Celular e Tecidual, Departamento de Ciências Morfofisiológicas, Pontifícia Universidade Católica do Rio Grande do Sul, Porto Alegre/RS, Brazil, **6** Laboratório de Pesquisa em Patologia, Departamento de Ciências Básicas da Saúde, Universidade Federal de Ciências da Saúde de Porto Alegre, Porto Alegre/RS, Brazil

Abstract

Lesions with great loss of skin and extensive burns are usually treated with heterologous skin grafts, which may lead rejection. Cell therapy with mesenchymal stem cells is arising as a new proposal to accelerate the healing process. We tested a new therapy consisting of sodium carboxymethylcellulose (CMC) as a biomaterial, in combination with adipose-derived stem cells (ADSCs), to treat skin lesions in an *in vivo* rat model. This biomaterial did not affect membrane viability and induced a small and transient genotoxicity, only at the highest concentration tested (40 mg/mL). In a rat wound model, CMC at 10 mg/mL associated with ADSCs increased the rate of cell proliferation of the granulation tissue and epithelium thickness when compared to untreated lesions (Sham), but did not increase collagen fibers nor alter the overall speed of wound closure. Taken together, the results show that the CMC is capable to allow the growth of ADSCs and is safe for this biological application up to the concentration of 20 mg/mL. These findings suggest that CMC is a promising biomaterial to be used in cell therapy.

Editor: Gobardhan Das, University of KwaZulu-Natal, South Africa

Funding: C. Rodrigues was the recipient of a Master fellowship from CAPES-REUNI program (Coordenação de aperfeiçoamento de Pessoal de Nível Superior); M.R. Wink, L.L. Xavier, and J. Saffi are recipients of research fellowships from CNPq (Conselho Nacional de Desenvolvimento Científico e Tecnológico). This study was supported by PRONEX and PRONEM - FAPERGS (Fundação de Amparo à Pesquisa do Estado do Rio Grande do Sul) and PROBITEC-CAPES. The funders had no role in study design, data collection and analysis, decision to publish, or preparation of the manuscript.

Competing Interests: The authors have declared that no competing interests exist.

* E-mail: mwink@ufcspa.edu.br

Introduction

The skin is the largest organ of the human body and its integrity allows body functions such as elimination of toxins, resistance to mechanical stress, prevention of water loss, participation in immune responses and protection against external agents [1,2,3]. However, many factors and systemic conditions may compromise the dermal function, such as, diabetes mellitus, trauma, inflammatory diseases, metabolic abnormalities, bleeding disorders, smoking, immunosuppression, malnutrition and obesity [1]. Therefore, promotion of skin lesion healing, regardless of the nature of the causative agent, is of utmost importance [4].

After injury to skin, a new epithelium has to form in order to close the wound and restore the barrier function. This requires the proliferation and directional migration of keratinocytes, fibroblasts and mesenchymal stem cells (MSCs) to wound edges [5,6]. MSCs are essential for skin homeostasis and repair of the lesions, by promoting cell differentiation, immune-modulation, secretion of growth factors to drive re-epithelialization and neovascularization and modulating resident stem cells [7,8]. Both, proliferation and migration of these cells are controlled by extracellular hormones, e.g., glucocorticoids and paracrine factors [5,6], which provides attractive opportunities for therapeutic interventions.

In the last decade, studies have indicated that human adipose tissue contains populations of MSCs, called adipose-derived stem cells (ADSCs) [9]. In contrast to stem cells obtained from bone marrow (BM-MSCs), whose access can be painful for the patient, requiring general or spinal anesthesia and most often resulting in low numbers of MSCs after processing [10], ADSCs can be collected from adipose tissue, as the waste material after liposuction in aesthetic surgeries [11]. It is estimated that 1 mL of lipoaspirate contains at least 7.3×10^3 fibroblast colony forming (CFU-F), corresponding to 0.5–5% of the total population of stromal vascular fraction (SVF) (4.04 ± 3.26 on average). Therefore, the SVF adipose tissue contains a quantity of CFU-F, 400 times higher than the bone marrow, where the frequency of CFU-F is estimated to be about 1 per 10^5 cells [12].

Thus, the possibility of deriving ADSCs from adult adipose tissues, the relatively easy protocols for their isolation and expansion in culture, and the fact that they are immunologically well tolerated, allowing the transplant from one individual to another of the same species, makes them attractive candidates for new studies in translational clinical research [9].

Besides the availability and abundance of ADSCs in the body, several steps remain to be solved before these cells become effectively and safely used as an established therapy. One of the obstacles is to find the correct substances or molecules capable to offer the correct environment to these cells, allowing their safe deployment in tissues or lesions. The long-term success of these therapies requires biocompatibility of the biomaterial that hosts the tissue while retaining the functionality of the cells within the implanted tissue [13]. Sodium carboxymethylcellulose (CMC) is synthesized by the hydration of the cellulose with sodium hydroxide and from an alkaline pulp catalyzed reaction with chloroacetic acid [13]. This synthetic polymer has been shown to be suitable for the protection and treatment of skin wounds, including partial thickness burns [14]. It has a strong ability to absorb and transport fluids and protect from bacterial exposure of the external environment [14]. These attributes make CMC suitable for use as a polymer matrix for wound healing, although its toxicity was not yet fully evaluated [15]. Over the years, CMC has been used as a single material with or without combination with drugs and co-excipients in partial thickness wounds, diabetic ulcers, lesions of the foot and deep dermal filler [15].

Considering that ADSCs have a great potential for clinical use in tissue repair and regeneration, but its efficacy depends of biomaterials that allow the proper tissue-engineering products, the aim of this study was to test the applicability of CMC as a scaffold for the ADSCs, to treat dermal lesions in a pre-clinical model.

Materials and Methods

Ethical Aspects

The protocols used in this study were approved by the Ethics Committee on Animal Use (CEUA) of Universidade Federal de Ciências da Saúde de Porto Alegre (UFCSPA), under the number **10-009**, following the resolutions of the the CONCEA (Conselho Nacional de Controle de Experimentação Animal). The NIH "Guide for the Care and Use of Laboratory Animals" (NIH publication N° 80–23, revised 1996) was followed in all experiments. All surgery was performed under ketamine and xylazine anesthesia and all efforts were made to minimize the animals suffering.

Chemicals

Dulbecco's modified Eagle medium low glucose (DMEM), Ca^{2+} and Mg^{2+} free Hank's balanced salt solution containing 10 mM sodium HEPES (HB-CMF-HBSS), collagenase Type I, trypsin–EDTA, penicillin/streptomycin and carboxymethylcellulose sodium (medium viscosity) were obtained from Sigma Chemical Co. (St. Louis, M.O., USA); fetal bovine serum (FBS) from Cultilab (São Paulo, S.P., Brazil); trypan blue (TB) was purchase from GIBCO (Grand Island, N.Y., USA). The measurement of the lactate dehydrogenase activity was determined by commercial kit (Labtest, Lagoa Santa, M.G., BR). Low-melting point agarose and agarose were obtained from Invitrogen (Carlsbad, C.A., USA).

Extraction and Cultivation of ADSCs

The ADSCs were extracted from visceral adipose tissue of Wistar rats (6–8 weeks), as previously described [16]. Briefly, the abdominal fat tissue from 2 rats was transferred to a petri dish, washed with Hanks solution and dissociated mechanically with a scalpel. The fat fragments were transferred to a Falcon tube containing 3 mg/mL collagenase Type I solubilized in DMEM medium without FBS and incubated in a water bath (37°C) for 45 minutes (being vortexed every 15 minutes). After this time twice the volume of DMEM supplemented with 10% FBS was added, to inactivate the collagenase. Cells were centrifuged for 10 minutes at 1000 rpm. The supernatant was discarded and cells were resuspended in a known volume of DMEM. Cell count was performed microscopically, and cell viability was assessed by trypan blue exclusion. 5×10^6 cells/cm^2 were seeded in six-well plates and cultured at 37°C in an incubator with 5% CO_2 in the presence of 3 mL DMEM low glucose medium supplemented with FCS 10%, penicillin (100 U/mL), streptomycin (100 mg/mL). After 24 hours of incubation, half of the culture medium was removed and replaced with fresh medium. At the end of 72 hours, the culture medium was removed completely and the non-adherent fraction was subtracted. The attached fraction was maintained until confluence, generally the fourth passage, when cells were harvested to be used in experiments.

Osteogenic and Adipogenic Differentiation of ADSCs

Cell differentiation was performed according the technique already described [17]. Osteogenic differentiation was induced by culturing ADSCs for up to 8 weeks in DMEM supplemented with 10^{-8} M dexamethasone, 5 μg/mL ascorbic acid 2-phosphate and 10 mM β-glycerophosphate. In order to observe calcium deposition, cultures were washed once with PBS, fixed with 4% paraformaldehyde in PBS for 15–30 minutes, and stained for 5 minutes with Alizarin Red S stain at pH 4.2. Excess of stain was removed by several washes with distilled water. To induce adipogenic differentiation, ADSCs were cultured for up to 8 weeks in DMEM supplemented with 10^{-8} M dexamethasone, 2.5 μg/mL insulin, 100 μM indomethacin and, in some experiments, 3.5 μM rosiglitazone. Adipocytes were easily discerned from the undifferentiated cells by phase-contrast microscopy. To further confirm their identity, cells were fixed with 4% paraformaldehyde in PBS for 1 hour at RT and stained with Oil Red O solution (three volumes of 3.75% Oil Red O in isopropanol plus two volumes of distilled water).

Immunophenotyping Analysis

The ADSCs were characterized to confirm the presence or absence of MSC surface markers using flow cytometry. The cells were trypsinized, washed twice with PBS, centrifuged and incubated for 25 minutes at 4°C with the antibodies described in the Table 1. Excess of antibody was removed by washing with PBS. The cells were analyzed using a FACScalibur cytometer equipped with 488 nm argon laser (Becton Dickinson, San Diego, CA) and with the CellQuest software. Gating was set by using unstained cells and at least 10,000 events were collected [18].

Scaffold Preparation

Sterile sodium carboxymethylcellulose (CMC) salt of medium viscosity was diluted in laminar flow with the aid of magnetic stirrer and glass rod in low glucose DMEM medium, previously supplemented with 3.7 g of sodium bicarbonate and 2.5 g of HEPES per liter and 10% FBS. The CMC was diluted to achieve the concentrations of 10 mg/mL, 20 mg/mL and 40 mg/mL and was used to test different viscosities and behavior of cells in culture.

Table 1. List of Antibodies.

Primary antibody	Host	Dilution	Supplier	Cat. N°
CD45 - PE	Mouse	1:1000	Invitrogen	MR6404
CD29 - PE	Mouse	1:1000	Invitrogen	A14886
CD90.1 - PerCP Cy5.5	Mouse	1:1000	Invitrogen	A14798
CD11b*	Mouse	1:1000	Invitrogen, USA	MR6200

*The unconjugated antibody CD11b was incubated with the secondary Anti-Mouse IgG – FITC antibody from Sigma Chemical Co. (St. Louis, M.O., USA) for 25 minutes

Lactate Dehydrogenase Activity Assay

MSCs on the 5th passage were plated with 3 mL of DMEM without phenol red in each well of 6-well plates (in triplicate), and allowed to growth until reaching a 70% confluence. Then, cells were treated with CMC at the concentrations of 10, 20 and 40 mg/mL. As a negative control we used cells cultured with standard culture medium, and as a positive control, cells lysed with Triton X-100 1%. After 24 hours or five days of growth in contact with CMC, the LDH cytotoxicity assay was performed following the manufacture protocol. The LDH activity was quantified at absorbance of 500 nm by a multiwell spectrophotometer (SpectraMax Plus Microplate Spectrophotometer, Molecular Devices, US). LDH activity in each cell supernatant was expressed by comparing it to the total enzyme activity in cells lysed with 1% of Triton X-100.

The Alkaline Comet Assay

The alkaline comet assay was performed as described in literature with minor changes [19,20]. After treatment of the cells with the biomaterial, cells were washed with ice-cold PBS, trypsinized with 100 µL trypsin (0.15%) and re-suspended in complete medium. Next, 20 µL of cell suspension (~10^6 cells/mL) was suspended in 0.75% low-melting point agarose and immediately spread onto a glass microscope slide pre-coated with a layer of 1% normal melting point agarose. The slides were incubated in ice-cold lysis solution (2.5 M NaCl, 10 mM Tris, 100 mM EDTA, 1% Triton X-100 and 10% DMSO, pH 10.0) at 4°C for a minimum of 1 hour to remove cellular proteins. The slides were then placed in a horizontal electrophoresis box containing a freshly prepared alkaline buffer (300 mM of NaOH and 1 mM of EDTA, pH ~13.0) at 4°C for 20 minutes to allow DNA unwinding. DNA electrophoresis was performed with a 300 mA and 25 V (0.90 V/cm) electric current for 20 minutes. The slides were then neutralized (0.4 M of Tris, pH 7.5), stained with silver nitrate as described in literature and analyzed using an optical microscope [21]. Images of 100 randomly selected cells (50 cells from each of two replicate slides) were analyzed for each test of different substance concentration. Cells were scored visually into five classes, according to tail size (from undamaged - 0, to maximally damaged - 4), and a damage index (DI) value was assigned to each comet according to its class. Visual scoring of comets is a valid evaluation method determined by international guidelines and recommendations for the comet assay [22]. The DI is an arbitrary score calculated for cells in different damage classes that were visually scored by measuring the DNA migration length and the amount of DNA in the tail. The DI ranged from 0 (no tail: 100 cells × 0) to 400 (with maximum migration: 100 cells × 4).

Micronucleus Assay

The micronucleus assay was performed according to Bonacker et al. (2004) with minor modifications. After treatment, cultures were washed twice with the media, and cytochalasin B (Cyt-B) was added at a final concentration of 2 µg/mL. Cultures were harvested 36 hours after Cyt-B addition. Cells were trypsinized and the cellular suspension was centrifuged at 2000 rpm for 5 minutes. Cells were then resuspended in 75 mM KCl solution and maintained at 4°C for 3 minutes (mild hypotonic treatment). Subsequently, the cells were centrifuged and a methanol/acetic acid (3:1) solution was gently added. This fixation step was repeated twice and cells were then resuspended in a small volume of methanol/acetic acid and dropped onto clean slides. The fixed cells were hydrolyzed in HCl and stained according to the Feulgen method [23]. Slides were mounted and coded prior to analysis. Micronuclei were counted in 1000 binucleated cells (BNCs) with well-preserved cytoplasms. The identification of micronuclei was carried out according to described in literature [24].

Skin Wound Model and Tissue Preparation

We used a total of 16 adult male Wistar rats (8 weeks), to avoid the influence of hormonal changes that could cause interferences in the healing process. The dorsal region was chosen to avoid animal access to their own wound and animals were housed one per box to prevent cross-access to the lesions. The animals were anesthetized with an intraperitoneal injection of ketamine and xylazine (90 mg/kg and 10 mg/kg, respectively), followed by shaving the dorsal region [25]. Then, with a punch, the animals were subjected to four circular lesions, measuring 7 mm of diameter, two upper and two lower, comprising dermis and epidermis. Immediately afterwards, the animals received topical treatment in the following order: Injury upper left (lesion 1): 100 µL of CMC 20 mg/mL+26×10^6 of ADSCs; injury upper right (lesion 2): 100 µL of CMC 10 mg/mL+ 26×10^6 of ADSCs; injury bottom left (lesion 3): 100 µL of CMC 20 mg/mL and injury bottom right (lesion 4): Sham, untreated. After the procedure, to ensure that the biomaterial stays housed in the lesion, the wounds were bandaged with polyurethane from 3 M (St. Paul, M.N., USA), permeable to oxygen, but acting as a barrier to liquids and bacteria. This bandage normally fell off naturally in the first or second day after surgery. After topical treatment, to analyze the wound healing, the animals were randomly organized in 4 groups, with 4 rats each: Group 1, the animals were euthanized four days after surgery; group 2, eight days after surgery; group 3, twelve days after surgery and group 4 were euthanized sixteen days after surgery. Animals were euthanized in CO_2 chamber and the lesions were removed with a scalpel, keeping 2 mm of intact skin around. Tissues were stored in phosphate buffered formaldehyde 4% for subsequent histological analysis.

Determination of Lesion Area

Lesions were photographed every day, next to one scale, using Sony Cyber Shot camera (Full HD 1080, 7.2 mega pixels, DSC-W120) and the area of lesion was measured using Image Pro Plus 6.1 software (IPP 6.1) (Media Cybernetics, CA, USA), after calibration.

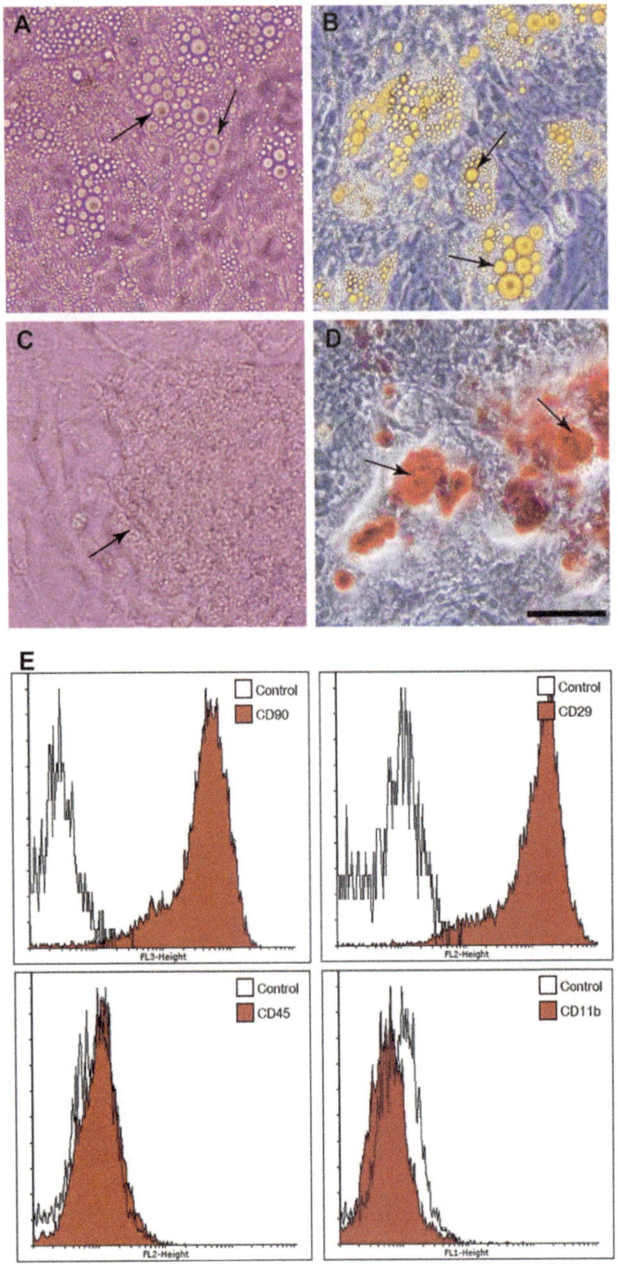

Figure 1. Analysis of the stem cell character of ADSCs extracted from rats. The differentiation of ADSCs is shown in light microscopy at 200X magnification. A. Cells differentiated into adipocytes showing the lipid vacuoles unstained. B. Cells differentiated into adipocytes showing the lipid vacuoles stained with Oil Red. C. Cells differentiated into osteoblasts showing the bone matrix in phase contrast. D. Cells differentiated into osteoblasts showing the bone matrix stained by Alizarin Red. A–D: Scale bar = 100 μm. E. Immunophenotyping analysis of surface marker expression in ADCSs. Flow cytometry histograms show the expression (shaded) of selected molecules (CD90, CD29, CD11b and CD45) by ADSC populations compared with controls (unshaded peaks).

Immunostaining of Cytokeratin

The tissue was included in paraffin being sectioned into slices of 4 μM and de-paraffinized. After, the slides were placed in sodium citrate pH 6.0, heated at 92°C for 40 minutes, with drawn to cool to room temperature, for 20 minutes and then passed into distilled water for antigen retrieval. The endogenous peroxidase was blocked by placing the slides in a mixture of 95 mL of methanol plus 5 mL of hydrogen peroxide 30 V for 5 minutes, protected from light. The tissues were washed twice with distilled water and then with PBS for 5 minutes, under stirring. To block the unspecific bindings, the samples were incubating with a 1% BSA, for 30 minutes. Subsequently, each slide was incubated with a volume of 50 mL of PBS plus a primary antibody at a dilution of 1:1000. The incubation occurred for 30 minutes at room temperature, and after at 4°C overnight. The next day, the slides were washed in PBS twice for 5 minutes and incubated with the secondary antibody (DAKO kit ADVANCE HPR), applied on each slide for 40 minutes at room temperature. Samples were washed with PBS twice for 5 minutes and incubated with tertiary antibody of the same kit. Revelation was performed with chromogen substrate (DAB), dehydration twice for 5 minutes, with xylene and assembled with Entellan. Reaction product of immunohistochemistry for cytokeratin was measured semi-quantitatively, by densitometry, using a Olympus BX-50 microscope, optical lens (10X/0.30 Ph1-UplanFI), coupled to a Motican 2500 camera and Image Pro Plus Software 6.1 (IPP 6.1) (Media Cybernetics, C.A., USA). The background correction and background staining subtraction were performed in accordance with a previously described protocol [26]. To measure optical density, at least two sections were analyzed in each case. Four images were obtained per section and four areas of interest (AOIs) were laid over the cytoplasmatic portions of epithelium in each digitized image. The epithelium thickness was measured in the same images, using an IPP 6.1 tool. At least four measurements were done in each image.

Epithelium Thickness

Slides with hematoxylin-eosin staining were also prepared in order to measure the epithelium thickness. This measurement was performed using Image Pro Plus 6.1 software (IPP 6.1) (Media Cybernetics, C.A., USA), after adjustment of calibration blade. For acquisition was used BX-50 Olympus microscope with optical lens (10X/0.30 Ph1-UplanFI), coupled to a camera Motican 2500. For each wound of each animal, two images were made, and four measurements for each image.

Collagen Fibers Analysis

After the preparation of histological slides, we used the staining of picrosirius, in order to examine whether the treatment used for healing was able to change the number of collagen fibers [27]. The images were taken with optical lens Olympus BX-50 microscope (10X/0.30 Ph1-UplanFI), using polarizing filter to highlight the collagen Type I and Type III on a black background. For each lesion of each animal, three fields of six sections were captured, using microscope images software DP2-BSW ver. 2.2 (Build 6212). The analysis was performed by the Image Pro Plus 6.1 software, proceeding with the marking of pixels according to their birefringence pattern (greenish/yellow-greenish or orange, orange-reddish). Quantification was performed on the percentage histogram of the software itself, totaling 16 points through injury.

Statistics

The toxicity experiments were repeated three times independently. The results were expressed as mean ± SD. Data were analyzed by one-way variance (ANOVA) for the toxicological analysis and two-way for histological parameters. The means were compared by Tukey's test for multiple comparisons, with $p<0.05$ considered statistically significant. Statistical analysis was performed using Prism 4.0 software (GraphPad, La Jolla, CA).

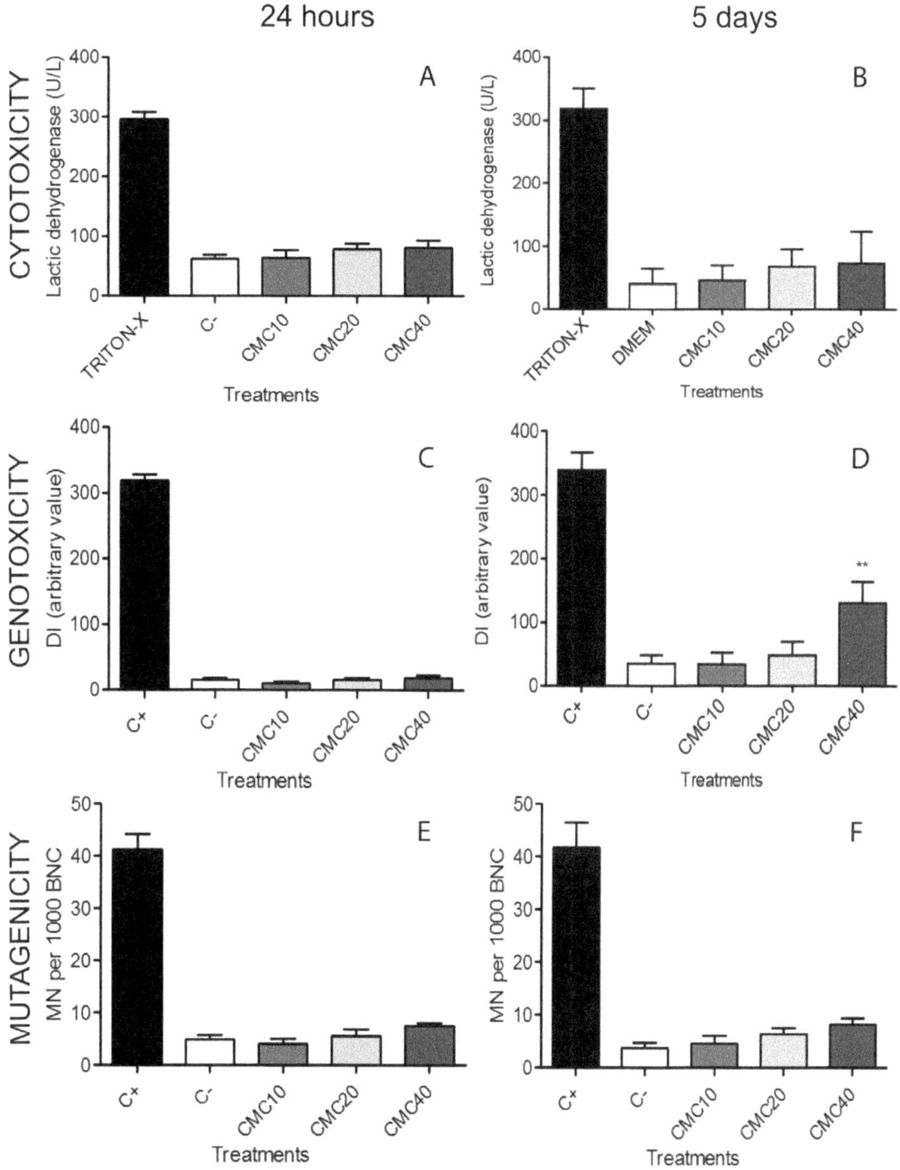

Figure 2. Determination of CMC toxicity on ADSCs cells. The cytotoxicity of CMC (A and B) was evaluated by measuring the LDH enzymatic activity on supernatant of ADSCs. The results are expressed in U/L and were analyzed by one-way ANOVA, followed by Tukey's multiple comparisons test. The values were considered significantly different from controls, when $P<0.05$. The potential genotoxicity of CMC (C and D) was analyzed by alkaline comet assay, as described in material and methods. Results are expressed as mean of damage index (DI) $\pm$ standard deviation of three independent experiments performed in triplicate. There was statistical significance when comparing CMC in the concentration of 40 mg/mL with the negative control (**$P<0.01$/One-way ANOVA Tukey's multiple comparison test). The potential of mutagenicity of CMC (E and F) was evaluated by the *in vitro* cytokinesis-block micronucleus method. The quantification was performed by counting the micronuclei in 1000 binucleated cells (BNCs) with well-preserved cytoplasms. The results were analyzed by one-way ANOVA, followed by Tukey's multiple comparisons test. The values were considered significantly different from controls, when $P<0.05$. All toxicity tests were performed in short and long periods of exposition of ADSCs to biomaterial, 24 hours (A, C and E) and 5 days (B, D and F), respectively. CMC: Sodium carboxymethylcellulose; ADSC: Adipose-derived stem cell. MN: micronucleus.

Results

Confirmation of the ADSC Stem Cell Character

Adherent cells, derived from samples of adipose tissue of rats, showed elongated fusiform morphology with a central nuclei. Cells could be maintained in culture for at least 18 passages. The multipotency of the populations studied was demonstrated by incubating the cells in media promoting differentiation into the osteogenic or adipogenic lineage. ADSCs differentiated into osteoblasts or adipocytes, as evidenced by calcium-rich mineralized matrix deposit or by the presence of intracellular lipid droplets, respectively (Figure 1A–D). The analysis of phenotype of ADSCs used in this study, by flow cytometry, showed that these cells express CD90 (glycosylphosphatidylinositol-anchored glycoprotein) and CD29 (Integrin β1 chain) and were negative for hematopoietic markers, CD11b and CD45 (Figure 1E). According to the literature, the capacity of differentiation and self-replication,

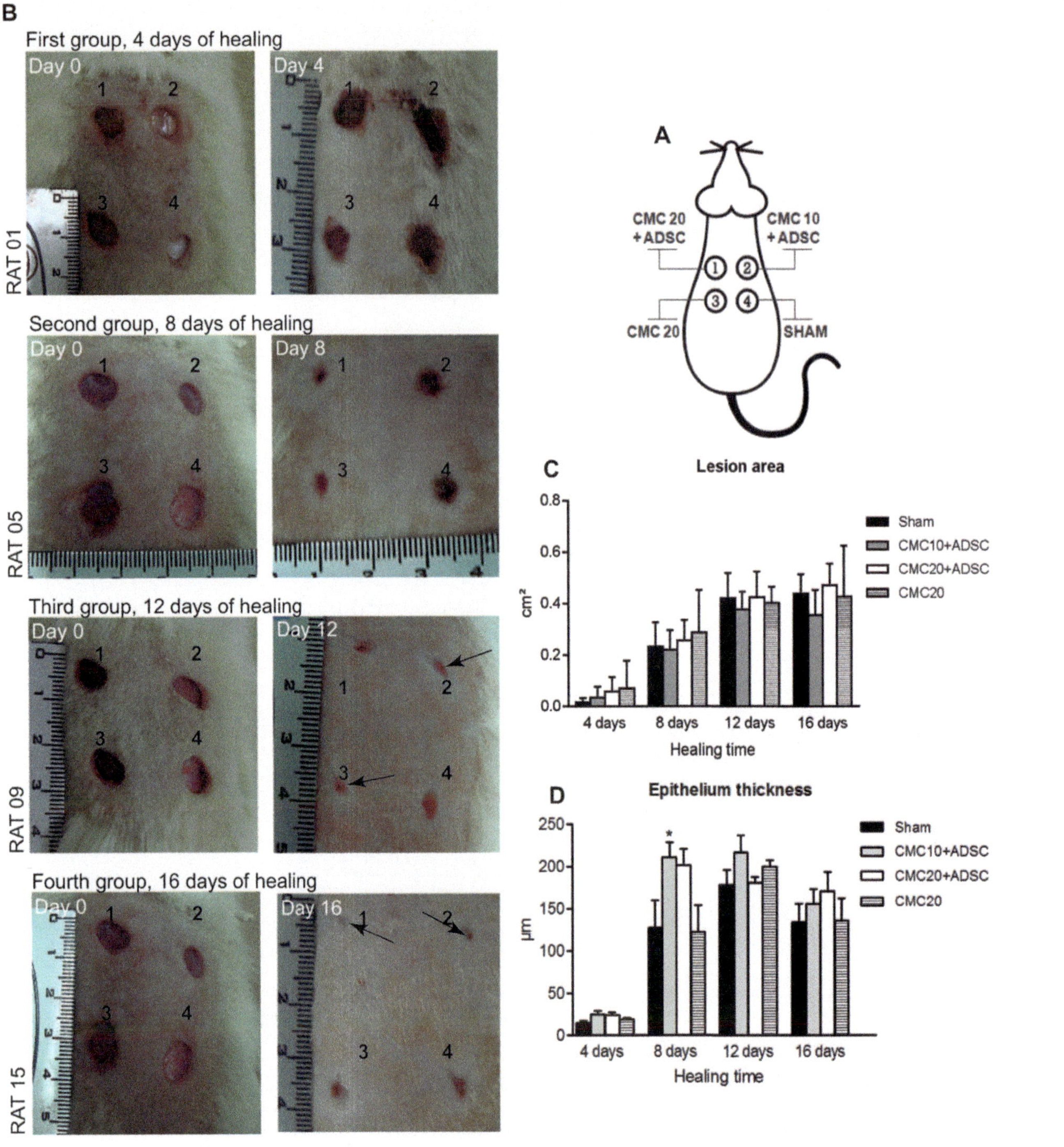

Figure 3. Progression of wound healing. A. Wound model. The scheme shows the four back injuries in a Wistar rat model, comprising the dermis and epidermis layers. The treatments were, in the upper left injury (1): 100 μL of CMC 20 mg/mL+26×10^6 of ADSCs; upper right (2): 100 μL CMC 10 mg/mL+26×10^6 of ADSCs; bottom left (3): 100 μL of CMC 20 mg/mL; bottom right (4): Sham. B. The figure is showing the image of one animal, representative of each group, from day 0 to 16, where the closure of wounds was followed and documented daily. Photographs in the left column were taken in day zero, immediately after surgery. In the right column, in the euthanasia day. C. Evaluation of the area of lesions. The graphic shows, in cm^2, the difference between the size of the initial lesion and the size after different days of healing. There was no difference among the groups. D. Determination of epithelium thickness. Values are expressed in micrometers (μM) and demonstrate a significant increase in the epithelium thickness in the group of animals treated with 10 mg/mL of CMC + ADSCs (lesion 2), with eight days of healing, when compared with control group (injury 4, Sham). (*$P<0.05$, two-way ANOVA test). CMC: Sodium carboxymethylcellulose; ADSC: Adipose-derived stem cell; Sham: Without treatment.

as well as adherence on plastic, are parameters acceptable to the confirmation of the identity of MSCs [16,17,18].

CMC Toxicity

In order to evaluate whether the CMC could damage the cell membrane, we measured the activity of LDH in the supernatant of

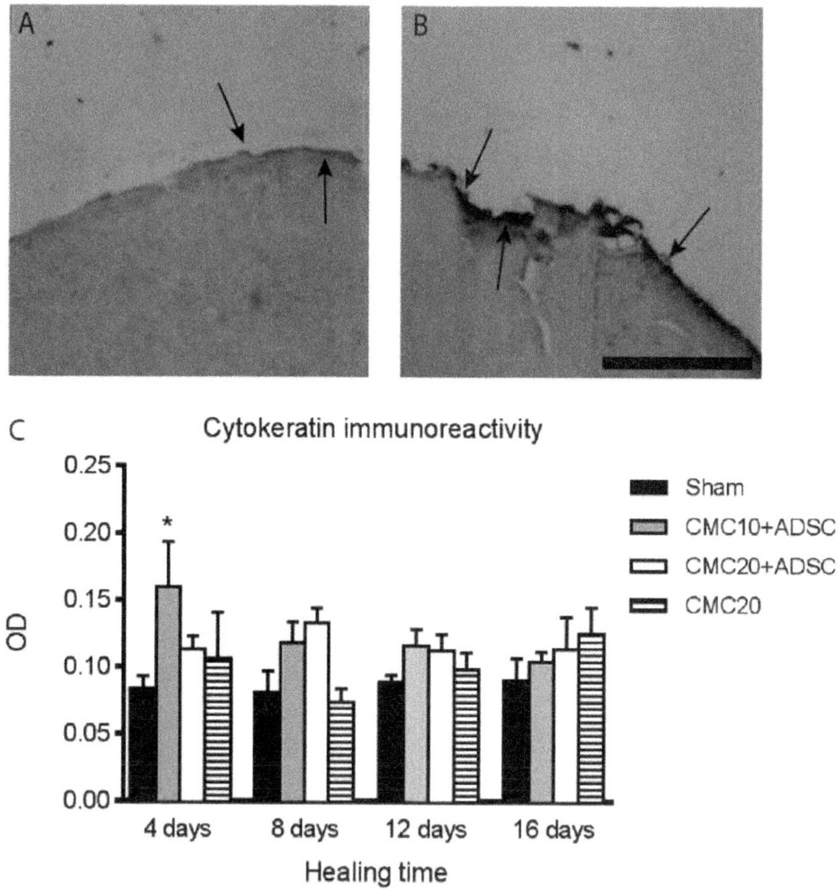

Figure 4. Cytokeratin immunoreactivity in the lesions. A and B. Show representative photomicrographs of the cytokeratin expression in the granulation tissue at the lesion site. The arrows indicate the points where the expression was higher in the lesions: A. Control lesion (Sham); and B. Treated with CMC 10 mg/mL+26×10^6 ADSCs, with four days of healing. Optical microscopy (400x). C. Quantification of cytokeratin expression. The animal groups were separated by healing time and the different treatments are represented in the legend at right. Analyzes demonstrated statistically significant increase in the immunoreactivity in the first group of animals which had four days of healing, (treated with CMC 10 mg/mL + ADSCs) when compared to Sham. Results are expressed as mean of optical density (OD) (*$P<0.05$, two-away ANOVA test). CMC: Sodium carboxymethylcellulose; ADSC: Adipose-derived stem cell; Sham: Without treatment. A–B: Scale bar = 10 μm.

treated cells for 24 hours or 5 days (Figure 2A and 2B). None of the CMC concentrations tested on ADSCs promoted a significant increase in LDH release. Cells that were cultured for as long as five days in contact with the biomaterial, remained adhered to the plate, similarly to cells grown in normal medium, despite the increased density of the culture medium, as a consequence of the presence of CMC. Cells replicated until reaching 100% of confluence.

The induction of single-strand, double-strand and alkali-labile breaks in ADSCs cells was analyzed using the alkaline comet assay (Figure 2C and 2D). The DNA damage index induced in ADSCs that were grown for 5 days in the presence of CMC was increased at the concentration of 40 mg/mL, when compared to the negative control (DMEM) (Figure 2D). Importantly, this concentration of CMC did not alter the proportions of micronuclei in short (24 hours) or long (5 days) periods (Figures 2E and 2F), suggesting that the small damage observed with the comet assay did not lead to a permanent damage of the cells.

Healing Parameters

The evaluation of the healing process of lesions treated with CMC in the presence or absence of ADSCs was determined by different histological parameters. The observation of lesions during the sixteen days of treatment showed that there was no visible suppurative inflammation. Also, none of the animals became sick or died during the treatment. The representative images of lesions of one animal in each group, at baseline (day zero) and end of healing (day of euthanasia) can be seen in Figure 3B. There was no statistical significance when comparing the sizes of the area of the lesions, treated with CMC, when compared with control lesions (Sham) (Figure 3C). The analysis of histological sections from animals that after treatment had 4 days of healing (first group of animals), showed an epithelium not completely formed to be measured. Among the groups, there was a statistical significant increase, in the second group of animals (8 days in the wound healing treated with CMC 10 mg/mL + ADSCs), when compared with lesions in the control group (Sham) (Figure 3D). The animals euthanized on day 4 (first group) showed a higher expression of cytokeratin with the treatment of CMC 10 mg/mL + ADSCs, when compared with the lesion in negative controls (Sham) (Figures 4 A–C).

The collagen fibers production increased progressively in the lesions during the sixteen days of treatment. There was a slight increase in collagen with the treatment of CMC 20 mg/mL associated with ADSCs, during the healing process. However the

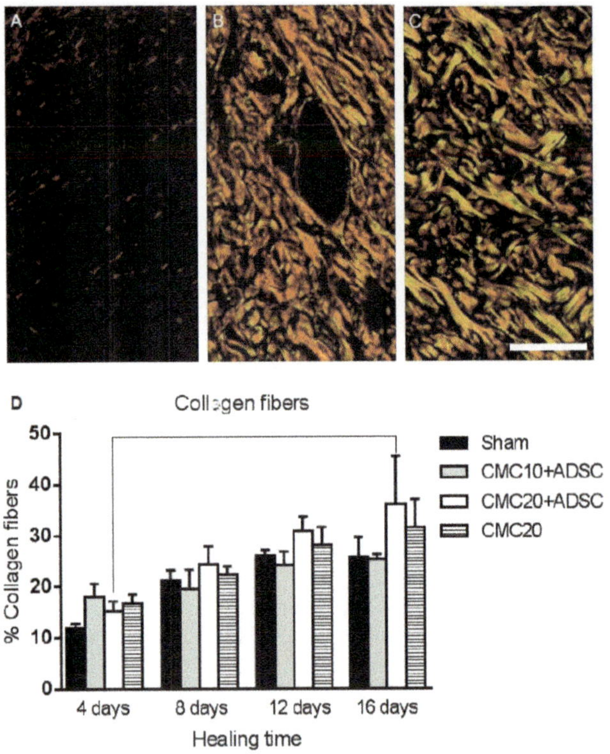

Figure 5. Evaluation of collagen fibers from scar tissues by picrosirius staining. Representative photomicrographs from scar tissues that were subjected to polarization microscopy for visualization of collagen fibers: A. Representative image from the first group with four days of healing. B. From the fourth group with sixteen days of healing, both having received treatment CMC 20 mg/mL+26×10^6 of ADSCs. C. Shows the collagen fibers in a biopsy of normal skin, not subjected to injury. Optical microscopy. Picrosirius staining (200x). D. The statistical analysis showed significance only in the progression of fibers, when were compared the groups of animals with 4 and 16 days of healing, with the treatment of CMC 20 mg/mL associated with ADSCs. There was no statistical difference in the amount of collagen fibers in any group treated, when compared with control lesions (Sham). CMC: Sodium carboxymethylcellulose; ADSCs: Adipose-derived stem cells; Sham: Without treatment. A–C: Scale bar = 50 μm.

quantification of microphotographs stained by picrosirius technique, showed that there was no statistical difference in the amount of collagen fibers in any treated groups, when compared with Sham (Figure 5D).

Discussion

CMC is a low cost biomaterial that has been used as dressings for wounds, with the purpose of protection and as an excipient for some drugs [28]. It has been tested as a biomaterial for growing cells [29,30], however, there is no data evaluating its potential risks and the safe concentrations to be used in therapies. Thus, the first concern in our study was to evaluate the range of safe CMC concentrations to use as a scaffold in cell therapy. For this purpose we performed cytotoxic, genotoxic and mutagenic assays, before testing it in the injured area of an animal. The results showed that there was no significant toxicity on cells cultured with CMC for five days, considering that concentrations of 10 to 40 mg/mL did not cause a significant membrane rupture.

The alkaline comet and micronucleus have been the most commonly used assays for the routine screening of potential genotoxic agents [22]. While the alkaline version of the comet assay [31] detects primary and repairable DNA damage, the micronucleus test detects DNA lesions after their fixation into chromosome mutations. Our results have shown that only at the highest concentration of CMC tested (40 mg/mL), after five days the ADSCs presented increase in DNA damage. However, this damage seems to be repaired, considering that the micronucleus assay did not detect chromosomal damage in ADSCs with the CMC concentrations from 10–40 mg/mL, even with five days of treatment. These results suggest that CMC should be used up to 20 mg/mL.

The epithelization process is an important event for wound healing, ensuring that the new epithelium is formed [32]. In this study, the epithelization rates were improved by CMC containing ADSCs. The cell differentiation and secretion of paracrine factors have been reported to justify the mechanisms by which MSCs contributed to repair damaged tissues [33]. Different molecules are involved in signaling pathways of ADSCs, such as, basic FGF (fibroblast growth factors), VEGF (vascular endothelial growth factor) and PDGF-A (platelet-derived growth factor) [12]. The early development of granulation tissue suggests that the biological events that characterize the dynamic healing are occurring more rapidly, in the group that received treatment of CMC 10 mg/mL associated with ADSCs. An increase in cytokeratin expression, between treated and untreated lesions, suggests a highly active biochemical and bioactive epithelial response, indicative of a migratory epithelium and migratory keratinocyte phenotype, common to the cutaneous wound healing response [34].

MSCs can modulate the host immune responses by changing macrophage, inducing regulatory T cells and inhibiting TH17 formation by both, direct cell-cell contact and indirect mechanisms [35]. This immunomodulatory property could be indirectly related to the epithelization process, since activated macrophages release expressive amounts of FGF [36], a cytokine involved in epithelial proliferation process [37]. A significant increase in cytokeratin expression was observed only in the group of animals, where the healing process occurred by four days. Considering that the therapeutic intervention occurred only once, after the surgery, these results could explain why the animals whose time of healing was 8, 12 and 16 days did not present a significant increase.

The primary route of wound healing in human beings is the formation of granulation tissue and re-epithelialization. Granulation tissue is essential, since it is formed on the surface of the lesion, protecting, nourishing and supporting fibroblasts, new capillaries, and cells from the immune system [38]. The histological slices of lesions stained with HE from animals in the first group (animals with four days of healing), showed a mix of cells, consisting mainly in fibroblasts and leukocytes, which characterize the granulation tissue in its first stage. In the second group, (animals with eight days of healing) the new epithelium observed inside of the wound, was clearly different from the first group, presenting the stratified epithelium. This group also presented an increased epithelium thickness in the lesions that received a concentration of 10 mg/mL of CMC associated with the ADSCs. Interestingly, this result is in accordance with the cytokeratin expression being higher in the group where the healing time was 4 days, with the same treatment. This suggests that during the first 4 days there was an increase in the epithelial cell proliferation, which in the next four days originated a new epithelium layer, capable of being measured and significantly different from the Sham group. Considering that four days of healing were not enough to form a new epithelium, the eight days healing group was the best for this analysis, because the other groups with longer healing time (12 and 16 days), already presented the new epithelium completely formed, independently of treatments.

Although there was a significant increase in the proliferation of epithelial cells in the lesions, with four days of healing, and increased epithelium thickness, in lesions with eight days of healing, when treated with CMC 10 mg/mL plus ADSC, there was no significant difference in the measurements in the injury area when compared to Sham. This improvement of healing at the histological level may be important for large wounds, which would be the ones in which this kind of therapy would be used in the future.

Collagen plays a key role in the wound healing process. The control of its production is a complex process involving a fine balance between its synthesis and degradation [39]. Our analysis by picrosirus stain showed that the treatments did not increase the number of these fibers in comparison with controls (Sham), even in the groups where the epithelium thickness and cytokeratin expression were improved by the treatment. This could be important, because in the healing process, when there is an overproduction of collagen, a fibrosis injury occurs, leading to loss of tissue function and leaving a compromised aesthetic aspect [40,41]. It is possible that ADSCs regulate the synthesis of collagen in the lesion, by releasing anti-fibrogenic molecules [39]. Taken together, the results suggest that CMC in association with the ADSCs contributed to wound healing in the *in vivo* model studied.

Despite promising, this new therapy will need further studies to better investigate the basic mechanisms behind this process, and thus be adapted to clinical applications.

Finally, the results presented here demonstrate an important application for CMC as a biomaterial that presents the properties required for ADSCs growth at concentrations of 20 mg/mL or below. In addition, we showed that CMC at 40 mg/mL has some genotoxic effects, but which do not translate into severe chromosomal damage leading to micronuclei. This can open new avenues for the exploration of the potential pharmaceutical uses of CMC in therapies.

Acknowledgments

The authors would like to thank Terezinha Stein and Rosalva Thereza Meurer (Laboratório de Pesquisa em Patologia, UFCSPA) for excellent technical assistance with histological analysis and Dr. Guido Lenz (Laboratório de sinalização e plasticidade celular, UFRGS) for critical editing of the manuscript.

Author Contributions

Conceived and designed the experiments: CR AMA DJM MCF MRW. Performed the experiments: CR AMA DJM. Analyzed the data: CR DJM GH LLX MCF MRW. Contributed reagents/materials/analysis tools: JS MCF MRW. Wrote the paper: CR MRW.

References

1. Chen M, Przyborowski M, Berthiaume F (2009) Stem cells for skin tissue engineering and wound healing. Crit Rev Biomed Eng 37: 399–421.
2. Simpson CL, Patel DM, Green KJ (2011) Deconstructing the skin: cytoarchitectural determinants of epidermal morphogenesis. Nat Rev Mol Cell Biol 12: 565–580.
3. Wendt H, Hillmer A, Reimers K, Kuhbier JW, Schafer-Nolte F, et al. (2011) Artificial skin-culturing of different skin cell lines for generating an artificial skin substitute on cross-weaved spider silk fibres. PLoS ONE 6: e21833.
4. Neuman MG, Nanau RM, Oruna L, Coto G (2011) In vitro anti-inflammatory effects of hyaluronic acid in ethanol-induced damage in skin cells. J Pharm Pharm Sci 14: 425–437.
5. Kirschner N, Houdek P, Fromm M, Moll I, Brandner JM (2010) Tight junctions form a barrier in human epidermis. Eur J Cell Biol 89: 839–842.
6. Balbino CA, Pereira LM, Curi R (2005) Mechanisms involved in wound healing: a revision. Rev Bras Cienc Farm 41: 27–51.
7. Balaji S, Keswani SG, Crombleholme TM (2012) The Role of Mesenchymal Stem Cells in the Regenerative Wound Healing Phenotype. Advances in wound care 1: 159–165.
8. Fuchs E, Horsley V (2008) More than one way to skin. Genes Dev 22: 976–985.
9. Zuk PA, Zhu M, Ashjian P, De Ugarte DA, Huang JI, et al. (2002) Human adipose tissue is a source of multipotent stem cells. Mol Biol Cell 13: 4279–4295.
10. Zuk PA, Zhu M, Mizuno H, Huang J, Futrell JW, et al. (2001) Multilineage cells from human adipose tissue: implications for cell-based therapies. Tissue Eng 7: 211–228.
11. Sterodimas A, de Faria J, Nicaretta B, Pitanguy I (2009) Tissue engineering with adipose-derived stem cells (ADSCs): current and future applications. J Plast Reconstr Aesthet Surg 63: 1886–1892.
12. Peroni D, Scambi I, Pasini A, Lisi V, Bifari F, et al. (2008) Stem molecular signature of adipose-derived stromal cells. Exp Cell Res 314: 603–615.
13. Sudhakar Y, Kuotsu K, Bandyopadhyay AK (2006) Buccal bioadhesive drug delivery–a promising option for orally less efficient drugs. J Control Release 114: 15–40.
14. Shimamura MK, Kamata K, Yao A, Deguchi T (2005) Scattering functions of knotted ring polymers. Phys Rev E Stat Nonlin Soft Matter Phys 72: 041804.
15. Ramlia NA, Wonga TW (2011) Sodium carboxymethylcellulose scaffolds and their physicochemical effects on partial thickness wound healing. Int J Pharm 403: 73–82.
16. Meirelles L S, Nardi NB (2009) Methodology, biology and clinical applications of mesenchymal stem cells. Front Biosci 14: 4281–4298.
17. Phinney DG, Kopen G, Isaacson RL, Prockop DJ (1999) Plastic adherent stromal cells from the bone marrow of commonly used strains of inbred mice: variations in yield, growth, and differentiation. J Cell Biochem 72: 570–585.
18. Gutiérrez-Fernández M, Rodríguez-Frutos B, Ramos-Cejudo J, Vallejo-Cremades TM, Fuentes B, et al. (2013) Effects of intravenous administration of allogenic bone marrow- and adipose tissue-derived mesenchymal stem cells on functional recovery and brain repair markers in experimental ischemic stroke. Stem Cell Res Ther. 4: 11.
19. Speit G, Hartmann A (1999) The comet assay (single-cell gel test). A sensitive genotoxicity test for the detection of DNA damage and repair. Methods Mol Biol 113: 203–212.
20. Tice RR, Strauss GH (1995) The single cell gel electrophoresis/comet assay: a potential tool for detecting radiation-induced DNA damage in humans. Stem Cells 13: 207–214.
21. Nadin SB, Vargas-Roig LM, Ciocca DR (2001) A silver staining method for single-cell gel assay. J Histochem Cytochem 49: 1183–1186.
22. Hartmann A, Speit G (1997) The contribution of cytotoxicity to DNA-effects in the single cell gel test (comet assay). Toxicology Letters 90: 183–188.
23. Thomas P, Holland N, Bolognesi C, Kirsch-Volders M, Bonassi S, et al. (2009) Buccal micronucleus cytome assay. Nat Protoc 4: 825–837.
24. Fenech M (2000) The in vitro micronucleus technique. Mutat Res. 455: 81–95.
25. Kim CH, Lee JH, Won JH, Cho MK (2011) Mesenchymal stem cells improve wound healing in vivo via early activation of matrix metalloproteinase-9 and vascular endothelial growth factor. J Korean Med Sci 26: 726–733.
26. Xavier LL, Viola GG, Ferraz AC, Da Cunha C, Deonizio JM, et al. (2005) A simple and fast densitometric method for the analysis of tyrosine hydroxylase immunoreactivity in the substantia nigra pars compacta and in the ventral tegmental area. Brain Res Brain Res Protoc 16: 58–64.
27. de Souza DB, Silva D, Cortez CM, Costa WS, Sampaio FJ (2012) Effects of chronic stress on penile corpus cavernosum of rats. J Androl 33: 735–739.
28. van Zuijlen PP, Vloemans JF, van Trier AJ, Suijker MH, van Unen E, et al. (2001) Dermal substitution in acute burns and reconstructive surgery: a subjective and objective long-term follow-up. Plast Reconstr Surg 108: 1938–1946.
29. Lipp C, Kirker K, Agostinho A, James J, Stewart P (2010) Testing wound dressings using an in vitro wound model. J Wound Care 19: 220–226.
30. Liuyun J, Yubao L, Chengdong X (2009) Preparation and biological properties of a novel composite scaffold of nano-hydroxyapatite/chitosan/carboxymethyl cellulose for bone tissue engineering. J Biomed Sci 16: 65.
31. Singh NP, McCoy MT, Tice RR, Schneider EL (1988) A simple technique for quantitation of low levels of DNA damage in individual cells. Exp Cell Res 175: 184–191.
32. Albuquerque-Júnior RLC, Barreto ALS, Pires JA, Reis FP, Ribeiro MAG, et al. (2009) Effect of bovine type-I collagen-based films containing red propolis on dermal wound healing in rodent model. Int J Morphol 27: 1105–1110.
33. Hocking AM, Gibran NS (2010) Mesenchymal stem cells: paracrine signaling and differentiation during cutaneous wound repair. Exp Cell Res 316: 2213–2219.
34. Wong P, Coulombe PA (2003) Loss of keratin 6 (K6) proteins reveals a function for intermediate filaments during wound repair. J Cell Biol 163: 327–337.
35. Khosrotehrani K (2013) Mesenchymal stem cell therapy in skin: why and what for? Exp Dermatol 22: 307–310.
36. Logan A, Frautschy SA, Gonzalez AM, Baird A (1992) A time course for the focal elevation of synthesis of basic fibroblast growth factor and one of its high-affinity receptors (flg) following a localized cortical brain injury. J Neurosci 12: 3828–3837.

37. Werner S, Grose R (2003) Regulation of wound healing by growth factors and cytokines. Physiol Rev 83: 835–870.
38. Harding KG, Moore K, Phillips TJ (2005) Wound chronicity and fibroblast senescence - implications for treatment. Int Wound J 2: 364–368.
39. Motawi TM, Atta HM, Sadik NA, Azzam M (2013) The therapeutic effects of bone marrow-derived mesenchymal stem cells and simvastatin in a rat model of liver fibrosis. Cell Biochem Biophys 13: 9698–9701.
40. Zhou S, Salisbury J, Preedy VR, Emery PW (2013) Increased collagen synthesis rate during wound healing in muscle. PLoS One 8: e58324.
41. Johnson A, DiPietro LA (2013) Apoptosis and angiogenesis: an evolving mechanism for fibrosis. FASEB J 27: 3893–3901.

21

Modified Extracorporeal Photopheresis with Cells from a Healthy Donor for Acute Graft-versus-Host Disease in a Mouse Model

Holger Budde[1]*, Susanne Kolb[1], Laura Salinas Tejedor[1], Gerald Wulf[2], Holger M. Reichardt[3], Joachim Riggert[1], Tobias J. Legler[1]

1 Department of Transfusion Medicine, University Medical Center Göttingen, Göttingen, Germany, **2** Department of Hematology and Oncology, University Medical Center Göttingen, Göttingen, Germany, **3** Institute for Cellular and Molecular Immunology, University Medical Center Göttingen, Göttingen, Germany

Abstract

Background: Graft-versus-host disease (GvHD) is a major challenge after hematopoietic stem cell transplantation but treatment options for patients are still limited. In many cases first-line treatment with glucocorticoids is not successful. Among second-line therapies the extracorporeal photopheresis (ECP) is frequently performed, due to induction of selective tolerance instead of general immunosuppression. However, for some patients with severe acute GvHD the leukapheresis step of the ECP procedure is physically exhausting and limits the number of ECP cycles.

Methods: We hypothesized that leukocytes from healthy cell donors could be used as a replacement for ECP leukocytes gained from the GvHD patient. For this purpose we used a well established mouse model of acute GvHD. The ECP therapy was based on cells with the genetic background of the initial donor of the stem cell transplantation. As a precondition we developed a protocol representing conventional ECP in mice equivalent to clinical used ECP setup.

Results: We could demonstrate that conventional, clinically derived ECP setup is able to alleviate acute GvHD. By using leukocytes obtained from healthy mice with the bone marrow donor's genetic background we could not observe a statistically significant therapeutic effect.

Conclusions: Conventional human ECP setup is effective in the mouse model of severe acute GvHD. In addition we could not prove that ECP cells from healthy mice with bone marrow donor's genetic background are as effective as ECP cells derived from GvHD mice. Based on our findings, new questions arise for further studies, in which the cellular characteristics for ECP mediated immune tolerance are a matter of investigation.

Editor: Vassiliki A. Boussiotis, Beth Israel Deaconess Medical Center, Harvard Medical School, United States of America

Funding: This work was supported by grants from the Deutsche José Carreras Leukämie Stiftung e.V. to H.B. (grant number: DJCLS R 12/34, http://www.carreras-stiftung.de/). The funders had no role in study design, data collection and analysis, decision to publish, or preparation of the manuscript.

Competing Interests: The authors have declared that no competing interests exist.

* Email: holger.budde@med.uni-goettingen.de

Introduction

Patients with malignant diseases of the hematopoietic system such as leukemia still have limited therapeutic options. Allogeneic hematopoietic stem cell transplantation (HSCT) is often performed to obtain a long-term disease-free survival [1]. However, HSCT is associated with severe side effects, of which some can be fatal. The graft-versus-host disease (GvHD) belongs to these side effects and is one of the major limiting factors in hematopoietic stem cell transplantation [2]. In brief, GvHD is based on alloreactive T-cells from the stem cell graft, which attack several organs of the recipient and cause a generalized tissue rejection [3]. Especially acute GvHD is associated with high morbidity and mortality [4]. First-line therapy of acute GvHD is mainly based on glucocorticoids [5]. However, after HSCT patients often suffer from reduced immunocompetence due to the myeloablative conditioning regimen. The use of glucocorticoids in these cases increases the risk of opportunistic infections. Furthermore, in steroid-resistant or steroid-refractory patients there is a need for alternative therapeutic interventions [6–8].

Extracorporeal photopheresis (ECP) is a widespread used second-line treatment for GvHD. It is frequently used for chronic GvHD but also effective for the treatment of acute GvHD with response rates up to 80% depending on the affected organ [9,10]. In comparison to other immunosuppressive therapies, ECP is rarely associated with side effects [11]. One major advantage of ECP therapy is the induction of immunotolerance without general immunosuppression. The frequency of opportunistic infections is not increased by the ECP [12,13]. Due to its efficacy and the low risk of side-effects, some groups suggest using ECP as part of the first-line therapy for acute GvHD [14]. During the ECP procedure, patient's own leukocytes are isolated by apheresis. The cells are treated *ex vivo* with 8-Methoxypsoralen (8-MOP) and UV-A light, a process, which causes cellular apoptosis [15]. Subsequently, the cells are transfused to the patient. Although up to 10% of the peripheral T-cells are eliminated by a single ECP procedure, the reduction of allo-reactive T-cells is probably not the key mechanism of ECP. More likely, 8-MOP/UV-A treated apoptotic cells are phagocytized and processed by antigen-presenting cells, a mechanism which initiates immune tolerance [16]. An increased number of regulatory T-cells (Tregs) was frequently observed after ECP therapy and is supposed to be a major cause of ECP mediated immune tolerance [17].

However, not all GvHD patients respond to ECP therapy and especially for patients suffering from severe acute GvHD the apheresis procedure can be a physical burden. In particular for these patients, modifications of the ECP procedure could be beneficial in the context of ECP application comfort. By replacing patient's leukocytes by the one's from blood donors, the patient apheresis would be omitted and ECP could be applied more frequently even to patients with severe GvHD.

Considering this, the aim of our study was to investigate whether 8-MOP/UV-A treated leukocytes from the donor of the original stem cell graft are able to alleviate severe acute GvHD *in vivo*. Therefore, we used a well known mouse model of acute GvHD and established a protocol for a standard murine ECP therapy, which is comparable to conventional human clinical ECP protocols. The therapeutic cells were taken from mice with GvHD after bone marrow transplantation, which had the same genetic background as the ECP treated mice. By using this protocol we could show for the first time that conventional ECP leads to reduced mortality rates in a MHC mismatched mouse model of severe acute GvHD. In a second step we used a modified ECP protocol, in which cells from the untreated bone marrow donor strain were used instead of cells from transplanted GvHD mice.

Altogether, the goal of our study was to examine whether a modified ECP setup with cells derived from healthy non-GvHD mice is able to improve the course of acute GvHD. Although we could not prove that such a modified ECP setup is therapeutically effective, our observations are important for the investigation of further modifications to improve the efficacy of the ECP therapy.

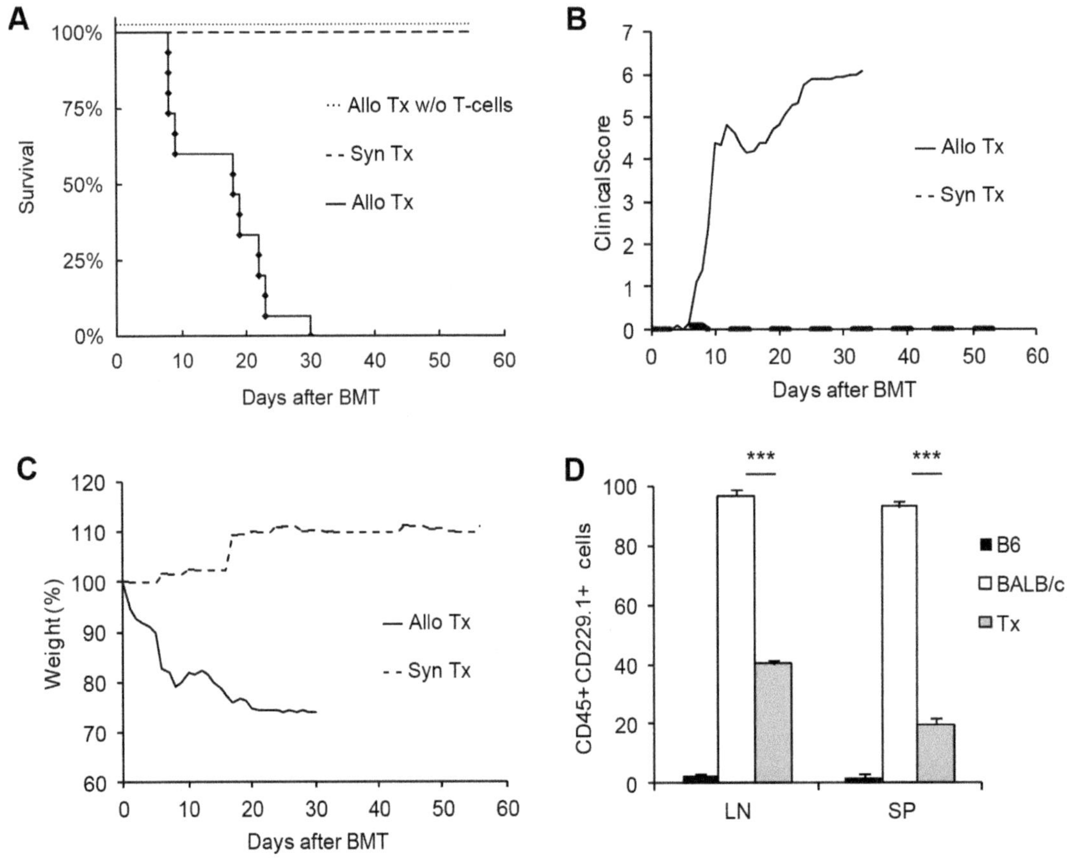

Figure 1. Major histocompatibility complex (MHC) mismatched mouse model of allogeneic BMT for GvHD induction. Survival after allogeneic BMT in a C57BL/6 (B6)→BALB/c mouse model (n = 15) of acute GVHD (—♦) (A). Both, syngeneic BALB/c→BALB/c BMT (n = 11) and allogeneic T-cell depleted B6→BALB/c BMT (n = 3) were performed as transplantation controls. Both control groups are represented as dashed lines with a survival rate of 100%. Daily assessed clinical score of allogeneic (including T-cells) vs. syngeneic transplanted mice (B). Weight changes in allogeneic (including T-cells) vs. syngeneic transplanted mice (C). Assessment of engraftment by cellular chimerism analysis 56 days after BMT in the T-cell depleted B6→BALB/c mouse model (D). CD45 positive leukocytes were stained for expression of CD229.1 antigen which is expressed on BALB/c leukocytes but not on B6 cells. Percentages of $CD45^+$ $CD229.1^+$ leukocytes from all splenocytes (SP) or lymph node cells (LN) are shown. Stainings were performed for wild-type B6 (black bars), wild-type BALB/c (white bars) and B6→BALB/c transplanted mice (Tx, grey bars). Chimerism analysis was performed in 3 animals from each group (mean ± SD, ***p<0.001).

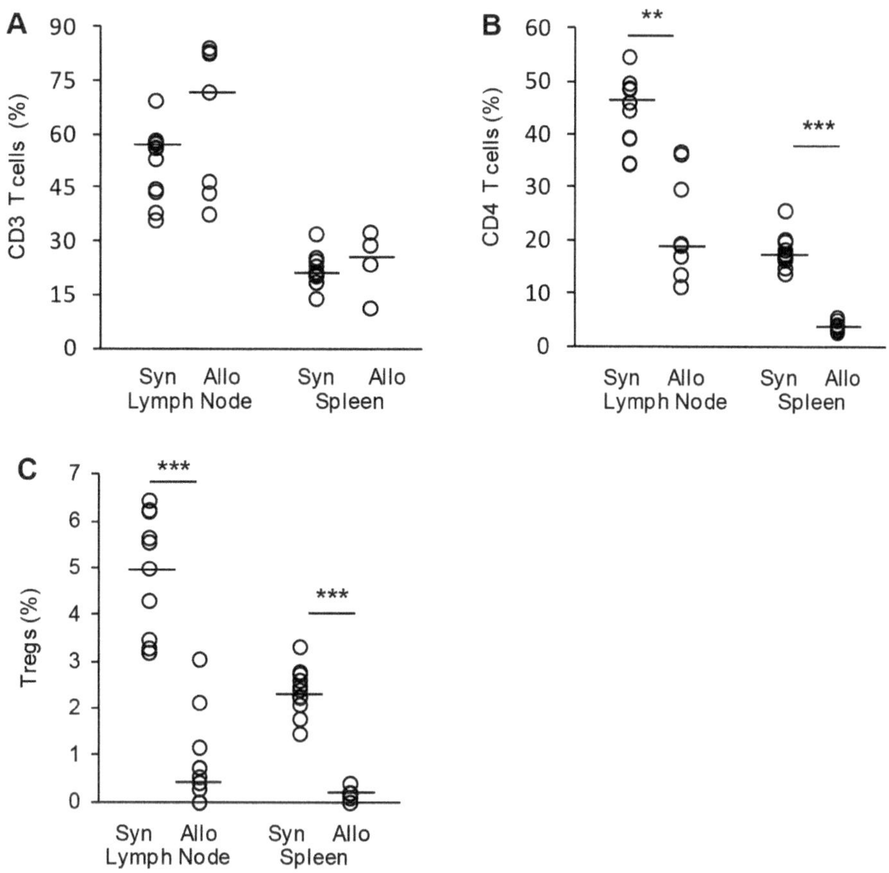

Figure 2. Evaluation of T-cell populations after syngeneic and allogeneic BMT. Analysis of T-cell subpopulations were performed after BMT in allogeneic C57BL/6→BALB/c (Allo) vs. syngeneic control transplantations (Syn). Percentage of CD3 positive T-cells (A), CD4 positive T-helper cells (B) and CD4 CD25 FoxP3 positive regulatory T-cells (C) from all splenocytes or lymph node cells were shown. Cells were analyzed from 7 allogeneic and 11 syngeneic transplanted mice. Animals were sacrificed when reaching clinical GvHD score ≥6 or when reaching day 56 after BMT (dot-plot with median, $^{**}p<0.01$, $^{***}p<0.001$).

Methods

Mice

Male BALB/c recipient mice were purchased from Charles River Laboratories (Sulzfeld, Germany). Male C57BL/6J (B6) donor mice were purchased from Janvier (St Berthevin Cedex, France). Mice were between 10 and 12 weeks of age when BMT was performed. Animals were housed under pathogen-free conditions in individual ventilated cages.

Ethics statement

All animal experiments were conducted according ethical regulations and approved by the corresponding authorities (Niedersächisches Landesamt für Verbraucherschutz und Lebensmittelsicherheit, permit number 33.9-42502-04-11/0472). Animals were treated with metamizole analgesics and comprehensive efforts were performed to minimize suffering. For further details about the animal experiments see Table S1 in the supporting information.

Bone marrow transplantation

Resembling human HSCT, allogeneic bone marrow transplantation (BMT) was performed in mice for induction of acute GvHD (modified after Tischner et al., [18]). Conditioning regimen consisted of a single total body irradiation (TBI) with a dose of 9.5 Gy. Intravenous transfusion of 10^7 syngeneic or allogeneic bone marrow (BM) cells and 2×10^6 T-cells were performed 24 h after TBI. BM cells were prepared from femur and tibia and T-cells were depleted from BM with CD90.2 MicroBeads (Miltenyi Biotec, Bergisch Gladbach, Germany). T-cells were isolated from spleen and lymph nodes by magnetic beads (Pan T-cell Isolation Kit, Miltenyi Biotec). To avoid bacterial infection, neomycin was added to the drinking water.

Chimerism analysis

Molecular chimerism was assessed 8 weeks after T-cell depleted BMT. $CD45^+/CD229.1^+$ cells were counted as BALB/c leukocytes and $CD45^+/CD229.1^-$ cells as leukocytes with B6 background [19].

GvHD pathology analysis

Animals were scored daily for GvHD symptoms. The disease score consisted of 4 criteria including posture, activity, fur/skin and diarrhea with a maximum of 2 points each. Maximum score was 8 but animals reaching a score of 6 due to disease severity were sacrificed. Animal weight was assessed daily.

ECP

Splenocytes were prepared from a second cohort of B6→BALB/c transplanted animals after showing first clinical

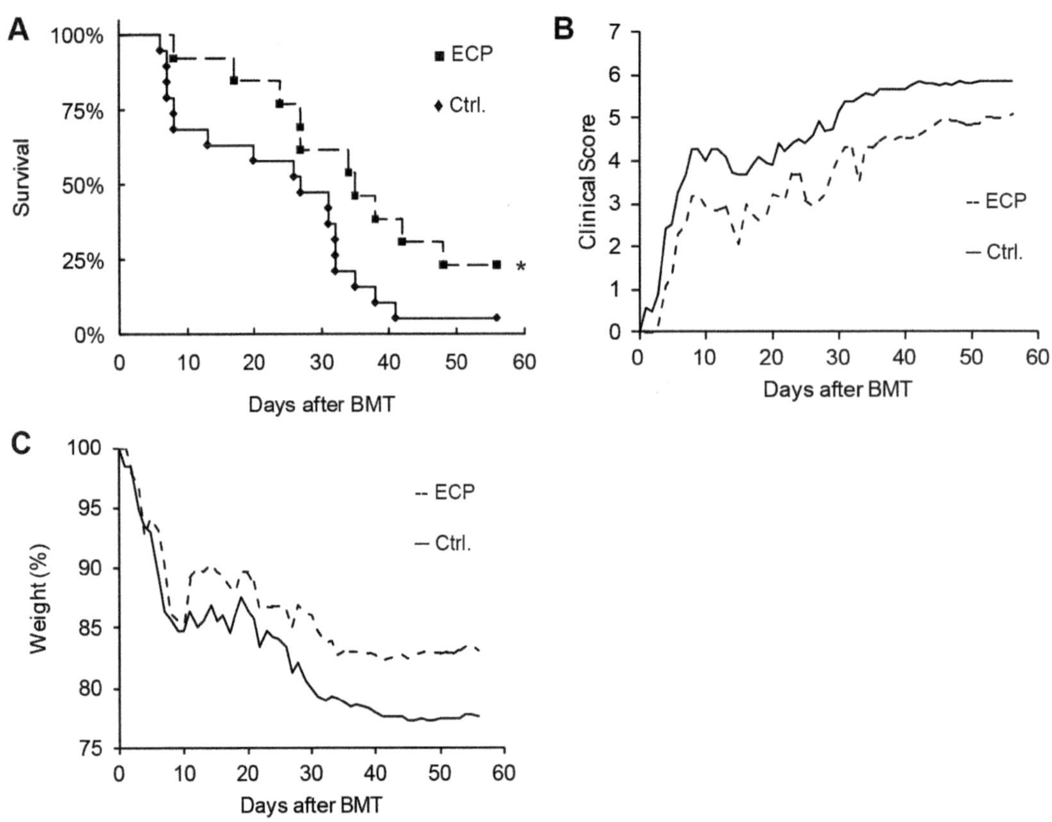

Figure 3. Conventional ECP therapy in the mouse model of acute GvHD. The therapeutic impact of conventional extracorporeal photopheresis (ECP) on GvHD severity in the C57BL/6→BALB/c mouse model was assessed. Animals of the ECP group (---■) were injected with 8-Methoxypsoralen and UV-A treated splenocytes from another cohort of C57BL/6→BALB/c transplanted GvHD animals with identical genetic chimerism. Animals received 3 ECP treatments in weekly intervals, started at day 1 after BMT. Control animals (—♦) received PBS injections at the same time points. Kaplan-Meier plot showing survival rates of the ECP treated vs. PBS control group (A). The ECP group consisted of 13 and the control group of 19 animals (*$p<0.05$, log rank test). Clinical GvHD score (B) and weight loss (C) were assessed daily.

signs as a model for "conventional ECP" or from healthy untreated B6 animals as a model for the "modified ECP". After red blood cell lysis, cells were incubated with 8-MOP (0.2 µg/ml) for 30 min prior to UV-A irradiation (2 J/cm^2). Afterwards cells were washed and diluted in PBS for intravenous tail vein injection. ECP treatment was performed weekly for three times for conventional ECP and, due to the lack of significant therapeutic response, four times for modified ECP. T-cell dose was 1×10^6 cells for conventional ECP and the modified ECP was performed with 1×10^6 and 5×10^6 cells in two different series, respectively. Control animals were injected with PBS only. In every experimental series, an own PBS control group was carried in parallel for precise analysis excluding shifts in GvHD score severity due to performing therapy and control series at different time points. Additionally, in each experimental series ECP treated animals and PBS control mice were housed together. Each experiment was performed in at least two independent series.

Flow cytometric analysis

Cells were analyzed when animals reached a disease score of 6 or at day 56 after BMT if score 6 was not reached before. Fc receptors were blocked with CD16/32 Fc receptor block (eBioscience, San Diego, CA, USA). Anti-mouse CD3e (Pacific Blue, clone 145-2C11, Armenian Hamster IgG) was purchased from BioLegend (San Diego, CA, USA). CD4 (FITC, clone RM4-5, Rat IgG2a) and CD45 (FITC, clone 30-F11, Rat IgG2b) were purchased from eBioscience and CD229.1 (APC, clone 30C7, IgG2a) from BD Bioscience-Pharmingen (San Diego, CA, USA). Cells were incubated with antibodies for 20 min at 4°C, washed with PBS and analysed by flow cytometry using a FACS Canto II flow cytometer (BD Biosciences, San Jose, CA, USA). Mouse Regulatory T-cell Staining Kit 2 was purchased from eBioscience and used according to the manufacturer's recommendations.

Statistical analysis

Survival rates were compared by using Kaplan-Meier plots and log-rank test was used for calculation of the corresponding p-values. Flow cytometry data from different groups were compared by using the non-parametric Mann-Whitney U test. All calculations were performed with "STATISTICA Version 10" software (StatSoft Inc., Tulsa, USA). Graphs are shown as bars with mean ± SD or, in case of high variance, as dot-plots with median.

Results

Allograft T-cell depletion completely prevents GvHD in MHC mismatched mouse model

We used a well known mouse model of acute GvHD with B6 (H2^b) as donor and BALB/c (H2^d) as recipient mouse strain. The model is based on MHC I and MHC II mismatches and GvHD is mediated by both CD4$^+$ and CD8$^+$ T-cells [20]. Due to the reproducible and robust development of GvHD symptoms, it is

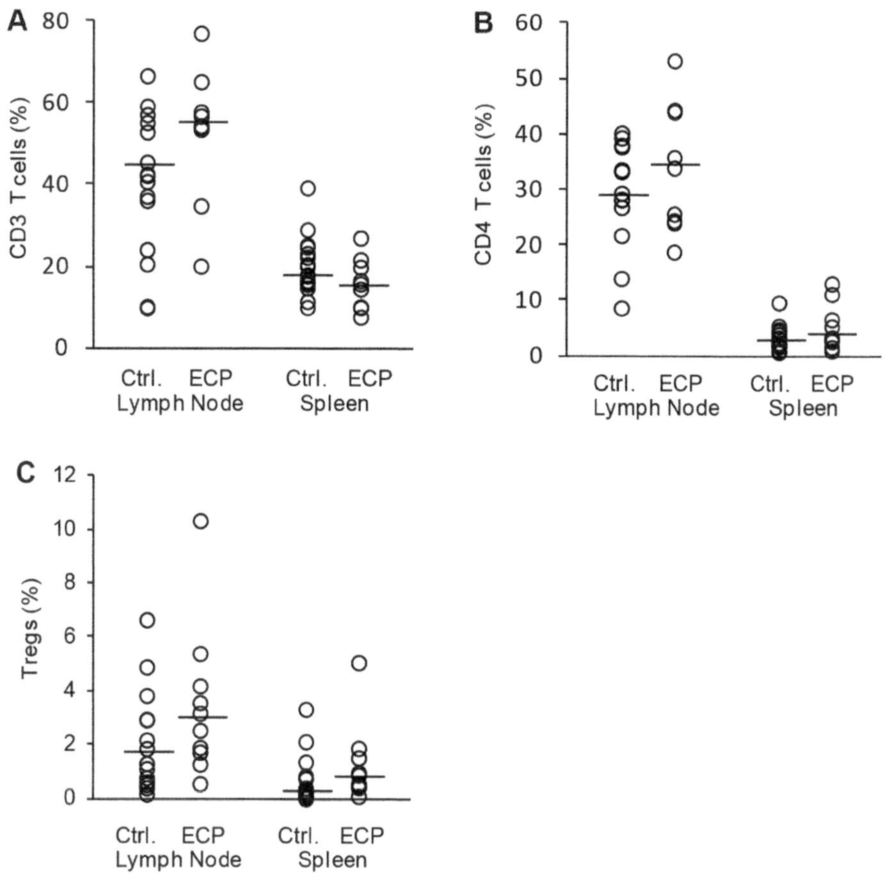

Figure 4. Analysis of T-cell subpopulations in GvDH mice treated with conventional ECP. GvHD was induced by allogeneic C57BL/6→BALB/c BMT and ECP treatment was performed weekly starting at day 1 after BMT (ECP). GvHD induction in control animals (Ctrl.) was performed simultaneously and animals received weekly PBS injections instead of ECP treatments. Percentage of CD3 positive T-cells (A), CD4 positive T helper cells (B) and CD4 CD25 FoxP3 positive Tregs (C) from all lymph node cells or splenocytes. Analysis is based on 10 ECP treated and 14 control mice. Animals were sacrificed when reaching clinical GvHD score ≥6 or at day 56 after BMT (dotplot with median).

one of the most studied mouse models for acute GvHD [20,21]. In most cases, allogeneic transplanted mice did not survive the experimental observation time of 56 days (Fig. 1A). Animals showed clinical symptoms such as diarrhea, rough fur, hunchback posture and reduced activity accompanied by weight loss (Fig. 1B and C). GvHD outcome and lethality was most distinct after the first and after the third week, respectively. In contrast, all control animals after syngeneic BALB/c→BALB/c transplantation survived without any symptoms (Fig. 1A–C). Furthermore, BALB/c mice transplanted with allogeneic T-cell depleted B6 bone marrow without T-cells from spleen or lymph node did not develop any disease symptoms (Fig. 1A), highlighting the pivotal role for allogeneic T-cells in GvHD induction. To control successful engraftment, we analyzed the chimerism level after T-cell depleted BMT. For this purpose we used a CD229.1 antibody, which reacts with BALB/c leukocytes but not with cells from the B6 strain. At day 56 after BMT we found ~40% of $CD45^+$ $CD229.1^+$ cells in lymph nodes and ~20% in spleen which corresponds to B6 donor leukocyte background in transplanted BALB/c mice of about 60% in lymph nodes and 80% in spleen (Fig. 1D).

GvHD is linked to reduced regulatory T-cell level

Since T-cells are key players in GvHD induction, we looked for different T-cell levels in lymph nodes and spleen after BMT. Allogeneic transplanted mice were analyzed at individual experimental break-off when reaching GvHD clinical score of 6 and syngeneic transplanted animals were analyzed 56 days after BMT. There was no difference in overall CD3 T-cell percentage between allogeneic transplanted animals suffering from GvHD compared to healthy syngeneic transplanted mice (Fig. 2A). In contrast, CD4 T-helper cell percentages of GvHD animals were significantly reduced in both lymph nodes and spleen (Fig. 2B). Furthermore, we could observe a massive reduction of $CD4^+$ $CD25^+$ $FoxP3^+$ regulatory T-cells (Tregs) in GvHD animals (Fig. 2C).

Conventional ECP ameliorates acute GvHD in a murine MHC major mismatched model

ECP is a widespread used clinical therapy for GvHD. We transferred the clinical used human ECP setup in our mouse model. Splenocytes from another cohort of allogeneic transplanted B6→BALB/c mice with GvHD symptoms were prepared and treated with 8-MOP and UV-A irradiation. Following 8-MOP/UV-A treatment, 1×10^6 splenocytes were injected intravenously in ECP recipient mice. The number of 1×10^6 splenocytes is about 10% of the mouse's peripheral leukocytes. This corresponds to the clinical ECP procedure in which about 10% of the peripheral leukocytes are treated *ex vivo* after apheresis [22]. Our murine ECP therapy was initiated one day after BMT and performed three times at weekly intervals. Compared to the control group injected with PBS only (n = 19), ECP treated animals (n = 13) had

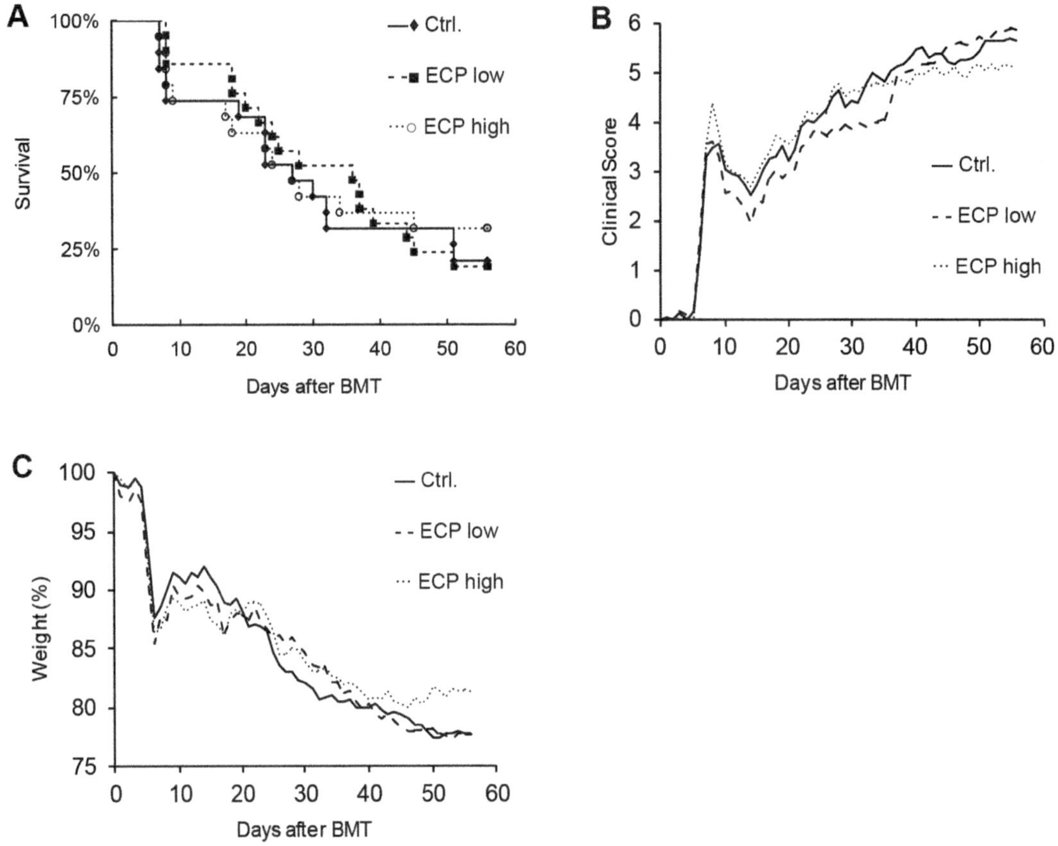

Figure 5. Modified ECP therapy for acute GvHD. Modified ECP therapy was performed as treatment for acute GvHD induced by C57BL/6→BALB/c bone marrow transplantation. ECP cells were isolated from wild type B6 spleens and treated with 8-Methoxypsoralen (8-MOP) and UV-A light. Treated cells were injected 4 times in C57BL/6→BALB/c transplanted GvHD mice at weekly intervals starting from day 1 after BMT. For each therapeutic injection, 8-MOP/UV-A treated cells were used either in low concentration of 1×10^6 cells (---■) or in high concentration of 5×10^6 (···○) cells. Control animals (—♦) were transplanted simultaneously and injected with PBS only at therapeutic time points. Therapeutic impact and GvHD severity were analyzed by comparing survival rates (A), clinical GvHD symptom score (B) and weight loss (C).

a significantly higher survival rate, demonstrated by Kaplan-Meier curves and log-rank test ($p<0.05$, Fig. 3A). Two months after BMT, three (23%) ECP treated animals were still alive, whereas only one (5%) control animal survived. Notably, during the first GvHD wave which occurs in between day 7 and 10 after BMT, survival was markedly reduced in the control group. In the ECP group only 1 of 13 (8%) animals did not survive the first GvHD wave compared to 6 of 19 (32%) animals in the control group. During the observation time of 56 days, the average clinical score of the ECP treated animals was around 1 point less than of the control animals (Fig. 3B). Furthermore, control animals had slightly but not significant higher final average weight loss (−22.4%) than animals from ECP treatment group (−16.9%, Fig. 3C). Concerning T-cell percentages of all splenocytes or lymph node cells, we could not find significant differences in $CD3^+$ cytotoxic T-cells, $CD4^+$ T-helper cells or Treg fractions between control and ECP treated animals (Fig. 4A–C). In comparison to the positive therapeutic effects described above, there was no effect on GvHD outcome and survival rates when performing just one single ECP treatment only (data not shown), suggesting that repeated ECP treatments are required for a therapeutic effect.

Modified ECP setup is less effective than conventional ECP

Conventional ECP is based on patient's own leukocytes, which are treated *ex vivo* with 8-MOP/UV-A and subsequently retransfused to the patient. A common clinical setup for the treatment of acute GvHD consists of two ECP procedures on consecutive days followed by further treatments at different intervals. However, a high frequency of apheresis cycles is not possible in all patients with severe GvHD. To maintain ECP frequency and apoptosis level in these patients, other cell sources for 8-MOP/UV-A treatment would be a major improvement for ECP application. The most standardized cell source would consist of allogeneic leukocytes of the healthy stem cell donor of the preceding stem cell transplantation. In addition, in case of previous related stem cell transplantation, donors would probably be well motivated for further leukocyte donations. To test whether mouse leukocytes with genetic background of the BMT donor are functional in ECP therapy, we used splenocytes from healthy B6 mice for 8-MOP/UV-A treatment and subsequent injection into B6→BALB/c transplanted GvHD mice. As well as for conventional ECP, we used 1×10^6 8-MOP/UV-A treated splenocytes at weekly intervals. In the modified ECP we observed a slight but not significant improvement in the survival rate at least within the first 35 days after BMT (Fig. 5A). The clinical score was similarly reduced but there was no difference in body weight loss between

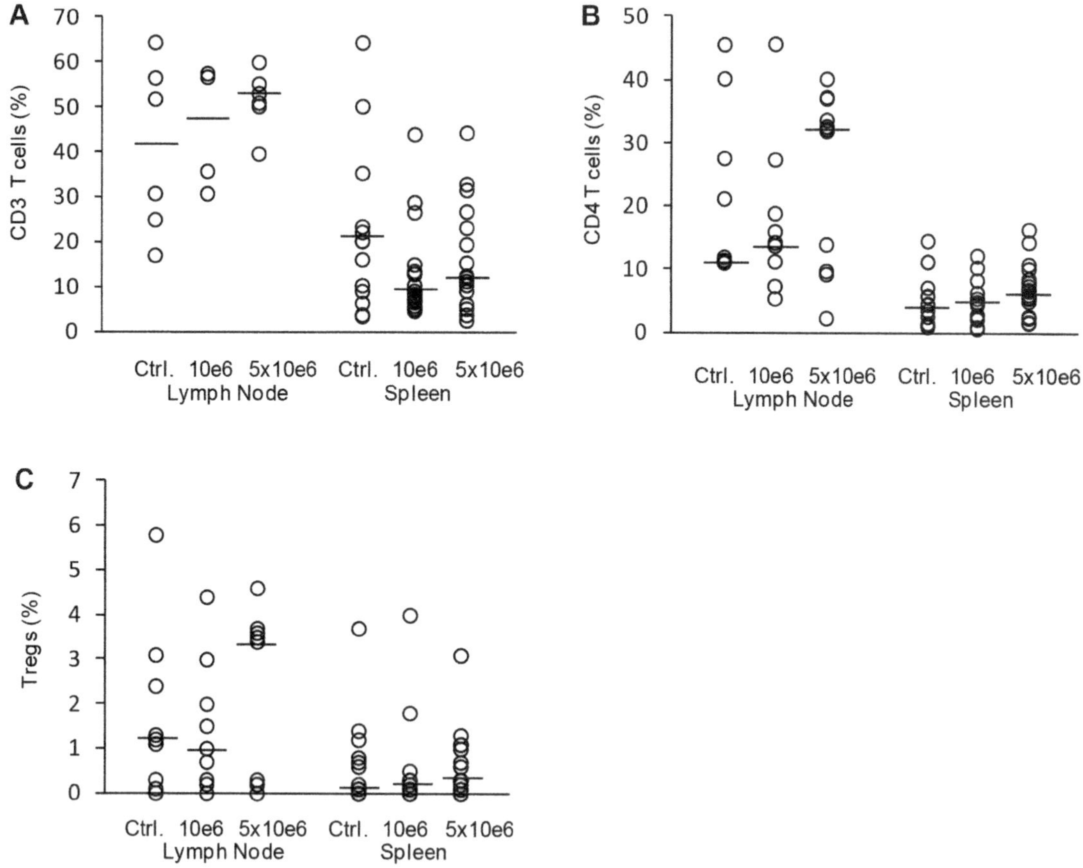

Figure 6. Assessment of T-cell populations after modified ECP therapy. T-cell subpopulations in GvHD mice treated by modified ECP were analyzed. ECP cells were isolated from wild type C57BL/6 spleens and treated with 8-Methoxypsoralen and UV-A light. Subsequently splenocytes were injected in GvHD mice after allogeneic B6→BALB/c bone marrow transplantation. ECP was performed 4 times at weekly intervals with cell concentrations of either 1×10^6 cells or 5×10^6 cells. Control animals were treated with PBS in parallel (Ctrl.). Percentage of CD3 positive T-cells (A), of CD4 positive T-helper cells (B) and CD4 CD25 FoxP3 positive Tregs (C) from all lymph node cells or splenocytes. Analysis is based on 9 animals treated with 1×10^6 cells (10e6), 11 animals treated with 5×10^6 cells (5×10e6) and 9 control mice (Ctrl.). Animals were sacrificed when reaching clinical GVHD score ≥6 or at day 56 after BMT (dot-plot with median).

control and ECP group despite of an additional fourth ECP cycle (Fig. 5B and C). To analyze, whether more cells are required in the modified ECP setup, we also tested the therapeutic effect of 5×10^6 8-MOP/UV-A treated B6 splenocytes in B6→BALB/c transplanted mice. In this treatment group four ECP cycles were performed at weekly intervals. On the one hand, increasing the therapeutic cell concentration to 5×10^6 cells did not result in improved clinical score, increased body weight or higher survival rate (Fig. 5A–C). On the other hand, we could not observe any negative or alloreactive effect using increased numbers of B6 splenocytes, suggesting that B6 T-cells were sufficiently inactivated by 8-MOP/UV-A treatment. Accordingly, cellular analysis of splenocytes and lymph node cells showed no difference in $CD3^+$ T-cell, $CD4^+$ T-cell or $CD4^+$ $CD25^+$ $FoxP3^+$ Treg percentages between therapy and control groups (Fig. 6A–C).

Discussion

Appropriate *in vivo* models are essential for the development of new GvHD therapies. In order to investigate the impact of ECP on severe acute GvHD, we chose the B6→BALB/c full MHC mismatched model. By using this model we observed severe GvHD signs and poor survival of untreated control mice, which is common for this model [23]. In contrast, T-cell depleted BMT does not lead to any GvHD symptoms due to the lack of alloreactive T-cells from the B6 donor strain, an observation which proves the validity of our model. These findings are in concordance with less frequent GvHD after T-cell depleted allogeneic HSCT in humans [24]. Additionally, we could proof by chimerism analysis of T-cell depleted animals that engraftment occurred in our BMT mouse model. For all subsequent therapeutic ECP experiments we performed transplantations with bone marrow and donor T-cells. Since T-cells are important for improved engraftment in allogeneic transplant settings, the chimerism of GvHD mice transplanted with both bone marrow and T-cells should be even higher towards donor cell population.

After having set up the *in vivo* model of acute GvHD we established a model of treating mice with ECP previously described by Gatza et al. [25]. In this model, cells for ECP were taken out of a second cohort of GvHD mice after allogeneic BMT to treat the first cohort of GvHD mice. Gatza and co-workers found that conventional ECP had a therapeutic effect in a MHC matched but minor histocompatibility mismatched mouse model of GvHD. Looking for the impact of ECP in the well known full MHC I and II mismatched B6→BALB/c model, we could demonstrate for the first time in mice that this conventional ECP setup is also effective in acute GvHD after BMT. In contrast to the work of Gatza et al. [25] and the findings in the contact hypersensitivity mouse model of Maeda et al. [26], we could not find increased Treg levels in mice treated with ECP. However, many of our ECP treated animals survived the observation time of 56 days. Hence, we analyzed cellular samples four weeks after the last ECP procedure. Thus, it is possible that Treg levels were higher directly after ECP but not after the full observation time. Further studies on the kinetics of Treg after ECP treatment could reveal new information for the development of the best ECP schedule.

From clinical experience it is well known that frequent repetition of the ECP procedure is required for a positive therapeutic outcome [27]. In our mouse model we could confirm that a single ECP cycle is not able to alleviate GvHD. These observations are in concordance with the immunological finding that immune tolerance can be induced by steady-state levels of apoptotic cells [28]. In the conventional ECP setup of our study, GvHD animals were injected with genetically identical 8-MOP/UV-A treated cells from another cohort of B6→BALB/c transplanted GvHD mice. This approach corresponds to the conventional human ECP setup with the major difference that the cells were not obtained from the ECP treated mouse itself. Thus, important evidence arises from our observation that the main mechanism of ECP is not the reduction of alloreactive T-cells, but rather the application of 8-MOP/UV-A treated apoptotic cells in general. Hence, it is maybe not necessary that cells for extracorporeal treatment are from the patient's origin. Therefore, our initial hypothesis was that cells from another source could perhaps replace the patient's apheresis procedure to avoid this physical intervention in patients with severe acute GvHD.

In a MHC matched and minor mismatched mouse model for GvHD, Capitini and co-workers found that 8-MOP/UV-A treated splenocytes from healthy non-GvHD mice with the background of BMT donor origin were able to reduce GvHD associated weight loss [29]. We wondered whether this modified ECP setup in which ECP cells are not patient-derived is able to reduce GvHD clinical score or even enhance the survival rate in an MHC mismatched model for acute GvHD. Furthermore we compared the efficacy of this modified ECP setup with our conventional ECP setup described above. Interestingly, we could observe only subtle therapeutic effects performing the modified ECP with 8-MOP/UV-A treated splenocytes from healthy B6 mice representing the genetic background of BMT donor origin. Marginal improvements in terms of clinical symptoms and survival rates were visible in the first 35 days after BMT but did not persistent until the end of the observation time at day 56. Injection of increased numbers of 8-MOP/UV-A treated splenocytes or prolongation of the therapeutic period by performing four instead of three ECP applications led to the same results.

Our findings are consistent with the hypothesis, that alloreactive apoptotic T-cells are essential for the induction of ECP mediated tolerance [30]. Since 8-MOP/UV-A treated cells in our modified ECP setup became apoptotic before getting into contact with the patient's MHC antigens, the cells were not alloreactive at any time point. Due to the lack of reactivity, these cells were possibly not able to induce tolerance in a sufficient and long-lasting manner. However, since we could observe subtle therapeutic effects of the modified setup with ECP cells derived from healthy B6 mice this approach could perhaps be successful in other settings e.g. in chronic GvHD. Notably, we could not observe any increase of GvHD score or mortality after injection of allogeneic 8-MOP/UV-A treated leukocytes. This is a crucial point because no additional alloreactions in the GvHD patient would be tolerable. In concordance with the well known effect of apoptosis induction by ECP treatment [15,31,32], we suggest that standard 8-MOP/UV-A treatment of allogeneic donor leukocytes is sufficient to prevent alloreactions of these cells in the recipient.

Since we could not find any negative effects of ECP with allogeneic 8-MOP/UV-A treated cells, the concept of replacing patient's cells by different types of allogeneic cells from healthy donors for the treatment of acute and chronic GvHD should be investigated in future studies.

Supporting Information

Table S1 ARRIVE Guidelines Checklist. Overview and background information about the *in vivo* experiments according to the ARRIVE Guidelines.

Acknowledgments

We thank Margret Rave-Fränk and Hendrik Wolff (Department of Radiation Oncology, University Medical Center Göttingen, Göttingen, Germany) for operating the irradiation system.

Author Contributions

Conceived and designed the experiments: HB GW HMR JR TJL. Performed the experiments: HB SK LST. Analyzed the data: HB SK JR TJL. Contributed to the writing of the manuscript: HB.

References

1. Li HW, Sykes M (2012) Emerging concepts in haematopoietic cell transplantation. Nat Rev Immunol 12: 403–16.
2. Shlomchik WD (2007) Graft-versus-host disease. Nat Rev Immunol 7: 340–52.
3. Korngold R, Sprent J (1978) Lethal graft-versus-host disease after bone marrow transplantation across minor histocompatibility barriers in mice. Prevention by removing mature T cells from marrow. J Exp Med 148: 1687–98.
4. Pasquini MC, Wang Z, Horowitz MM, Gale RP (2010) 2010 report from the Center for International Blood and Marrow Transplant Research (CIBMTR): current uses and outcomes of hematopoietic cell transplants for blood and bone marrow disorders. Clin Transpl: 87–105.
5. Wolff D, Steiner B, Hildebrandt G, Edinger M, Holler E (2009) Pharmaceutical and cellular strategies in prophylaxis and treatment of graft-versus-host disease. Curr Pharm Des 15: 1974–97.
6. Perfetti P, Carlier P, Strada P, Gualandi F, Occhini D, et al. (2008) Extracorporeal photopheresis for the treatment of steroid refractory acute GVHD. Bone Marrow Transplant 42: 609–17.
7. Wolf D, von Lilienfeld-Toal M, Wolf AM, Schleuning M, von Bergwelt-Baildon M, et al. (2012) Novel treatment concepts for graft-versus-host disease. Blood 119: 16–25.
8. Blazar BR, Murphy WJ, Abedi M (2012) Advances in graft-versus-host disease biology and therapy. Nat Rev Immunol 12: 443–58.
9. Greinix HT, Worel N, Knobler R (2010) Role of extracorporeal photopheresis (ECP) in treatment of steroid-refractory acute graft-versus-host disease. Biol Blood Marrow Transplant 16: 1747–8.
10. Marshall SR (2006) Technology insight: ECP for the treatment of GvHD–can we offer selective immune control without generalized immunosuppression? Nat Clin Pract Oncol 3: 302–14.
11. Hart JW, Shiue LH, Shpall EJ, Alousi AM (2013) Extracorporeal photopheresis in the treatment of graft-versus-host disease: evidence and opinion. Ther Adv Hematol 4: 320–34.
12. Greinix HT, Volc-Platzer B, Kalhs P, Fischer G, Rosenmayr A, et al. (2000) Extracorporeal photochemotherapy in the treatment of severe steroid-refractory acute graft-versus-host disease: a pilot study. Blood 96: 2426–31.
13. Suchin KR, Cassin M, Washko R, Nahass G, Berkson M, et al. (1999) Extracorporeal photochemotherapy does not suppress T- or B-cell responses to novel or recall antigens. J Am Acad Dermatol 41: 980–6.
14. Merlin E, Paillard C, Rochette E, David A, Isfan F, et al. (2010) Extracorporeal photochemotherapy as second- or first-line therapy of acute GVHD? Bone Marrow Transplant 45: 963–5.
15. Heshmati F (2003) Mechanisms of action of extracorporeal photochemotherapy. Transfus Apher Sci 29: 61–70.
16. Albert ML (2004) Death-defying immunity: do apoptotic cells influence antigen processing and presentation? Nat Rev Immunol 4: 223–31.

17. Xia CQ, Campbell KA, Clare-Salzler MJ (2009) Extracorporeal photopheresis-induced immune tolerance: a focus on modulation of antigen-presenting cells and induction of regulatory T cells by apoptotic cells. Curr Opin Organ Transplant 14: 338–43.
18. Tischner D, Theiss J, Karabinskaya A, van den Brandt J, Reichardt SD, et al. (2011) Acid sphingomyelinase is required for protection of effector memory T cells against glucocorticoid-induced cell death. J Immunol 187: 4509–16.
19. Tawara I, Sun Y, Lewis EC, Toubai T, Evers R, et al. (2012) Alpha-1-antitrypsin monotherapy reduces graft-versus-host disease after experimental allogeneic bone marrow transplantation. Proc Natl Acad Sci USA 109: 564–9.
20. Schroeder MA, DiPersio JF (2011) Mouse models of graft-versus-host disease: advances and limitations. Dis Model Mech 4: 318–33.
21. Reddy P and Ferrara JLM (2009) Mouse models of graft-versus-host disease. StemBook, ed. The Stem Cell Research Community. doi/10.3824/stembook.1.36.1.
22. Wolf CE, Meyer M, Riggert J. (2005) Leukapheresis for the extraction of monocytes and various lymphocyte subpopulations from peripheral blood: product quality and prediction of the yield using different harvest procedures. Vox Sang 88: 249–55.
23. Reddy P, Negrin R, Hill GR (2008) Mouse models of bone marrow transplantation. Biol Blood Marrow Transplant 14: 129–35.
24. Devine SM, Carter S, Soiffer RJ, Pasquini MC, Hari PN, et al. (2011) Low risk of chronic graft-versus-host disease and relapse associated with T cell-depleted peripheral blood stem cell transplantation for acute myelogenous leukemia in first remission: results of the blood and marrow transplant clinical trials network protocol 0303. Biol Blood Marrow Transplant 17: 1343–51.
25. Gatza E, Rogers CE, Clouthier SG, Lowler KP, Tawara I, et al. (2008) Extracorporeal photopheresis reverses experimental graft-versus-host disease through regulatory T cells. Blood 112: 1515–21.
26. Maeda A, Schwarz A, Bullinger A, Morita A, Peritt D, et al. (2008). Experimental extracorporeal photopheresis inhibits the sensitization and effector phases of contact hypersensitivity via two mechanisms: generation of IL-10 and induction of regulatory T cells. J Immunol. 181: 5956–62.
27. Greinix HT, Socié G, Bacigalupo A, Holler E, Edinger MG, et al. (2006) Assessing the potential role of photopheresis in hematopoietic stem cell transplant. Bone Marrow Transplant 38: 265–73.
28. Birge RB, Ucker DS (2008) Innate apoptotic immunity: the calming touch of death. Cell Death Differ 15: 1096–102.
29. Capitini CM, Davis JP, Larabee SM, Herby S, Nasholm NM, et al. (2011) Extracorporeal photopheresis attenuates murine graft-versus-host disease via bone marrow-derived interleukin-10 and preserves responses to dendritic cell vaccination. Biol Blood Marrow Transplant 17: 790–9.
30. Edelson RL (2013) Mechanistic insights into extracorporeal photochemotherapy: Efficient induction of monocyte-to-dendritic cell maturation. Transfus Apher Sci: doi:10.1016/j.transci.2013.07.031.
31. Gasparro FP, Dall'Amico R, Goldminz D, Simmons E, Weingold D (1989) Molecular aspects of extracorporeal photochemotherapy. Yale J Biol Med 62: 579–93.
32. Bladon J, Taylor PC (2006) Extracorporeal photopheresis: a focus on apoptosis and cytokines. J Dermatol Sci 43: 85–94.

Berberis libanotica Ehrenb Extract Shows Anti-Neoplastic Effects on Prostate Cancer Stem/Progenitor Cells

Rabih El-Merahbi[1]☯, Yen-Nien Liu[2]☯, Assaad Eid[1], Georges Daoud[1], Leina Hosry[3], Alissar Monzer[1], Tarek H. Mouhieddine[1], Aline Hamade[4]*, Fadia Najjar[5]*, Wassim Abou-Kheir[1]*

1 Department of Anatomy, Cell Biology and Physiological Sciences, Faculty of Medicine, American University of Beirut, Beirut, Lebanon, **2** Graduate Institute of Cancer Biology and Drug Discovery, College of Medical Science and Technology, Taipei Medical University, Taipei, Taiwan, **3** Faculty of Pharmacy, Lebanese University, Hadath, Lebanon, **4** Department of Biology, Faculty of Sciences II, Lebanese University, Fanar, Lebanon, **5** Department of Chemistry and Biochemistry, Faculty of Sciences II, Lebanese University, Fanar, Lebanon

Abstract

Cancer stem cells (CSCs), including those of advanced prostate cancer, are a suggested reason for tumor resistance toward conventional tumor therapy. Therefore, new therapeutic agents are urgently needed for targeting CSCs. Despite the minimal understanding of their modes of action, natural products and herbal therapies have been commonly used in the prevention and treatment of many cancers. *Berberis libanotica* Ehrenb (BLE) is a plant rich in alkaloids which may possess anti-cancer activity and a high potential for eliminating CSCs. We tested the effect of BLE on prostate cancer cells and our data indicated that this extract induced significant reduction in cell viability and inhibited the proliferation of human prostate cancer cell lines (DU145, PC3 and 22Rv1) in a dose- and time-dependent manner. BLE extract induced a perturbation of the cell cycle, leading to a G0-G1 arrest. Furthermore, we noted 50% cell death, characterized by the production of high levels of reactive oxidative species (ROS). Inhibition of cellular migration and invasion was also achieved upon treatment with BLE extract, suggesting a role in inhibiting metastasis. Interestingly, BLE extract had a major effect on CSCs. Cells were grown in a 3D sphere-formation assay to enrich for a population of cancer stem/progenitor cells. Our results showed a significant reduction in sphere formation ability. Three rounds of treatment with BLE extract were sufficient to eradicate the self-renewal ability of highly resistant CSCs. In conclusion, our results suggest a high therapeutic potential of BLE extract in targeting prostate cancer and its CSCs.

Editor: Maria Cristina Vinci, Centro Cardiologico Monzino, Italy

Funding: This research was supported by funding from the Medical Practice Plan-MPP (AUB-FM) and the Lebanese National Council for Scientific Research (CNRS). The funders had no role in study design, data collection and analysis, decision to publish, or preparation of the manuscript.

Competing Interests: The authors have declared that no competing interests exist.

* Email: wa12@aub.edu.lb (WAK); Fadia.najjar@gmail.com (FN); alinehamade@gmail.com (AH)

☯ These authors contributed equally to this work.

Introduction

Prostate cancer (PC) is the most commonly diagnosed non-cutaneous malignancy and the third most common cause of cancer mortality in the Western male population [1,2]. Primary PC is androgen-dependent in nature and is usually treated with androgen deprivation therapy (ADT). Most frequently, however, hormonal therapy leads to recurrence in a few years and PC eventually progresses to an androgen-independent state or, a so-called castrate resistant PC (CRPC). CRPC is an aggressively metastatic and lethal form of PC and currently, there is no known effective treatment for it.

Prostate cancer stem cells (CSCs) share properties with normal stem cells as they tend to express high levels of: aldehyde dehydrogenase (ALDH) - a detoxifying enzyme - [3], multidrug resistance (MDR) efflux pumps and ABC transporters [4–6]. These defensive strategies render conventional therapy ineffective, due to the presence of fast proliferative cells in the tumor bulk and a great potential for sparing the putative cancer stem/progenitor cells [7]. In addition, it has been indicated that prostate CSCs do not express androgen receptors (AR) [8,9] and may not respond to ADT as mature tumor cells do. Following ADT, cancer stem cells may frequently manage to repopulate the tumor mass with androgen-independent PC, which is an aggressively metastatic and lethal form of PC.

A wide range of strategies have been employed for the discovery of novel drugs that might carry beneficial effects for cancer patients. A targeted therapy is urgently needed to eradicate not only the cancer bulk, but also the CSC pool found within the tumor. Previous work in our laboratory has demonstrated the ability to enrich a population of PC stem/progenitor cells by growing them in 3D spheres-forming culture conditions, namely called prostaspheres [10,11]. Several studies have recently shown that a number of bioactive food compounds may have an anti-CSC effect. For example, it has been recently reported that palm oil-extracted gamma-tocotrienol [12], polysaccharide-P (PSP), an active component extracted from the mushroom Turkey tail [13], and Genistein, a major isoflavone constituent of soybeans and soy products [14], exhibit potent inhibitory activity on protosphere formation ability and tumorigenicity of PC cells. These substances have been shown to target prostate CSCs *in vitro*, and suppress tumor formation *in vivo*. These findings highlight the potential use of herbal bioactive compounds for the production of therapeutic agents targeting CSC, either for the prevention or the treatment of PC.

Berberis libanotica, specifically its roots, has been used in traditional herbal Lebanese remedies for rheumatic and neuralgic diseases [15,16]. *B. vulgaris* and *B. stata* are the two most studied species of *Berberis* [17–20]. Alkaloids constitute the major class of compounds reported to exist in *Berberis* species and represent a very wide range of secondary metabolites with important biological activities [21,22]. Various herbal alkaloids exhibit *in vitro and in vivo* anti-proliferative and anti-metastatic effects on various types of cancers. Alkaloids, such as camptothecin [23] and vinblastine [24], have already been successfully developed into anti-cancer drugs. Berberine, a major alkaloid characterizing *Berberis* species, has been intensively investigated for its pharmacological properties. It was shown to inhibit the migration of melanoma cancer cells [25], and the growth of human tongue squamous carcinoma tumors in a murine xenograft model [26], enhance tumor necrosis factor-related apoptosis-inducing ligand in breast cancer [27], and exert a cytotoxic effect against many cell lines [25,28,29]. To date, the biological and phytochemical properties of *B. libanotica* extracts have only been investigated in two publications reporting the inhibition of adult T-cell leukaemia viability via ethanol fraction [30], and the inhibition of key enzymes linked to Alzheimer's disease [15]. However, little is known about the effect of BLE extract on PC and the proliferation and viability of CSCs. Therefore, the aim of this study is to investigate the therapeutic efficacy of *B. libanotica* root extract (via ammonia-dichloromethane extraction) in inhibiting the proliferation and reducing the viability of human PC cell lines grown in a conventional 2D monolayer model. Furthermore, this study will focus on an advanced assay (sphere formation assay) to investigate the effect of BLE extract in targeting an enriched population of prostate CSCs.

Materials and Methods

Cell Culture

Human prostate cancer cell lines, PC3, DU145, and 22Rv1 were obtained from the American Type Culture Collection (Manassas, VA), and were sent to us as a gift by Dr. Kathleen Kelly at the National Institutes of Health (Bethesda, MD). Cells were cultured in RPMI-1640 AQ medium (Sigma, USA) supplemented with 10% FBS (Sigma, USA), 1% penicillin-streptomycin (Sigma, USA) and 1% non-essential amino acids (Sigma, USA) and incubated at 37°C in a humidified incubator (95% air, 5% CO_2). Cells were harvested by trypsin-EDTA at 37°C (Sigma, USA). Cells were pelleted, re-suspended and transferred into new-75 cm^2 tissue culture flasks for maintenance, or seeded for experiments at a concentration of 1×10^4 cells/cm^2 for PC3 and Du145 and 1.5×10^4 cells/cm^2 for the 22Rv1 cell line.

BLE Extraction

The roots of *Berberis libanotica* Ehrenb were collected in September 2011 from the Cedars area in North Lebanon (altitude 1400–1700 m; approximate GPS location: N 34°14'41"; E 36°2'15"). No specific permissions were required for these locations/activities as they did not involve endangered or protected species. The plant was identified by Pr. G. Tohmé (CNRS, Lebanon). The voucher specimen (code number BLCS12) was deposited in the herbarium of the Faculty of Sciences II, Lebanese University, Beirut, Lebanon.

The plant roots were shade-dried and powdered using an electric blender. The ammonia-dichloromethane extract was prepared as follows: 10 g of plant powder were moistened for 2 h with NH_4OH solution followed by dichloromethane addition (100 ml). The mixture was macerated for 24 h under magnetic stirring, followed by filtration. The organic phase was concentrated at 40°C under reduced pressure and cryo-dried. The obtained extracts were kept in a cool place in dark containers. The rotary evaporator used was IKA RV 10 BASIC and the lyophiliser was LYOVAC GT4.

MTT/Cytotoxicity Assay

Anti-proliferative and cytotoxic effects of BLE were measured *in vitro* via MTT ([3-(4, 5-dimethylthiazol-2-yl)-2, 5-diphenyltetrazolium bromide]) assays. DU145 and PC3 cells were seeded in 100 µl complete medium in 96-well culture plates at a density of 7500 cells/well and 22Rv1 cells at a density of 13250 cells per well. Cells were incubated overnight in the incubator and then treated in triplicates with various extract concentrations diluted in 100 µl complete media for 24, 48 and 72 h. In case of the ROS scavenger N-acetyl-L-cysteine (NAC; Sigma), DU145 cells were pretreated with 5 mM NAC for 30 minutes and cells were then treated with 30 µg/ml of BLE extract for 24 and 48 h as above. For each time point, 20 µl of 5 mg/ml - in 1 x phosphate buffered saline (PBS) - MTT reagent was added to each well and incubated at 37°C for 4 h. Finally, 100 µl of solubilization solution was added into each well to dissolve the formazan crystals. The reduced MTT optical density (OD) was measured at a wavelength of 595 nm using an ELISA reader (Multiskan Ex). The percentage of cell viability was presented as an OD ratio between the treated and untreated cells at the indicated concentrations.

Trypan Blue Exclusion Method

Supernatants containing dead cells were collected and attached live cells were harvested by trypsin EDTA and added to the supernatant. The cell pellet was re-suspended in 100 µl media and 50 µl of cell suspension was mixed with 50 µl of trypan blue and then live/dead cells were counted using a hemocytometer.

Cell Cycle Analysis by Flow Cytometry

DU145 and PC3 cells were seeded in duplicates into 6-well plates at a density of 1×10^5 cells/well and were incubated for 24 h prior to drug treatment for 24, 48 or 72 h. Cells were then harvested, washed twice with PBS, centrifuged at 1500 rpm for 5 min at 4°C, re-suspended in 1 mL of cold PBS, fixed in 4 mL of cold absolute ethanol and then stored at −20C° until staining and analysis. Fixed cells were then treated for 1 h with 200 µg/ml DNase-free RNase A, stained with 1 mg/ml propidium iodide (PI) (Sigma, USA) and incubated for 10 min in the dark in a flow tube (BD Flacon). Fluorescence of PI, a measure of DNA content in a cell population, was done using flow cytometry (FACScan, Becton Dickinson). A total of 10,000 gated events were acquired in order to assess the proportions of cells in different stages of the cell cycle. Analysis of cell cycle distribution was performed using FlowJo Software.

Lactate Dehydrogenase (LDH) Assay

The cytotoxicity of BLE extract on the DU145 cell line was determined by measuring the activity of lactate dehydrogenase (LDH) released from the cytosol of damaged cells using a Cytotoxicity Detection Kit PLUS (LDH) (Roche Diagnostics

GmbH). The assay was conducted following the manufacturer's instructions. Briefly, medium supernatants of control and treated cells were collected and equal volume of reaction mixture was added. The reaction mixtures were incubated for 20 min at room temperature in the dark. Following incubation, the absorbance was determined at 490 nm using an ELISA plate reader. The percentage of cytotoxicity was calculated as: (exp.value - spontaneous control release) ÷ (max.control release - spontaneous control release) $\times 100$.

ROS detection by Nitroblue Tetrazolium Assay

DU145 cells were cultured in triplicate in a 96-well plate at a density of 1×10^4 cells/cm^2 and were then incubated for 24 h prior to BLE extracts treatment for 24, 48 or 72 h. In each condition, ROS production was evaluated by a nitroblue tetrazolium assay (NBT; Sigma). Culture media were aspirated and cells were incubated in PBS containing 1 mg/ml NBT at 37°C in the dark for 60 minutes. NBT was reduced by ROS to a dark-blue insoluble form of NBT called formazan that could be visualized. To quantify the formazan product, the intracellular formazan was solubilized in KOH 2 M and DMSO 5 M solution, and absorbance was determined at 630 nm using an ELISA plate reader.

Apoptosis Assay

Apoptosis was assayed using the cellular DNA fragmentation ELISA (Roche Diagnostics GmbH, Mannheim, Germany) for the detection of BrdU-labeled DNA fragments in culture supernatants and cell lysates. The assay was used according to the manufacturer's protocol and as described previously [31]. Cells were incubated with 30 μg/ml of BLE extract for 48 h in the presence or absence of 5 mM N-acetyl-L-cysteine that was added 30 minutes before BLE treatment. Apoptosis in the control untreated cells was adjusted to 100% and the results were plotted as percent of control.

Wound healing assay

For wound healing, or scratch assay, cells were seeded in a six-well plate with the same numbers, and incubated until they reached 80–90% confluence. Cells were then treated with 10 μg/ml mitomycin C (Sigma) for 2 h in order to block cellular proliferation. A sterile 200 μl tip was used to scratch wounds of the same width on each monolayer, the plates were then washed twice with PBS to remove the detached cells, and the remaining cells were cultured in complete media with or without treatment. Photos were subsequently taken at 0, 12, 30, and 48 h. The distance traveled by the cells into the wounded area enumerated the closure of the wounds. The experiment was repeated three times with duplicate measurements in each experiment.

Sphere Formation Assay

In this study, all three cell lines, PC3, DU145 and 22Rv-1, were able to form spheroids in non-adherent culture, suggesting the presence of cancer stem-like cells within these cell lines. After counting the single cell suspension of prostate cancer cell lines, a concentration of 1,000 cells/well was suspended in cold Matrigel^{TM}/serum-free RPMI-1640 (1:1) in a total volume of 50 μl. Cells were seeded uniformly in a circular manner around the bottom rim of a well in a 24-well plate and allowed to solidify in the incubator at 37°C for 45 minutes, before 0.5 ml of complete RPMI-1640 −2% FBS media (with or without treatment) was added gently in the middle of each well. Spheres were replenished with warm media as in the original seeding (with or without treatment) every two to three days. Spheres were counted after 13 days. To propagate spheres, the medium was aspirated and Matrigel^{TM} was digested with 0.5 ml Dispase solution (Invitrogen, Carlsbad, CA, 1 mg/ml, dissolved in RPMI-1640 incomplete medium) for 60 minutes at 37°C. Spheres were collected, incubated in 1 ml warm Trypsin- EDTA at 37°C for 5 minutes, and then passed through a 27-gauge syringe 5 times. Cells were counted by a hemocytometer and re-seeded.

Transwell Invasion Assay

For the invasion assay, 2.5×10^5 cells were seeded in a serum-free medium with or without treatment in the top chamber onto the Matrigel^{TM}-coated membrane (24-well insert; pore size, 8 μm; Falcon), and a medium supplemented with serum was used as a chemo-attractant in the lower chamber. Each well was freshly coated with 100 μl of Matrigel^{TM} (BD Bioscience) at a dilution of 1:10 in cold PBS and was then air-dried overnight before starting the invasion assay. Cells were allowed to migrate through the membrane coated with Matrigel^{TM} at 37°C in a 5% CO2 incubator for 24 and 48 h. Non-migratory cells in the upper chamber were then gently scraped off with a cotton-tip applicator. Invading cells on the lower surface of the membrane were fixed and stained with Hematoxylin and Eosin. After staining, the total number of invading cells was counted under the light microscope ($\times10$ objective) from 6 consecutive fields for each well.

RNA extraction and quantitative real time PCR (qRT-PCR)

Total RNA was extracted from cells, treated or non-treated with 30 μg/ml of BLE extract for 48 h, using the RNeasy Micro Kit (Qiagen) according to the manufacturer's instructions. cDNA was generated from total RNA using the Super Script III First Strand Synthesis System for RT-PCR (Invitrogen), and PCR was performed using Platinum Taq Polymerase (Invitrogen). For quantitative RT-PCR, the amplification step was done using the SYBR green PCR master mix (Applied Biosystems, Bedford, MA). All reactions were run in duplicate using the specific primers listed below and all values were normalized to the house keeping gene GAPDH. The primers used were: GAPDH-F: 5′ ACCTGGC-TAGCGAAAAGCAA3′; GAPDH-R: 5′CCACTTTGTCAAG-CTCATTTCCT3′; Sox2-F: 5′AACCCCAAGATGCACAAC-TC3′; Sox2-R: 5′GCTTAGCCTCGTCGATGAAC3′; Oct4-F: 5′AGAACATGTGTAAGCTGCGG3′; Oct4-R: 5′GTTGCC-TCTCACTCGGTTC3′; Nanog-F: 5′ACCTCAGCTACAAAC-AGGTGAA3′; Nanog-R: 5′AAAGGCTGGGGTAGGTAGG-T3′; CD44-F: 5′TTTGCATTGCAGTCAACAGTC3′; CD44-R: 5′GTTACACCCCAATCTTCATGTCCAC3′; CD166-F: 5′TGGCAATATCACATGGTACAGG3′; CD166-R: 5′AGCC-TTGGTTGTCTTGTACTC3′.

Data Analysis

Statistical analysis was performed using Microsoft Excel 2013. The significance of the data was analyzed using a Student's t-test, and differences between two means with $p<.05$ (*) and $p<.01$ (#) were considered significant and very significant, respectively.

Results

BLE extract inhibited PC cell proliferation in a dose- and time-dependent manner

Our first objective was to investigate the *in vitro* effects of the BLE extract on PC cell growth and viability. Using MTT assays (Fig. 1A), the proliferation activity of PC cell lines Du145, PC3 and 22Rv-1 was found to be inversely correlated with the extract concentration in the culture medium. After 72 h of treatment, the inhibitory effect commenced at low concentrations and significantly inhibited proliferation by 50% when PC cells were grown in media supplied with 30 μg/ml of BLE extract. However, at high extract concentration (100 μg/ml), around 75% of the anti-proliferative effect was significantly achieved. These results were compatible with confluency changes in culture (Figure S1) and verified by trypan blue exclusion assays (Fig. 1B). At high concentrations (60 and 100 μg/ml), increased cytotoxicity and cell death were observed.

BLE extract induced cellular toxicity on DU145 cell line

BLE was cytotoxic and induced significant cell death in the DU145 cell line in a dose- and time-dependent manner (Fig. 2). The cytotoxic potential of the BLE extract was determined based on the measurement of intracellular LDH enzyme, which is a stable cytosolic enzyme released into the cell culture medium upon plasma membrane damage. A 30% cytotoxic effect was significantly achieved at 48 h with 100 μg/ml and at 72 h with 60 μg/ml. However, the maximum percentage of cell death was around 50% after 72 h of treatment at a concentration of 100 μg/ml of BLE extract.

BLE pro-oxidative activity induced DU145 cell death

Intracellular reactive oxygen species (ROS) production was measured using nitroblue tetrazolium (NBT). The intensity of the dye is inversely propotional to the presence of ROS. Results from Figure 3 indicate a dose- and time-dependent reduction of NBT. NBT reduction was initiated at 24 h with 100 μg/ml, while the maximum percentage of NBT reduction was after 72 h of treatment with 100 μg/ml of BLE extract. Interestingly, in the presence of a ROS scavenger (N-acetyl-L-cysteine, NAC), BLE failed to affect cell viability at 24 h and 48 h. Furthermore, BLE-induced cellular DNA fragmentation, indicative of apoptosis and cell death, was not seen in the presence of NAC (Fig. S2). This suggests that the BLE extract induces cell death by the induction of ROS production.

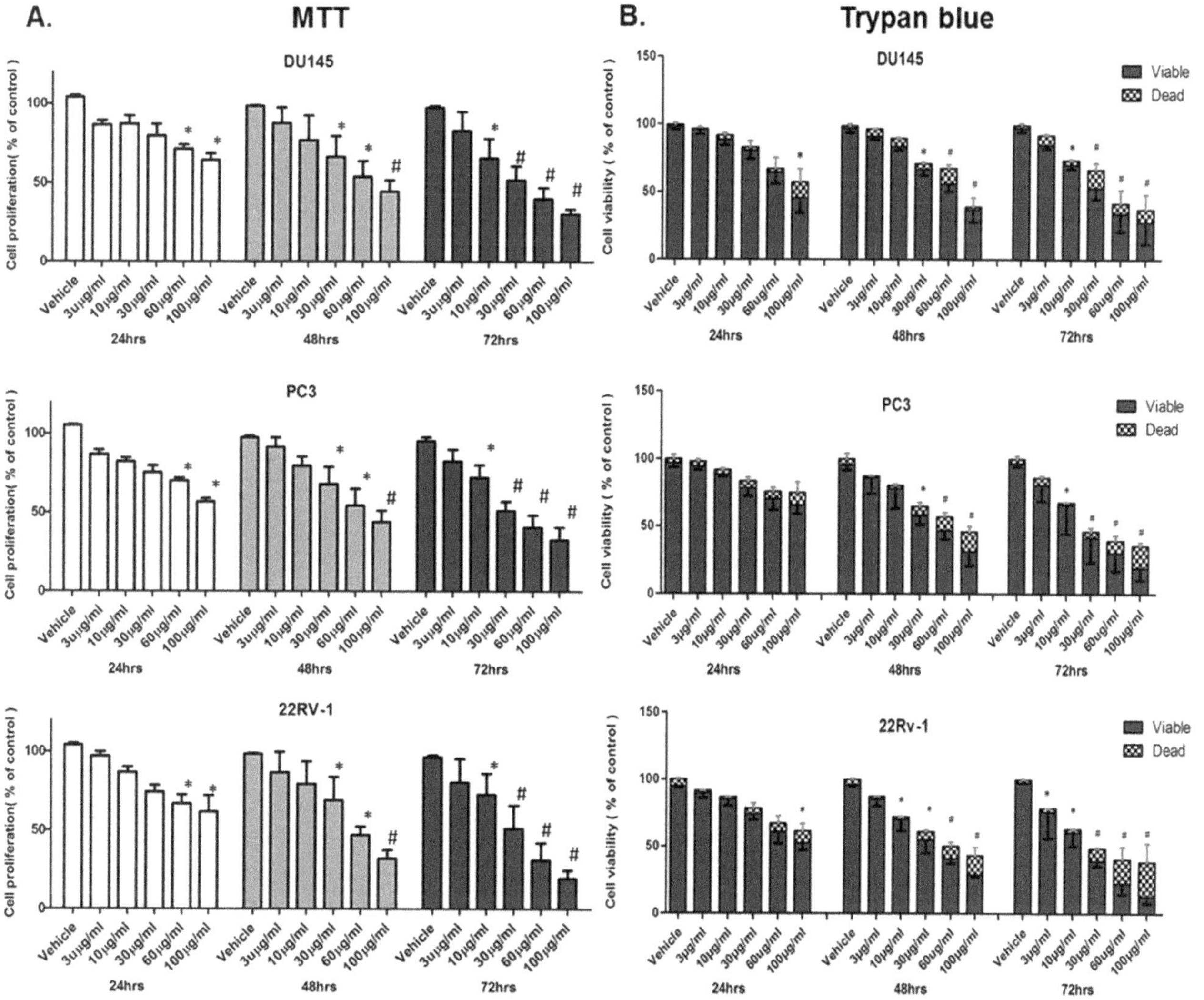

Figure 1. The effect of various concentrations of BLE extract on prostate cancer cell line proliferation and viability by using (A) MTT assay and (B) Trypan blue dye exclusion assay. After incubation of the three prostate cancer cell lines (DU145, PC3 and 22Rv-1) for 24, 48 and 72 h with or without treatment with BLE extract, cell proliferation was determined. Results are expressed as a percentage of the studied group compared to its control. Data represent an average of five independent experiments. The data are reported as mean ± SD (* P<0.05; # P<0.01).

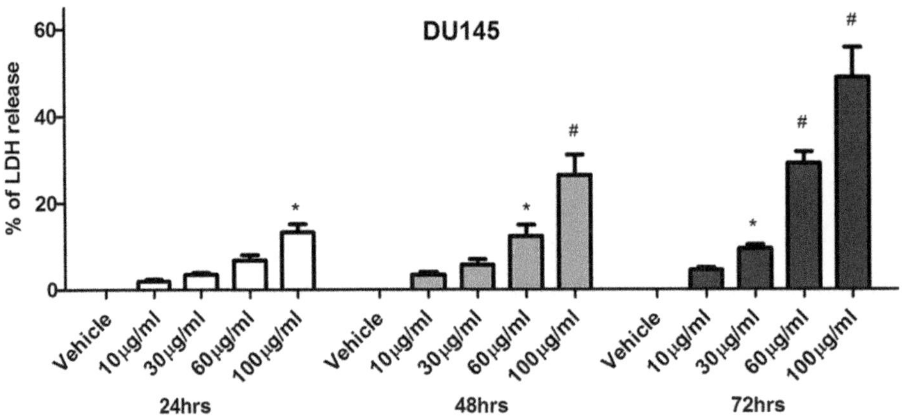

Figure 2. BLE extract induces cell death in DU145 prostate cancer cell lines. Cell death was determined by lactate dehydrogenase (LDH) assay treated with or without indicated concentrations of BLE extract for 24, 48 and 72 h. Cell death was calculated as the percentage of LDH release according to a previously mentioned formula. Data represent an average of three independent experiments. The data are reported as mean ± SD (* $P<0.05$; # $P<0.01$).

BLE extracts induced cell cycle arrest at G0-G1 phase

In an attempt to understand how the BLE extract is able to decrease the cell count and inhibit cell proliferation, we sought to analyse the cell cycle. We examined cellular DNA content distribution by flow cytometry after 24, 48 and 72 h of treatment. Cell cycle analysis revealed that DNA content distribution is altered in BLE-treated PC3 (Table 1) and DU145 (Table 2) cells. As summarized in Table 1, BLE treatment of PC3 cells with either 10 μg/ml or 30 μg/ml caused an increase, within 24 h, in the number of cells in the G1 phase of the cell cycle, providing evidence of G1 arrest. At 48 h and 72 h post-treatment, the G1 population increased from less than 50% in the control to more than 60% in PC3 cells treated with BLE extract at 10 μg/ml or 30 μg/ml. Interestingly, this was associated with a concomitant decrease in the percentage of S phase cells and an accumulation of pre-G1 (G0) phase cells. The same profile was seen when DU145 cells were subjected to the same treatment and analysis (Table 2). These data suggest that BLE treatment results in a perturbation of the cell cycle.

BLE extract greatly reduced the migration and invasion abilities of DU145 cells

Next, we investigated the effect of the BLE extract on cell migration and invasion,, two phenotypes that are associated with the progression to metastasis. Using a wound-healing assay, in the presence of mitomycin C to inhibit cell division, BLE treatment significantly suppressed the cell migration ability of DU145 cells compared to the control, whereby the wound was completely healed after 48 h (Fig. 4A and B). These data were compatible with the invasion assay where upon treatment with 30 μg/ml of BLE extract, invasion ability, in response to FBS, was significantly reduced by more than three folds compared to the control (Fig. 4 C and D). Interestingly, BLE treatment significantly reduced the basal invasion (no FBS) of the cells when compared to the control. Moreover, and after correcting for the effect on proliferation/ apoptosis, the effect of BLE on invasion remained significant between treatment and control group. This suggests that BLE extract has a high inhibitory effect on PC cell migration and invasion.

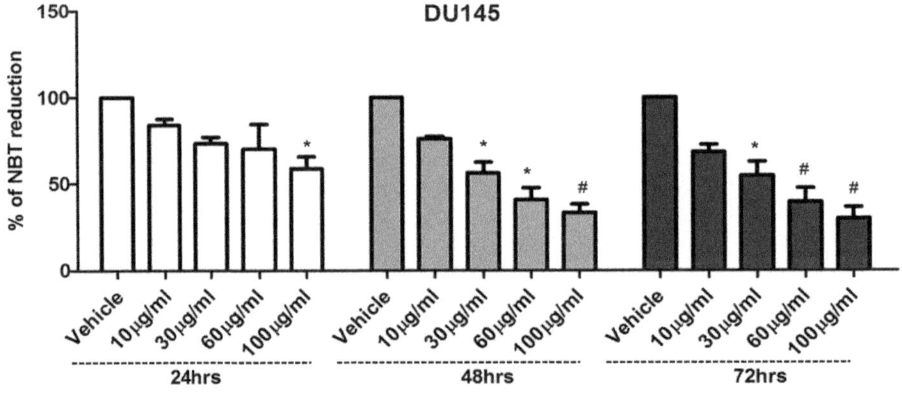

Figure 3. Effect of BLE extract on the production of reactive oxygen species (ROS). ROS production was measured by nitroblue tetrazolium (NBT) reduction in DU145 cell line with or without treatment with indicated concentrations of BLE extract for 24, 48 and 72 h. NBT reduction was calculated as the percentage of the control of the same group. Data represent an average of three independent experiments. The data are reported as mean ± SD (* $P<0.05$; # $P<0.01$).

Table 1. Summary of cell cycle analysis of PC3 cells by flow cytometry.

	PC3	Pre G0-G1	G0-G1	S	G2-M
24 hr	Control	1.42±0.3	46.44±1.5	21.19±1.3	30.64±2.1
	Vehicle	1.32±0.5	49.32±2.3	19.4±2.2	29.69±3.1
	10 µg/ml	1.51±0.4	57.24±2.6	16.31±3.1	**24.61±3.3***
	30 µg/ml	**2.22±0.2***	**61.81±3.2***	**11.04±1.2***	**24.97±3.2***
48 hr	Control	1.40±0.42	47.29±4.14	18.65±1.23	32.24±2.55
	Vehicle	1.49±0.32	49.11±1.64	20.64±3.21	29.19±2.23
	10 µg/ml	**2.23±1.3***	**63.96±3.2***	**10.85±2.92***	**22.86±2.52***
	30 µg/ml	**2.25±1.33***	**70.49±3.24***	**11.87±2.1***	**15.68±2.45***
72 hr	Control	1.541±0.4	44.89±4.88	19.41±2.21	31.20±3.16
	Vehicle	1.69±0.53	47.78±3.33	17.87±2.14	31.18±3.21
	10 µg/ml	**2.11±1.13***	**59.98±3.21***	**14.87±2.33***	**22.73±3.12***
	30 µg/ml	**2.51±1.23***	**73.34±3.21***	**9.44±1.22***	**14.32±3.12***

Table showing the distribution of phases of the cell cycle under the effect of the BLE extract at 24, 48 and 72 h in PC3 cells. Cells that were left untreated with either vehicle or BLE (10 µg/ml and 30 µg/ml) are referred to as control. Data represent an average of three independent experiments. The data are reported as mean ± SD (**Bold*** P<0.05).

BLE extract targeted an enriched population of PC stem/progenitor cells using 3D spheres-formation assay

The ability to form and propagate spheres in non-adherent culture is one of the characteristics of CSCs. To confirm that BLE treatment can inhibit prostate CSC properties, formation of prostaspheres from PC cell lines was studied in the presence or absence of BLE extract. In this assay, prostaspheres were generated by embedding 1000 single cells per well in Matrigel^{TM} at low serum concentration (2% FBS). After culturing DU145 cells for 13 days, a low number of these cells were able to form tumor spheres in untreated conditions with a sphere-forming unit (SFU) of 6.8%. However, treatment with 20 µg/ml or 30 µg/ml of BLE extract significantly reduced the number of prostaspheres by more than 50% (Fig. 5A). In addition, no spheres were observed in wells treated with BLE extract at a concentration above 30 µg/ml. Interestingly, the average size of BLE extract-treated prostaspheres was significantly smaller compared to control (Fig 5B and D) and this could be explained by the significant reduction in the average number of cells per prostasphere as shown in figure 5C.

Self-renewal activity is considered to be one of the major hallmarks of stem/progenitor cells. Thus, to ensure that our extract is able to target the self-renewing CSC pool, sphere formation activity was assessed over five generations. DU145 cells that were able to form spheres in the first generation (G1) were collected and propagated by dissociating spheres into single cells and then the same number of cells (1000 cell/well) was re-seeded. The assay was performed untill the fifth generation (G5), following an experimental design illustrated in figure 6B. Treatment with 20 µg/ml of BLE extracts was sufficient to significantly reduce the sphere formation ability to around 50% whenever treated once and consistently over five generations (Fig. 6A). Interestingly, upon successive treatment and propagation of cells that were able to

Table 2. Summary of cell cycle analysis of DU145 cells by flow cytometry.

	DU145	Pre G0-G1	G0-G1	S	G2-M
24 hr	Control	1.35±0.58	45.44+4.82	22.37+2.04	30.63±2.23
	Vehicle	1.34±0.34	47.95±3.84	21.63±3.03	29.08±3.96
	10 µg/ml	1.55±0.43	**56.3±1.17***	18.02±2.74	24.07±2.38
	30 µg/ml	**1.894±1.01***	**61.36±2.53***	**14.40±1.91***	**22.61±2.76***
48 hr	Control	1.618±0.69	48.02±1.17	17.42±3.33	32.8±2.98
	Vehicle	1.75±0.92	48.27±2.36	20.64±1.91	29.34±1.63
	10 µg/ml	1.83±0.52	**60.22±2.23***	**14.65±2.83***	**23.49±3.74***
	30 µg/ml	1.71±0.21	**65.42±2.82***	**14.50±1.02***	**18.71±2.82***
72 hr	Control	1.60±0.34	44.7±4.25	21.15±2.92	32.66±2.57
	Vehicle	1.82±0.66	48.42±2.23	18.90±3.77	31.10±2.56
	10 µg/ml	**2.45±0.86***	**58.91±2.26***	**15.43±3.23***	**23.43±2.22***
	30 µg/ml	**2.89±0.61***	**71.27±2.74***	**10.01±1.53***	**16.28±1.12***

Table showing the distribution of phases of the cell cycle under the effect of the BLE extract at 24, 48 and 72 h in DU145 cells. Cells that were left untreated with either vehicle or BLE (10 µg/ml and 30 µg/ml) are referred to as control. Data represent an average of three independent experiments. The data are reported as mean ± SD (**Bold** * P<0.05).

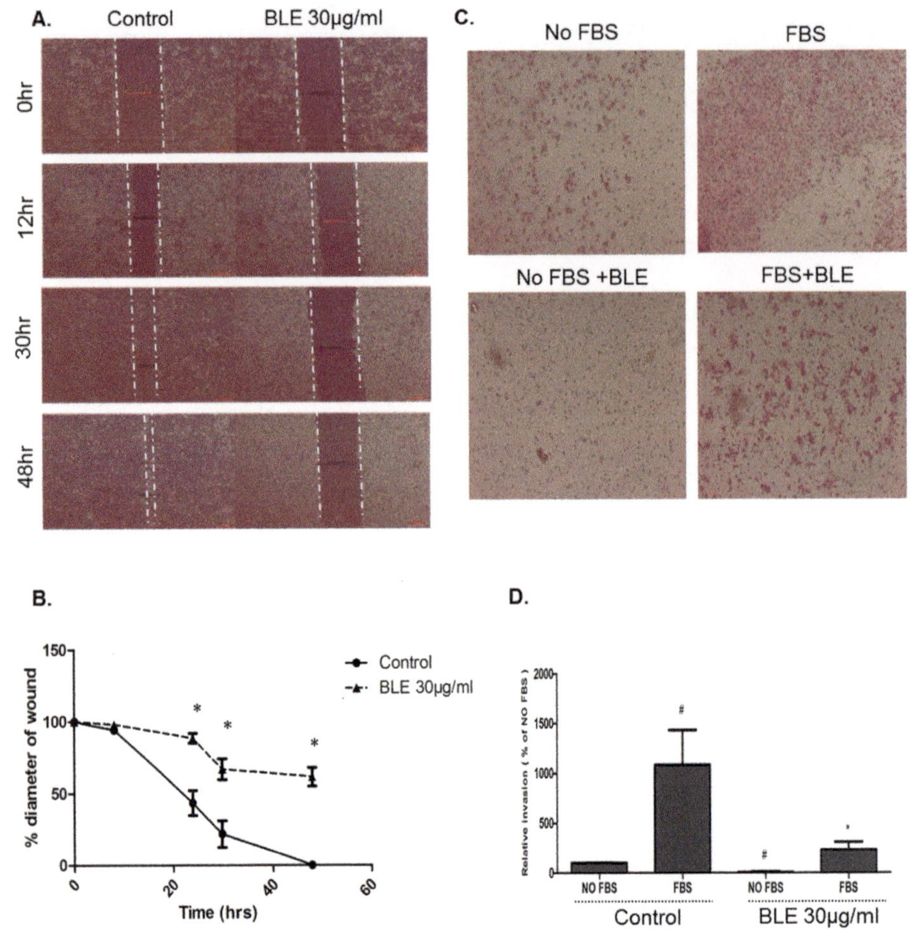

Figure 4. BLE extract reduces the invasive potential of human prostate cancer cells. (**A**) A scratch, in DU145 cells, was made using a yellow tip, and images were taken at T=0 h, 12 h, 30 h and 48 h with or without treatment, and quantification of the diameter of the wound closure was assessed over time (**B**). Data represent an average of three independent experiments. The data are reported as mean ± SD (* $P<0.05$; # $P<0.01$). (**C**): representative images of trans-well invasion assay. DU145 cells were seeded onto the Matrigel^{TM}-coated membrane in the top chamber of the trans-well and were either treated or not treated with 30 µg/ml of BLE extract in the presence or absence of FBS in the lower chamber. Cells that invaded to the lower chamber after 48 h were fixed with methanol, stained with H&E, counted and represented as a percentage of invaded cells compared to the control (**D**). Data represent an average of three independent experiments. The data are reported as mean ± SD (* $P<0.05$; # $P<0.01$).

resist and survive the first BLE treatment at G1, extremely demoted SFUs were produced at G2, and no prostaspheres were formed at G3. Notably, cells that were treated once at G1 were able to regain their sphere-forming ability upon further propagation of the prostaspheres in the absence of any treatment (Fig. 6B). This suggests that serial treatment with BLE extract is needed and is able to target the highly resistant CSCs, and thus extinguishing their self-renewal ability. Notably, cells treated with 30 µg/ml of BLE extract showed a significant reduction in the expression levels of Sox2, Oct4, Nanog, CD44 and CD166, all markers known to be expressed in prostate cancer stem cells, compared to control non-treated cells (Fig. S3).

Discussion and Conclusions

The overall goal of cancer therapy is to target cancer cells only, while leaving viable normal cells unscathed. In this study, *Berberis libanotica* extract exhibited anti-neoplastic effects by reducing the viability and proliferation of three PC cell lines in a time- and dose-dependent manner, whereby the 22RV-1 cell line showed a higher sensitivity to treatment than the other two cell lines (PC3 and DU145). A concentration of 30 µg/ml of BLE extract was successful in achieving IC50, while a concentration of 60 µg/ml induced a 70% proliferation inhibition and cell death was achieved at 100 µg/ml. Several other studies that investigated the effect of alkaloids extracted from the roots of other plants such as Glaucium flavum [32], Sanguinarine [33] and Rauwolfia vomitoria [34], showed similar results in terms of their cytotoxicity and anti-proliferative effects. Previous and current work, including this study, present alkaloids as new promising agents for cancer chemotherapy.

Our data revealed a high increase in the G0-G1 peak in response to treatment with the BLE extract, suggesting an arrest at the G0-G1 phase due to the treatment. Since other studies showed that the anti-proliferative effect of alkaloid-rich plant extracts is due to G2-M arrest rather than G0-G1 arrest [35–37], as in the BLE extract, more studies should be done at the molecular level to better understand the altered pathways in response to BLE treatment.

The ability of cancer cells to metastasize to vital organs is the most life-threatening stage of cancer, and thus, targeting the rapidly proliferating cancer cells is critical for any chemothera-

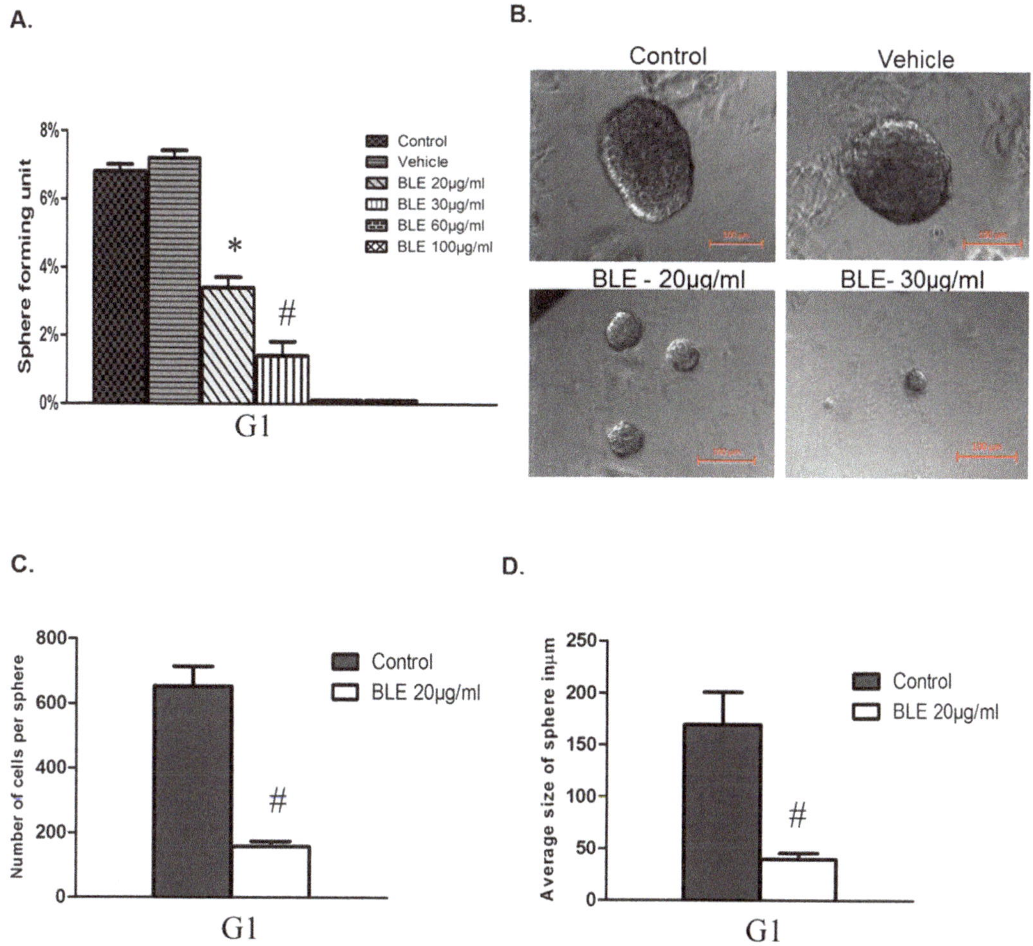

Figure 5. The effects of BLE extract on sphere-forming ability of DU145 cells. (**A**) Sphere-Forming Unit (SFU) is shown with or without BLE treatment (20, 30, 60 and 100 µg/ml). Generated spheres are referred to as G1 (Generation 1) spheres. SFU is calculated according to the following formula: SFU = (number of spheres counted ÷ number of input cells) ×100. Data represent an average of three independent experiments. The data are reported as mean ± SD (* $P<0.05$; # $P<0.01$). (**B**) Representative images of DU145 prostaspheres with or without BLE treatment. Images were visualized by Carl Zeiss microscope at 10x magnification and analyzed by Carl Zeiss Zen 2012 image software. (**C**) Quantification of the average diameter of DU145 prostaspheres with or without treatment conditions. Data represent an average of three independent experiments. The data are reported as mean ± SD (* $P<0.05$; # $P<0.01$). (**D**) Quantification of the average number of cells per DU145 prostasphere with or without treatment. Data represent an average of three independent experiments. The data are reported as mean ± SD (* $P<0.05$; # $P<0.01$).

peutic drug. Unfortunately, recurrence is still observed in many patients that undergo conventional chemotherapy, such as Docetaxel [38,39]. This is because most of these drugs are likely to target the rapidly proliferating cells in the tumor bulk, ultimately missing the quiescent or the slowly dividing CSCs. CSCs that survive chemotherapy would re-enter the cell cycle and produce highly-proliferative and rapidly-dividing progenitor cells that can re-establish the tumor. It is even probable that successive cycles of chemotherapy would strengthen a tumor by generating therapy-resistant CSCs that will give rise to a resistant progeny of rapidly proliferating cells. Therefore, it is crucial to develop a drug that affects the whole tumor, including CSCs that are found within. Recent studies highlighted novel drugs with the ability to target prostate CSCs [12–14]. That is why we aimed to study the effect of the BLE extract on this sub-population of stem/progenitor cells using the sphere-formation assay and the extract's ability to inhibit metastasis. Our results showed a strong inhibitory effect on the sphere-formation ability in response to BLE treatment, which is translated into a decrease in sphere size and number of cells per sphere. Interestingly, after three consecutive treatments with the same concentration of BLE extract, spheres lose their self-renewing pool (i.e. CSCs), which eventually results in their death after the 3rd generation. Moreover, our results from the wound-healing and invasion assays showed a significant decrease in the ability of PC cells to migrate and invade the artificial basement membrane suggesting a role in inhibiting metastasis. These observations suggest that the therapeutic effect of the BLE extract is in its ability to target the resistive CSCs pool and ultimately prevent recurrence.

In conclusion, treatment with BLE extract at a low concentration was able to inhibit proliferation, cell migration, cell invasion and reduce viability. Moreover, the BLE extract was able to target the highly resistive CSCs population and inhibit their self-renewing ability. We believe that this one-of-a-kind approach, using the sphere-formation assay, is very useful in evaluating the potential of novel targeted therapeutics *in vitro*, which in turn, might dictate *in vivo* testing. The next step would be to evaluate the *ex vivo* efficacy of the BLE extract on primary cells extracted from human or animal PC tumors and the *in vivo* efficacy on animal models of prostate cancer.

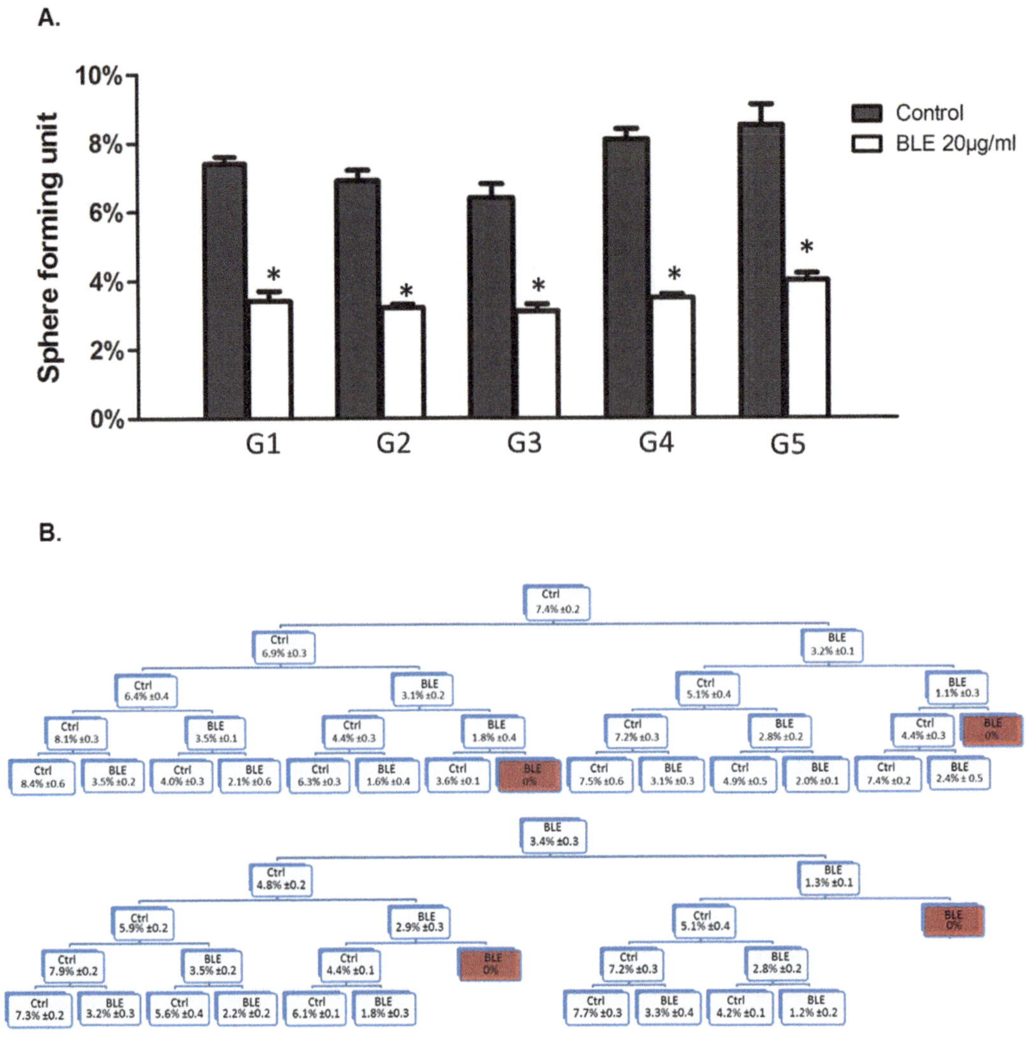

Figure 6. BLE treatment inhibits the self-renewal capacity of prostate cancer stem/progenitor cells. (A) SFU obtained from serially passaged prostaspheres over five generations is shown under untreated conditions (always control) and with 20 µg/ml of BLE extract-treated condition (treated once after propagation of control). (B) Prostate CSC were enriched from DU145 cell line and treated with either BLE extracts (20 µg/ml) or media (control) (G1). After each propagation, cells that were initially treated with the BLE extract (20 µg/ml) or media (control) were seeded into separate wells. Spheres were propagated for five generations in duplicates of each condition. SFU is counted and an average of two independent experiments is represented. The data are reported as mean ± SD (* P<0.05; # P<0.01).

Supporting Information

Figure S1 Cell Morphology and Confluency changes of prostate cancer cell lines in culture. Representative images of (A) DU145, (B) PC3 and (C) 22Rv-1 cells were taken after 72 h in culture with or without BLE treatment. Cells were visualized by Carl Zeiss inverted image microscope at 4x magnification.

Figure S2 ROS scavenger decreased the effect of BLE extract on prostate cancer cell line proliferation and DNA fragmentation. After incubation of DU145 for 24 and 48 h, with or without 30 µg/ml of BLE extract, with one condition being in the presence of 5 mM NAC, cell proliferation (A) and DNA fragmentation (B) were determined. Results are expressed as a percentage of the studied group compared to its control. The data are reported as mean ± SD (* P<0.05).

Figure S3 BLE extract reduces the expression of prostate cancer stem cell markers. The expression levels, using qRT-PCR analysis, of SOX2, Oct4, Nanog, CD44, and CD166 were determined in DU145 cells control or treated with 30 µg/ml of BLE extract for 48 h. The values were normalized to GAPDH and expressed relative to control. The data are reported as mean ± SD (* P<0.05).

Acknowledgments

We would like to thank all the members of the Abou-Kheir's Laboratory for their help on this work. We would like to thank Dr. Kathleen Kelly for her generous gift of the cells used in this study. This research was partly supported by funding from the Medical Practice Plan-MPP (AUB-FM) and the Lebanese National Council for Scientific Research (CNRS).

Author Contributions

Conceived and designed the experiments: WAK AH FN YNL REM. Performed the experiments: REM LH AM TM. Analyzed the data: REM TM AM WAK. Contributed reagents/materials/analysis tools: WAK YNL AE GD AH FN. Wrote the paper: REM YNL TM AE GD AH FN WAK.

References

1. Jemal A, Bray F, Center MM, Ferlay J, Ward E, et al. (2011) Global cancer statistics. CA: a cancer journal for clinicians 61: 69–90.
2. Bubendorf L, Schöpfer A, Wagner U, Sauter G, Moch H, et al. (2000) Metastatic patterns of prostate cancer: an autopsy study of 1,589 patients. Human pathology 31: 578–583.
3. Ma I, Allan AL (2011) The role of human aldehyde dehydrogenase in normal and cancer stem cells. Stem Cell Reviews and Reports 7: 292–306.
4. Lou H, Dean M (2007) Targeted therapy for cancer stem cells: the patched pathway and ABC transporters. Oncogene 26: 1357–1360.
5. Foster BA, Gangavarapu KJ, Mathew G, Azabdaftari G, Morrison CD, et al. (2013) Human prostate side population cells demonstrate stem cell properties in recombination with urogenital sinus mesenchyme. PLoS One 8: e55062.
6. Fetsch PA, Abati A, Litman T, Morisaki K, Honjo Y, et al. (2006) Localization of the ABCG2 mitoxantrone resistance-associated protein in normal tissues. Cancer Lett 235: 84–92.
7. Maitland NJ, Collins AT (2008) Prostate cancer stem cells: a new target for therapy. J Clin Oncol 26: 2862–2870.
8. Sharifi N, Kawasaki BT, Hurt EM, Farrar WL (2006) Stem cells in prostate cancer: resolving the castrate-resistant conundrum and implications for hormonal therapy. Cancer Biol Ther 5: 901–906.
9. Schroeder A, Herrmann A, Cherryholmes G, Kowolik C, Buettner R, et al. (2013) Loss of androgen receptor expression promotes a stem-like cell phenotype in prostate cancer through STAT3 signaling. Cancer research.
10. Abou-Kheir WG, Hynes PG, Martin PL, Pierce R, Kelly K (2010) Characterizing the contribution of stem/progenitor cells to tumorigenesis in the Pten−/− TP53−/− prostate cancer model. Stem Cells 28: 2129–2140.
11. Abou-Kheir W, Hynes PG, Martin P, Yin JJ, Liu Y-N, et al. (2011) Self-renewing Pten−/−TP53−/−protospheres produce metastatic adenocarcinoma cell lines with multipotent progenitor activity. PloS one 6: e26112.
12. Luk SU, Yap WN, Chiu YT, Lee DT, Ma S, et al. (2011) Gamma-tocotrienol as an effective agent in targeting prostate cancer stem cell-like population. International Journal of Cancer 128: 2182–2191.
13. Luk S-U, Lee TK-W, Liu J, Lee DT-W, Chiu Y-T, et al. (2011) Chemopreventive effect of PSP through targeting of prostate cancer stem cell-like population. PloS one 6: e19804.
14. Zhang L, Li L, Jiao M, Wu D, Wu K, et al. (2012) Genistein inhibits the stemness properties of prostate cancer cells through targeting Hedgehog–Gli1 pathway. Cancer Letters 323: 48–57.
15. Bonesi M, Loizzo MR, Conforti F, Passalacqua NG, Saab A, et al. (2013) Berberis aetnensis and B. libanotica: a comparative study on the chemical composition, inhibitory effect on key enzymes linked to Alzheimer's disease and antioxidant activity. Journal of Pharmacy and Pharmacology 65: 1726–1735.
16. Marc EB, Nelly A, Annick D-D, Frederic D (2008) Plants used as remedies antirheumatic and antineuralgic in the traditional medicine of Lebanon. Journal of ethnopharmacology 120: 315–334.
17. Imanshahidi M, Hosseinzadeh H (2008) Pharmacological and therapeutic effects of Berberis vulgaris and its active constituent, berberine. Phytotherapy research 22: 999–1012.
18. Zovko Končić M, Kremer D, Karlović K, Kosalec I (2010) Evaluation of antioxidant activities and phenolic content of *Berberis vulgaris* L. and *Berberis croatica* Horvat. Food and chemical toxicology 48: 2176–2180.
19. Kim S, Choi JH, Kim JB, Nam SJ, Yang J-H, et al. (2008) Berberine suppresses TNF-α-induced MMP-9 and cell invasion through inhibition of AP-1 activity in MDA-MB-231 human breast cancer cells. Molecules 13: 2975–2985.
20. Singh J, Kakkar P (2009) Antihyperglycemic and antioxidant effect of *Berberis aristata* root extract and its role in regulating carbohydrate metabolism in diabetic rats. Journal of ethnopharmacology 123: 22–26.
21. Roberts MF, Wink M (1998) Alkaloids: biochemistry, ecology, and medicinal applications: Springer.
22. Aniszewski T (2007) Alkaloids-Secrets of Life:: Aklaloid Chemistry, Biological Significance, Applications and Ecological Role: Elsevier.
23. Huang M, Gao H, Chen Y, Zhu H, Cai Y, et al. (2007) Chimmitecan, a novel 9-substituted camptothecin, with improved anticancer pharmacologic profiles in vitro and in vivo. Clinical cancer research 13: 1298–1307.
24. Li W, Shao Y, Hu L, Zhang X, Chen Y, et al. (2007) BM6, a new semi-synthetic vinca alkaloid, exhibits its potent in vivo anti-tumor activities via its high binding affinity for tubulin and improved pharmacokinetic profiles. Cancer Biology & Therapy 6: 787–794.
25. Singh T, Vaid M, Katiyar N, Sharma S, Katiyar SK (2011) Berberine, an isoquinoline alkaloid, inhibits melanoma cancer cell migration by reducing the expressions of cyclooxygenase-2, prostaglandin E2 and prostaglandin E2 receptors. Carcinogenesis 32: 86–92.
26. Ho Y-T, Yang J-S, Lu C-C, Chiang J-H, Li T-C, et al. (2009) Berberine inhibits human tongue squamous carcinoma cancer tumor growth in a murine xenograft model. Phytomedicine 16: 887–890.
27. Refaat A, Abdelhamed S, Yagita H, Inoue H, Yokoyama S, et al. (2013) Berberine enhances tumor necrosis factor-related apoptosis-inducing ligand-mediated apoptosis in breast cancer. Oncology letters 6: 840–844.
28. Liu B, Wang G, Yang J, Pan X, Yang Z, et al. (2011) Berberine inhibits human hepatoma cell invasion without cytotoxicity in healthy hepatocytes. PloS one 6: e21416.
29. Peng P-L, Hsieh Y-S, Wang C-J, Hsu J-L, Chou F-P (2006) Inhibitory effect of berberine on the invasion of human lung cancer cells via decreased productions of urokinase-plasminogen activator and matrix metalloproteinase-2. Toxicology and applied pharmacology 214: 8–15.
30. Esseily F, El Ezzy M, Gali-Muhtasib H, Safi S, Esseily J, et al. (2012) The ethanol fraction from the stem of Berberis libanotica inhibits the viability of adult T cell leukemia. Minerva Biotecnologica 24: 129–133.
31. Eid AA, Ford BM, Bhandary B, de Cassia Cavaglieri R, Block K, et al. (2013) Mammalian target of rapamycin regulates Nox4-mediated podocyte depletion in diabetic renal injury. Diabetes 62: 2935–2947.
32. Bournine L, Bensalem S, Peixoto P, Gonzalez A, Maiza-Benabdesselam F, et al. (2013) Revealing the anti-tumoral effect of Algerian *Glaucium flavum* roots against human cancer cells. Phytomedicine.
33. Slaninová I, Pěnčíková K, Urbanová J, Slanina J, Táborská E (2013) Antitumour activities of sanguinarine and related alkaloids. Phytochemistry Reviews: 1–18.
34. Yu J, Ma Y, Drisko J, Chen Q (2013) Antitumor Activities of *Rauwolfia vomitoria* Extract and Potentiation of Carboplatin Effects Against Ovarian Cancer. Current Therapeutic Research 75: 8–14.
35. Kaseb AO, Chinnakannu K, Chen D, Sivanandam A, Tejwani S, et al. (2007) Androgen receptor and E2F-1 targeted thymoquinone therapy for hormone-refractory prostate cancer. Cancer Res 67: 7782–7788.
36. Wang Y, Liu Q, Liu Z, Li B, Sun Z, et al. (2012) Berberine, a genotoxic alkaloid, induces ATM-Chk1 mediated G2 arrest in prostate cancer cells. Mutation Research/Fundamental and Molecular Mechanisms of Mutagenesis 734: 20–29.
37. Rafi MM, Vastano BC, Zhu N, Ho C-T, Ghai G, et al. (2002) Novel polyphenol molecule isolated from licorice root (Glycrrhiza glabra) induces apoptosis, G2/M cell cycle arrest, and Bcl-2 phosphorylation in tumor cell lines. Journal of agricultural and food chemistry 50: 677–684.
38. Corcoran C, Rani S, O'Brien K, O'Neill A, Prencipe M, et al. (2012) Docetaxel-resistance in prostate cancer: evaluating associated phenotypic changes and potential for resistance transfer via exosomes. PloS one 7: e50999.
39. Domingo-Domenech J, Vidal SJ, Rodriguez-Bravo V, Castillo-Martin M, Quinn SA, et al. (2012) Suppression of acquired docetaxel resistance in prostate cancer through depletion of notch-and hedgehog-dependent tumor-initiating cells. Cancer cell 22: 373–388.

A Novel Model of Human Skin Pressure Ulcers in Mice

Andrés A. Maldonado[1]*, Lara Cristóbal[1], Javier Martín-López[2], Mar Mallén[3], Natalio García-Honduvilla[4], Julia Buján[4]

1 Department of Plastic and Reconstructive Surgery and Burn Unit, University Hospital of Getafe, Madrid, Spain, **2** Department of Pathology, University Hospital of Puerta de Hierro, Madrid, Spain, **3** Department of Genetics, University Hospital Central de la Defensa, Madrid, Spain, **4** Department of Medical Specialties, Faculty of Medicine, University of Alcalá, Networking Research Centre on Bioengineering, Biomaterials and Nanomedicine (CIBER-BBN), Madrid, Spain

Abstract

Introduction: Pressure ulcers are a prevalent health problem in today's society. The shortage of suitable animal models limits our understanding and our ability to develop new therapies. This study aims to report on the development of a novel and reproducible human skin pressure ulcer model in mice.

Material and Methods: Male non-obese, diabetic, severe combined immunodeficiency mice (n = 22) were engrafted with human skin. A full-thickness skin graft was placed onto 4×3 cm wounds created on the dorsal skin of the mice. Two groups with permanent grafts were studied after 60 days. The control group (n = 6) was focused on the process of engraftment. Evaluations were conducted with photographic assessment, histological analysis and fluorescence in situ hybridization (FISH) techniques. The pressure ulcer group (n = 12) was created using a compression device. A pressure of 150 mmHg for 8 h, with a total of three cycles of compression-release was exerted. Evaluations were conducted with photographic assessment and histological analysis.

Results: Skin grafts in the control group took successfully, as shown by visual assessment, FISH techniques and histological analysis. Pressure ulcers in the second group showed full-thickness skin loss with damage and necrosis of all the epidermal and dermal layers (ulcer stage III) in all cases. Complete repair occurred after 40 days.

Conclusions: An inexpensive, reproducible human skin pressure ulcer model has been developed. This novel model will facilitate the development of new clinically relevant therapeutic strategies that can be tested directly on human skin.

Editor: Amit Gefen, Tel Aviv University, Israel

Funding: This study was supported by a Grant from the Fundación MAPFRE (SA/11/AYU/444). The funders had no role in study design, data collection and analysis, decision to publish, or preparation of the manuscript.

Competing Interests: The authors have declared that no competing interests exist.

* Email: mail@andresmaldonado.es

Introduction

Pressure ulcers (PU) are a high-prevalence problem in our society. It is estimated that 1.3 million to 3 million adults have a PU, with an estimated cost of $500 to $40,000 to heal each ulcer. The incidence varies greatly by clinical setting; in the hospital, for example, the incidence is estimated to be 0.4% to 38.0% [1].

Although conservative management is conducted in clinical practice (e.g., postural changes, dressing care), there is great disparity in the approach and management of these patients [2,3]. According to the American and European Pressure Ulcer Advisory Panel guidelines, nutrition is an important aspect of a comprehensive care plan for prevention and treatment of pressure ulcers (although limited evidence-based research is available) [4]. Moreover, according to Thomas [5], prescriptions should be individually tailored to persons with pressure ulcers with regard to both macro- and micronutrients. Surgical treatment is used in only a small numbers of patients. However Larson and others advocate good results with a surgical approach (without consideration of nutritional status or osteomyelitis) [6], while other authors have reported a high recurrence rate with this method [7,8].

Approaches incorporating cellular therapy and growth factors are thought to be on the horizon. The combined clinical evidence on platelet-derived growth factor (PDGF) suggests that PDGF-BB may improve healing of pressure ulcers. However, the evidence is not sufficient to recommend this treatment for routine use [9]. According to Akita et al. [10] adipose-derived stem cells can promote human dermal fibroblast proliferation by directly contacting cells and via paracrine activation in the re-epithelialization phase of wound healing. Moreover, skin substitutes were made by employing advanced tissue-engineering approaches and have been used for clinical applications, promoting the healing of acute and chronic wounds [11]. For example, bilayered bioengineered human skin equivalent (Apligraf, Novartis) has been shown to be efficacious in a case study of patients with heel PUs (level IV evidence) [12]. Additionally, growth factors could be another alternative to stem cells. According to Yang et al. [13], the expression of VEGF and bFGF in PU tissue is decreased. This leads to a reduction in angiogenesis, which may be a crucial factor in the formation of PUs.

Regarding the etiology of PUs, external pressure is viewed as the main factor. Other patient-specific factors leading to derangement in tissue perfusion may account for an observed development of a pressure ulcer [14]. It is well known that ischemia reperfusion injury contributes to the pathophysiology of PUs more significantly than a single, prolonged ischemic insult [15]. The animal models described in the literature employ a variety of devices to apply localized pressure on the back of mouse skin, and many of these models use external magnets; this technique is based on the repetition of ischemia-reperfusion cycles [16–19]. However, all of these models are based on mouse skin, which could be a potential limitation to studying the effect of human stem cells or growth factors in the PU environment.

From our point of view, the shortage of suitable animal models together with the ethical and practical considerations for humans limits our understanding of PUs and the development of new therapies. This study aims to report a novel and reproducible PU model of human skin graft. Cell therapy, growth factors and other techniques could be applied directly to human skin instead of mouse skin.

Material and Methods

Animals

Three-week-old, male, non-obese diabetic/severe combined immunodeficiency (NOD.CB17-Prkdscid/NCrHsd) mice (n = 22) (Harlan Laboratories S.r.l. Barcelona, Spain) were used in this study. All mice were caged under standard light and temperature conditions with free access to food and water throughout the study. All experimental procedures were made to minimize suffering and they were approved by the local committee for animal welfare and were conducted in accordance with the European Community Council Directive (86/609/EEC). The ethical Committee at University of Alcalá (Madrid, Spain) approved this research.

Human skin grafts

Mice were engrafted under general anesthesia (Ohmeda, BOC Health Care) with female human skin. All human skin came from abdominoplasty or breast reduction procedures. Ethics committee from University of Alcalá (Madrid, Spain) was approved and written informed consent from all patients was obtained. Full-thickness skin grafts (FTSGs) were placed onto a 4×3 cm wound created with scalpel on the dorsal skin of the animals. Mice skin was incised down to the muscle and removed, exposing muscular layer. FTSGs were sutured in place with 4/0 nylon. Postoperative analgesia (meloxicam and buprenorphine) was provided for 3 days and the dressing was tied on for the first 5 days. Special care was taken with sterilization and postoperative animal handling due to the immunodeficient status of the mice. A total of 4 mice died in the immediate postoperative period.

Mice were classified into two groups after 60 days with permanent human skin grafts. In the control group (n = 6), mice were sacrificed after 190 days and the FTSG was removed to study the engraftment process.

Pressure ulcers

In the PU group (n = 12), mice were placed in a compression device (Fig. 1A) following the method described by Stadler et al. [17] and modified by our group in terms of the device and the timing. A modified pressure device (7 mm×5 mm) that delivered a pressure of 150 mmHg to the human FTSG was used. The exerted pressure was measured with a dynamometer (Fig. 1B). Three cycles of compression-release (8 hours of clamping after 16 hours of no compression) were applied to the human FTSG. This group was subdivided into two groups. From the first one (n = 6), biopsies were taken at 5, 25, 45 and 130 days post-cycle. From the second subgroup (n = 6), animals were used to assess PU evolution by photographic analysis. All animals were sacrificed after 190 days.

Macroscopic analysis

At the macroscopic level, the behavior of the graft was evaluated every 15 days for 60 days in the control group. In the PU group, the PU was evaluated every 7 days for 130 days after compression cycles. The evolution of the graft was analyzed morphometrically (ImageJ for Windows XP NIH Image) using photographs from days 0 to 190. These measurements were taken by 2 independent researchers who were blinded to the treatment group. The values are expressed as the means ± standard deviation.

Microscopic evaluation

At the end of the experiments, tissue specimens were collected for different studies and placed in 10% buffered formaldehyde, Bouin and Carnoy. Then the samples were dehydrated and embedded in paraffin. Tissue sections (5 μm-thick) passing through the center plane of each wound were stained with hematoxylin-eosin and Masson's trichrome for morphological assessment. Cytogenetic analysis using fluorescence in situ hybridization (FISH) for the X and Y chromosomes (sonde XA X/Y, D-5608-100-OG, MetaSystems GmbH) was also performed.

Statistical analysis

Areas of the FTSG were compared among treatment groups by ANOVA followed by the Mann-Whitney U test. Differences with $p<0.05$ were considered statistically significant.

Results

Human skin grafts

Stable human FTSGs from the control groups took successfully, as demonstrated by photographic assessment (Fig. 2A). Macro-

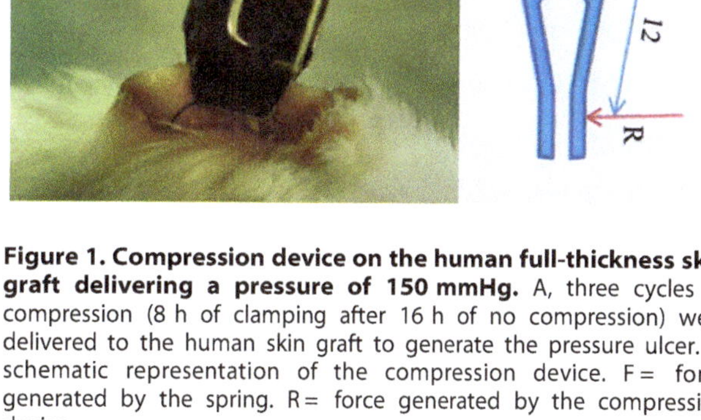

Figure 1. Compression device on the human full-thickness skin graft delivering a pressure of 150 mmHg. A, three cycles of compression (8 h of clamping after 16 h of no compression) were delivered to the human skin graft to generate the pressure ulcer. B, schematic representation of the compression device. F = force generated by the spring. R = force generated by the compression device.

scopically, the graft was soft and pliable after 60 days, resembling normal human skin (Fig. 1A–d). Human FTSGs showed uniform behavior in all studied animals. The FTSGs featured normal coloration (Fig. 2A–a) during the first few days but successively transformed into crusty skin over the first 30 days (Figs. 2A–b and 2A–c). This crust progressively disappeared and was replaced by a fully re-epithelialized area resembling normal and fine human skin after 60 days (Fig. 2A–d). Morphometrical analysis showed the effect of the contraction and retraction of the grafted skin. An important retraction of 33% of the graft was observed in the first 7 days. A progressive decrease with a retraction of 71% after 120 days was reached (Fig. 3).

In the histological analysis, a panoramic view showed (Fig. 4A) a clear delimitation between the human skin and the mouse skin. The presence of well-preserved human skin could be observed in the center of the graft which was formed by a stratified epidermis on a dermis with many papillae, resembling human skin. At higher magnification, keratinized squamous epithelium was observed on a papillae dermis with pressure corpuscles. A lymphocytic infiltrate was randomly distributed without specific accumulations. Human dermis (superficial and deep level) was well vascularized, which implies graft stability. The human skin was surrounded by a discrete and well-vascularized dermal layer from the receptor tissue (mouse). The host tissue presented the usual features of mouse skin. This animal model had a fine skin, two or three layers of keratinocytes, a papillary dermis with well-preserved hair follicles and a deep dermis with great development of adipose tissue.

The human tissue over the mouse receptors was assessed using the FISH technique (Fig. 5). Chromosomes XX from the human skin and chromosomes XY from the mouse skin were found as green-green dots (X-X) and green-red dots (X-Y). Most cells showing the Y chromosome (red dot) appeared to correspond to mouse cells infiltrating the human tissue.

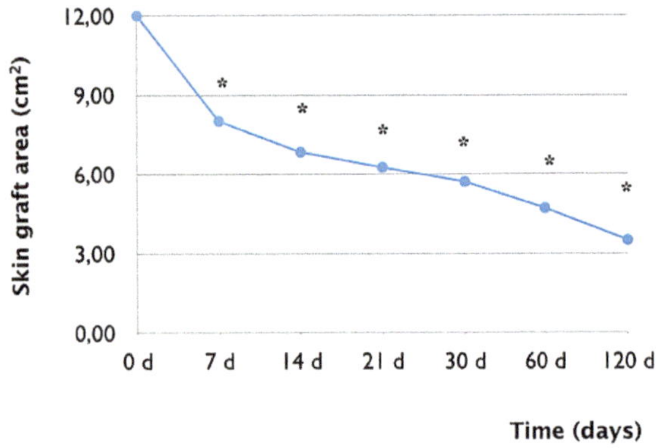

Figure 3. Area in cm^2 of the human full thickness skin graft on the back of the mice during the first 120 days after surgery. A significant reduction of the skin graft can be observed during the first week after the transplantation.

Pressure ulcers

Placing the compression device on the human skin graft for three cycles induced irreversible damage, characterizing a PU (Fig. 2B). A dermoepidermal necrotic fold covering the entire longitudinal extension under the site of the compression device was observed day 1 post-cycle (Fig. 2B–b). The ischemia produced

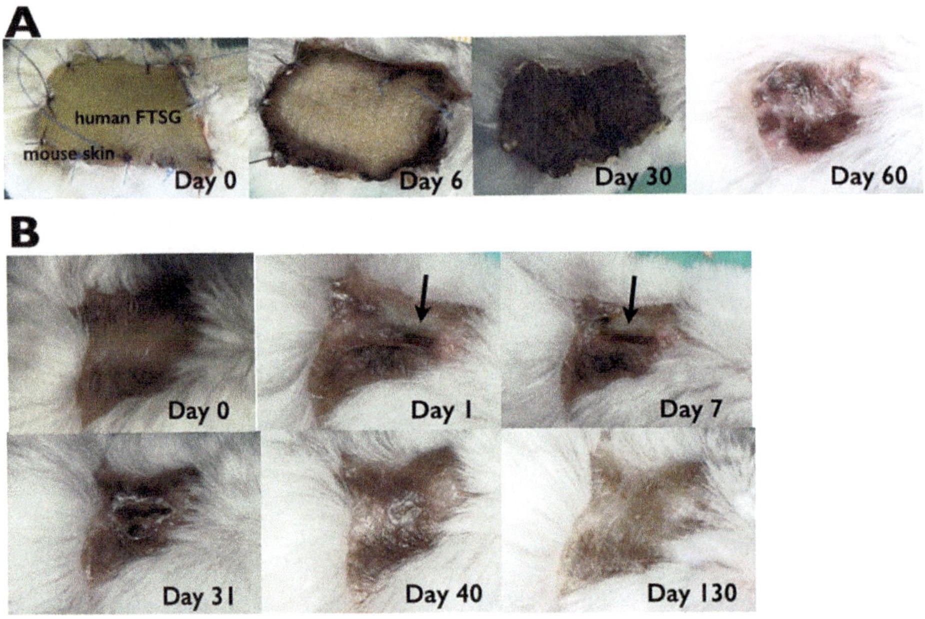

Figure 2. Photographic evolution of the human skin graft in mice. A, human full-thickness skin graft evolution from the control group. (a) Day of surgery. (b) Day 6 after surgery. (c) Day 30 after surgery. (d) Day 60 after surgery. Macroscopically, the graft was soft and pliable, resembling normal human skin. Note that the stable human full-thickness skin graft from the control groups took successfully. B, human full-thickness skin graft evolution after placing the compression device for three cycles. (a) Day before the compression device was applied. (b) Day 1 after the three compression cycles. The human skin graft remained folded with a hemorrhagic area in the center of the fold (see arrow). (c) Day 7 after compression cycles. Irreversible damage characterizing a PU can be observed. The center of the fold was occupied by a necrotic and hemorrhagic area (see arrow). (d) Day 31 after the compression cycles. Crusty remnants were observed. (e) Day 40 after compression cycles. Only a small central crusted island could be observed. (f) Day 130 after compression cycles. Complete regeneration of the graft was noticed.

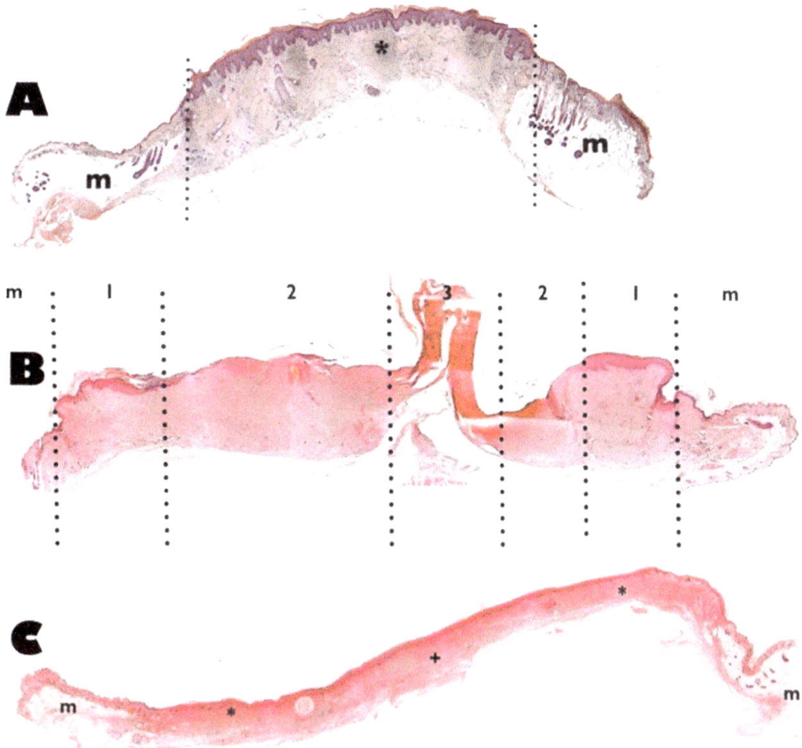

Figure 4. Histological analysis (H&E at 5x, panoramic view) of the mouse and human skin. A, human skin taken 60 days after placement of the human full-thickness skin graft. A clear delimitation between the normal human skin (*) and the mouse skin (m) can be observed. The transition between human and mouse skin has been marked with dotted lines. B, pressure ulcer over the human full-thickness skin graft. 7 days post-compression cycles. Pressure ulcer tissue can be observed as a consequence of mechanical damage. Four zones can be differentiated: m = receptor mouse skin; zone 1 = normal human skin; zone 2 = medium damage human skin; zone 3 = maximal damage human skin. C, 130 days post-compression cycles. A central area with a stratified epithelium over a non-papillary neodermis (+), a homogeneous and uniform human skin (*) and receptor mouse skin tissue (m) can be observed.

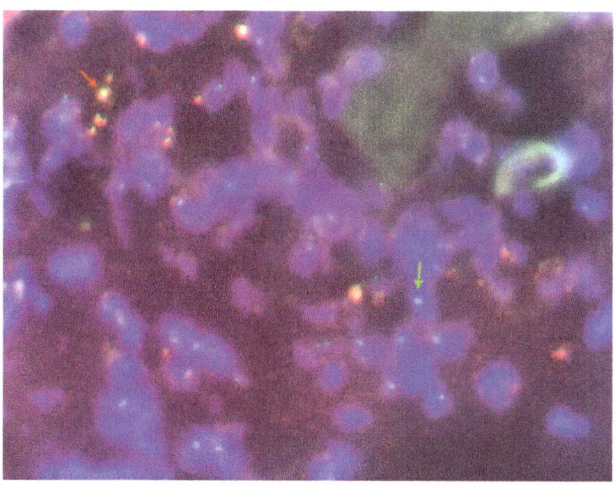

Figure 5. Fluorescence in situ hybridization (FISH) for chromosomes XX and XY. Chromosomes XX from the female human skin and chromosome XY from the male mouse skin were found as green-green dots (X-X, see green arrow) and green-red dots (X-Y, see red arrow). Most cells showing the Y chromosome (red dot) appeared to correspond to mouse cells infiltrating the human tissue.

over the next few days continued to bleed on day 7 post-cycle (Fig. 2B–c). Then, a further evolution towards a retracted and elevated lesion with crusty edges with underlying granulation tissue was observed (Fig. 2B–d). The appearance of the lesion changed completely on day 40 due to the loss of the crust and the formation of a sclerotic surface (Fig. 2B–e). Fully recovered dermoepidermal tissue then emerged. The stability of the human skin graft even after recovery of the PU was striking; the human skin had completely recovered after 130 days (Figs. 2B–a vs. 2B–f).

Histological analysis of the injured human skin was performed systematically in mice at various intervals (days 5, 25, 45 and 130 days post-cycle) after the three ischemic periods of 8 hours. Tissue ischemia was observed during the first days after the cycles of compression as a consequence of mechanical damage (Fig. 4B), reminiscent of PU tissue. The effect of the compression device induced potent tissue degeneration, necrosis in the center and full thickness skin loss involving subcutaneous tissue damage (stage III pressure ulcer). Tissue with a "U" shape was formed by dermal collagen residues. The edges of the tissue were also affected by loss of epithelialization, vascularization and cell population (Fig. 4B, zone 3). These edges were followed by skin areas showing denuded surface, epithelial desquamation and small surface areas covered with crust. The superficial dermis was composed of thick bundles of acellular collagen, while the deep dermis showed signs of cell habitability (Fig. 4B, zone 2). Compression effects were not observed in remote areas (Fig. 4B, zone 1). Finally, this area was bordered by the recipient mouse tissue (Fig. 4B, zone m). The evolution of this tissue damage was studied. Total tissue recovery was observed after 130 days. Fig. 4C shows a sagittal section of the human FTSG. The repair capacity of the human skin could be assessed; zones 1 and 2 of the previous figure (Fig. 4B) were replaced by a homogeneous and uniform human skin, showing the characteristics expected for a reparative epithelium after a significant injury. The central area of Figure 4C (zone 3 in Fig. 4B) showed a stratified epithelium over a non-papillary dermis, which is characteristic of neodermis.

Discussion

Human skin grafts

In our initial experiments, human FTSG rejection was observed in all mice using semiathymic and athymic nude mice (unpublished results). Although several papers have reported human skin transplantation to the nude mice, [20–24] hypertrophic scarring of the tissue developed instead of viable normal human skin. Yang et al. [24] postulated that although nude mice are immunologically defective of T cells, there might be other immune mechanisms involved, most likely related to the strong antigenicity of the skin, remnant T cells in nude mice (extrathymic lymphocyte populations) and the enhanced role of macrophages and natural killer cells. In the same way, Lin et al. [25] published that NK cells and macrophages could be activated in the absence of T cells or xenoantibodies to directly reject xenografts. To solve this problem, male NOD.CB17-Prkdscid/NCrHsd mice were used in our model. No graft rejection was observed and a viable normal human skin was achieved after 60 days.

Other models of human skin transplantation on mice were used in order to study the rejection process. Murray et al. [26,27] transplanted 7×7 mm skin graft on a C.B-17 SCID mice. Changes that resembled skin rejection in humans were observed after 2 weeks. Waldron-Lynch et al. [28] used neonatal NOD/SCID/IL2Rc cnull mice, reconstituted with human CD34+ hematopoietic stem cells. A murine skin transplant model in humanized mice was used to test human monoclonal antibody therapy. Racki et al. [29] studied human skin transplantation on immunodeficient mice and rejection following engraftment of allogeneic peripheral blood mononuclear cells. They transplanted 1.5 cm^2 human skin graft onto NOD-scid IL2rγ^{null} and CB17-scid bg for this purpose. Although all these models are useful for studying the immunology system and the rejection process, non of them provide enough dimensions of normal human skin as our model. Even when a fully new re-epithelialized area resembling normal and fine human skin was available after 60 days, the graft dimensions were about 5 cm^2. Although it is a 60% less than the initial dimensions (12 cm^2), this area should be enough to study the PU process or other possible skin damage.

Other models have used genetically engineered human skin on the backs of NOD/SCID mice [30–32]. Selected keratinocytes were assembled in a live fibroblast-containing fibrin dermal matrix orthotopically grafted onto mice. Although the authors presented stable human bilayer skin, we could not apply this to our model because human dermis is a more complex structure with a key role in PU generation and resolution.

Our model is based on a four-month stable human FTSG with complete dermal and epidermal layers. Studies have focused on the progress and shrinking of the graft and on the re-epithelialization after the generation of a PU. The graft has shown a high rate of shrinkage in the implant site, specially during the first 7 days. This process, which is known as primary contraction, is the immediate recoil of freshly harvested grafts as a result of the elastin in the dermis. The more dermis the graft has, the more primary the contraction that will be experienced [33]. In our model, more than half of the original size of the graft was observed after the FTSG reached a stable phase. The maximum dimensions of the back of the mice (4×3 cm) were used to achieve enough human graft after 60 days to develop the PU. The evolution of the graft was towards the initial formation of a crust related to the dermal and epidermal surface area of the graft. After 60 days, the FTSG site was denuded of a crusty layer and a new intensely keratinized and vascularized human stratified epithelium was formed.

Therefore, based on these results, we have used this experimental model to induce PUs in human skin. Skin stability was observed up to 190 days after the FTSG surgery, which demonstrates the effectiveness of our experimental model.

Pressure ulcers

PU models have been developed using mouse skin [17–19,34,35]. Our model presents a true human skin pressure ulcer model for the first time (see Table 1).

Based on the PU model presented by Stadler et al. [17], we used an original modification, replacing magnets with a mechanical compression device delivering a known, constant and controlled pressure. A modification of the magnet model was published by Wassermann et al. [18] A steel disk was implanted under the gluteus maximus muscle and pressure cycles were applied in conjunction with a magnet. This method could potentially damage the tissue underlying the implanted steel disk and, in particular, the area where the human FTSG was placed in our model. Our model differs from other models because the pressure exerted by the magnets depends on the thickness of the fold and on the position of the magnet. Moreover, we found it more reproducible to put forceps in the same position after every cycle than magnetic disks. After 3 cycles of 8 hours of clamping, irreversible tissue ischemia led to tissue necrosis, which was visible a week after the last cycle. This tissue damage was observed as an ulcer on the edges of the compression device and in the underlying deep dermal and subcutaneous tissue (stage III pressure ulcer).

Ulceration and bleeding could be observed for the rest of the first month. In all cases, recovery of the tissue continuity was evident after 40 days, which corroborated the notion that our model of ischemia had been successful in obtaining an experimental transient PU over human skin. Regenerated epidermis was characterized as a keratinized epidermis without melanocytes and with no papillary dermis, as in human skin defects. The underlying dermis demonstrated parallel bundles of mature collagen with new vascular components. Lymphocytic infiltrates were limited to the interface between the host (mouse) and the grafted (human) tissue.

This study has two main limitations. First, mice immunosuppression does not allow a normal inflammation process. However, as mentioned above, human PUs may be not only a pressure problem [14]. Systemic factors as diabetes, may produce decreased tissue perfusion, poor wound healing, slower epithelialization and immunosuppression [36]. We think that previous models [17–19,34] using completely healthy mice are distant from a clinical setting. Our immunosuppressed model would be more similar to the overall condition that many patients with PUs have: delayed wound healing, altered healing environment, etc. Moreover, from a histological point of view, our model would be more analogous to human PUs than previous models. Nevertheless, further improvements of the model remain to be performed. The second limitation involves the damaged adipose tissue under the human skin. Adipocyte cells come from the mouse, not from the human

Table 1. Characteristics of the different published pressure ulcer models.

Model	Animal	Compression device	Cycles	Tissue
Stadler et al. (2004){Stadler:2004jx}	BALB/c mouse	magnetic plates	12 h compression - 12 h release	mouse skin
Reid et al. (2004){Reid:2004ee}	C57BL/6J mouse	magnetic plates	1, 2 or 5 cycles/day of 0.5 h	mouse skin
Wassermann et al. (2009){Wassermann:2009ba}	Balb/c nu/nu nude mouse	neodymium magnet	2 h compression - 1 h release	mouse skin
Garza-Rodea et al. (2011){delaGarzaRodea:2011 kq}	NOD-LtSz-scid/scid/J mouse	magnetic disks	Only one cycle (4/8/12/14/20 h of clamping)	mouse skin
Maldonado et al.	NOD.CB17-Prkdscid/NCrHsd mouse	forceps	8 h compression - 16 h release	human skin

graft. Although that could be a limitation, we think that it should not affect the cicatrization process as the dermis and epidermis play the main role.

We believe this PU model could represent the most similar situation to human PUs at this time. Further research into any kinds of cellular (i.e. stem cells) or molecular therapies (i.e. growth factors) could be tested directly on damaged human skin after pressure without the ethical issues involved with research in humans. As we described above, the cicatrization time of this model is well-known. Improving this process on our model with new therapies, could potentially mean a direct application in the healing of human PUs.

Conclusions

To our knowledge, this is the first model where PU has been developed over human skin. In comparison with other mouse skin PU models, future therapies applied in our human skin model could be more realistically extrapolated to a human PU. We think it opens up prospects for testing different cellular or molecular therapies directly over human skin.

Acknowledgments

We thank Dr. Purificación Holguín and all the staff of the Department of Plastic Surgery and Burns, University Hospital of Getafe, Madrid, Spain; Dr. Mario Arenillas (Fundación para la Investigación Biomédica of Getafe, Spain) and Dr. José Carrasco (Department of Materials, School of Mines of Madrid, Spain).

Author Contributions

Conceived and designed the experiments: AAM LC NGH JB. Performed the experiments: AAM LC JML MM. Analyzed the data: AAM LC. Contributed reagents/materials/analysis tools: AAM LC. Wrote the paper: AAM LC JB.

References

1. Lyder CH (2003) Pressure ulcer prevention and management. JAMA 289: 223–226.
2. Levine SM, Stadler I, Sinno S, Zhang R-Y, Oskoui P, et al. (2013) Current thoughts for the prevention and treatment of pressure ulcers: using the evidence to determine fact or fiction. Ann Surg 257: 603–608.
3. Moore Z, Cowman S (2008) A systematic review of wound cleansing for pressure ulcers. J Clin Nurs 17: 1963–1972.
4. Dorner B, Posthauer ME, Thomas D, National Pressure Ulcer Advisor Panely (2009) The role of nutrition in pressure ulcer prevention and treatment: National Pressure Ulcer Advisory Panel white paper. Adv Skin Wound Care 22: 212–221.
5. Thomas DR (2014) Role of Nutrition in the Treatment and Prevention of Pressure Ulcers. Nutr Clin Pract 29: 466–472.
6. Larson DL, Hudak KA, Waring WP, Orr MR, Simonelic K (2012) Protocol management of late-stage pressure ulcers: a 5-year retrospective study of 101 consecutive patients with 179 ulcers. Plast Reconstr Surg 129: 897–904.
7. Kierney PC, Engrav LH, Isik FF, Esselman PC, Cardenas DD, et al. (1998) Results of 268 pressure sores in 158 patients managed jointly by plastic surgery and rehabilitation medicine. Plast Reconstr Surg 102: 765–772.
8. Keys KA, Daniali LN, Warner KJ, Mathes DW (2010) Multivariate predictors of failure after flap coverage of pressure ulcers. Plast Reconstr Surg 125: 1725–1734.
9. Dealey C, Clark M, Defloor T, Schoonhoven L, Vanderwee K, et al. (2009) European Pressure Ulcer Advisory Panel and National Pressure Ulcer Advisory Panel. Treatment of pressure ulcers: Quick Reference Guide. Washington DC: National Pressure Ulcer Advisory Panel.
10. Akita S, Yoshimoto H, Akino K, Ohtsuru A, Hayashida K, et al. (2012) Early experiences with stem cells in treating chronic wounds. Clin Plast Surg 39: 281–292.
11. Groeber F, Holeiter M, Hampel M, Hinderer S, Schenke-Layland K (2012) Skin Tissue Engineering—In Vivo and In Vitro Applications. Clin Plast Surg 39: 33–58.
12. Karr J (2008) Utilization of living bilayered cell therapy (Apligraf) for heel ulcers. Adv Skin Wound Care 21: 270–274.
13. Yang J-J, Wang X-L, Shi B-W, Huang F (2013) The angiogenic peptide vascular endothelial growth factor-basic fibroblast growth factor signaling is up-regulated in a rat pressure ulcer model. Anat Rec (Hoboken) 296: 1161–1168.
14. Thomas DR (2010) Does pressure cause pressure ulcers? An inquiry into the etiology of pressure ulcers. J Am Med Dir Assoc 11: 397–405.
15. Tsuji S, Ichioka S, Sekiya N, Nakatsuka T (2005) Analysis of ischemia-reperfusion injury in a microcirculatory model of pressure ulcers. Wound Repair Regen 13: 209–215.
16. Peirce SM, Skalak TC, Rodeheaver GT (2000) Ischemia-reperfusion injury in chronic pressure ulcer formation: a skin model in the rat. Wound Repair Regen 8: 68–76.
17. Stadler I, Zhang R-Y, Oskoui P, Whittaker MBS, Lanzafame RJ (2004) Development of a Simple, Noninvasive, Clinically Relevant Model of Pressure Ulcers in the Mouse. J Invest Surg 17: 221–227.
18. Wassermann E, van Griensven M, Gstaltner K, Oehlinger W, Schrei K, et al. (2009) A chronic pressure ulcer model in the nude mouse. Wound Repair Regen 17: 480–484.
19. De la Garza-Rodea de AS, Knaän-Shanzer S, van Bekkum DW (2011) Pressure Ulcers: Description of a New Model and Use of Mesenchymal Stem Cells for Repair. Dermatology 223.
20. Kischer CW, Sheridan D, Pindur J (1989) Use of nude (athymic) mice for the study of hypertrophic scars and keloids: vascular continuity between mouse and implants. Anat Rec 225: 189–196.
21. Kischer CW, Wagner HN, Pindur J, Holubec H, Jones M, et al. (1989) Increased fibronectin production by cell lines from hypertrophic scar and keloid. Connect Tissue Res 23: 279–288.
22. Shetlar MR, Shetlar CL, Kischer CW, Pindur J (1991) Implants of keloid and hypertrophic scars into the athymic nude mouse: changes in the glycosaminoglycans of the implants. Connect Tissue Res 26: 23–36.
23. Wang J, Ding J, Jiao H, Honardoust D, Momtazi M, et al. (2011) Human hypertrophic scar-like nude mouse model: Characterization of the molecular and cellular biology of the scar process. Wound Repair Regen 19: 274–285.

24. Yang DY, Li SR, Wu JL, Chen YQ, Li G, et al. (2007) Establishment of a Hypertrophic Scar Model by Transplanting Full-Thickness Human Skin Grafts onto the Backs of Nude Mice. Plast Reconstr Surg 119: 104–109.
25. Lin Y, Vandeputte M, Waer M (1997) Natural killer cell- and macrophage-mediated rejection of concordant xenografts in the absence of T and B cell responses. J Immunol 158: 5658–5667.
26. Murray AG, Petzelbauer P, Hughes CC, Costa J, Askenase P, et al. (1994) Human T-cell-mediated destruction of allogeneic dermal microvessels in a severe combined immunodeficient mouse. Proc Natl Acad Sci USA 91: 9146–9150.
27. Murray AG, Schechner JS, Epperson DE, Sultan P, McNiff JM, et al. (1998) Dermal microvascular injury in the human peripheral blood lymphocyte reconstituted-severe combined immunodeficient (HuPBL-SCID) mouse/skin allograft model is T cell mediated and inhibited by a combination of cyclosporine and rapamycin. Am J Pathol 153: 627–638.
28. Waldron-Lynch F, Deng S, Preston-Hurlburt P, Henegariu O, Herold KC (2012) Analysis of human biologics with a mouse skin transplant model in humanized mice. Am J Transplant 12: 2652–2662.
29. Racki WJ, Covassin L, Brehm M, Pino S, Ignotz R, et al. (2010) NOD-scid IL2rgamma(null) mouse model of human skin transplantation and allograft rejection. Transplantation 89: 527–536.
30. Río M, Larcher F, Serrano F, Meana A, Muñoz M, et al. (2002) A Preclinical Model for the Analysis of Genetically Modified Human Skin In Vivo. Hum Gene Ther 13: 959–968.
31. García M, Escamez MJ, Carretero M, Mirones I, Martínez-Santamaría L, et al. (2007) Modeling normal and pathological processes through skin tissue engineering. Mol Carcinog 46: 741–745.
32. Carretero M, Guerrero-Aspizua S, Del Río M (2011) Applicability of bioengineered human skin: From preclinical skin humanized mouse models to clinical regenerative therapies. Biobugs 2: 203–207.
33. Thorne C, Grabb WC, Beasley RW (2007) Techniques and principles in Plastic Surgery. Grabb and Smith's Plastic Surgery. 6th ed. Philadelphia: Lippincott Williams & Wilkins. pp. 7.
34. Reid RR, Sull AC, Mogford JE, Roy N, Mustoe TA (2004) A novel murine model of cyclical cutaneous ischemia-reperfusion injury. J Surg Res 116: 172–180.
35. Salcido R, Popescu A, Ahn C (2007) Animal models in pressure ulcer research. J Spinal Cord Med 30: 107–116.
36. Gantwerker EA, Hom DB (2012) Skin: Histology and Physiology of Wound Healing. Clin Plast Surg 39: 85–97.

Permissions

All chapters in this book were first published in PLOS ONE, by The Public Library of Science; hereby published with permission under the Creative Commons Attribution License or equivalent. Every chapter published in this book has been scrutinized by our experts. Their significance has been extensively debated. The topics covered herein carry significant findings which will fuel the growth of the discipline. They may even be implemented as practical applications or may be referred to as a beginning point for another development.

The contributors of this book come from diverse backgrounds, making this book a truly international effort. This book will bring forth new frontiers with its revolutionizing research information and detailed analysis of the nascent developments around the world.

We would like to thank all the contributing authors for lending their expertise to make the book truly unique. They have played a crucial role in the development of this book. Without their invaluable contributions this book wouldn't have been possible. They have made vital efforts to compile up to date information on the varied aspects of this subject to make this book a valuable addition to the collection of many professionals and students.

This book was conceptualized with the vision of imparting up-to-date information and advanced data in this field. To ensure the same, a matchless editorial board was set up. Every individual on the board went through rigorous rounds of assessment to prove their worth. After which they invested a large part of their time researching and compiling the most relevant data for our readers.

The editorial board has been involved in producing this book since its inception. They have spent rigorous hours researching and exploring the diverse topics which have resulted in the successful publishing of this book. They have passed on their knowledge of decades through this book. To expedite this challenging task, the publisher supported the team at every step. A small team of assistant editors was also appointed to further simplify the editing procedure and attain best results for the readers.

Apart from the editorial board, the designing team has also invested a significant amount of their time in understanding the subject and creating the most relevant covers. They scrutinized every image to scout for the most suitable representation of the subject and create an appropriate cover for the book.

The publishing team has been an ardent support to the editorial, designing and production team. Their endless efforts to recruit the best for this project, has resulted in the accomplishment of this book. They are a veteran in the field of academics and their pool of knowledge is as vast as their experience in printing. Their expertise and guidance has proved useful at every step. Their uncompromising quality standards have made this book an exceptional effort. Their encouragement from time to time has been an inspiration for everyone.

The publisher and the editorial board hope that this book will prove to be a valuable piece of knowledge for researchers, students, practitioners and scholars across the globe.

List of Contributors

Michael Andrew Yu, Alan Kiang, Vicky Yu, Elham Rahimy, Kevin T. Brumund, Robert A. Weisman and Weg M. Ongkeko
Division of Otolaryngology-Head and Neck Surgery, Department of Surgery, University of California San Diego, La Jolla, California, United States of America

Jessica Wang-Rodriguez and Lesley G. Ellies
Veterans Administration Medical Center and Department of Pathology, University of California San Diego, La Jolla, California, United States of America

Martin Haas
Moores Cancer Center, University of California San Diego, La Jolla, California, United States of America

Jing Chen and Jian-Bing Fan
Illumina Inc., San Diego, California, United States of America

Mutsuko Minata, Kaushal Joshi, Christopher Hong
Department of Neurological Surgery, The Ohio State University Medical Center, Columbus, Ohio, United States of America

Mariko Nakano-Okuno
Department of Internal Medicine, The Ohio State University Medical Center, Columbus, Ohio, United States of America

Chi-Hung Nguyen
Pharmaco-chemistry, UMR 176 CNRS-Institut Curie, Orsay, France

Harley I. Kornblum
Departments of Psychiatry and Molecular and Medical Pharmacology, University of California Los Angeles, Los Angeles, California, United States of America

Annie Molla
Institut Albert Bonniot, Universite´ Joseph Fourier, Grenoble Cedex 9, France

Chunyu Gu
Department of Neurological Surgery, The Ohio State University Medical Center, Columbus, Ohio, United States of America
Departments of Neurosurgery, Beijing Sanbo Brain Hospital, Capital Medical University, Beijing, China

Ichiro Nakano
Department of Neurological Surgery, The Ohio State University Medical Center, Columbus, Ohio, United States of America
James Comprehensive Cancer Center, The Ohio State University Medical Center, Columbus, Ohio, United States of America

Zhen Ma, Huaxiao Yang, Honghai Liu and Bruce Z. Gao
Department of Bioengineering, Clemson University, Clemson, South Carolina, United States of America

Meifeng Xu
Department of Pathology and Laboratory Medicine, University of Cincinnati Medical Center, Cincinnati, Ohio, United States of America

Raymond B. Runyan
Department of Cellular and Molecular Medicine, University of Arizona, Tucson, Arizona, United States of America

Carol A. Eisenberg
New York Medical College/Westchester Medical Center Stem Cell Laboratory, New York Medical College, Valhalla, New York, United States of America

Roger R. Markwald and Thomas K. Borg
Department of Regenerative Medicine and Cell Biology, Medical University of South Carolina, Charleston, South Carolina, United States of America

Sindhu G. Rajendra and Edi Levi
Veterans Affairs Medical Center, Wayne State University, Detroit, Michigan, United States of America

Yingjie Yu, Anita Vasudevan and Lulu Farhana
Veterans Affairs Medical Center, Wayne State University, Detroit, Michigan, United States of America
Department of Internal Medicine, Wayne State University, Detroit, Michigan, United States of America

Pratima Nangia-Makker and Adhip P. N. Majumdar
Veterans Affairs Medical Center, Wayne State University, Detroit, Michigan, United States of America
Karmanos Cancer Institute, Wayne State University, Detroit, Michigan, United States of America
Department of Internal Medicine, Wayne State University, Detroit, Michigan, United States of America

Dong-feng Lu, Ying Wang, Zi-zhuo Su, Zhao-hua Zeng, Xiao-wen Xing and Zhi-yu He
Department of Cardiology, the First Affiliated Hospital of Guangzhou Medical University, Guangzhou, Guangdong, China

Chunxiang Zhang
Department of Pharmacology and Cardiovascular Research Center, Rush Medical college of Rush University, Chicago, Illinois, United States of America

Tony Tung-Yin Lee, Yu-Chih Wang, Mi-Chun Kao and Jau-Nan Lee
Department of Obstetrics and Gynecology and Center of Excellence for Environmental Medicine, Kaohsiung Medical University Hospital, Kaohsiung, Taiwan

Cheng-Fang Tsai
Graduate Institute of Medicine, Kaohsiung Medical University College of Medicine, Kaohsiung, Taiwan

Tsung-Hsun Hsieh and Jia-Jin Jason Chen
Institute of Biomedical Engineering, National Cheng Kung University, Tainan, Taiwan

Ruey-Meei Wu
Department of Neurology, National Taiwan University Hospital, Taipei, Taiwan

Sher Singh
Department of Life Science, National Taiwan Normal University, Taipei, Taiwan

Eing-Mei Tsai
Department of Obstetrics and Gynecology and Center of Excellence for Environmental Medicine, Kaohsiung Medical University Hospital, Kaohsiung, Taiwan
Graduate Institute of Medicine, Kaohsiung Medical University College of Medicine, Kaohsiung, Taiwan

Stephen Davis, Jill Waibel, Arsalan Shabbir and Audrey Cox
Department of Dermatology and Cutaneous Surgery, University of Miami Miller School of Medicine, Miami, Florida, United States of America

Luis Rodriguez-Menocal, Marcela Salgado and Evangelos V. Badiavas
Department of Dermatology and Cutaneous Surgery, University of Miami Miller School of Medicine, Miami, Florida, United States of America
Interdisciplinary Stem Cell Institute, University of Miami Miller School of Medicine, Miami, Florida, United States of America

D. Ryan Ormond, Craig Shannon, Julius Oppenheim, Kaushik Das, Raj Murali and Meena Jhanwar-Uniyal
Department of Neurosurgery, New York Medical College, Valhalla, New York, United States of America

Richard Zeman
Department of Cell Biology and Anatomy, New York Medical College, Valhalla, New York, United States of America

Baowan Lin and Philip S. Garza
Cerebral Vascular Disease Research Laboratories, Leonard M. Miller School of Medicine, University of Miami, Miami, Florida, United States of America

Chuanhui Dong, Weizhao Zhao, Tatjana Rundek and Ralph L. Sacco
Department of Neurology, Leonard M. Miller School of Medicine, University of Miami, Miami, Florida, United States of America

Erika B. Rangel and Ian McNiece
Interdisciplinary Stem Cell Institute, Leonard M. Miller School of Medicine, University of Miami, Miami, Florida, United States of America

Ami P. Raval and Miguel Perez-Pinzon
Cerebral Vascular Disease Research Laboratories, Leonard M. Miller School of Medicine, University of Miami, Miami, Florida, United States of America

Department of Neurology, Leonard M. Miller School of Medicine, University of Miami, Miami, Florida, United States of America

Joshua M. Hare
Interdisciplinary Stem Cell Institute, Leonard M. Miller School of Medicine, University of Miami, Miami, Florida, United States of America
The Department of Medicine, Leonard M. Miller School of Medicine, University of Miami, Miami, Florida, United States of America

Dileep R. Yavagal
Cerebral Vascular Disease Research Laboratories, Leonard M. Miller School of Medicine, University of Miami, Miami, Florida, United States of America
Department of Neurology, Leonard M. Miller School of Medicine, University of Miami, Miami, Florida, United States of America
Interdisciplinary Stem Cell Institute, Leonard M. Miller School of Medicine, University of Miami, Miami, Florida, United States of America

Wing-Hon Lai, Yau-Chi Chan, Joyce H. L. Ng, Ka-Wing Au and Lai-Yung Wong
Cardiology Division, Department of Medicine, Queen Mary Hospital, the University of Hong Kong, Hong Kong, HKSAR, China

Jenny C. Y. Ho, Chung-Wah Siu and Hung-Fat Tse
Cardiology Division, Department of Medicine, Queen Mary Hospital, the University of Hong Kong, Hong Kong, HKSAR, China
Research Center of Heart, Brain, Hormone and Healthy Aging, Li Ka Shing Faculty of Medicine, the University of Hong Kong, Hong Kong, HKSAR, China

Soo Jin Choi and Won Il Oh
Biomedical Research Institute, MEDIPOST Co., Ltd., Seoul, Korea

So Yoon Ahn, Se In Sung and Hye Soo Yoo
Department of Pediatrics, Samsung Medical Center, Sungkyunkwan University School of Medicine, Seoul, Korea

Dong Kyung Sung
Samsung Biomedical Research Institute, Sungkyunkwan University School of Medicine, Seoul, Korea

Yun Sil Chang, Won Soon Park
Department of Pediatrics, Samsung Medical Center, Sungkyunkwan University School of Medicine, Seoul, Korea
Samsung Biomedical Research Institute, Sungkyunkwan University School of Medicine, Seoul, Korea

Esther Wolfs, Bryan Holvoet, Cindy Casteels, Koen Van Laere and Christophe M. Deroose
Nuclear Medicine and Molecular Imaging, Department of Imaging and Pathology, KU Leuven, Leuven, Belgium

Rik Gijsbers, Michael Maris, Abdelilah Ibrahimi and Zeger Debyser
Laboratory of Molecular Virology and Gene Therapy, Department of Pharmaceutical and Pharmacological Sciences, KU Leuven, Leuven, Belgium, and Leuven Viral Vector Core, KU Leuven, Leuven, Belgium

Scott J. Roberts
Skeletal Biology and Engineering Research Center, Department of development and regeneration, KU Leuven, Leuven, Belgium

Tom Struys
Lab of Histology, Department of morphology, Biomedical Research Institute, Universiteit Hasselt, Hasselt, Belgium

Catherine M. Verfaillie
Stem Cell Institute Leuven, Department of development and regeneration, KU Leuven, Leuven, Belgium

Wooseok Im, Jaejun Ban, Jiyeon Lim, Mijung Lee, Soon-Tae Lee, Kon Chu and Manho Kim
Department of Neurology, Biomedical Research Institute, Seoul National University Hospital, Seoul, Korea

Nicholas Bryan, Fiona C. Lewis and John A. Hunt
Clinical Engineering, United Kingdom Centre for Tissue Engineering, Institute of Ageing and Chronic Disease, University of Liverpool, Liverpool, United Kingdom

Damian Bond and Christopher Stanley
CellCap Technologies Ltd., The Greens Farm, Todmorden, Bacup, United Kingdom

Kotturathu Mammen Cherian
Stem Cells and Tissue Engineering Laboratory, International Centre for Cardiothoracic and Vascular Diseases, Frontier Lifeline, Chennai, India

Soma Guhathakurta
Department of Engineering Design, Indian Institute of Technology Madras, Chennai, India

Rama Shanker Verma
Stem Cells and Molecular Biology Laboratory, Department of Biotechnology, Indian Institute of Technology Madras, Chennai, India

Dillip Kumar Bishi and Santosh Mathapati
Stem Cells and Molecular Biology Laboratory, Department of Biotechnology, Indian Institute of Technology Madras, Chennai, India
Stem Cells and Tissue Engineering Laboratory, International Centre for Cardiothoracic and Vascular Diseases, Frontier Lifeline, Chennai, India

Jyun-Yi Wu and Ming-Long Yeh
Institute of Biomedical Engineering, National Cheng Kung University, Tainan, Taiwan, Republic of China

Yan- Hsiung Wang
Orthopaedic Research Center, College of Medicine, Kaohsiung Medical University, Kaohsiung, Taiwan, Republic of China
School of Dentistry, College of Dental Medicine, Kaohsiung Medical University, Kaohsiung, Taiwan, Republic of China

Chia-Hsin Chen
Department of Physical Medicine and Rehabilitation, Kaohsiung Medical University Hospital, Kaohsiung, Taiwan, Republic of China
Department of Physical Medicine and Rehabilitation, Faculty of Medicine, College of Medicine, Kaohsiung Medical University, Kaohsiung, Taiwan, Republic of China
Department of Physical Medicine and Rehabilitation, Kaohsiung Municipal Ta-Tung Hospital, Kaohsiung, Taiwan, Republic of China

Chau-Zen Wang and Mei-Ling Ho
Orthopaedic Research Center, College of Medicine, Kaohsiung Medical University, Kaohsiung, Taiwan, Republic of China
Department of Physiology, College of Medicine, Kaohsiung Medical University, Kaohsiung, Taiwan, Republic of China
Graduate Institute of Medicine, College of Medicine, Kaohsiung Medical University, Kaohsiung, Taiwan, Republic of China

Mohamad Masoumya and Jack Yu
Department of Surgery, Georgia Regents University, Augusta, Georgia, United States of America

Jun Yao Liu and Mahmood S. Mozaffari
Department of Oral Biology, Georgia Regents University, Augusta, Georgia, United States of America

Nathan Yanasak
Department of Radiology and Imaging, Georgia Regents University, Augusta, Georgia, United States of America

Christopher Middleton
Cancer Center, Georgia Regents University, Augusta, Georgia, United States of America

Folami Lamoke
Department of Ophthalmology, Georgia Regents University, Augusta, Georgia, United States of America

Babak Baban
Department of Surgery, Georgia Regents University, Augusta, Georgia, United States of America
Department of Oral Biology, Georgia Regents University, Augusta, Georgia, United States of America

Monique Terwijn, Wendelien Zeijlemaker, Angéle Kelder, Arjo P. Rutten, Alexander N. Snel, Willemijn J. Scholten Sonja Zweegman, Gert J. Ossenkoppele and Gerrit J. Schuurhuis
Department of Hematology, VU University Medical Center, Amsterdam, The Netherlands

Thomas Pabst
Department of Medical Oncology, Inselspital, Bern University Hospital, University of Bern, Bern, Switzerland

Gregor Verhoef
Department of Hematology, University Hospital Leuven, Leuven, Belgium

Bob Löwenberg
Department of Hematology, Erasmus University Medical Center, Rotterdam, The Netherlands

Cheng-Cheng Hwang, Liang-Che Chang, Wen-Ying Chi, Hui-Ping Chien, Tsan-Yu Hsieh and Jim-Ray Chen
Department of Pathology, Chang Gung Memorial Hospital, Keelung, Taiwan

Shin Nieh
Department of Pathology and Graduate Institute of Pathology, National Defense Medical Center & Tri-Service General Hospital, Taipei, Taiwan

Chien-Hong Lai
Division of Hematology-Oncology, Department of Internal Medicine, Chang Gung Memorial Hospital, Keelung, Taiwan

Chien-Sheng Tsai
Department of Radiation Oncology, Chang Gung Memorial Hospital, Keelung, Taiwan

Chung-Ching Hua
Division of Pulmonary, Critical Care and Sleep Medicine, Chang Gung Memorial Hospital, Keelung, Taiwan

Chih-Wei Wang
Department of Anatomic Pathology, Chang Gung Memorial Hospital, Tao-Yuan, Taiwan

Siu-Cheung Chan
Department of Diagnostic Radiology, Chang Gung Memorial Hospital, Keelung, Taiwan

Cristiano Rodrigues and Márcia Rosângela Wink
Laboratórios de Pesquisa em Biologia Celular, Departamento de Ciências Básicas da Saúde, Universidade Federal de Ciências da Saúde de Porto Alegre, Porto Alegre/RS, Brazil

Dinara J. Moura and Jenifer Saffi
Genética Toxicológica, Departamento de Ciências Básicas da Saúde, Universidade Federal de Ciências da Saúde de Porto Alegre, Porto Alegre/RS, Brazil

Graziele Halmenschlager
Estresse Oxidativo e Poluição Atmosférica, Departamento de Ciências Básicas da Saúde, Universidade Federal de Ciências da Sau´de de Porto Alegre, Porto Alegre/RS, Brazi

Léder Leal Xavier
Laboratório de Biologia Celular e Tecidual, Departamento de Ciências Morfofisiológicas, Pontifícia Universidade Católica do Rio Grande do Sul, Porto Alegre/RS, Brazil

Marilda da Cruz Fernandes
Laboratório de Pesquisa em Patologia, Departamento de Ciências Básicas da Saúde, Universidade Federal de Ciências da Saúde de Porto Alegre, Porto Alegre/RS, Brazil

Holger Budde, Susanne Kolb, Laura Salinas Tejedor, Joachim Riggert and Tobias J. Legler
Department of Transfusion Medicine, University Medical Center Göttingen, Göttingen, Germany

Gerald Wulf
Department of Hematology and Oncology, University Medical Center Göttingen, Göttingen, Germany

Holger M. Reichardt
Institute for Cellular and Molecular Immunology, University Medical Center Göttingen, Göttingen, Germany

Rabih El-Merahbi, Assaad Eid, Georges Daoud, Tarek H. Mouhieddine and Wassim Abou-Kheir
Department of Anatomy, Cell Biology and Physiological Sciences, Faculty of Medicine, American University of Beirut, Beirut, Lebanon

Yen-Nien Liu
Graduate Institute of Cancer Biology and Drug Discovery, College of Medical Science and Technology, Taipei Medical University, Taipei, Taiwan

Leina Hosry
Faculty of Pharmacy, Lebanese University, Hadath, Lebanon

Aline Hamade
Department of Biology, Faculty of Sciences II, Lebanese University, Fanar, Lebanon

Fadia Najjar
Department of Chemistry and Biochemistry, Faculty of Sciences II, Lebanese University, Fanar, Lebanon

Andrés A. Maldonado and Lara Cristóbal
Department of Plastic and Reconstructive Surgery and Burn Unit, University Hospital of Getafe, Madrid, Spain

Javier Martín-López
Department of Pathology, University Hospital of Puerta de Hierro, Madrid, Spain

Mar Mallén
Department of Genetics, University Hospital Central de la Defensa, Madrid, Spain

Natalio García-Honduvilla and Julia Buján
Department of Medical Specialties, Faculty of Medicine, University of Alcalá, Networking Research Centre on Bioengineering, Biomaterials and Nanomedicine (CIBER-BBN), Madrid, Spain

Index